Genomics, Proteomics, and Biotechnology

High-throughput molecular technologies ("omics") can help to decipher the contributions of different physiological systems and identify candidate molecules that are representative of different physiological pathways thereby allowing the discovery of biomarkers. Notably, the omics technologies along with computational methods, bioprospecting, and artificial intelligence will continue to lead to better understanding of biological mechanisms that are responsible for physical attributes, or phenotypes. Research breakthroughs obtained through these technologies can be used to enhance productivity of food animals, meet the increasing demand for animal-sourced foods, enhance high-quality nutrient availability, ensure nutrient safety, mitigate the effects of climate variability, and result in new technologies that provide continued improvement in food security worldwide. Such breakthroughs are an urgent necessity because over the past 50 years, there has been an unprecedented increase in the world's population, which will reach ten billion by the year 2050. Innovative and technological advancements that enhance all aspects of food production will arise from basic, fundamental research. Besides food, animal by-products have found many applications in the fields of pharmaceuticals, cosmetics, and household and industrial products. Hence, the need to ameliorate the productivity, reproductivity, growth performance, and disease resistance in animals has created a worldwide interest in gaining a deeper understanding of animal biology, biotechnology and genomics, and proteomics.

The present volume thoroughly discusses the omics studies in domestic and non-domestic animals and their role in mitigation of various challenges ahead.

The volume thus focuses on

i. Omics (genomics, proteomics, transcriptomics, metabolonomics) technologies in identifying, characterizing biodiversity
ii. Role of molecular techniques for improvement of domestic and non-domestic organisms
iii. Animal and alternative model systems (using stem cells, tissue engineering, cell free systems, 3D platforms etc.) for studying life phenomena
iv. Genetically modified organisms as factories for the products

Translating Animal Science Research

Series Editor RC Sobti, Department of Biotechnology, Panjab University, Chandigarh, India

Biodiversity: Threats and Conservation
RC Sobti

Environmental Studies and Climate Change
RC Sobti, S.K. Malhotra, Kamal Jaiswal, and Sanjeev Puri

Genomics, Proteomics, and Biotechnology
RC Sobti, Manishi Mukesh, and Aastha Sobti

For more information, visit webpage: www.routledge.com/Translating-Animal-Science-Research/book-series/TASR?pd=published,forthcoming&pg=1&pp=12&so=pub&view=list?pd=published,forthcoming&pg=1&pp=12&so=pub&view=list

Genomics, Proteomics, and Biotechnology

Edited by RC Sobti,
Manishi Mukesh, and
Aastha Sobti

CRC Press
Taylor & Francis Group
Boca Raton London New York

CRC Press is an imprint of the
Taylor & Francis Group, an **informa** business

Cover image credit: ©Shutterstock

First edition published 2023
by CRC Press
6000 Broken Sound Parkway NW, Suite 300, Boca Raton, FL 33487–2742

and by CRC Press
2 Park Square, Milton Park, Abingdon, Oxon, OX14 4RN

CRC Press is an imprint of Taylor & Francis Group, LLC

Library of Congress Cataloging-in-Publication Data
Names: Sobti, R. C., editor. | Mukesh, Manishi, editor. | Sobti, Aastha, editor.
Title: Genomics, proteomics and biotechnology / edited by RC Sobti, Manishi Mukesh and Aastha Sobti.
Other titles: Translating animal science research (Series)
Description: First edition. | Boca Raton, FL : CRC Press, 2023. | Series: Translating animal science research | Includes bibliographical references and index.
Identifiers: LCCN 2022023880 (print) | LCCN 2022023881 (ebook) | ISBN 9781032116334 (hardback) | ISBN 9781032116341 (paperback) | ISBN 9781003220831 (ebook)
Subjects: MESH: Genomics | Proteomics | Genetic Techniques—veterinary | Genetic Engineering—veterinary | Biomedical Technology—methods
Classification: LCC QH447 (print) | LCC QH447 (ebook) | NLM QU 460 | DDC 572.8/629—dc23/eng/20220912
LC record available at https://lccn.loc.gov/2022023880
LC ebook record available at https://lccn.loc.gov/2022023881

ISBN: 978-1-032-11633-4 (hbk)
ISBN: 978-1-032-11634-1 (pbk)
ISBN: 978-1-003-22083-1 (ebk)

DOI: 10.1201/9781003220831

Typeset in Times New Roman
by Apex CoVantage, LLC

Content

Editor Biographies

Ranbir Chander Sobti, former Vice-Chancellor at Panjab University in Chandigarh and at Babasaheb Bhimrao Ambedkar University (Central University) in Lucknow, is a scientist, an able administrator, and dynamic institution builder. He has published more than 300 papers in journals of national and international repute, such as *Mutation Research*, *Carcinogenesis*, *Achieves of Toxicology*, *Cancer Genetics and Cytogenetics*, *Molecular Cell Biochemistry*, and *PLOS 1* to name a few. He has also published more than 50 books with international publishers such as Springer, Academic Press, CRC, Elsevier, and others. He is a Fellow of the Third World Academy of Sciences, National Academy of Sciences India, Indian National Science Academy, National Academy of Medical Sciences, National Academy of Agricultural Sciences, Canadian Academy of Cardiovascular Diseases, and others. He was the general president of Indian Science Congress for the 102nd session held at the University of Jammu in 2013. Dr. Sobti is the recipient of many prestigious awards, like the INSA Young Scientist Medal, UGC Career Award, Punjab Rattan Award, JC Bose Oration and Sriram Oration Awards, and of Life Time Achievement Awards of the Punjab Academy of Sciences, Zoological Society of India, and the Environment Academy of India, besides many other medals and awards at national and international levels.

Manishi Mukesh is Principal Scientist (Professor) at ICAR-National Bureau of Animal Genetic Resources in Karnal (India), and engaged in the characterization and conservation of indigenous livestock breeds. He has more than 25 years of experience as a research scientist in the area of cattle and buffalo genomics. He has been engaged in assessing the genome diversity of various cattle breeds of India using neutral and functional markers. He has recently co-authored *Advances in Animal Biotechnology* (Springer). Dr. Mukesh has been awarded the ICAR National Fellow position twice for two terms since May 2011 by Indian Council of Agricultural Research for his contribution in the area of animal sciences. He is also the recipient of the Dr PG Nair Memorial award, the Dr BS Chauhan Gold Medal, and an Eminent Scientist award from different national agencies.

Aastha Sobti has a master's degree in clinical dentistry and oro-facial surgery (UK) and works in the field of head and neck cancers using her intellect and her background in oral and maxillofacial surgery to bring reforms that are required in the present multifarious surgical and research areas. She has teaching and research experience of about seven years and has published a number of papers in reputed journals. She has been awarded a number of prizes and medals for her exceptional work. She has attended and presented papers in international conferences in Croatia, Brazil, Canada, USA, Japan, Switzerland, and other countries.

Contributors

Hemant Agrawal
Flowcytometry Solutions Pvt Ltd
Jaipur, India

Ahmad Ali
Department of Zoology
Panjab University, Sector 14
Chandigarh, India

Navneet Batra
Department of Biotechnology
Goswami Ganesh Dutta Sanatan Dharma College, Sector 32
Chandigarh, India

Sonu Bhatia
Department of Biotechnology
Goswami Ganesh Dutta Sanatan Dharma College, Sector 32
Chandigarh, India

Ekta Chaudhary
Department of Microbiology
Panjab University
Chandigarh, India

Archana Chauhan
Department of Zoology
Panjab University, Sector 14
Chandigarh, India

Manmohan Singh Chauhan
ICAR-Central Institute for Research on Buffaloes
Hisar, Haryana, India

Supriya Chhotaray
Animal Genetics and Breeding Division
ICAR-National Dairy Research Institute
Karnal, Haryana, India

Nisha Phour Dhull
Department of Microbiology
Panjab University
Chandigarh, India

Phuntsog Dolma
Department of Zoology
Panjab University, Sector 14
Chandigarh, India

Seema Dua
ICAR-Central Institute for Research on Buffaloes
Hisar, Haryana, India

Ravi Kr. Gupta
Department of Environmental Microbiology
Babasaheb Bhimrao Ambedkar University

Yudhvir Jakhar
Department of Microbiology
Panjab University
Chandigarh, India

Arashdeep Kaur
Department of Microbiology
Panjab University
Chandigarh, India

Bhawanpreet Kaur
College of Animal Biotechnology
Guru Angad Dev Veterinary and Animal Sciences University
Ludhiana, Punjab, India

Harjot Kaur
Department of Zoology
Panjab University
Chandigarh, India

Maninderjeet Kaur
Department of Microbiology
Panjab University
Chandigarh, India

Ravneet Kaur
Department of Zoology
Post Graduate Government College
Chandigarh, India

JK Kaushik
National Dairy Research Institute
Karnal, India

Abhishek Kumar
Department of Environmental Microbiology
Babasaheb Bhimrao Ambedkar University (A Central University)
VidyaVihar, Lucknow, India

Ashok Kumar
Centre for Systems Biology & Bioinformatics, UIEAST
Panjab University
Chandigarh, India

Dharmendra Kumar
CAR-Central Institute for Research on Buffaloes
Hisar, Haryana, India

Pradeep Kumar
ICAR-Central Institute for Research on Buffaloes
Hisar, Haryana, India

Ravinder Kumar
Department of Zoology
Panjab University
Chandigarh, India

Sudarshan Kumar
National Dairy Research Institute
Karnal, India

Venkatesh R Kumar
Department of Zoology
Babasaheb Bhimrao Ambedkar University
Lucknow, India

Vijay Kumar
Department of Zoology
Panjab University
Chandigarh, India

Vinay Kumar
Department of Environmental Microbiology
Babasaheb Bhimrao Ambedkar University
Lucknow, Uttar Pradesh, India

Anju Kumari
Department of Zoology
Panjab University
Chandigarh, India

Ritika Manhas
Department of Microbiology
Panjab University
Chandigarh, India

K Mohanty
National Dairy Research Institute
Karnal, India

Manishi Mukesh
ICAR-National Bureau of Animal Genetic Resources
Karnal, India

CS Mukhopadhyay
College of Animal Biotechnology
Guru Angad Dev Veterinary and Animal Sciences University
Ludhiana, India

Rajendra Mundkar
KSSRDI
Talaghattapura, Bangalore, India

SK Niranjan
ICAR-National Bureau of Animal Genetic Resources
Karnal, Haryana, India

Meeti Punetha
ICAR-Central Institute for Research on Buffaloes
Hisar, Haryana, India

Deepak K Rahi
Department of Microbiology
Panjab University
Chandigarh, India

Sonu Rahi
Department of Botany
Government Girls College
A.P.S. University
Rewa, India

Praveen Rishi
Department of Microbiology
Panjab University
Chandigarh, India

Ritika Sahi
Department of Microbiology
Panjab University
Chandigarh, India

Ruchi Sankhwar
Department of Environmental Microbiology
Babasaheb Bhimrao Ambedkar University
(A Central University)
VidyaVihar, Lucknow, India

Naresh L Selokar
ICAR-National Dairy Research Institute
Karnal, Haryana, India

Abhishek Sharma
Centre for Systems Biology & Bioinformatics
UIEAST, Panjab University
Chandigarh, India

Apurav Sharma
Department of Microbiology
Panjab University
Chandigarh, India

Neetu Sharma
Department of Biotechnology
Goswami Ganesh Dutta Sanatan Dharma
College, Sector 32
Chandigarh, India

Rakesh Sharma
ICAR-Central Institute for Research on
Buffaloes
Hisar, Haryana, India

Devika Shrivastava
KSSRDI
Thalgattapura, Bangalore, India

Manoj Kumar Singh
ICAR-National Dairy Research
Institute
Karnal, Haryana, India

RC Sobti
Department of Biotechnology
Panjab University
Chandigarh, India

Abhinashi Singh Sodhi
Department of Biotechnology
Goswami Ganesh Dutta Sanatan Dharma
College, Sector 32
Chandigarh, India

Monika Sodhi
ICAR-NBAGR
Karnal, India

Raman Soni
Department of Microbiology
Panjab University
Chandigarh, India

Sanjeev Kumar Soni
Department of Microbiology
Panjab University
Chandigarh, India

Ankita Srivastava
Department of Environmental Microbiology
Babasaheb Bhimrao Ambedkar University
Lucknow, Uttar Pradesh, India

Himanshu Tillu
Flowcytometry Solutions Pvt Ltd
Jaipur, India

Vivek Tiwari
Department of Biotechnology
Goswami Ganesh Dutta Sanatan Dharma
College, Sector 32
Chandigarh, Idia

Akanksha Vishwakarma
Department of Environmental Microbiology
Babasaheb Bhimrao Ambedkar University
Lucknow, Uttar Pradesh, India

Vikas Vohra
Animal Genetics and Breeding Division
ICAR-National Dairy Research Institute
Karnal, Haryana, India

Prem Singh Yadav
ICAR-Central Institute for Research on Buffaloes
Hisar, Haryana, India

Shilpi Yadav
Department of Biotechnology
Babasaheb Bhimrao Ambedkar University

1 Genomics, Transcriptomics and Proteomics for Computational Biology

Abhishek Sharma and Ashok Kumar

CONTENTS

1.1 BACKGROUND

The discovery of remarkable findings like the structure of DNA and human genome sequencing and assembly and its transcription into messenger RNA, which further translated into proteins the crucial part of cells and life, leads to the generation of a very huge amount of data at three steps: one is genomics, another transcriptomics, and the final is proteomics. To infer the information, a biologist needs very accurate and fast methods to analyze the data to understand the biological components of their systems and how they interact with each other; therefore computational approaches were developed. Computational biology, science that uses biological data to infer or develop mathematical algorithms, programs to understand various aspects of biology, in health and disease. This involves data processing its statistical analysis pathway and network information and at last modeling of molecules, e.g., the assembly of genome and its annotations, identification of differentially expressed genes at transcriptomics level and further protein-protein interactions. In the last few years, artificial intelligence and machine learning techniques have also gained popularity. Analyzing the biological data also uses computational techniques for prediction.

1.2 GENOMICS AND COMPUTATIONAL BIOLOGY

Genomics is a field in genetics that applies recombinant DNA technology, DNA sequencing strategies, and bioinformatics to sequence, collect, and examine the features and shape of the genome

DOI: 10.1201/9781003220831-1

[1, 2]. This leads to deciphering the entire DNA sequence of organisms and their genetic mapping. This includes whole genome, epigenetics, and functional genomics study [3].

There are many factors that have an effect on the state of an organism, and the primary crucial factor is genetic makeup. By analyzing the background makeup one can decipher the mutations and variants pathways that can discriminate health and disease [4].

Computational biology approaches make it easier to understand the key factors behind the processes.

1.2.1 Genome Analysis

Three levels are usually concerned in genome analysis: the sequencing of DNA, the assembly of that sequence to create an illustration of the original chromosome, accompanied via way of means of the annotation, and evaluation of that illustration [5]. A rough draft of the human genome was given by the Human Genome Project in 2001. By 2003, the project was completed; the entire genome of one specific person was sequenced. By 2007, this sequence project was declared "finished", with less than one error in 20,000 bases and all chromosomes assembled [6]. Sequence assembly is the next step in genome projects, which is the process of aligning and merging of fragments of much longer DNA sequences in order to reconstruct the original sequence [7]. This reconstruction is not possible without computational approaches.

The final step is Genome Annotation whereby biological information is assigned to several genes. Gene prediction is done here by using different available algorithms and software [8].

Next Generation Sequencing (NGS) is a very powerful platform that arose a few years back. NGS technology is distinctive from the Sanger approach in that it offers hugely parallel analysis, extraordinarily high-throughput from a couple of samples at reduced cost [9]. This results in a huge amount of data that needs to be analyzed to infer the important information. NGS data analysis unravels important points for treatment of diseased persons, improving agriculture varieties, and other scientific problems. Whole data analysis is done by efficient and accurate computational biology approaches and it is assumed that future treatments of diseased persons will be done on the basis of human genomes (Figure 1.1). The latest NGS applications used today are:

- Whole genome de-novo assembly
- miRNA-seq

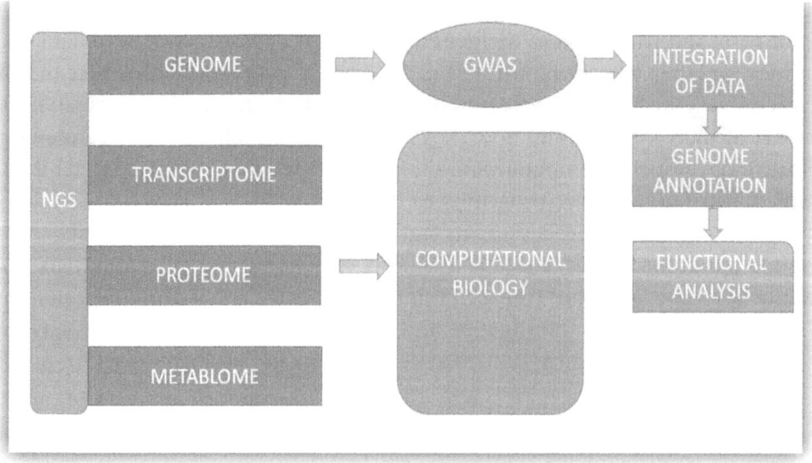

FIGURE 1.1 Data collection from different NGS sources.

- DNA-seq
- Whole genome metagenomics
- RNA-seq

A part from next generation sequencing, **nanopore sequencing** technology is more prevalent these days. When nucleic acid passes through protein nanopores it senses the electrical signals. The outcome signal is decoded further [70]. Nano-pipe is an effective tool with compatible web interface for nanopore data sequencing analysis [69] (Table 1.1).

1.2.1.1 Comparative Genomics

Comparative genomics involves comparing the genome arrangement of dissimilar organisms. Due to the presence of large genomic data, several computational tools have been developed for comparative analysis. It is impossible to visualize long genomic regions manually, hence web-based browsers are available with relevant information; these browsers are updated constantly [10].

1.2.1.2 Genome-Wide Association Studies

Genome-wide association studies (GWAS) are used in research to associate specific genetic variations with a particular disease. This is used to check the different people groups to identify the genetic markers for identification of disease. Without computational techniques it is not easy to perform such operations, hence different pipelines, like Hail pipeline, have been established for the study of GWAS [11].

TABLE 1.1
List of Latest Software/Databases Used in Genomics

Software/ Databases	Short Description	Environment	Reference
SPAdes	Genome assembly	Linux	[18]
MIRA	Genome assembly	Linux	[19]
VELVET	Genome assembly	Linux	[20]
MASURCA	Genome assembly	Linux	[21]
PROKKA	Gene prediction	Linux	[22]
Glimer3	Gene prediction	Online webserver	[23]
Gene-Mark-HMM	Gene prediction	Online webserver	[24]
CIRCOS	Genome Visualization	Linux	[25]
FASTQC	Read quality check	Linux	[26]
Cut-Adapt	Trimming of sequences	Linux	[27]
NCBI	Series of Databases	Online Database	[28]
UCSC	Genome Browser	Online Database	[29]
BLAST	Find similarity between sequences	Online similarity search tool	[30]
GATK	Variant detection	Linux	[31]
Ensemble VEP	Variant effect prediction	Linux	[32]
Hail Pipeline	GWAS analysis	Linux	[33]
MACS2	Identifying transcription factor binding sites used in epigenomics	Linux	[34]
CHIPseeker	ChIP-seeker is an R package for annotating ChIP-seq data analysis	Linux, Windows	[35]

1.2.1.3 Epigenomics

The study of modification of chromatin is epigenomics. It includes DNA and its associated proteins, which play crucial roles in gene coordination and environment response. ChIP-seq is also widely used to study patterns of histone modifications and chromatin modifiers [12]. With the help of computational biological tools, one can process different types of epigenetic data, predicting chromatin states and also their dynamics [13]. Computational advancement and studies in different cell types are found in the human Encyclopaedia of DNA Elements (ENCODE) project [14], the NIH Roadmap Epigenomics effort [15], the Human Epigenome Project [16], and, the most recent, BLUEPRINT, which claims to identify the epigenome of 100 different cell types [17].

1.3 TRANSCRIPTOMICS AND COMPUTATIONAL BIOLOGY

Transcriptomics is the study of the transcriptome, the complete set of RNA transcripts from the genome. The data is expressed through the genome inside the cell in the form of transcripts. Messenger RNA works as an intermediary molecule, while noncoding RNA, like miRNA, has other relevant functions. Transcriptomics means the total content of RNA in a given space and time in a cell. There are new advancements and new computational biology techniques evolving to understand the differentially expressed genes that are regulated or unregulated during a particular time and space. Two technologies that are highly relevant in the area of transcriptomics are Microarray, which quantifies a particular sequence, and RNA-seq, which uses high throughput sequencing. Microarray data stored as high-resolution images requires feature detection and statistical analysis [36]. A human gene expression could be accurately captured by using RNA-seq with 30 million 100 bp sequences per sample [36]. This requires around 2 GB of disk space per sample. Data from both the source microarray and RNA-seq are stored in public repositories like NCBI-SRA (sequence read archive) and Gene Expression Omnibus. RNA-seq is today's most advanced technique.

1.3.1 RNA-Seq Data Analysis

From the experiments a huge amount of raw data is generated, which has to be processed in the form of useful information (Figure 1.2). Generally, data evaluation is divided into four parts:

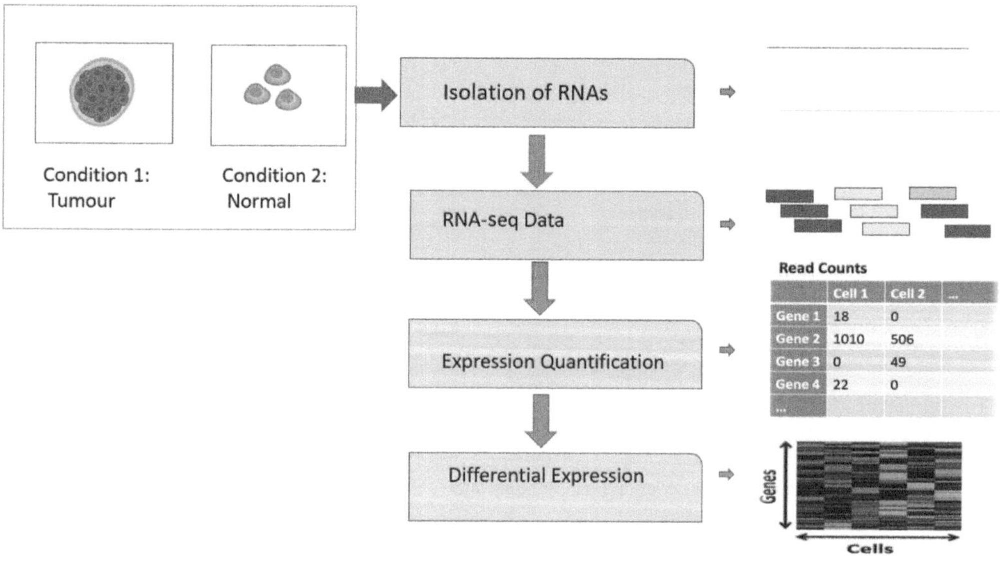

FIGURE 1.2 Overview of RNA-seq workflow.

TABLE 1.2
Transcriptomic Analysis Tools

Software	Description	Environment
Velvet [38]	RNA-seq assembly	Linux
CLC genomics workbench [39]	RNA-seq assembly	Windows, Linux
Newbler [40]	RNA-seq assembly	Linux
Cuffdiff2 [41]	Differential gene expression	Unix-based
EdgeR [42]	Differential gene expression	R/Bioconductor
DEseq2 [43]	Differential gene expression	R/Bioconductor
Limma/Voom [44]	Differential gene expression	R/Bioconductor

TABLE 1.3
Transcriptomic Databases

Databases	Host	Type of Data Stored
Gene Expression Omnibus [45]	NCBI	Microarray RNA-seq
ArrayExpress [46]	ENA	Microarray
Expression Atlas [47]	EBI	Microarray RNA-seq
NONCODE [48]	Noncode.org	RNA-seq

quality control, alignment, quantification, and differential expression [37]. Most of the programs and software are run through Linux Operating system-based application or R/Bioconductor package [90]. Other application tools required for analysis are described in Table1.2.

1.3.2 Transcriptome Databases

The large amount of raw data generated by experiments are deposited in databases from where one can easily access the data for analysis purposes (Table 1.3). Transcriptomics revolutionized the research with explaining the differential expression of the genome. This is possible due to new advanced computational techniques that led to deciphering the problems that were faced three decades ago. Low cost and availability of data make these technologies easy for the scientific community to understand various aspects.

1.4 PROTEOMICS AND COMPUTATIONAL BIOLOGY

Proteomics refers to the study of proteome on a large scale. Proteome shows functional part pf the biological systems which changes from cell to cell with respect to time. Proteomics answer the questions whether the protein involved in some biological process, interaction with other protein and subcellular localizations. The recent development in storing and analyzing large-scale proteomics data is a challenging task. Computational biology gives particular solutions to perform these tasks by integrating raw protein data and proteomics database searching to the functional interpretation of huge protein data sets.

1.4.1 Protein Motif and Domain Analysis

To know the functions of proteins that still lack information, and also their interactions and behavior in relevant pathways with other members proteins, it is helpful to identify them by the known amino acid sequence for specific domains and motifs. BLAST is a computational biology program search against the database of known protein sequences and results in the homology percentage between query and target protein sequences [30]. Pfam, Interpro, SMART, and DAVID are online webservers used in amino acid sequence to identified proteins that share a specific protein fold with other proteins [49–52]. These servers used Hidden Markov Model algorithm (HMM) at their back-end to predict a given amino acid sequence. Other algorithms, such as MotifX or PhosphoMotif, analyze the post-translational modifications sites [53,54].

1.4.2 Protein-Protein Interaction Study

Most of the proteins do not work independently but require other proteins to form complexes for important cell functionality. For the analysis of the protein complexes – their formation and dissociation and their interaction with one another – the data are stored in biological repositories such as MINT, BioGRID, IntAct, or HRPD [55–58]. Data stored in these databases related to protein interaction are analyzed on an experimental basis. The most used database for interaction data is STRING, which uses literature mining from the PubMed database and other resources for the prediction of interactions and networks. This also creates intensive layout networks by providing gene lists or protein lists to show higher degrees of connectivity between hub proteins [59]. Cystoscope, a visualization and graphical tool, creates networks of interaction (Figure 1.3) [60]. These tools and databases make understanding the interactions between proteins very easy.

1.4.3 Mass Spectra

Mass spectra (MS) is currently the main technique for large-scale whole-proteome study. Today's advanced techniques produce data sets around one million spectra, i.e., approximately 100 GB of

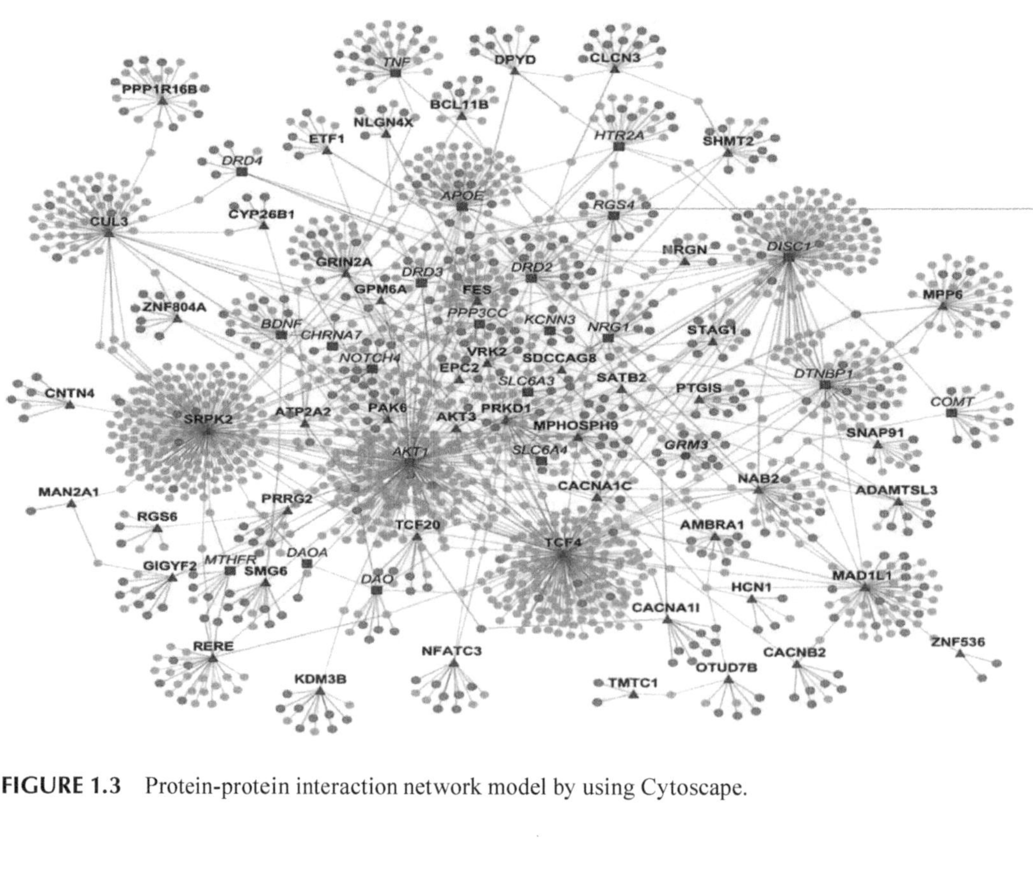

FIGURE 1.3 Protein-protein interaction network model by using Cytoscape.

data. Initial processing of data contains noise detection, subtraction, peak detection, and alignment of LC/MS maps. For analysis, generally four steps are followed: (1) identification of amino-acid sequences, peptides, and proteins in Peptide-Spectrum Match (PSM), and detection, quantification, annotation, and alignment of features; (2) peptide and protein significance analysis; (3) class discovery and prediction; and (4) pathway detection. The biggest repositories – PRIDE, Proteome Commons, and Peptide Atlas – inside the proteome exchange task allow entry to a greater number of the saved proteomic data sets and afford beneficial analysis [61]. neXtProt is online platform for human proteins that contains information such as their function, subcellular location, expression, interactions, and role in diseases [62]. For MS analysis there is one software, the Trans-Proteomic Pipeline, that statistically validates peptide-spectra-matches (PSM) using the results of search engines by estimating a false discovery rate (FDR) on PSM level [63].

1.4.4 Molecular Dynamic Simulations

A computational biology technique is used to mimic the physical motion of atoms in protein molecules with respect to actual environment. For a certain period of time these atoms of protein molecules are allowed to interact with each other to compute their trajectory with respect to time [64]. Potential energy of system of particles is used with the help of Molecular Mechanics Force Fields. MD simulation applications such as NAMD [65], GROMACS [66], and AMBER [67], are widely used, which constantly improves their algorithms and performance to give ~10 100ns/day/workstation/cluster [68]. Protein simulation is done in mimicking environment conditions where several factors like temperature and pressure are also considered. Advancement in MD simulation application is used to infer complex biological systems for longer time scale. When the simulation process runs, it produces big data in gigabytes. Data are related to protein atom motions, and conformational changes yield information that is not possible to analyze without computational techniques [64].

The advancement in computational biological tools and databases in the field of proteomics leads to the study of all proteins and their functions, interactions. Their atomic behavior on nanoscales with each other opens a new path of innovation in the field of human pathology and crop improvement (Figure 1.4).

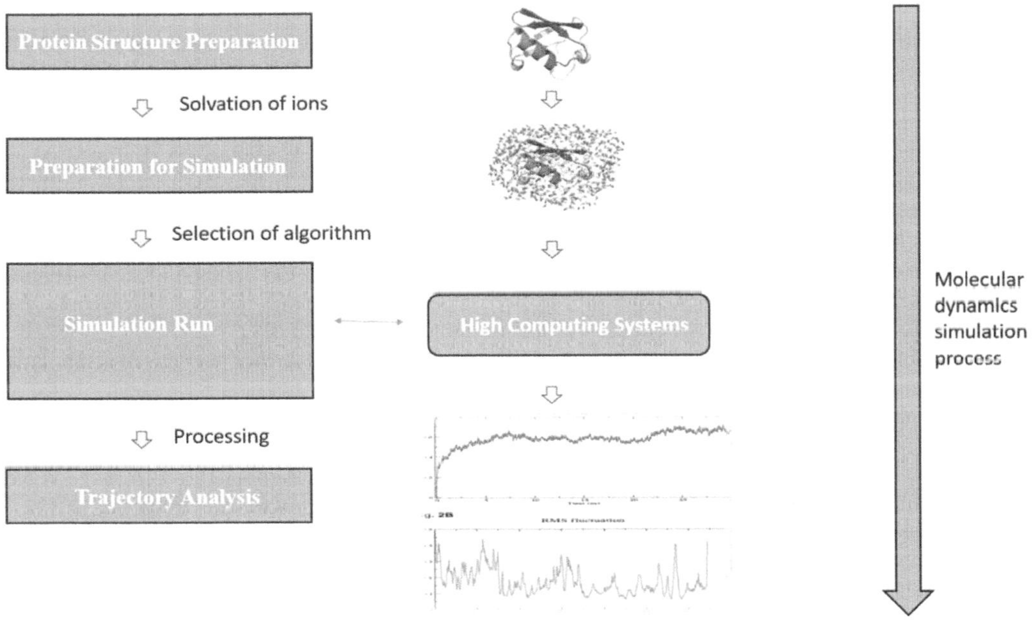

FIGURE 1.4 Molecular dynamic simulations process.

1.5 CONCLUSION

High-throughput data from genomics transcriptomics and proteomics studies in laboratories are often generated by large consortia. Omics data are proving very promising. Today's advanced techniques are very diverse and analyze complex biological data. Computational biology evolution will increase accuracy and robustness, and reduce costs. These innovations continue shifting from production of data to analysis of data. An understanding of these computational biological approaches will develop a good scale of analysis to infer the learning process in life sciences.

BIBLIOGRAPHY

Adams D, Altucci L, Antonarakis SE, Ballesteros J, Beck S, Bird A, Bock C, Boehm B, Campo E, Caricasole A, Dahl F, Dermitzakis ET, Enver T, Esteller M, Estivill X, Ferguson-Smith A, Fitzgibbon J, Flicek P, Giehl C, Graf T, Grosveld F, Guigo R, Gut I, Helin K, Jarvius J, Kuppers R, Lehrach H, Lengauer T, Lernmark A, Leslie D, Loeffler M, Macintyre E, Mai A, Martens JH, Minucci S, Ouwehand WH, Pelicci PG, Pendeville H, Porse B, Rakyan V, Reik W, Schrappe M, Schubeler D, Seifert M, Siebert R, Simmons D, Soranzo N, Spicuglia S, Stratton M, Stunnenberg HG, Tanay A, Torrents D, Valencia A, Vellenga E, Vingron M, Walter J, Willcocks S (2012) BLUEPRINT to decode the epigenetic signature-written in blood. *Nat Biotech.* 30(3):224–226

Ahmad Abu Turab Naqvi, Taj Mohammad, Gulam Mustafa Hasan, Md. Imtaiyaz Hassan. Advancements in docking and molecular dynamics simulations towards ligand-receptor interactions and structure-function relationships. *Curr Top Med Chem.* 2018;18(20). https://doi.org/10.2174/1568026618666181025114157.

Altschul SF, Gish W, Miller W, Myers EW, Lipman DJ. Basic local alignment search tool. *J Mol Biol.* 1990 Oct 5;215(3):403–410. doi: 10.1016/S0022-2836(05)80360-2. PMID: 2231712.

Amanchy R, Periaswamy B, Mathivanan S, Reddy R, Tattikota S, Pandey A: A curated compendium of phosphorylation motifs. *Nat Biotech.* 2007, 25(3):285–286.

Andrews S (2010) *FastQC: A quality control tool for high throughput sequence data.* Available online at: www.bioinformatics.babraham.ac.uk/projects/fastqc

Angarica, V E, del Sol A. (2017). Bioinformatics tools for genome-wide epigenetic research. *Neuroepigenom Aging Dis.* 489–512. doi: 10.1007/978-3-319-53889-1_25

Bankevich A, Nurk S, Antipov D, Gurevich AA, Dvorkin M, Kulikov AS, Lesin VM, Nikolenko SI, Pham S, Prjibelski AD, Pyshkin AV, Sirotkin AV, Vyahhi N, Tesler G, Alekseyev MA, Pevzner PA. SPAdes: A new genome assembly algorithm and its applications to single-cell sequencing. *J Comput Biol.* 2012 May;19(5):455–77. doi: 10.1089/cmb.2012.0021. Epub 2012 Apr 16. PMID: 22506599; PMCID: PMC3342519.

Bergman NH. (2007). Bergman, N.H. (ed.). *Comparative Genomics: Volumes 1 and 2.* Totowa, New Jersey: Humana Press. ISBN 978-193411-537-4. PMID 21250292.

Bernstein BE, Stamatoyannopoulos JA, Costello JF, Ren B, Milosavljevic A, Meissner A, Kellis M, Marra MA, Beaudet AL, Ecker JR, Farnham PJ, Hirst M, Lander ES, Mikkelsen TS, Thomson JA (2010) The NIH roadmap epigenomics mapping consortium. *Nat Biotechnol.* 28(10):10

Borodovsky M, Lomsadze A. Eukaryotic gene prediction using GeneMark.hmm-E and GeneMark-ES. *CurrProtoc Bioinformat.* 2011 Sep;Chapter 4:Unit 4.6.1–10. doi: 10.1002/0471250953.bi0406s35. PMID: 21901742; PMCID: PMC3204378.

Bradbury J (2003) Human epigenome project – up and running. *PloSBiol.* 1(3):E82.

Case DA, Cheatham TE 3rd, Darden T, Gohlke H, Luo R, et al. (2005) The Amber biomolecular simulation programs. *J Comput Chem.* 26:1668–1688.

Chatraryamontri A, Ceol A, Montecchi Palazzi L, Nardelli G, Schneider MV, Castagnoli L, Cesareni G: MINT: the Molecular INTeraction database. *Nucleic Acids Res.* 2007, 35(Database):D572–D574.

Chevreux B. (2005). MIRA: An automated genome and EST assembler. Zerbino DR, Birney E. Velvet: Algorithms for de novo short read assembly using de Bruijn graphs. *Genome Res.* 2008;18:821–829.

Chou M, Schwartz D. *Current Protocols in Bioinformatics*, 01 Sep 2011, Chapter 13:Unit 13.15–24. DOI: 10.1002/0471250953.bi1315s35 PMID: 21901740

Damian Szklarczyk, Annika L Gable, David Lyon, Alexander Junge, Stefan Wyder, Jaime Huerta-Cepas, Milan Simonovic, Nadezhda T Doncheva, John H Morris, Peer Bork, Lars J Jensen, Christian von Mering, STRING v11: protein–protein association networks with increased coverage, supporting functional discovery in genome-wide experimental datasets. *Nucl Acids Res.* 08 January 2019;47(D1):D607–D613, https://doi.org/10.1093/nar/gky1131

Dror RO, Dirks RM, Grossman JP, Xu H, Shaw DE. Biomolecular simulation: a computational microscope for molecular biology. *Annu Rev Biophys*. 2012;41:429–452.

Edgar R, Domrachev M, Lash AE. Gene Expression Omnibus: NCBI gene expression and hybridization array data repository. *Nucl Acids Res*. 2002;30:207–210. PMID: 11752295.

Hail Team. Hail 0.2.13–81ab564db2b4. https://Github.com/hail-is/hail/releases/tag/0.2.13 http://Doi.org/10.5281/zenodo.2646680

Hail Team. Hail 0.2.13–81ab564db2b4. https://github.com/hail-is/hail/releases/tag/0.2.13

Hannon P, Markiel A, Ozier O, et al. Cytoscape: a software environment for integrated models of biomolecular interaction networks. *Genome Res*. 2003;13(11):2498–2504. doi:10.1101/gr.1239303. PMC 403769. PMID 14597658.

Hunter S, Jones P, Mitchell A, Apweiler R, Attwood TK, Bateman A, Bernard T, Binns D, Bork P, Burge S, et al. Interpro in 2011: new developements in the family and domain prediction database. *Nucleic Acids Res*. 2012;40(D1):D306–D312.

Image citations: Ben Amar A, Daldoul S, Reustle G, Krczal G, Mliki A. *Curr Genomics*. 2016 Dec;17(6):460–475. doi: 10.2174/1389202917666160520102827.

International Human Genome Sequencing Consortium. Finishing the euchromatic sequence of the human genome. *Nature*. 2004;431:931–945.

Jiao X, Sherman B, Huang D, Stephens R, Baseler M, Lane C, Lempicki R. DAVID-WS: a stateful web service to facilitate gene/protein list analysis. *Bioinformatics*. 2012. Jul 1;28(13):1805–6. doi:10.1093/bioinfomratics/bts251

Kaur, S. Genomics, Editor(s). Stanley Maloy, Kelly Hughes, *Brenner's Encyclopaedia of Genetics*, 2nd ed. Academic Press, 2013: 310–312. ISBN 9780080961569.

Keller A, Nesvizhskii A, Kolker E, Aebersold R. Empirical statistical model to estimate the accuracy of peptide identifications made by MS/MS and database search. *Anal Chem*. 2002;74(20):5383–5392. doi:10.1021/ac025747h. PMID 12403597.

Kelley DR, Liu B, Delcher AL, Pop M, Salzberg SL. Gene prediction with Glimmer for metagenomic sequences augmented by classification and clustering. *Nucleic Acids Res*. 2012 Jan;40(1):e9. doi: 10.1093/nar/gkr1067. Epub 2011 Nov 18. PMID: 22102569; PMCID: PMC3245904.

Kerrien S, Aranda B, Breuza L, Bridge A, Broackes-Carter F, Chen C, Duesbury M, Dumousseau M, Feuermann M, Hinz U, et al. The IntAct molecular interaction database in 2012. *Nucleic Acids Res*. 2012;40(D1):D841–D846.

Keshava Prasad TS, Goel R, Kandasamy K, Keerthikumar S, Kumar S, Mathivanan S, Telikicherla D, Raju R, Shafreen B, Venugopal A, et al. Human Protein Reference Database – 2009 update. *Nucleic Acids Res*. 2009;37(Database):D767–D772.

Kolesnikov N, Hastings E, Keays M, Melnichuk O, Tang YA, Williams E, et al. ArrayExpress update simplifying data submissions. *Nucleic Acids Res*. 2015;43:D1113–D1116. https://doi.org/10.1093/nar/gku1057. PMID: 25361974.

Krzywinski M, Schein J, Birol I, et al. Circos: an information aesthetic for comparative genomics. *Genome Res*. 2009;19(9):1639–1645. doi:10.1101/gr.092759.109

Kumar S, Blaxter ML. Comparing de novo assemblers for 454 transcriptome data. *BMC Genomics*. 2010;11:571. https://doi.org/10.1186/1471-2164-11-571. PMID: 20950480

Letunic I, Doerks T, Bork P: SMART 7: recent updates to the protein domain annotaion resource. *Nucleic Acids Res*. 2012;40(D1):D302–D305.

Love MI, Huber W, Anders S. Moderated estimation of fold change and dispersion for RNA-seq data with DESeq2. *Genome Biol*. 2014;15:550. https://doi.org/10.1186/s13059-014-0550-8. PMID:25516281.

Lowe R, Shirley N, Bleackley M, Dolan S, Shafee T. Transcriptomics technologies. *PloSComputeBiol*. 2017;13(5):e1005457. https://doi.org/10.1371/journal.pcbi.1005457

Lydie L. "neXtProt: a knowledge platform for human proteins. *Nucleic Acids Res*. 2012;40(Database issue):D76–D83. doi:10.1093/nar/gkr1179. PMC 3245017. PMID 22139911.

Makałowski W, Shabardina V. Bioinformatics of nanopore sequencing. *J Hum Genet*. 2020 Jan;65(1):61–67. doi: 10.1038/s10038-019-0659-4

Mardi's ER: A decade's perspective on DNA sequencing technology. *Nature*. 2011;470:198–203. doi: 10.1038/nature09796

Margulies M, Egholm M, Altman WE, Attiya S, Bader JS, Bemben LA, et al. Genome sequencing in microfabricated high-density picolitre reactors. *Nature*. 2005;437:376–380. https://doi.org/10.1038/nature03959 PMID: 16056220.

MARTIN M. Cutadapt removes adapter sequences from high-throughput sequencing reads. *EMBnet.journal.* [S.l.] May 2011;17(1):10–12. ISSN 2226–6089. Available at: <https://journal.embnet.org/index.php/embnetjournal/article/view/200>. Date accessed: 10 mar. 2021. doi: https://doi.org/10.14806/ej.17.1.200

Mc Elheny V. *Drawing the map of life: Inside the human genome project.* New York, NY: Basic Books, 2010. ISBN 9780465043330.

McKenna A, Hanna M, Banks E, Sivachenko A, Cibulskis K, Kernytsky A, Garimella K, Altshuler D, Gabriel S, Daly M, DePristo MA. The Genome Analysis Toolkit: a MapReduce framework for analyzing next-generation DNA sequencing data. *Genome Res.* 2010 Sep;20(9):1297–303. doi: 10.1101/gr.107524.110. Epub 2010 Jul 19. PMID: 20644199; PMCID: PMC2928508.

McLaren W, Gil L, Hunt SE et al. The ensembl variant effect predictor. *Genome Biol.* 2016;17:122. https://doi.org/10.1186/s13059-016-0974-4

National Center for Biotechnology Information (NCBI)[Internet]. *Bethesda (MD): National Library of Medicine (US), National Center for Biotechnology Information* [1988] [cited 2017 Apr 06]. Available from: www.ncbi.nlm.nih.gov

National Human Genome Research Institute. *A brief guide to genomics*, 2010. www.genome.gov. Retrieved 3/12/2011.

Patodia S, Bagaria A, Chopra D. Molecular dynamics simulation of proteins: A brief overview. *J Phys Chem Biophys* 2014;4:166. doi: 10.4172/2161–0398.1000166

Petryszak R, Keays M, Tang YA, Fonseca NA, Barrera E, Burdett T, et al. Expression Atlas update an integrated database of gene and protein expression in humans, animals and plants. *Nucleic Acids Res.* 2016;44:D746–D752. https://doi.org/10.1093/nar/gkv1045 PMID: 26481351.

Pevsner J. *Bioinformatics and Functional Genomics*, 2nd ed. Hoboken, NJ: Willey Blackwell, 2009. ISBN97804700851.

Pevsner J. *Bioinformatics and Functional Genomics*, 2nd ed. Hoboken, NJ: Willey Blackwell, 2009. ISBN97804700851.

Phillips JC, Braun R, Wang W, Gumbart J, Tajkhorshid E, et al. Scalable molecular dynamics with NAMD. *J Comput Chem.* 2005;26:1781–1802.

Prokop A, Csukás D. (eds.). *Systems Biology.* Springer Dordrecht, https://doi.org/10.1007/978-94-007-6803-1

Pronk S, Pall S, Schulz R, Larsson P, Bjelkmar P, et al. GROMACS 4.5: A high-throughput and highly parallel open-source molecular simulation toolkit. *Bioinformatics* 2013;29: 845–854.

Punta M, Coggill P, Eberhardt R, Mistry J, Tate J, Boursnell C, Pang N, Forslund K, Ceric G, Clements J, et al. The Pfam protein families database. *Nucleic Acids Res.* 2012;40(Database):D290–D301.

Reverse Genetics and High Throughput Sequencing Methodologies for Plant Functional Genomics. *Curr Genom.* 17:1–1. doi: 10.2174/1389202917666160520102827. www.genengnews.com/insights/protein-protein-interactions-get-a-new-groove-on/; https://learn.gencore.bio.nyu.edu/single-cell-rnaseq/

Ritchie ME, Phipson B, Wu D, Hu Y, Law CW, ShiW, Smyth GK. limma powers differential expression analyses for RNA-sequencing and microarray studies. *Nucleic Acids Res.* 2015;43:e47. https://doi.org/10.1093/nar/gkv007. PMID: 25605792.

Robinson MD, McCarthy DJ, Smyth GK. edgeR: a Bioconductor package for differential expression analysis of digital gene expression data. *Bioinformatics.* 2010;26:139–140. https://doi.org/10.1093/bioinformatics/btp616. PMID:19910308.

San Francisco. *Concepts of Genetics*, 10th ed. Pearson Education, 2012. ISBN9780321724120.

Seemann T, Prokka: rapid prokaryotic genome annotation. *Bioinformatics.* 15 July 2014;30(14):2068–2069.

Shabardina V, Kischka T, Manske F, Grundmann N, Frith MC, Suzuki Y, et al. NanoPipe-a web server for nanopore MinION sequencing data analysis. *Gigascience.* 2019 Feb 1;8(2):giy169. PMID: 30689855.

Stark C, Breitkreutz BJ, Reguly T, Boucher L, Breitkreutz A, Tyers M: BioGRID: a general repository for interaction datasets. *Nucleic Acids Res.* 2006;34(suppl 1):D535–D539.

Stein L. Genome annotation: From Sequence to Biology. *Nat Rev Genet.* 2001;2(7):493–503.

Trapnell C, Hendrickson DG, Sauvageau M, Goff L, Rinn JL, Pachter L. Differential analysis of gene regulation at transcript resolution with RNA-seq. *Nat Biotechnol.* 2013;31:46–53. https://doi.org/10. 1038/nbt.2450. PMID: 23222703.

UCSC Genome Browser: Kent WJ, Sugnet CW, Furey TS, Roskin KM, Pringle TH, Zahler AM, Haussler D. The human genome browser at UCSC. *Genome Res.* 2002 Jun;12(6):996–1006.

Van Verk MC, Hickman R, Pieterse CM, Van Wees SC. RNA-Seq: revelation of the messengers. *Trends Plant Sci.* 2013;18:175–179. https://doi.org/10.1016/j.tplants.2013. 02.001. PMID: 23481128.

Yu, G, Wang, LG, He, QY. ChIPseeker: an R/Bioconductor package for ChIP peak annotation, comparison and visualization. *Bioinformatics*. 2015;31(14):2382–2383. doi: [10.1093/bioinformatics/btv145] (http://dx.doi.org/10.1093/bioinformatics/btv145)

Zerbino DR, Birney E. Velvet: algorithms for de novo short read assembly using de Bruijn graphs. *Genome Res*. 2008;18:821–829. https://doi.org/10.1101/gr.074492.107. PMID: 18349386.

Zhang Y, Liu T, Meyer CA, Eeckhoute J, Johnson DS, Bernstein BE, Nusbaum C, Myers RM, Brown M, Li W, Liu XS. (2008) Model-based Analysis of ChIP-Seq (MACS). *Genome Biol*. 2008;9(9):R137.

Zhao Y, Li H, Fang S, Kang Y, Wu W, Hao Y, et al. NONCODE 2016; An informative and valuable data source of long non-coding RNAs. *Nucleic Acids Res*. 2016;44:D203–D208. https://doi.org/10.1093/nar/gkv1252. PMID: 26586799.

Zhou VW, Goren A, Bernstein BE. Charting histone modifications and the functional organization of mammalian genomes. *Nat Rev Genet*. 2011;12(1):7–18

Zhu B, de Jong PJ. Identification and analysis of functional elements in 1% of the human genome by the ENCODE pilot project. *Nature*. 2007;447(7146):799–816

Zimin AV, Marçais G, Puiu D, Roberts R, Salzberg S L, Yorke J A. The MaSuRCA genome assembler. *Bioinformatics*. 1 Nov 2013;29(21):2669–2677.

2 Genomics
The Rise of Sequencing Technology

Archana Chauhan, Ahmad Ali and Phuntsog Dolma

CONTENTS

2.1 INTRODUCTION

Sanger's studies of insulin 75 years ago established the importance of sequence in biological macromolecules (Hutchison, 2007). For the very first time, the research demonstrated that proteins are made up of linear polypeptides formed by joining amino acid residues in a programmed sequence but apparently seem to be put together in a random sequence (Sanger, 1949; Sanger and Tuppy, 1951). Later, in 1953, when Watson and Crick demonstrated the double-helical structure of DNA working from crystallographic data produced by Rosalind Franklin and Maurice Wilkins, the implications for DNA base sequences were supposed to be considered (Watson and Crick, 1953; Zallen, 2003). As a result, the stage was set to stress the coding problem: how does the base sequence of the DNA gene that encodes a protein decide the amino acid sequence of that protein? (Hutchison, 2007). This contributed to the development of a conceptual framework for both DNA replication and nucleic

DOI: 10.1201/9781003220831-2

acid encoding proteins for a long time, however, the ability to "read" or sequence DNA remained elusive. Strategies for inferring the sequence of protein chains did not seem to be easily transferable to nucleic acid research: DNA molecules were much longer and made up of smaller, more similar units, making it more difficult to differentiate between them (Hutchison, 2007; Heather and Chain, 2016). The genetic and biochemical properties of terrestrial life are basically encoded in nucleic acid sequences in polynucleotide chains.

The introduction of sequencing technologies has had a major impact on the study of organism genomic sequences. Frederick Sanger published an important paper in 1977 that introduced the DNA sequencing method and made a significant contribution to the advancement of biological sciences over the last few decades (Sanger et al., 1977). The methods vastly improved existing DNA sequencing techniques pioneered by Maxam and Gilbert published in the same year, and the "plus and minus" system of Sanger and Coulson's published two years before (Sanger and Coulson, 1975; Maxam and Gilbert, 1977). The dideoxy chain terminator DNA sequencing method used automated gel electrophoresis and fluorescent terminator chemistry, first on slab gels and then on capillary gels (Kircher and Kelso, 2010). DNA sequences are stored in files generated by a DNA sequencer machine; these sequences are called reads on an alphabet based on the nitrogenous base A, T, C, G, and N. An ambiguity is represented by the letter N (Shendure and Ji, 2008). The benefit of handling less toxic chemicals, low radioactive substance and its high-efficiency Sanger sequencing has become the sole technique of sequencing used for almost 30 years (Pareek et al., 2011). Sanger provided a perfect tool for deciphering the nucleotide sequences of a full gene and later the whole genome of an organism, it has been automated and commercialized as the "Sanger Sequencing Technology." It was the core technology that allowed the Human Genome Project (HGP) to be completed successfully (International Human Genome Sequencing Consortium, 2004).

Before the advent of a new age of sequencing technologies that opened new perspectives for genome exploration and study, the most popular sequencing technologies used by biologists were Sanger and Maxam-Gilbert (Mehdi et al., 2017). Later, in 2005, Roche's 454 technology first commercialized the Next Generation Sequencing (NGS) technologies, which have the potential to generate high-throughput sequences (HTS) at a lesser charge than the traditional sequencing techniques (Shendure and Ji, 2008; Kchouk et al., 2017). Since then, the NGS technologies have advanced at a breakneck pace; new sequencing platforms have entered the market, and the amount of data generated by these new technologies has exploded (Figure 2.1; Metzker, 2010; Zhou et al.,

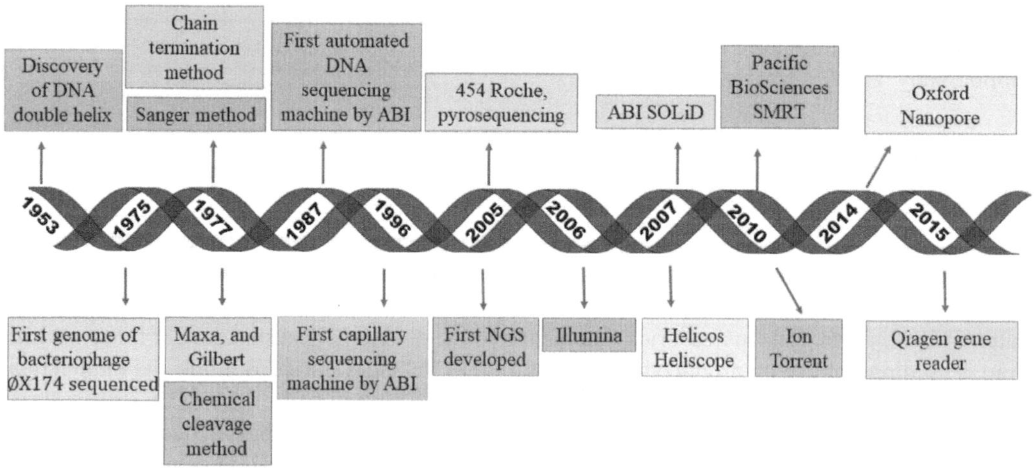

FIGURE 2.1 History of nucleotide sequencing technology.

2010). These accomplishments not only provided a vast volume of raw data for scientific research, but they also revealed new scientific theories and fundamentally altered the working methods in the field of genomics (Guo et al., 2010; Yu et al., 2010; Zhou et al., 2010). NGS technologies allow massively parallel sequencing, high-throughput analysis of multiple samples at a much lower cost (Eid et al., 2009). NGS technologies can sequence millions of reads per cycle, and the time it takes to produce GB-sized reads is just a few hours or a week, making it greater to first-generation sequencing techniques like Sanger sequencing (Wheeler et al., 2008; Korlach et al., 2010).

2.2 GENERATION OF SEQUENCING TECHNOLOGIES

The drastic advancement in sequencing technologies and the number of sequencers has increased exponentially in recent years (Shendure and Ji, 2008; Thudi et al., 2012). There are at least four generations of DNA sequencing methods that can be categorized based on their key characteristics (McGinn and Gut, 2013). Different sequencing methods and platforms are described in Table 2.1 and Figure 2.2.

2.2.1 First Generation Sequencing

With their publication in 1977, Sanger and Maxam-Gilbert introduced DNA sequencing in the field of life science, and these methods are categorized as first-generation sequencing technologies (Maxam and Gilbert, 1977; Liu et al., 2012; Thudi et al., 2012). Fragment ladders for first-generation sequencing are formed by enzymatically elongating a primer hybridized to a pool of template DNA and addition of unique adenine (A), thymine (T), guanine (G), and cytosine (C) terminators (Sanger and Coulson, 1975). Nucleotide base-specific cleavage was also announced in a system developed by Maxam and Gilbert (Maxam and Gilbert, 1977). Isolation of ladders is carried out through gel electrophoresis. Either radioactive labeling and film exposure is used to reveal the products or by the addition of fluorescent dyes to the termination reaction and fluorescent imaging.

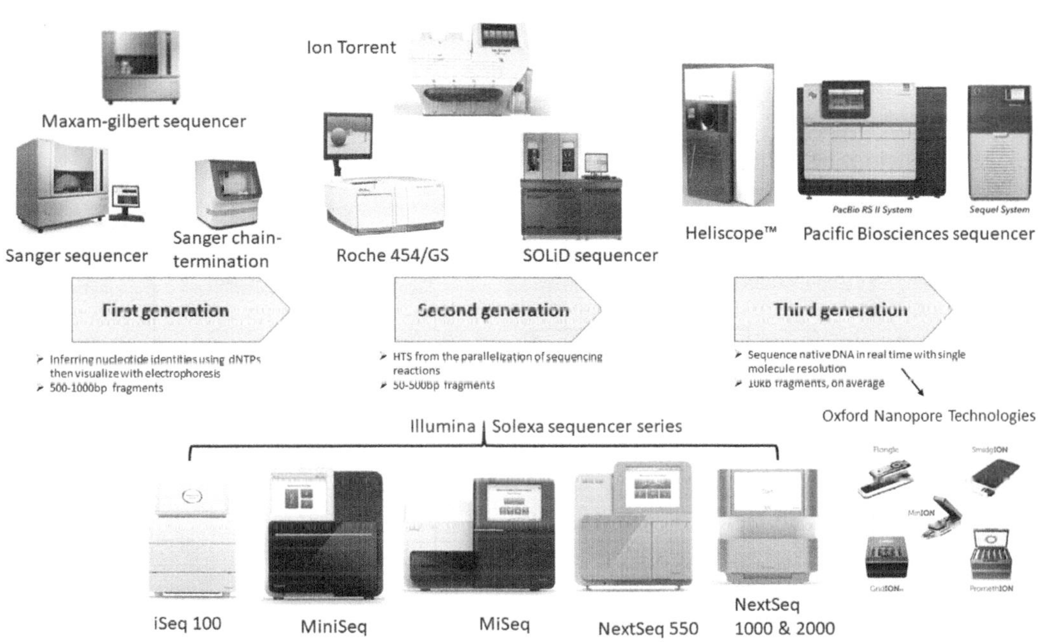

FIGURE 2.2 Evolution and generation of sequencing technology.

2.2.1.1 Sanger Sequencing

Sanger sequencing is a type of targeted sequencing that uses short nucleotide primers to detect a particular DNA region. It is also known as the dideoxynucleotide method or chain termination method. The first step in Sanger sequencing is the denaturation of the double-stranded DNA (dsDNA) because it uses one strand of the dsDNA as a template to be sequenced. After that, the single-stranded DNA (ssDNA) is annealed to short nucleotide primers and elongation takes place with a blend of deoxynucleotide triphosphates (dNTPs) that provide the required A, T, G, C nucleotides to build the new dsDNA. A minute amount of chain-terminating chemically modified nucleotides called dideoxynucleotide triphosphates (ddNTPs) is also included for each nucleotide. Further, the addition of dNTPs to the sequence is done before a ddNTP is attached. These dNTPs are labeled with ddA, ddT, ddG, and ddC for each nitrogenous base. The sequence will remain elongated with dNTPs until a ddNTP is incorporated. Once these ddNTPs are incorporated to the elongating nucleotide chain it stops further addition of dNTPs and terminates the elongation. The final products will be different-sized DNA fragments ended with dNTPs. Then these fragments are isolated by a gel sab electrophoresis based on their size and the resultant bands matching to nucleotide fragments are visualized by UV light (Sanger and Coulson, 1975; El-Metwally et al., 2014). A fluorescent marker is used in each ddNTP (ddATP, ddTTP, ddGTP, and ddCTP). The base can show fluoresce depending on the matching of the corresponding nucleotide when a ddNTP is added to the elongating sequence. Green fluorescence denotes A, red denotes T, black denotes G, and blue denotes C. Fluorescent intensity is detected by a laser within the system that reads the sequence, which is represented as a "peak." When a heterozygous mutation arises in a series, two fluorescent dyes of similar intensity will be used to capture loci. When a homozygous allele exists, the expected fluorescent dye color is entirely replaced by the color of the novel base pair (Figure 2.3).

In the mid-1990s, a new advancement in Sanger sequencing took place, in which the DNA fragments are sorted by size in capillary sequencing machines using a long and thin acrylic-fiber capillary rather than a gel slab electrophoresis. In capillary sequencing, a capillary is injected with a sample

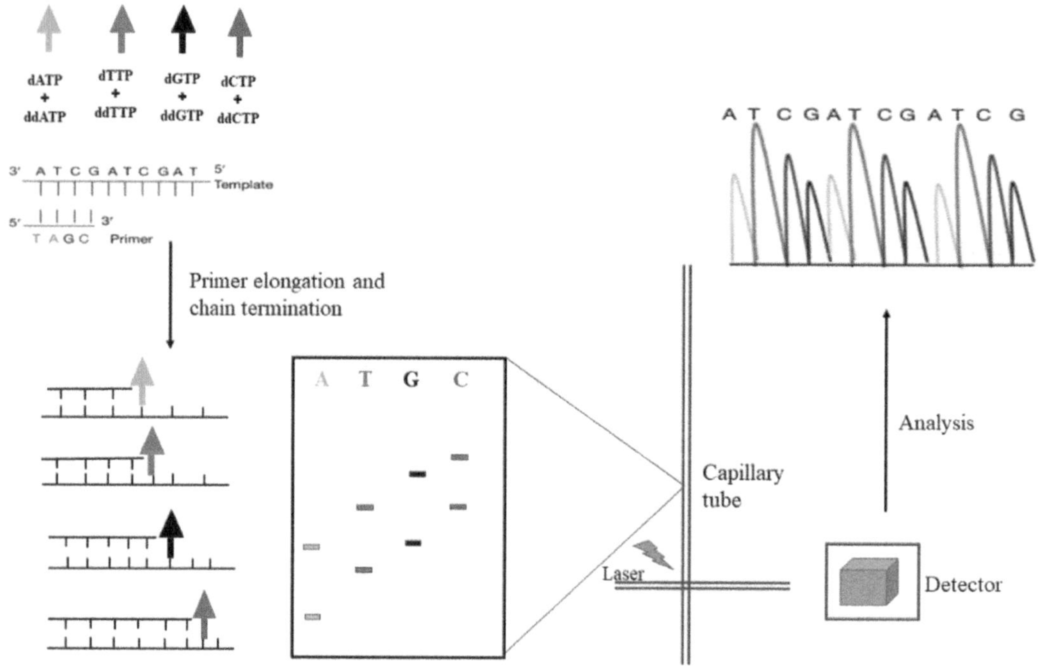

FIGURE 2.3 Sanger sequencing technology.

containing fragments of DNA. This is accomplished by immersing a capillary and an electrode in a sample solution and applying a short electric current. DNA moves to the capillary's end as a result of this. The fluorescence color tags on the nucleotide glow as a result of a fluorescence-detecting laser that is integrated within the machine shooting through the capillary fiber. A sensor detects the color of the fluorescent bases, and the sequencing machine records it. Finally, the basic colors are then presented as a graph of differences on a computer. Sanger sequencing is a commonly used approach for detecting small nucleotide variations (SNVs). During *in-vitro* DNA replication, the DNA polymerase selectively incorporates the chain-terminating dideoxynucleotides. Since the sequencing of both alleles of an autosomal locus and shown as analog electropherograms at the same time.

2.2.1.2 Maxam-Gilbert Sequencing

Maxam-Gilbert sequencing is a chemical degradation technique established by Allan Maxam and Walter Gilbert in 1977. It is also called chemical sequencing because it is formulated on the chemical modifications of the DNA molecule (Maxam and Gilbert, 1977). It's a chemical technique that partially splits a terminally labeled DNA molecule at every repetition of a base that can be used to sequence DNA (Figure 2.4). The technique involves using polynucleotide kinase to radioactively mark the 5′-phosphate ends of dsDNA with 32P-dATP. After that, denaturation of DNA at 90°C with DiMethyl SulfOxide (DMSO), and the ssDNA are separated by electrophoresis. The Adenosine (A), Thymidine (T), Guanosine (G), and Cytidine (C) residues are modified with nitrogenous base-specific reactions, enabling chemical cleavage at the 5′-P side of the ssDNA. In each of the four reactions, the chemical treatment causes breakage in a tiny fraction of one or two of the

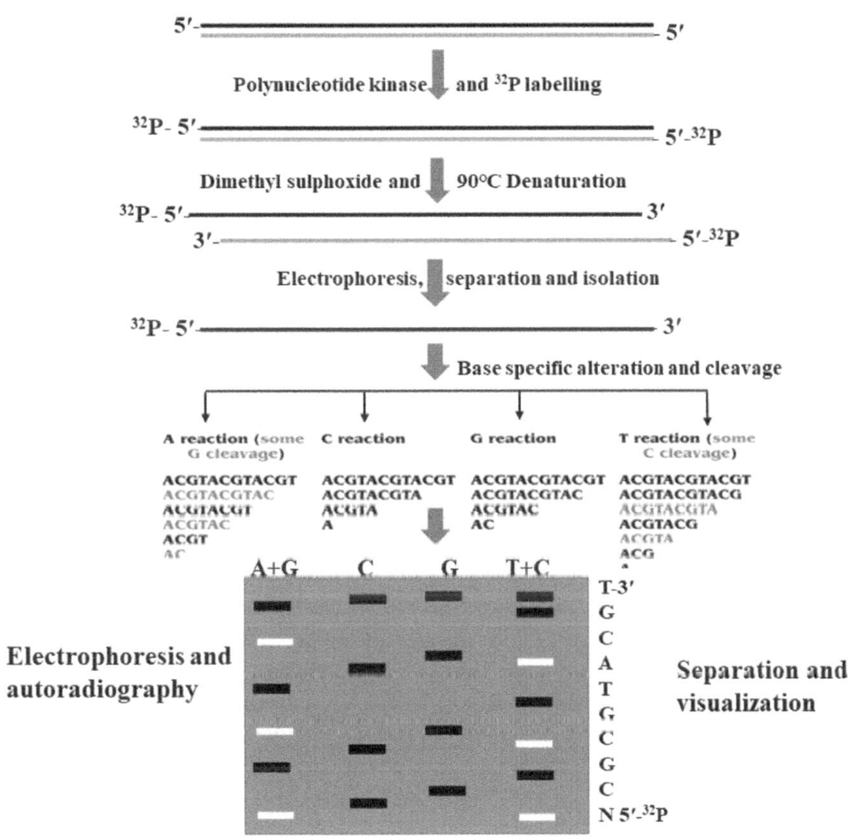

FIGURE 2.4 Overview of Maxam-Gilbert sequencing.

four nucleotide bases (G, A+G, C, T+C). The DNA sequence can be read from the pattern of radioactive bands of the products of these four reactions, the locations of that base are then determined by the lengths of the labeled fragments and separated by size using electrophoresis on a polyacrylamide gel (Maxam and Gilbert, 1977; Masoudi-Nejad et al., 2013). Since it enables direct sequencing of the purified DNA without the requirement of *in-vivo* cloning in ssDNA preparation steps, it was used robustly. However, later on, due to the significant advancement in sequencing technology, the Sanger sequencing approach became the methodology of choice.

2.2.2 Second-Generation Sequencing

Sanger sequencing was dominant in the area of genomics for almost 30 years, but the cost and time consumption were major shortcomings. In 2005, second-generation sequencing emerged to overcome the gaps in the first generation. The second-generation sequencing technology is based on the complete transformation of first-generation sequencing (Metzker, 2010; Margulies et al., 2005). Separate clonal DNA templates are created and sequenced in massive parallel by microfluidics; cycle enzymatic reactions are used to carry out sequencing reactions to develop a base-specific signal and finally captured by imaging (Gut, 2013). The main features of the second-generation sequencing technology are the increased speed of the sequencing process over the first generation, low cost of sequencing, the simultaneous generation of several millions of short reads, and the sequencing performance is directly detected without the use of electrophoresis (Nyrén, 1987; Ronaghi et al., 1996; Ronaghi, 1998). The three major second-generation sequencing platforms are Roche/454 (introduced in 2005), Illumina/Solexa (introduced in 2006), and ABI/SOLiD (introduced in 2007; Margulies et al., 2005; Bentley et al., 2008; Valouev et al., 2008; Rothberg et al., 2011). These widely used sequencing platforms are briefly discussed next.

2.2.2.1 Roche/454 GS (Pyrosequencing)

Pyrosequencing was introduced in 2005. It uses real-time sequencing by synthesis method. DNA polymerase combines the nucleotides, resulting in the formation of a pyrophosphate molecule. As a result of the action of the enzyme luciferase, a light will be released, which will start a cascade of downstream processes. The quantity of joined nucleotides is exactly proportional to the amount of light produced by the reactions (Margulies et al., 2005). The library fragments are restricted on Sepharose or Styrofoam beads, which have oligonucleotides complementary to the 454-specific adapter sequences ligated on both ends of the fragmented library on their surfaces (Vezzi, 2012; Liu et al., 2012). PCR reactions are present in oil micro-reactors and the emulsion PCR thermal cycling is utilized to amplify library fragments (El-Metwally et al., 2014). Each library fragment is amplified on the membrane of a single bead in a single micro-reactor. Such an amplification mechanism create billion copies of the same pieces (Figure 2.5). Single-stranded enriched beads are then produced and authenticated to a particular sequencing primer. After that, the beads are placed in a picotiter plate (PTP) with one million well capacity per plate. Each plate can hold only one amplified DNA bead. In the PTP, four layers of treated beads are placed. All the layers are deposited by centrifugation, from lowest to topmost: diluted pyrosequencing enzyme beads, amplification beads and finally PPiase enzyme beads. The PTP is then sequenced on a Roche/454 sequencer. A constant flow of nucleotide solutions (A, T, G, and C) is introduced throughout the sequencing stages. The light emanated from each bead, as well as the flow of the nucleotide solutions, are captured using a CCD camera. The GS FLX$^+$ machine delivers 200 nucleotide flow cycles to create up to 800 bp sequenced reads. The resulting raw reads are analyzed using software, which is subsequently filtered using various quality filters to eliminate low-quality reads (Huse et al., 2007; Mardis, 2008).

2.2.2.2 Illumina/Solexa

It was previously called Solexa and was developed in 2006. It is primarily utilized in most of the present sequencing applications because of its large capacity (Figure 2.6). It utilizes a sequencing

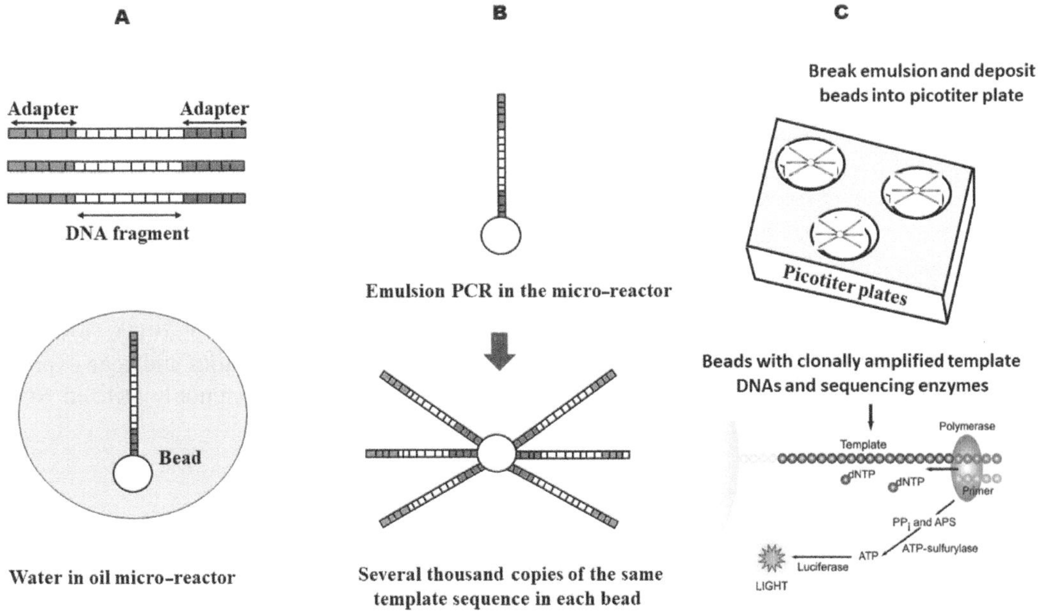

FIGURE 2.5 Sequencing by synthesis method on Roche/454 GS.

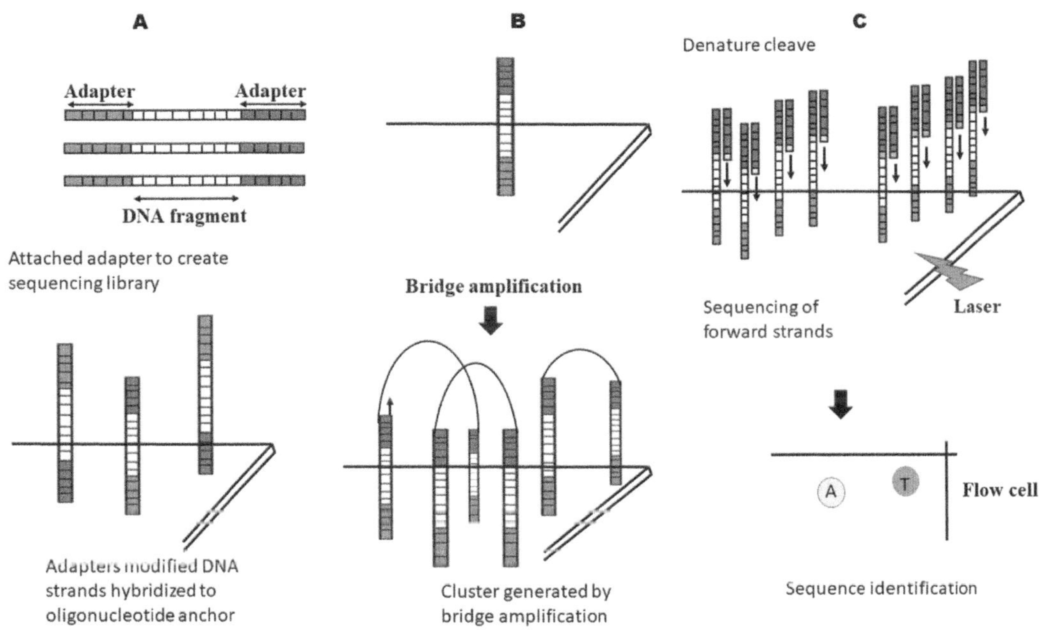

FIGURE 2.6 Bridge amplification sequencing on Illumina platform.

by synthesis method, similar to the Roche 454 genome sequencers, but it is coupled with bridge amplification on the surface of the flow cell, each separated into eight distinct ways. Adapters are connected to the interior surfaces of the flow cells that connect library fragments to oligos. On the flow cell, active heating and cooling stages are performed to aid the DNA fragment to oligo hybridization. The incubation, amplification reactants, and an iso-thermal polymerase method are used to create millions of clusters of library pieces (Shendure and Ji, 2008; Bentley et al., 2008; Heo, 2015).

The cluster is supplied with polymerase and four fluorescent nucleotides that are tagged individually with chemically inactivated OH during the sequencing process. This step acts as a blocking step, guaranteeing that only one base per-flow cycle is used. Subsequently, each nucleotide has been integrated, the stimulation and imaging steps are performed to isolate the inserted nucleotide in each cluster. A chemical deblocking step is employed to remove the fluorescent group, allowing the inserted nucleotide to flow freely during the following flow cycle (Balasubramanian, 2015; Reuter et al., 2015; Kulski, 2016). After computing the sequence of each cluster, the quality filtering phase separates the poor-quality reads (Shendure and Ji, 2008). The HiSeq 2000, HiSeq 1000, MiSeq, and Genome Analyzer IIx are the four commercially available variants. They have sequencing outputs of up to 600 GB, 300 GB, 15 GB, and 95 GB, respectively. In just 27 hours, the HiSeq2500 platform may create up to 120 GB of data (Buermans and Den, 2014). However, MiSeq is a cost-effective technique for targeted gene sequencing, metagenomics and gene expression research, among other things. MiSeq has become one of the most commonly utilized NGS systems as a result of these factors.

2.2.2.3 SOLiD Sequencing

Applied Biosystems presented the SOLiD sequencer in 2007. It uses sequencing by oligo ligation technique; it differs from Roche 454 or Illumina sequencers (Shendure and Ji, 2008). Oligonucleotide linked an adapter with corresponding oligos is sorted out on a 1 mm magnetic bead surface. To individually amplify the beads, an emulsion PCR is employed. The beads are then affixed to the surface of a specifically prepared glass slide, which is subsequently positioned in a sequencer on a fluidics cassette, where an appropriate sequencing primer is solidified to join the SOLiD-specific adapters to the library fragments (Alic et al., 2016). In the instrument, four 8-mer fluorescent semi-degenerate oligos coupled with DNA ligase are introduced consistently. The oligo is then combined with the sequence of the DNA segment adjacent to the universal primer. As a result, the DNA-ligase binds to the phosphate group and closes it. A fluorescence readout is then used to identify the ligated 8-mer oligo, which matches one of the four possible bases (Figure 2.7). To continue the ligation round, a chemical cleavage occurs in the 8-mer oligo between the fifth and sixth bases and the fluorescent group is eliminated. The next sequencing round, which includes the rounds of ligation arbitrated sequencing, begins with the hybridization of n-1 to n-4 placed universal primers (Goodwin et al., 2016; Mardis, 2008).

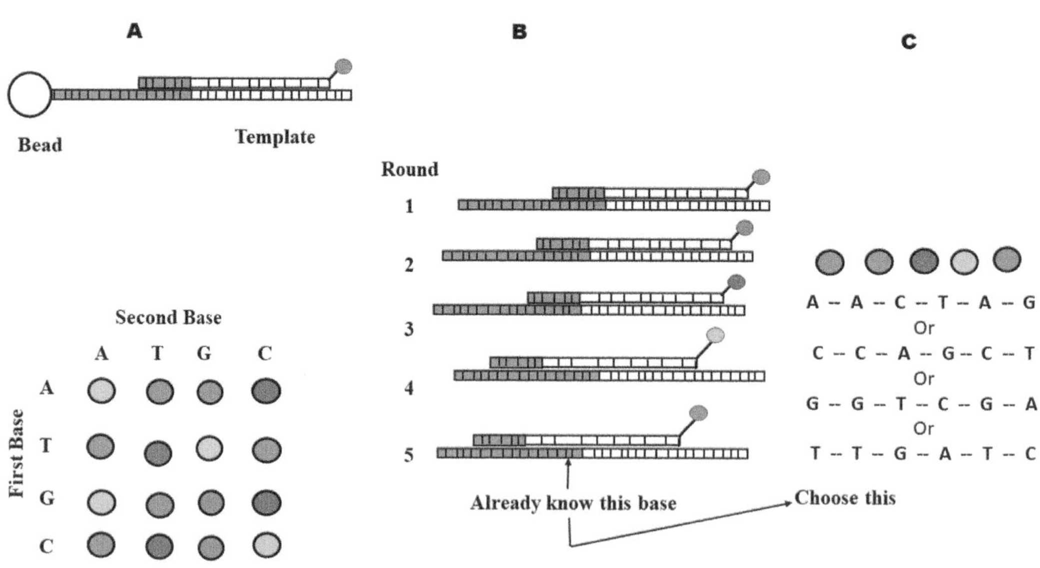

FIGURE 2.7 Sequencing by oligo ligation technique on SOLiD sequencer.

2.2.2.4 Life Technologies Ion Torrent

Ion Torrent was introduced by Life Technologies in 2010; it's also known as Personal Genome Machine (PGM). It utilizes post-light sequencing technology. The polymerase integrates a nucleotide into a strand of deoxyribose nucleotide (DNA). This causes the hydrogen ion to be released, allowing for detection (Rothberg et al., 2011; Reuter et al., 2015). This technique is carried out simultaneously and in huge numbers using a high-density array of micro-machined wells. DNA templates from the library are carried in the wells. The ion-sensitive layer beneath the wells and an exclusive ion sensor measures the variation in hydrogen ion concentration caused by the nucleotide incorporation (Figure 2.8; Dressman et al., 2003; Nakano et al., 2003; Dohm et al., 2008). Ion Torrent platform makes use of 314, 316, or 318 ion chips, which may yield up to 10 megabytes, 100 megabytes, or 1 gigabyte, depending on the needed sequencing coverage (Loman et al., 2012). There are two variants of the Ion Proton benchtop sequencer, that is, the Ion Proton I chip, which has 165 million wells (approximately 100 times greater than the Ion 314 chip), and the Ion Proton II chip, which has 660 million wells (around 1000-fold greater than the Ion 314 chip; Rothberg et al., 2011). The latest versions decode these chemical changes into digital data using metal oxide semiconductor chip technology (Merriman et al., 2012; Ma et al., 2013; Golan and Medvedev, 2013).

2.2.3 Third-Generation Sequencing

In comparison to the first generation of sequencing technologies, the second generation revolutionized DNA sequencing methods and data analysis and has been the most widely used. However, most second-generation sequencing technologies require a PCR amplification stage, which is a time-consuming procedure with a high sequencing cost. Furthermore, with the advancement in genomic technologies, it becomes evident that the genomes are extremely complex, with numerous short repetitive sequences, and second-generation sequencing technologies are unable to resolve them, and it makes the genome assembly more complicated by these short reads. To overcome these problems, researchers have developed an advanced new generation of sequencing known as third-generation sequencing. Direct reading of individual nucleic acid molecules characterizes the third generation. The sequence is not identified using a replication enzymatic system.

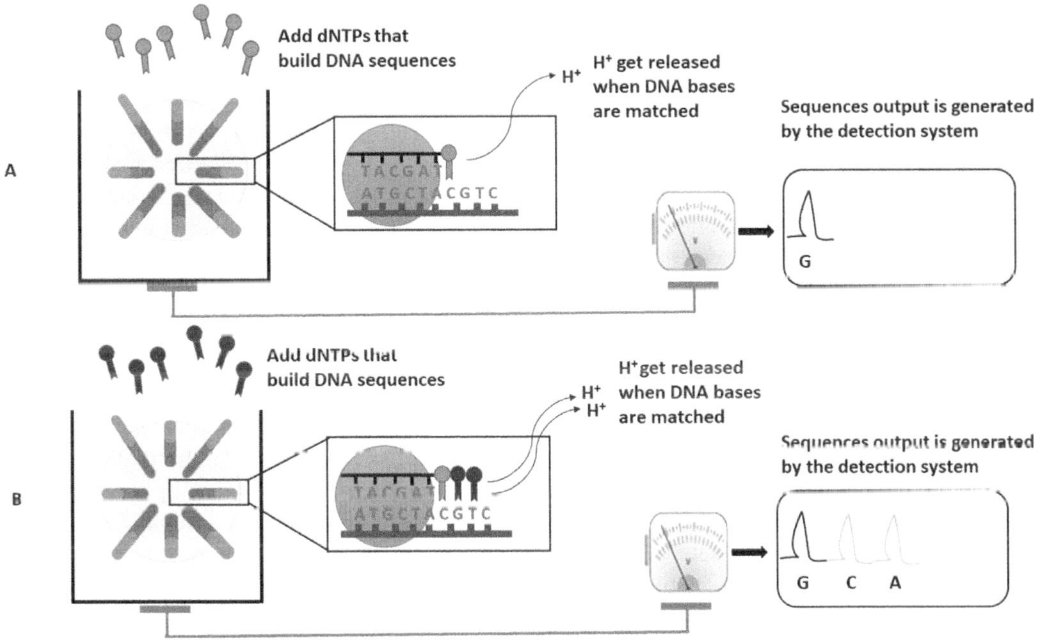

FIGURE 2.8 Post-light sequencing on Life Technologies Ion Torrent.

2.2.3.1 Heliscope™ Single Molecule Sequencer

Braslavsky et al. (2003) for the very first time developed methods for nucleotide sequencing from a single DNA molecule. Helicos biosciences patented it as the first commercial single-molecule fluorescent sequencing method in 2007 (Shendure and Ji, 2008; Thompson and Steinmann, 2010). Heliscope's sequencing concept is based on the "true single molecule sequencing" technology. It starts with the library preparation, which includes DNA shearing, poly-A tailing, and combining of DNA fragments to dT50 oligonucleotides linked on the surface of the flow cell for billions of molecules to be sequenced in parallel (Ozsolak and Milos, 2011). When fluorescent nucleotides are combined with the terminating nucleotide, the cyclical mechanism is slowed till an image of one nucleotide from every DNA sequence is recorded, and the procedure is only repeated when all of the fragments are sequenced (Eid et al., 2009; Travers et al., 2010). DNA elongation with one of the four fluorescently labeled nucleotides are accompanied by nucleotide identification with the HeliScope sequencer in the sequencing cycle. Following the chemical cleavage of the fluorophores, the subsequent step of DNA extension will continue with additional fluorescently labeled nucleotide, allowing the DNA sequence to be determined. This sequencing technique combines hybridization and synthesis sequencing with the use of a DNA polymerase (Fuller et al., 2009). Since no ligation or PCR amplification is needed, ample planning prevents the GC material and size prejudices seen in other technologies (Hart et al., 2010). Read lengths in HeliScope sequencing vary from 25 to more than 60 bases, with an average of 35 bases. The HeliScope sequencer will sequence up to 28 gigabytes per cycle, which takes about one week. It will produce fast reads up to 55 bases in length (Braslavsky et al., 2003; Pushkarev et al., 2009; Ozsolak and Milos, 2011).

2.2.3.2 Pacific Biosciences SMRT DNA Sequencing

Single Molecule Real-Time (SMRT) DNA sequencing was introduced by Pacific Biosciences in 2010. It makes use of a *single-molecule* sequencing platform that is based on SMRT fluorescence (Korlach et al., 2010). It is a synthesis-based single-molecule sequencing method that does not require an amplification step for sample preparation. This approach employs a nano-structure known as a Zero Mode Waveguide (ZMW) to see DNA polymerization in real time. A thin metal sheet supported by a transparent substrate was perforated by tens of thousands of sub wavelengths with ten-nanometer diameter holes. During sequencing, the DNA polymerase, which is located at the bottom level of the waveguides, synthesizes matching DNA strands from the single-stranded template. The platform utilizes four phosphorus linked multicolored nucleotides. The fluorescent label is held by the terminal 5-P group, not the nucleotide base. This aids in the release of the fluorescent component with nucleotide insertion, which sets this technique apart from others. This eliminates the washing phase in every nucleotide flow, reducing the time required for nucleotide incorporation (Figure 2.9). This method takes the use of DNA polymerase's inherent ability to sequence more than ten nucleotides per second in thousands of parallel ZMWs (Lundquist et al., 2008; Korlach et al., 2010; Zhou et al., 2010; Coupland et al., 2012; Chin et al., 2013).

2.2.3.3 Oxford Nanopore Technologies

Nanopore-based sequencing was first released in 2014. The PromethION and GridION is a small bench-top machine, while the MinION is a compact small (handheld) unit for DNA and RNA sequencing that is connected directly to a laptop/computer via a USB port (Bayley, 2015). Nanopore sequencing could take several methods, but it typically depends on individual nucleotides or DNA passing through the tiny channel (Wang et al., 2015). A sequencing flow cell with Oxford Nanopore's current technology consists of more than a hundred individual micro-wells, each comprising a synthetic double layer that is perforated by biologic nanopores (Wang et al., 2015). The bases are lead-in piece through the pore by a molecular motor protein, and the sequence is determined by monitoring distinctive variations in current. The small piece of DNA and adapters ligation are the only steps in the library preparation process. The library preparation approach is similar to SMRT sequencing; it can be done with or without PCR amplification. The first adapter is

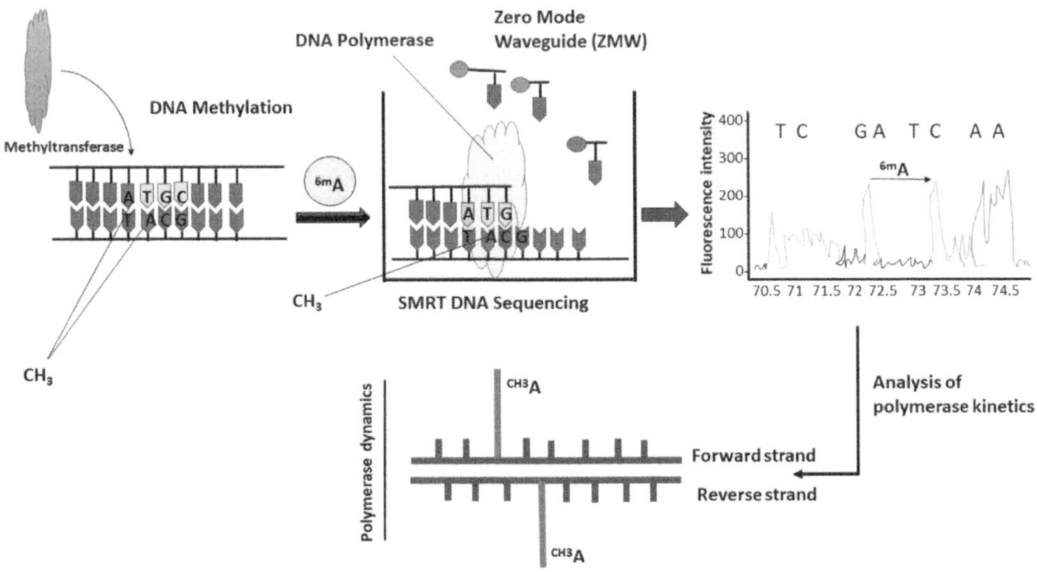

FIGURE 2.9 Single molecule real-time (SMRT) DNA sequencing.

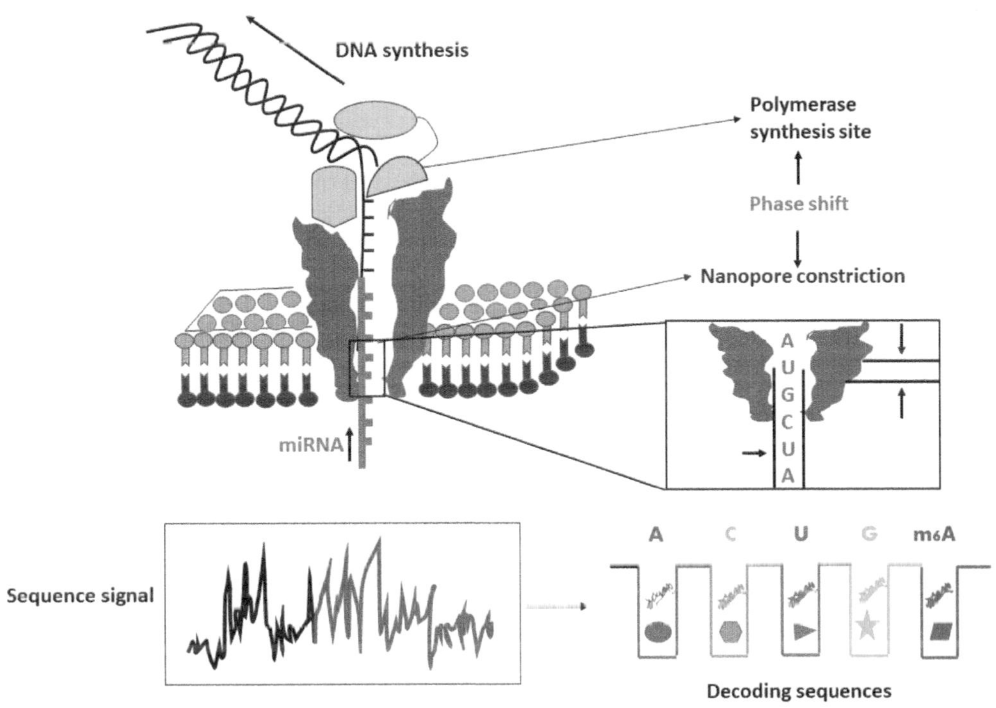

FIGURE 2.10 Oxford Nanopore sequencing process.

a hairpin oligonucleotide that is bound with the hairpin motor protein and molecular tether, while the second adaptor is a hairpin oligonucleotide that is attached with a proprietary motor enzyme as well as a molecular tether (Quick et al., 2014). This library configuration permits both strands of DNA to be sequenced from a single molecule to increasing accuracy (Figure 2.10; Quick et al., 2014; Ashton et al., 2015).

TABLE 2.1
Details of Various NGS Platform

Platform	Instrument	Avg Read length (bp)	Read Type	Error Type	Reads per Run	Error Rate (%)	Data Generated per Run (Gb)	Year
Third Generation								
Nanopore Technologies	PromethION	9846	2D	indel	NA	NA	2Tb	2018
Nanopore Technologies	GridION	NA	1D	indel	NA	12	35	2017
Nanopore Technologies	MinION	9545	1D	indel	100	15	7	2015
PacBio	RSIIP5C3	8500	SE	indel	528	13	0.5–1	2014
PacBio	RSIIP6C4	13500	SE	indel	660	12	0.5–1	2014
PacBio	RSII C2XL	4600	SE	indel	564	15	0.5–1	2013
PacBio	RSC2	2500	SE	indel	432	15	0.5–1	2012
PacBio	RSC1	1300	SE	indel	432	15	0.54	2011
PacBio	RSC2XL	4300	SE	indel	432	15	0.5–1	2012
Helicose Biosciences	HeliScope	>1000	SE	indel	500M	NA	15	2003
Second Generation								
454	GS Junior	400	SE, PE	indel	100	1	0.04	2010
IonTorrent	PGM314chipv2	400	SE	indel	400.000–550.000	1	0.06–0.1	2011
IonTorrent	PGM316chipv2	200	SE	indel	2M–3M	1	0.6–1	2011
Illumina	MiSeq	300	SE, PE	mismatch	25M (max)	0.1	15 (max)	2011
SOLiD	5500W	75	SE	mismatch	3B	0.1	160	2011
Illumina	HiSeq	150	SE, PE	mismatch	5B (max)	0.1	1.5Tb (max)	2012
IonTorrent	Ion Proton	200	SE	indel	60M–80M	1	10	2012
Illumina	MiniSeq	150	SE, PE	mismatch	25M (max)	1	7.5 (max)	2013
SOLiD	5500xlW	75	SE	mismatch	6B	~0.1	320	2013
IonTorrent	PGM318chipv2	400	SE	indel	4M–5.5M	1	1.2–2	2013
454	GS Junior+	700	SE, PE	indel	100	1	0.07	2014
Illumina	HiSeqX	150	SE, PE	mismatch	6B (max)	0.1	1.8Tb (max)	2014
Illumina	NextSeq	150	SE, PE	mismatch	400M (max)	1	120 (max)	2014
IonTorrent	IonS5/S5XL520	400	SE	indel	3M–5M	1	1.2–2	2015
IonTorrent	IonS5/S5XL530	400	SE	indel	15M–20M	1	03–05	2015
IonTorrent	IonS5/S5XL540	400	SE	indel	60M–80M	1	NA	2015
First Generation								
ABI Sanger	3730xl	400–900*	SE	NA	96	0.3	0.00069–0.0021	2002

* depends on the run module

2.3 SEQUENCING TECHNOLOGY IN GENOME RESEARCH

For the very first time using the paired-end sequencing approach on the first automated fluorescent DNA sequencing machine, a complete gene locus for the hypoxanthine-guanine phosphoribosyltransferase (HPRT) gene was sequenced (Edwards et al., 1990). However, the speed and range of sequencing have vastly increased since the first complete bacterial genomes 2 megabases (Mb) was sequenced in 1995. Later, in 2003, the first Human Genome Project (HGP) was completed using the automated DNA sequencer of the first-generation sequencing technology. In subsequent years, another breakthrough in sequencing was accomplished by the DNA sequencing of the first small phage genome of almost 5386 bases in length and sequencing of the human genome of up to ~3 billion bases (Lander et al., 2001; Venter et al., 2001). Surprisingly, such advancements have been made using techniques that are refinements of Sanger's original dideoxy technique, which was published in 1977. Two groups that publicly financed the process were the HGP and the Celera Groups, successfully accomplished the historic sequencing of the human genome. Both participants used various approaches. The HGP group used a map-based method to create a working draught of the human genome, while Celera used a whole-genome shotgun sequencing (WGS) technique to sequence the human genome. The HGP strategy was focused on the localization of bacterial artificial chromosomes (BACs) containing vast fragments of human DNA within the structure of a landmark-based physical map, which was originally developed by a publicly funded initiative. In an ideal world, sequencing would have been performed clone by clone, with clones chosen from the shortest BAC tiling path. The subsequent 'mapping' phase, in which the BACs were each located on the genome's chromosomes by searching for distinctive marker sequences, known as sequence-tagged sites (STSs), whose location had already been pinpointed, was crucial to the HGP's strategy.

The first NGS system in the market developed by 454 Life Sciences was effectively validated by combining single-molecule emulsion PCR with pyrosequencing (shotgun sequencing procedure) of the entire 580,070 bp of the *Mycoplasma genitalia* genome at 99.96% accuracy and 96% coverage in a single run (Tawfik and Griffiths, 1998; Margulies et al., 2005). The NGS technologies also have a significant effect on mammalian genomics (Schuster et al., 2008). The pyrosequencing technique is based on the concept of "sequencing by synthesis." It differs from Sanger sequencing in that it relies on the identification of pyrophosphate release on nucleotide integration rather than dideoxynucleotide chain termination (Nyrén, 1987; Ronaghi et al., 1996; Ronaghi et al., 1998).

Significant progress has been made in the fields of genome decoding in the last two decades. Apart from advances in sequencing methods, the last ten years will be known as the decade of genome science. Next-generation sequencing has been fully incorporated into genomics and genetics. Its ability to rapidly sequence massive quantities of deoxyribonucleic acid (DNA) at high resolution allows it to analyze several genes at the same time, which is particularly useful for atypical or poorly described phenotypes or where genetic heterogeneity exists. Technical advances in a new age of genome applications, tailored genomes, and unusual variant identification using genome sequencing to deal with cancer science, mammalian evolution, and population development. The technological advancement in the arena of genomics, particularly the HTS have increased a series of applications such as structural dissimilarity, mutation finding and carrier screening, whole genome sequencing (WGS), *de novo* assembling and re-assembling of the genome, chromatin immunoprecipitation coupled to DNA microarray (ChIP-chip), sequencing of the mitochondrial genome, detection of hereditary syndromes and multifaceted diseases, DNA library construction, paired ends and genomic captures and RNA sequencing (RNA-seq), etc. A detailed overview of the potential impact of sequencing technologies on genomics studies is shown in Table 2.2 and Figure 2.11.

2.3.1 Gene Expression Analysis

The microarray technologies revolutionized genome research or enabling for the first-time classification of total gene expression. However, there were some shortcomings in the technology that limited sensitivity and accuracy, which were resolved by sequencing. The sensitivity of sequencing-based research is primarily restricted by the depth of sequencing, whereas identification of recessive genes using microarrays is restricted by background noise. RNA-based gene expression analysis can be done in different ways, according to recent studies. Primarily cost-directed, SAGE-like methods were common (Hoen et al., 2008). However, falling costs are now driving research toward full RNA sequencing (RNAseq), which requires far deeper sequencing to reveal the complete illustration, including differential splicing and allelic expression, thorough investigation of the entire transcript (Lappalainen et al., 2013; Westra et al., 2013). HTS has also been commonly used to investigate the in-depth analysis of RNA structure, with particular methods built to describe transcription initiation using the 3′-end of transcripts (de Klerk et al., 2012) or the 5′-start (cap site) (Valen et al., 2009). It can also be used to study small RNAs in a variety of ways, for instance, a differential gene expressed of miRNA binding sites for leukemia was identified through NGS sequencing (Ramsingh et al., 2010).

Numerous effective procedures to carry out RNA-seq on the Illumina have been introduced recently with certain technical issues such as the construction of a complex miRNA database, preparation of small RNA libraries and analyses of the raw data for measuring miRNAs (Buermans et al., 2010; Lee et al., 2010; Marguerat and Bähler, 2010; Nagalakshmi et al., 2010; Yang et al., 2010). Similarly, in cancer research (cell lines and tissues), RNA sequencing using Illumina and 454 technologies is an effective method for identifying novel gene fusions (Maher et al., 2009). Moreover, many transcriptomic studies are conducted using a newly established deep sequencing method that is not time-consuming and is cheaper (Wang et al., 2009). Short reads generated by HTS, especially Illumina and SOLiD sequencers are potentially useful for gene expression studies. RNA-seq has been used to reliably track the gene expression profile of a particular gene to assess differential splicing, allele-specific transcript expression, and various other biological concerns (Costa et al., 2010).

2.3.2 Epigenetics

The NGSTs have the potential to remarkably speed up the examination of inheritable gene regulation that does not include the DNA sequence but rather it alters the higher-order structures (HOS) of protein, such as transcription factor interactions with their immediate targets, histone post-translational modifications, and nucleosome positioning on the entire genome as well as the identification of the epigenetics patterns of DNA methylation (Buermans et al., 2010; Fouse et al., 2010; Bhaijee et al., 2011). As a post-translational modification of histone and DNA methylation are the two essential epigenetic pathways that control how genes are expressed. Histone modification and the position of transcription factors (TFs) can be analyzed at the entire genome level using chromatin immunoprecipitation (ChIP-Seq) and direct sequencing technology, while bisulfite and methylated DNA immunoprecipitation (meDIP) procedures can be used to study DNA methylation patterns (Neff and Armstrong, 2009).

2.3.2.1 ChIP-Seq

ChIP-sequencing is an important tool for determining the DNA binding sites for TFs and other proteins. ChIP-seq methods are used to immunoprecipitate DNA-bound proteins with a specific antibody. The bound DNA is then purified, sequenced, and co-precipitated. The use of NGS on ChIP has uncovered new information about gene regulation events involved in a variety of diseases and biological processes, including cancer growth and progression. ChIP-seq allows for a comprehensive study of protein-nucleic acid interactions on a genome-wide scale (Blat and Kleckner, 1999). For instance, the binding sites for a TFs and a human growth-associated binding protein were directly sequenced using ChIP-seq on the NGS platform instead of being hybridized on a chip-array reveals the broad and complicated gene pathways controlled by the Peroxisome Proliferator Activated Receptor Gamma (PPARG) gene (Costa et al., 2010) and predicting *de novo* motif discovery (Jiao et al., 2010). Initially combining ChIP-chip with DNA microarrays has been used to investigate protein-DNA interactions. The ChIP-seq technique, on the other hand, receives two benefits from the HTS platforms; first, it is not constrained by microarray material, and second, it is not dependent on probe hybridization performance. However, due to the limited content of the array, decisions had to be made on which sequences to examine. Generally, CpG-island and promoter regions of genes are selected. The signals obtained were always poor because it was hard to enrich the targeted sequences to utmost purity. The advantage of sequencing the enhanced ChIP-products is that all sequences bound are known, and the number of sequence reads can be used to calculate sensitivity; even if the enrichment is minute, rising the quantity of reads can help achieve a strong indication. In human cells, the ChIP-seq method was used to characterized binding sites for transcription factors such as signal transducers and activators of transcription (STAT1) and the Neuron Restrictive Silencer Factor (NRSF) (Robertson et al., 2007; Euskirchen et al., 2007). They compared their results to those produced by ChIP-chip and concluded that ChIP-seq provided improved resolution and allowed less samples.

2.3.2.2 Methylation

Like ChIP-seq, another application advanced by HTS technologies was the investigation of genome-wide DNA methylation. The methylation of DNA can be studied using different techniques such as methylated DNA immunoprecipitation (meDIP), methylated CpG island recovery assay (MIRA-Seq), and whole-genome shotgun bisulfite protocols (WGSBS) (Popp et al., 2010; Chung et al., 2010; Fouse et al., 2010; Bhaijee et al., 2011). NGS analysis specifically shows both enriched sequences regardless of the method used to extract the methylated sequences, such as MIRA-Seq, MeDIP-seq, or WGSBS (Laird, 2010; Neff and Armstrong, 2009; Jiao et al., 2010). Moreover, chemical methods may be used to classify methylated nucleotides in addition to enriching methylated DNA sequences. When sequencing both unprocessed and bisulfite treated DNA, the cytosines that are methylated but not chemically transformed are highlighted, resulting in thymine. Although, the bisulfite-treatment reduced complexity of the DNA (most C changed to T) will pose some issues during data analysis (determining the origin of the reads obtained). Single-molecule sequencing technology, such as those developed by Pacbio, are appealing alternatives because they can detect all DNA changes, not just C-residue methylation. Several NGS methylation studies have been published, including the ENCODE consortium's (ENCODE Project Consortium, 2004; Herb et al., 2012).

2.3.3 Mutation Detection and Diseased Gene Identification

Using practical genomic fingerprinting, NGS platforms are being used to study high-throughput mutation identification and carrier testing of fibroblast growth factor (FGF)-related genetic disorders. In addition to the detection of mutations for syndromes/diseases and drug response, the strategy suggests careful enhancement of functional genomic regions (the promoterome, exome, and exon splice enhancers) (Senapathy et al., 2010). The microarray target enrichment with HTS allows the study of full genomic areas for many genes in a disease cascade and several samples simultaneously, making it a more useful tool for detailed mutation screening (Amstutz et al., 2011). Similarly, by finding all the rare and common genetic diversity in human populations, NGS aims to make structural variation studies easier on the genome-wide scale of the human population (Xi et al., 2010; Henn et al., 2010; Bowne et al., 2011). A combination of genome-wide association studies (GWAS) and selective targeting by sequence capture of the genomic areas discovered is now frequently employed to identify and classify the variation that functionally connects the DNA with the phenotype. Likewise, genome sequencing can be used to assess haplotype composition, characterize the genetic difference in a population, and use this information to impute alleles and improve GWAS performance. To date, NGS has created a complete genetic map of all human variation. Scientists will be able to do more systematic trials to diagnose genetic variation underlying the reaction to drugs (1000 Genomes Project Consortium, 2010). Even though only one or a few cases are open, NGS studies are much more effective, where multiple unrelated cases are possible or destructive variations are present, parent-child trio research has proven to be very useful in detecting dominant *de novo* diseases, whereas recessive diseases can be detected when many unrelated cases are existing or when destructive variations are present (Hoischen et al., 2010; Ng et al., 2010; Vissers et al., 2010).

2.3.4 Non-invasive Prenatal Testing and Cancer Research

Traditionally, prenatal diagnosis usually requires the expensive and unsafe sampling of both chorionic villi and the amniotic fluid to extract the DNA of a fetus. While it has long been understood that fetal DNA can be present in the mother's blood (blood cell-free serum), it is rare and of poor quality, making it difficult to distinguish fetal from maternal DNA (Buermans et al., 2014). The immense potential of NGS technologies, on the other hand, seems to be especially appealing for non-invasive prenatal monitoring (NIPT). A very basic but successful and very efficient approach was developed to diagnose trisomies, specifically 21-trisomy or Down's syndrome. In this brute force approach, the DNA sequencing, mapping, and counting are carried out, then the reads are mapped to the human genome and count each chromosome after DNA from maternal serum is sequenced (Vermeer et al., 2010; Bell et al., 2011). Eventually, trisomies can be identified after 5–10 million reads have been mapped, with a significantly increased number of reads mapping to a single chromosome. Another fascinating use of NGS technology is in the field of cancer research (Kitzman et al., 2012; Norton et al., 2012). The use of NGS technologies has revolutionized cancer genomics, creating broad data sets that can be studied in a variety of methods to address a variety of queries regarding the disease's genomic modifications: point mutations, such as insertion, substitutions and deletions, significant structural modifications, and change in copy number. Various diagnostic and therapeutic tools for assessing genetic variations in cancer based on heuristic and probabilistic models have been established and applied. While several important discoveries have been made, the majority of these techniques are still in the early stages of production. The task of detecting and evaluating genetic phenomena, such as somatic mutations, copy number alterations (CANs), and single nucleotide variant (SNVs), is still unsolved (Ding et al., 2010; McBride et al., 2010; Pfeifer and Hainaut, 2011). As a result, the optimal balance of adaptation and accuracy for mutation detection in cancer research is always the use of multiple algorithms accompanied by confirmation (Meyerson et al., 2010; Ding et al., 2010).

2.3.5 Metagenomics

Metagenomics is the study of the whole genetic material recovered directly from the environmental samples or diverse consortia (e.g., within a waste-treatment system or bioreactor). In certain natural habitats, such as chemically polluted soils, obtaining DNA that is truly reflective of an entire population can be difficult. Any bacteria and spores are much more difficult to lyse to release their nucleic acids. During the DNA extraction and purification process, it's also critical to eliminate co-extracted particles that could compete with the downstream strategies (Brady and Salzberg, 2009). Unlike conventional single-genomics methods, the metagenomics approach can identify and sequencing all genes and genomes from a mixed species at once, rather than singularizing human bacterial clones from diverse microbial mixtures (Schloss et al., 2005; Huson et al., 2007). Metagenomics can be done in two ways: by sequencing all or by concentrating on certain constantly conserved sequences, such as ribosomal RNA (rRNA) genes. In this approach, the data collected is much less abstract and more sequences may be attributed to a single individual or a group of similar organisms. The single-isolate technique is effective in identifying and analyzing diseases caused by basically a single genotype (Schloss et al., 2009). However, choosing a single colony for sequencing, as is used in epidemiological research, can obscure the likelihood that a population of highly related but yet distinguishable individual genotypes are to blame for the disease phenotype or epidemic (Venter et al., 2004; Cox-Foster et al., 2007; Wang et al., 2009; Bos et al., 2011). Metagenomics is a way to analyze what is present in a sample, of what is residing where. For

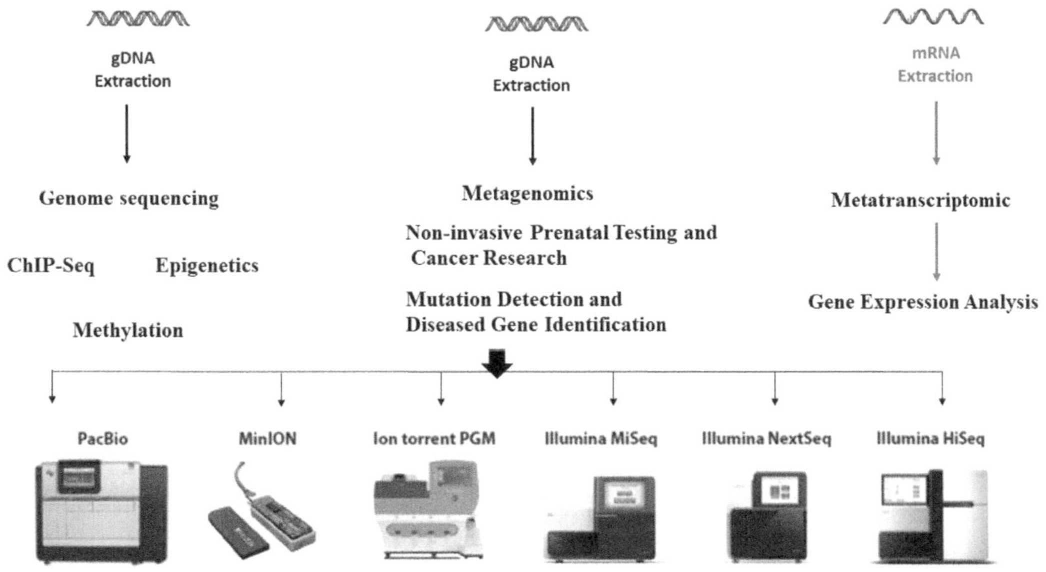

FIGURE 2.11 Sequencing technologies in genome research.

TABLE 2.2
Detailed Overview of the Sequencing Technologies in Genome Research

Applications	Description	References
ChIP-seq	The study conducted on the gene pathways regulated by PPARG using RNA-seq recognized coding as well as regulatory regions of PPARγ gene-novel nucleotide differences and haplotypes associated with human diseases, defining a PPARG binding map by ChIP-seq, and solving the broad and complex gene pathways of the same.	Costa et al. (2010)
	Advanced statistical technique to predict the *de-novo* motif detection in the peak environments by examining the human GABPα on ChIP-seq.	Jiao et al. (2010)
Cancer research	Introduction to the development of an integrated database for various and tumor cancers genomes to provide a comprehensive understanding of cancer's genetic origin.	Ding et al. (2010)
	A complete analysis of NGS applications to cancer genomes, with a focus on glioblastoma multiforme (GBM), found that the gene codes for isocitrate dehydrogenase 1 (IDH1), which is a candidate for cancer-causing mutations.	Pfeifer and Hainaut (2011)
	Rapidly expanding the ability to recognize the causal mutations that drive cancer initiation and metastasis with NGS technologies increases substantial hopes for enhancing oncologic outcomes.	Katsios et al. (2012)
	NGS Chip-seq technology was used to identify 4883 SOX2 binding regions in the glioblastoma multiformecancer genome.	Fang et al. (2011)
	NGS approaches for whole-exome, whole-transcriptome, and whole-genome studies. Sequencing includes somatic genome mutations, cancer genetics, diagnosis, and treatment.	Meyerson et al. (2010)

(*Continued*)

TABLE 2.2
(Continued)

Applications	Description	References
Epigenetics	A study was conducted via the NGS platform to show the DNA methylation where the comparison between two different bisulfite conversions using whole methylome sequencing methods.	Chung et al. (2010)
	The study conducted on profiling methylation patterns at single-base resolution showed new findings in the divergence of DNA methylation in eukaryotes and their gene regulation.	Bhaijee et al. (2011)
	Investigation revealed the methylation detection reagents and their wide range of applications with advanced molecular approaches such as sequencing and microarray. They also planned international coordination to standardize methylome platforms so as to create a database of methylome maps from unique cell type and tissues.	Fouse et al. (2010)
	NGS methylation repository/database for mouse, Arabidopsis, and the human genome, including a wider variety of tissues methylation or the variations happening along with pathological conditions. (http://bioinfo2.ugr.es/NGSmethDB/gbrowse/)	Hackenberg et al. (2010)
De Novo assembling and re-assembling of the human genome	Experiments proposed a new method for the *de-novo* assembly of human genomes from short-read sequences. A study accomplished the assembly of N50 contig size of 7.4 and 5.9 kb and scaffold of 446.3 and 61.9 kb of the human genome.	Li et al. (2010)
	The investigation recommended re-sequencing of the genome by DNA pool which was a new statistical method, CRISP (Comprehensive Read analysis for Identification of SNPs from Pooled sequencing) that can recognize both common and rare variations. This method can detect 80–85% of SNPs identified using individual sequencing.	Bansal (2010)
	Studies conducted on the complete genome of a 4,000-year-old human with 20-fold coverage which provided a fresh look at human population history.	Shapiro and Hofreiter (2010)
	A comprehensive study on recent advancements in software packages and algorithms in analyzing NGS data.	Nagarajan and Pop (2010)
	Research has been taken to recognize the relative performance of the Roche 454 and Illumina Genome Analyzer and for the re-sequencing of 16 genes linked with hypertrophic cardiomyopathy. They resolved that the practicability of combination PCR with NGS platforms for targeted re-sequencing of hypertrophic cardiomyopathy linked genes are more feasible and yield better inference.	Dames et al. (2010)
Structural variation detection in the Human population	NGS sequencing revealed the genetic variations in the genome that cause disease and structural change in the diverse human genome.	Henn et al. (2010)
	A study was conducted to detect the disease-causing mutations in patients with monogenic congenital diseases using NGS. It's an effective method for detecting new and rare mutations causing heterogeneous monogenic diseases. The disease-causing mutations can now be identified in 65% of autosomal dominant cases due to the advancement of sequencing technology.	Bowne et al. (2011)
	The research was conducted on WGS of 179 samples from four different populations, to detect and develop diverse approaches for genome-wide sequence variation using NGS.	1000 Genomes Project Consortium (2010)
	A comprehensive review on the identification of structural variations in the human genome using NGS.	Xi et al. (2010)

Applications	Description	References
Sequencing of the mitochondrial genome	The researchers suggested a HTS and bioinformatics pipeline for mitochondrial genome, which has implications for the annotation and interpretation of organelles, long contiguous regions in nuclear and viral genomes.	Jex et al. (2010)
	A novel method for sequencing and *de-novo* assembly of multiple mtDNA without the costs of indexing was proposed in this study.	McComish et al. (2010)
	The mitochondrial genomes were compared and found to have relatively little nucleotide variation within the population, which may have implications for environmental protection strategies.	Soroka and Burzyński (2010)
Mutation detection and carrier screening	Experiments were conducted to propose a selective enrichment of functional genomic regions such as the promoterome, exome, or exon splice enhancers through functional genomics fingerprint approach on the NGS platform to discover the causal mutations and drug response.	Senapathy et al. (2010)
	The investigation recognized a gene with mutation and has shown its linkage with autosomal-recessive cerebellar ataxia, using the combination of SNP array-based linkage analysis and targeted resequencing of pertinent sequences in the linkage interval on the NGS platform.	Vermeer et al. (2010)
	A cost-effective method of carrier screening using NGS is now open to the public with extreme recessive juvenile conditions.	Bell et al. (2011)
	A thorough investigation of entire genomic regions for various molecular pathways and multiple samples concurrently thereby provides an effective method for systematic diagnostic testing of mutations.	Amstutz et al. (2011)
Detection of inherited disorders	Through exon study of the entire human genome, researchers were able to identify human monogenic diseases.	Kuhlenbäumer et al. (2011)
	Comprehensive analysis of the NGS on brain science across the past two decades, covering a wide range of psychological and mental conditions.	Zoghbi and Warren (2010)
	To better elucidate the genetic basis of human sexual differentiation and determination, a combination method of comparative genomic hybridization and sequencing using NGS was proposed.	Bashamboo et al. (2010)
Complex human diseases	In mapping genetic features using NGS, a thorough analysis of experimental setup deliberations, data management problems, and analytical creation tools are required.	Day-Williams and Zeggini (2011)
	NGS is being used to classify the heritable risks that contribute to or predispose to multifaceted diseases.	Singleton et al. (2010)
	For selective re-sequencing of genes involved in hypertrophic cardiomyopathy, principles of next-generation sequencing library planning, sequencing chemistries, and raw data analysis were used.	Voelkerding et al. (2010)
	A systematic analysis of the discovery of causal variants usually includes protein coding, regulatory, and structural sequences in the surrounding of disease linked SNPs.	Kingsley (2011)
Library preparation paired ends and genomic captures for NGS platforms	The research used a special 6bp DNA barcode to present a comprehensive and lucrative preprocessing technique for library development, allowing composite sample processing and sequencing of almost 32 libraries in a single run-on SOLiD sequencing machine.	Farias-Hesson et al. (2010)
	This research gives a novel sequence insertion pipeline to find out different genetic variations. (http://compbio.cs.sfu.ca/strvar.htm)	Hajirasouliha et al. (2010)

(*Continued*)

TABLE 2.2 (Continued)

Applications	Description	References
	Researchers in this examination compared NGS with CAGE and SAGE; they discovered 196 new regulatory regions that are used more often in multiplying cells or already differentiated cells. These libraries have stable expression levels and could add tissue-specific promoters and alternate 3′-UTR use to existing genome annotations.	Hestand et al. (2010)
	The genetic assay used to detect, sequence, and diagnose mutations in 21 genes, the research established a precise, detailed, and cost-effective isolation of heritable mutations for breast and ovarian cancer.	Walsh et al. (2010)
Whole-genome genotyping	A simple method of human leukocyte antigen typing was developed. First, the genes were amplified with PCR. Then the resulting amplicons were sequenced using the Roche 454 platform. Finally, the sequenced data were analyzed with different NGS software using different pipelines.	Lind et al. (2010)
	A new study determines the utero disease screening of whole genome embryonic sequencing using non-invasive NGS of the mother's blood.	Burgess (2011)
	A complete assessment and analysis on the effect of NGS on prenatal diagnosis screening.	Raymond et al. (2010)
RNA sequencing	An improved method for generating advance miRNA sequencing libraries for the Illumina sequencer was proposed in this report.	Buermans et al. (2010)
	The Illumina sequencing platform was used in this analysis to demonstrate an effective method for conducting RNA-seq.	Nagalakshmi et al. (2010)
	A new database (The deepBase) was proposed in the study to aid in the systematic annotation and detection of noncoding RNAs from RNA transcript data.	Yang et al. (2010)
	Introduction of a time-saving protocol for preparing non coding RNA libraries for sequencing and analyzing the data to determine miRNA abundance.	Morin et al. (2010)
	A study on the classification of small RNAs and demonstrated an alignment of pattern matrices score. Scoring scheme for classifying small RNAs that merely use the comparative positions and lengths of reads from NGS results. (www.bio.ifi.lmu.de/ALPS)	Erhard and Zimmer (2010)
	An in-depth investigation on RNA-seq for transcriptome analyses using HTS. The investigation also observed how to reliably assess the expression levels of individual genes, selective splicing, allele-specific transcript expression, and a variety of other biological concerns that come up in RNA-seq analysis.	Costa et al. (2010)
	A thorough examination of RNA-seq, including the technical problems that arise during the generation and analysis of RNA-seq data.	Marguerat and Bähler (2010)
	Using next-generation sequencing systems, a systematic survey of micro-RNA sequence variants was conducted from mouse and human samples. The study devised a tool for constructing a database that indexes the entire catalog of miRNA sequences to decide the maximum number of sequences and the level of heterogeneity for each particular miRNA species. (http://galas.systemsbiology.net/cgi-bin/isomir/find.pl)	Lee et al. (2010)

Applications	Description	References
	A systematic overview of the role of next-generation sequencing (NGS) in functional genomics research, with an emphasis on transcription factor binding sites.	Werner (2010)
	The research recognized genetic variations of micro-RNA genes and screened for variations in micro-RNA binding sites in an individual with acute myeloid leukemia, making it the first full analysis of the "miRNAome" in a human cancer.	Ramsingh et al. (2010)

example, analyzing and comparing all DNA from healthy people to people that have been affected reveals the "extra" DNA that most probable comes from the infectious agent (McHardy et al., 2007; Finn et al., 2010).

2.4 CONCLUSION AND FUTURE PROSPECT

The advancement in sequencing technologies provided cost-effective and timely sequencing has aided in the advancement of various genome research that had previously been overlooked in genomics. The accessibility of deep sequencing of DNA transformed extraordinarily in the arenas of cancer research, proteomics, genomics, and other omics studies. Furthermore, ultra-deep sequencing in the near future can become a propitious instrument in the analysis of whole-genome genotyping, de novo assembling and re-assembling of genomes, structural difference, mutation finding and carrier testing, detection of congenital disease and complex human disorder diagnostics and medicine. The decreasing cost of this technology has opened the door to the prospect of sequencing whole genomes of organisms within a species as well as understanding various pathways underlying gene expression, and diversity analysis through *de-novo* sequencing, which was previously impossible due to a lack of reference genomes.

NGS technology can be widely applied for coding and small noncoding RNA sequencing, besides that it can be used to find out DNA methylation patterns in bisulfite sequencing, to aid in the elucidation of all related biological development from these entities to transcriptional changes. While this is a chance to extensively investigate the full range of DNA and RNA modifications in all main cancer cells, further, in addition to identifying major molecular variations in cancer cells, the incorporation efforts will also find out the basic interactions of such molecular variations and their associations with definite clinical features.

Moreover, to obtain fetal genome-wide sequence data currently conducting a diagnostic invasive technique to collect fetal DNA, which is most typically done via amniocentesis or CVS. A variety of studies have shown that noninvasive sequencing of the fetal genome from cell-free DNA circulating in the maternal plasma is technically feasible. Working on isolating intact fetal cells, mainly of trophoblastic origin, from maternal blood holds a lot of promise as a noninvasive source of foetal DNA that can be amplified and exposed to NGS to diagnose single-gene defects, chromosomal copy number variations, and aneuploidy.

The use of NGS technology will help researchers better understand unknown organisms and diseases. Since various industries use different platforms with different features and advantages, assembly efficiency and accuracy are determined by the number of reads and read the time. As a result, one of the most critical areas of future study will be the advancement of methods for analyzing the massive amounts of data generated by NGS. The objectives would be to improve assembly sequencing precision, reduce processing time, and fine-tune the performance of research algorithms. To make the most of NGS data, state-of-the-art bioinformatics pipelines to extract mean values must be designed.

BIBLIOGRAPHY

1000 Genomes Project Consortium. (2010). A map of human genome variation from population-scale sequencing. *Nature*, *467*(7319), 1061.

Alic, A. S., Ruzafa, D., Dopazo, J., & Blanquer, I. (2016). Objective review of de novo stand-alone error correction methods for NGS data. *Wiley Interdisciplinary Reviews: Computational Molecular Science*, *6*(2), 111–146.

Amstutz, U., Andrey-Zürcher, G., Suciu, D., Jaggi, R., Häberle, J., & Largiadèr, C. R. (2011). Sequence capture and next-generation resequencing of multiple tagged nucleic acid samples for mutation screening of urea cycle disorders. *Clinical Chemistry*, *57*(1), 102–111.

Ashton, P. M., Nair, S., Dallman, T., Rubino, S., Rabsch, W., Mwaigwisya, S., . . . & O'grady, J. (2015). MinION nanopore sequencing identifies the position and structure of a bacterial antibiotic resistance island. *Nature Biotechnology*, *33*(3), 296–300.

Balasubramanian, S. (2015). Solexa sequencing: Decoding genomes on a population scale. *Clinical Chemistry*, *61*(1), 21–24.

Bansal, V. (2010). A statistical method for the detection of variants from next-generation resequencing of DNA pools. *Bioinformatics*, *26*(12), i318–i324.

Bashamboo, A., Ledig, S., Wieacker, P., Achermann, J. C., & McElreavey, K. (2010). New technologies for the identification of novel genetic markers of disorders of sex development (DSD). *Sexual Development*, *4*(4–5), 213–224.

Bayley, H. (2015). Nanopore sequencing: from imagination to reality. *Clinical Chemistry*, *61*(1), 25–31.

Bell, C. J., Dinwiddie, D. L., Miller, N. A., Hateley, S. L., Ganusova, E. E., Mudge, J., . . . & Kingsmore, S. F. (2011). Carrier testing for severe childhood recessive diseases by next-generation sequencing. *Science Translational Medicine*, *3*(65), 65ra4–65ra4.

Bentley, D. R., Balasubramanian, S., Swerdlow, H. P., Smith, G. P., Milton, J., Brown, C. G., . . . & Roe, P. M. (2008). Accurate whole human genome sequencing using reversible terminator chemistry. *Nature*, *456*(7218), 53–59.

Bhaijee, F., Pepper, D. J., Pitman, K. T., & Bell, D. (2011). New developments in the molecular pathogenesis of head and neck tumors: a review of tumor-specific fusion oncogenes in mucoepidermoid carcinoma, adenoid cystic carcinoma, and NUT midline carcinoma. *Annals of Diagnostic Pathology*, *15*(1), 69–77.

Blat, Y., & Kleckner, N. (1999). Cohesins bind to preferential sites along yeast chromosome III, with differential regulation along arms versus the centric region. *Cell*, *98*(2), 249–259.

Bos, K. I., Schuenemann, V. J., Golding, G. B., Burbano, H. A., Waglechner, N., Coombes, B. K., . . . & Krause, J. (2011). A draft genome of Yersinia pestis from victims of the Black Death. *Nature*, *478*(7370), 506–510.

Bowne, S. J., Sullivan, L. S., Koboldt, D. C., Ding, L., Fulton, R., Abbott, R. M., . . . & Daiger, S. P. (2011). Identification of disease-causing mutations in autosomal dominant retinitis pigmentosa (adRP) using next-generation DNA sequencing. *Investigative Ophthalmology & Visual Science*, *52*(1), 494–503.

Brady, A., & Salzberg, S. L. (2009). Phymm and PhymmBL: Metagenomic phylogenetic classification with interpolated Markov models. *Nature Methods*, *6*(9), 673–676.

Braslavsky, I., Hebert, B., Kartalov, E., & Quake, S. R. (2003). Sequence information can be obtained from single DNA molecules. *Proceedings of the National Academy of Sciences*, *100*(7), 3960–3964.

Buermans, H. P. J., Ariyurek, Y., van Ommen, G., den Dunnen, J. T., & AC't Hoen, P. (2010). New methods for next generation sequencing based microRNA expression profiling. *BMC Genomics*, *11*(1), 1–16.

Buermans, H. P. J., & Den Dunnen, J. T. (2014). Next generation sequencing technology: advances and applications. *Biochimica et Biophysica Acta (BBA)-Molecular Basis of Disease*, *1842*(10), 1932–1941.

Burgess, D. J. (2011). Next-generation sequencing of the next generation. *Nature Reviews Genetics*, *12*(2), 78–79.

Chin, C. S., Alexander, D. H., Marks, P., Klammer, A. A., Drake, J., Heiner, C., . . . & Korlach, J. (2013). Nonhybrid, finished microbial genome assemblies from long-read SMRT sequencing data. *Nature Methods*, *10*(6), 563.

Chung, C. A. B., Boyd, V. L., McKernan, K. J., Fu, Y., Monighetti, C., Peckham, H. E., & Barker, M. (2010). Whole methylome analysis by ultra-deep sequencing using two-base encoding. *PLoS One*, *5*(2), e9320.

Costa, V., Gallo, M. A., Letizia, F., Aprile, M., Casamassimi, A., & Ciccodicola, A. (2010). PPARG: gene expression regulation and next-generation sequencing for unsolved issues. *PPAR Research*, *2010*.

Coupland, P., Chandra, T., Quail, M., Reik, W., & Swerdlow, H. (2012). Direct sequencing of small genomes on the Pacific Biosciences RS without library preparation. *Biotechniques*, *53*(6), 365–372.

Cox-Foster, D. L., Conlan, S., Holmes, E. C., Palacios, G., Evans, J. D., Moran, N. A., . . . & Lipkin, W. I. (2007). A metagenomic survey of microbes in honey bee colony collapse disorder. *Science*, *318*(5848), 283–287.

Dames, S., Durtschi, J., Geiersbach, K., Stephens, J., & Voelkerding, K. V. (2010). Comparison of the illumina Genome Analyzer and Roche 454 GS FLX for resequencing of hypertrophic cardiomyopathy-associated genes. *Journal of Biomolecular Techniques: JBT*, *21*(2), 73.

Day-Williams, A. G., & Zeggini, E. (2011). The effect of next-generation sequencing technology on complex trait research. *European Journal of Clinical Investigation*, *41*(5), 561–567.

de Klerk, E., Venema, A., Anvar, S. Y., Goeman, J. J., Hu, O., Trollet, C., . . . & 't Hoen, P. A. (2012). Poly (A) binding protein nuclear 1 levels affect alternative polyadenylation. *Nucleic Acids Research*, *40*(18), 9089–9101.

Ding, L., Wendl, M. C., Koboldt, D. C., & Mardis, E. R. (2010). Analysis of next-generation genomic data in cancer: accomplishments and challenges. *Human Molecular Genetics*, *19*(R2), R188–R196.

Dohm, J. C., Lottaz, C., Borodina, T., & Himmelbauer, H. (2008). Substantial biases in ultra-short read data sets from high-throughput DNA sequencing. *Nucleic Acids Research*, *36*(16), e105.

Dressman, D., Yan, H., Traverso, G., Kinzler, K. W., & Vogelstein, B. (2003). Transforming single DNA molecules into fluorescent magnetic particles for detection and enumeration of genetic variations. *Proceedings of the National Academy of Sciences*, *100*(15), 8817–8822.

Edwards, A., Voss, H., Rice, P., Civitello, A., Stegemann, J., Schwager, C., . . . & Ansorge, W. (1990). Automated DNA sequencing of the human HPRT locus. *Genomics*, *6*(4), 593–608.

Eid, J., Fehr, A., Gray, J., Luong, K., Lyle, J., Otto, G., . . . & Turner, S. (2009). Real-time DNA sequencing from single polymerase molecules. *Science*, *323*(5910), 133–138.

El-Metwally, S., Ouda, O. M., & Helmy, M. (2014). New horizons in next-generation sequencing. In *Next Generation Sequencing Technologies and Challenges in Sequence Assembly* (pp. 51–59). Springer, New York, NY.

ENCODE Project Consortium. (2004). The ENCODE (ENCyclopedia of DNA elements) project. *Science*, *306*(5696), 636–640.

Erhard, F., & Zimmer, R. (2010). Classification of ncRNAs using position and size information in deep sequencing data. *Bioinformatics*, *26*(18), i426–i432.

Euskirchen, G. M., Rozowsky, J. S., Wei, C. L., Lee, W. H., Zhang, Z. D., Hartman, S., . . . & Snyder, M. (2007). Mapping of transcription factor binding regions in mammalian cells by ChIP: comparison of array-and sequencing-based technologies. *Genome Research*, *17*(6), 898–909.

Fang, X., Yoon, J. G., Li, L., Yu, W., Shao, J., Hua, D., . . . & Lin, B. (2011). The SOX2 response program in glioblastoma multiforme: an integrated ChIP-seq, expression microarray, and microRNA analysis. *BMC Genomics*, *12*(1), 1–17.

Farias-Hesson, E., Erikson, J., Atkins, A., Shen, P., Davis, R. W., Scharfe, C., & Pourmand, N. (2010). Semi-automated library preparation for high-throughput DNA sequencing platforms. *Journal of Biomedicine and Biotechnology*, *2010*, 617469.

Finn, R. D., Mistry, J., Tate, J., Coggill, P., Heger, A., Pollington, J. E., . . . & Bateman, A. (2010). The Pfam protein families database. *Nucleic Acids Research*, *38*(suppl_1), D211–D222.

Fouse, S. D., Nagarajan, R. P., & Costello, J. F. (2010). Genome-scale DNA methylation analysis. *Epigenomics*, *2*(1), 105–117.

Fuller, C. W., Middendorf, L. R., Benner, S. A., Church, G. M., Harris, T., Huang, X., . . . & Vezenov, D. V. (2009). The challenges of sequencing by synthesis. *Nature Biotechnology*, *27*(11), 1013–1023.

Golan, D., & Medvedev, P. (2013). Using state machines to model the Ion Torrent sequencing process and to improve read error rates. *Bioinformatics*, *29*(13), i344–i351.

Goodwin, S., McPherson, J. D., & McCombie, W. R. (2016). Coming of age: ten years of next-generation sequencing technologies. *Nature Reviews Genetics*, *17*(6), 333.

Guo, J., Yu, L., Turro, N. J., & Ju, J. (2010). An integrated system for DNA sequencing by synthesis using novel nucleotide analogues. *Accounts of Chemical Research*, *43*(4), 551–563.

Gut, I. G. (2013). New sequencing technologies. *Clinical and Translational Oncology*, *15*(11), 879–881.

Hackenberg, M., Barturen, G., & Oliver, J. L. (2010). NGSmethDB: a database for next-generation sequencing single-cytosine-resolution DNA methylation data. *Nucleic Acids Research*, *39*(suppl_1), D75–D79.

Hajirasouliha, I., Hormozdiari, F., Alkan, C., Kidd, J. M., Birol, I., Eichler, E. E., & Sahinalp, S. C. (2010). Detection and characterization of novel sequence insertions using paired-end next-generation sequencing. *Bioinformatics*, *26*(10), 1277–1283.

Hart, C., Lipson, D., Ozsolak, F., Raz, T., Steinmann, K., Thompson, J., & Milos, P. M. (2010). Single-molecule sequencing: Sequence methods to enable accurate quantitation. *Methods in Enzymology*, *472*, 407–430.

Heather, J. M., & Chain, B. (2016). The sequence of sequencers: The history of sequencing DNA. *Genomics*, *107*(1), 1–8.

Henn, B. M., Gravel, S., Moreno-Estrada, A., Acevedo-Acevedo, S., & Bustamante, C. D. (2010). Fine-scale population structure and the era of next-generation sequencing. *Human Molecular Genetics*, *19*(R2), R221–R226.

Heo, Y. (2015). *Improving quality of high-throughput sequencing reads* (Doctoral dissertation, University of Illinois at Urbana-Champaign).

Herb, B. R., Wolschin, F., Hansen, K. D., Aryee, M. J., Langmead, B., Irizarry, R., . . . & Feinberg, A. P. (2012). Reversible switching between epigenetic states in honeybee behavioral subcastes. *Nature Neuroscience*, *15*(10), 1371–1373.

Hestand, M. S., Klingenhoff, A., Scherf, M., Ariyurek, Y., Ramos, Y., van Workum, W., . . . & 't Hoen, P. A. (2010). Tissue-specific transcript annotation and expression profiling with complementary next-generation sequencing technologies. *Nucleic Acids Research*, *38*(16), e165–e165.

Hoen, T. PAC et al. (2008). Deep sequencing-based expression analysis shows major advances in robustness, resolution and inter-lab portability over five microarray platforms. *Nucleic Acids Research*, *36*, e141.

Hoischen, A., van Bon, B. W., Gilissen, C., Arts, P., van Lier, B., Steehouwer, M., . . . & Veltman, J. A. (2010). De novo mutations of SETBP1 cause Schinzel-Giedion syndrome. *Nature Genetics*, *42*(6), 483–485.

Huse, S. M., Huber, J. A., Morrison, H. G., Sogin, M. L., & Welch, D. M. (2007). Accuracy and quality of massively parallel DNA pyrosequencing. *Genome Biology*, *8*(7), 1–9.

Huson, D. H., Auch, A. F., Qi, J., & Schuster, S. C. (2007). MEGAN analysis of metagenomic data. *Genome Research*, *17*(3), 377–386.

Hutchison III, C. A. (2007). DNA sequencing: Bench to bedside and beyond. *Nucleic Acids Research*, *35*(18), 6227–6237.

International Human Genome Sequencing Consortium. (2004). Finishing the euchromatic sequence of the human genome. *Nature*, *431*(7011), 931.

Jex, A. R., Hall, R. S., Littlewood, D. T. J., & Gasser, R. B. (2010). An integrated pipeline for next-generation sequencing and annotation of mitochondrial genomes. *Nucleic Acids Research*, *38*(2), 522–533.

Jiao, S., Bailey, C. P., Zhang, S., & Ladunga, I. (2010). Probabilistic peak calling and controlling false discovery rate estimations in transcription factor binding site mapping from ChIP-seq. In *Computational Biology of Transcription Factor Binding* (pp. 161–177). Humana Press, Totowa, NJ.

Katsios, C., Ziogas, D. E., Liakakos, T., Zoras, O., & Roukos, D. H. (2012). Translating cancer genomes sequencing revolution into surgical oncology practice. *Journal of Surgical Research*, *173*(2), 365–369.

Kchouk, M., Gibrat, J. F., & Elloumi, M. (2017). Generations of sequencing technologies: From first to next generation. *Biology and Medicine*, *9*(3).

Kingsley, C. B. (2011). Identification of causal sequence variants of disease in the next generation sequencing era. *Disease Gene Identification*, 37–46.

Kircher, M., & Kelso, J. (2010). High-throughput DNA sequencing – concepts and limitations. *Bioessays*, *32*(6), 524–536.

Kitzman, J. O., Snyder, M. W., Ventura, M., Lewis, A. P., Qiu, R., Simmons, L. E., . . . & Shendure, J. (2012). Noninvasive whole-genome sequencing of a human fetus. *Science Translational Medicine*, *4*(137), 137ra76–137ra76.

Korlach, J., Bjornson, K. P., Chaudhuri, B. P., Cicero, R. L., Flusberg, B. A., Gray, J. J., . . . & Turner, S. W. (2010). Real-time DNA sequencing from single polymerase molecules. *Methods in Enzymology*, *472*, 431–455.

Kuhlenbäumer, G., Hullmann, J., & Appenzeller, S. (2011). Novel genomic techniques open new avenues in the analysis of monogenic disorders. *Human Mutation*, *32*(2), 144–151.

Kulski, J. K. (2016). Next-generation sequencing – an overview of the history, tools, and "omic" applications. *Next Generation Sequencing-Advances, Applications and Challenges*, 3–60.

Laird, P. W. (2010). Principles and challenges of genome-wide DNA methylation analysis. *Nature Reviews Genetics*, *11*(3), 191–203.

Lander, E. S., Linton, L. M., Birren, B., Nusbaum, C., Zody, M. C., Baldwin, J., . . . & Rowen, L. (2001). Erratum: initial sequencing and analysis of the human genome: International human genome sequencing consortium (nature (2001) 409 (860–921)). *Nature*, *412*(6846), 565–566.

Lappalainen, T., Sammeth, M., Friedländer, M. R., Ac't Hoen, P., Monlong, J., Rivas, M. A., . . . & Dermitzakis, E. T. (2013). Transcriptome and genome sequencing uncovers functional variation in humans. *Nature*, *501*(7468), 506–511.

Lee, L. W., Zhang, S., Etheridge, A., Ma, L., Martin, D., Galas, D., & Wang, K. (2010). Complexity of the microRNA repertoire revealed by next-generation sequencing. *Rna*, *16*(11), 2170–2180.

Li, R., Zhu, H., Ruan, J., Qian, W., Fang, X., Shi, Z., . . . & Wang, J. (2010). De novo assembly of human genomes with massively parallel short read sequencing. *Genome Research*, *20*(2), 265–272.

Lind, C., Ferriola, D., Mackiewicz, K., Heron, S., Rogers, M., Slavich, L., . . . & Monos, D. (2010). Next-generation sequencing: the solution for high-resolution, unambiguous human leukocyte antigen typing. *Human Immunology*, *71*(10), 1033–1042.

Liu, L., Li, Y., Li, S., Hu, N., He, Y., Pong, R., . . . & Law, M. (2012). Comparison of next-generation sequencing systems. *Journal of Biomedicine and Biotechnology*, *2012*, 251364.

Loman, N. J., Misra, R. V., Dallman, T. J., Constantinidou, C., Gharbia, S. E., Wain, J., & Pallen, M. J. (2012). Performance comparison of benchtop high-throughput sequencing platforms. *Nature Biotechnology*, *30*(5), 434–439.

Lucy Raymond, F., Whittaker, J., Jenkins, L., Lench, N., & Chitty, L. S. (2010). Molecular prenatal diagnosis: the impact of modern technologies. *Prenatal Diagnosis*, *30*(7), 674–681.

Lundquist, P. M., Zhong, C. F., Zhao, P., Tomaney, A. B., Peluso, P. S., Dixon, J., . . . & Zaccarin, D. (2008). Parallel confocal detection of single molecules in real time. *Optics Letters*, *33*(9), 1026–1028.

Ma, Z., Lee, R. W., Li, B., Kenney, P., Wang, Y., Erikson, J., . . . & Lao, K. (2013). Isothermal amplification method for next-generation sequencing. *Proceedings of the National Academy of Sciences*, *110*(35), 14320–14323.

Maher, C. A., Kumar-Sinha, C., Cao, X., Kalyana-Sundaram, S., Han, B., Jing, X., . . . & Chinnaiyan, A. M. (2009). Transcriptome sequencing to detect gene fusions in cancer. *Nature*, *458*(7234), 97–101.

Mardis, E. R. (2008). The impact of next-generation sequencing technology on genetics. *Trends in Genetics*, *24*(3), 133–141.

Marguerat, S., & Bähler, J. (2010). RNA-seq: from technology to biology. *Cellular and Molecular Life Sciences*, *67*(4), 569–579.

Margulies, M., Egholm, M., Altman, W. E., Attiya, S., Bader, J. S., Bemben, L. A., & Rothberg, J. M. (2005). Genome sequencing in microfabricated high-density picolitre reactors. *Nature*, *437*(7057), 376–380.

Masoudi-Nejad, A., Narimani, Z., & Hosseinkhan, N. (2013). *Next generation Sequencing and Sequence Assembly: Methodologies and Algorithms* (Vol. 4). Springer, Science & Business Media.

Maxam, A. M., & Gilbert, W. (1977). A new method for sequencing DNA. *Proceedings of the National Academy of Sciences*, *74*(2), 560–564.

McBride, D. J., Orpana, A. K., Sotiriou, C., Joensuu, H., Stephens, P. J., Mudie, L. J., . . . & Campbell, P. J. (2010). Use of cancer-specific genomic rearrangements to quantify disease burden in plasma from patients with solid tumors. *Genes, Chromosomes and Cancer*, *49*(11), 1062–1069.

McComish, B. J., Hills, S. F., Biggs, P. J., & Penny, D. (2010). Index-free de novo assembly and deconvolution of mixed mitochondrial genomes. *Genome Biology and Evolution*, *2*, 410–424.

McGinn, S., & Gut, I. G. (2013). DNA sequencing – spanning the generations. *New Biotechnology*, *30*(4), 366–372.

McHardy, A. C., Martín, H. G., Tsirigos, A., Hugenholtz, P., & Rigoutsos, I. (2007). Accurate phylogenetic classification of variable-length DNA fragments. *Nature Methods*, *4*(1), 63–72.

Mehdi, K., Gibrat, J. F., & Elloumi, M. (2017). Generations of sequencing technologies: From first to next generation. *Electromagnetic Biology and Medicine*, *9*(3), 8.

Merriman, B., R&D Team, I. T., & Rothberg, J. M. (2012). Progress in ion torrent semiconductor chip based sequencing. *Electrophoresis*, *33*(23), 3397–3417.

Metzker, M. L. (2010). Sequencing technologies – the next generation. *Nature Reviews Genetics*, *11*(1), 31–46.

Meyerson, M., Gabriel, S., & Getz, G. (2010). Advances in understanding cancer genomes through second-generation sequencing. *Nature Reviews Genetics*, *11*(10), 685–696.

Morin, R. D., Zhao, Y., Prabhu, A. L., Dhalla, N., McDonald, H., Pandoh, P., . . . & Marra, M. (2010). Preparation and analysis of microRNA libraries using the Illumina massively parallel sequencing technology. In *RNAi and microRNA-Mediated Gene Regulation in Stem Cells* (pp. 173–199). Humana Press, Totowa, NJ.

Nagalakshmi, U., Waern, K., & Snyder, M. (2010). RNA-Seq: A method for comprehensive transcriptome analysis. *Current Protocols in Molecular Biology, 89*(1), 4–11.

Nagarajan, N., & Pop, M. (2010). Sequencing and genome assembly using next-generation technologies. *Computational Biology*, 1–17.

Nakano, M., Komatsu, J., Matsuura, S. I., Takashima, K., Katsura, S., & Mizuno, A. (2003). Single-molecule PCR using water-in-oil emulsion. *Journal of Biotechnology, 102*(2), 117–124.

Neff, T., & Armstrong, S. A. (2009). Chromatin maps, histone modifications and leukemia. *Leukemia, 23*(7), 1243–1251.

Ng, S. B., Buckingham, K. J., Lee, C., Bigham, A. W., Tabor, H. K., Dent, K. M., . . . & Bamshad, M. J. (2010). Exome sequencing identifies the cause of a mendelian disorder. *Nature Genetics, 42*(1), 30–35.

Norton, M. E., Brar, H., Weiss, J., Karimi, A., Laurent, L. C., Caughey, A. B., . . . & Song, K. (2012). Non-Invasive Chromosomal Evaluation (NICE) study: Results of a multicenter prospective cohort study for detection of fetal trisomy 21 and trisomy 18. *American Journal of Obstetrics and Gynecology, 207*(2), 137–e1.

Nyrén, P. (1987). Enzymatic method for continuous monitoring of DNA polymerase activity. *Analytical Biochemistry, 167*(2), 235–238.

Ozsolak, F., & Milos, P. M. (2011). RNA sequencing: advances, challenges and opportunities. *Nature Reviews Genetics, 12*(2), 87–98.

Pareek, C. S., Smoczynski, R., & Tretyn, A. (2011). Sequencing technologies and genome sequencing. *Journal of Applied Genetics, 52*(4), 413–435.

Pfeifer, G. P., & Hainaut, P. (2011). Next-generation sequencing: emerging lessons on the origins of human cancer. *Current Opinion in Oncology, 23*(1), 62–68.

Popp, C., Dean, W., Feng, S., Cokus, S. J., Andrews, S., Pellegrini, M., . . . & Reik, W. (2010). Genome-wide erasure of DNA methylation in mouse primordial germ cells is affected by AID deficiency. *Nature, 463*(7284), 1101–1105.

Pushkarev, D., Neff, N. F., & Quake, S. R. (2009). Single-molecule sequencing of an individual human genome. *Nature Biotechnology, 27*(9), 847–850.

Quick, J., Quinlan, A. R.& Loman, N. J. (2014). A reference bacterial genome dataset generated on the MinION™ portable single-molecule nanopore sequencer. *Gigascience, 3*(1), 22.

Ramsingh, G., Koboldt, D. C., Trissal, M., Chiappinelli, K. B., Wylie, T., Koul, S., . . . & Link, D. C. (2010). Complete characterization of the microRNAome in a patient with acute myeloid leukemia. *Blood, The Journal of the American Society of Hematology, 116*(24), 5316–5326.

Reuter, J. A., Spacek, D. V., & Snyder, M. P. (2015). High-throughput sequencing technologies. *Molecular Cell, 58*(4), 586–597.

Robertson, G., Hirst, M., Bainbridge, M., Bilenky, M., Zhao, Y., Zeng, T., . . . & Jones, S. (2007). Genome-wide profiles of STAT1 DNA association using chromatin immunoprecipitation and massively parallel sequencing. *Nature Methods, 4*(8), 651–657.

Ronaghi, M. (1998). Real-time pyrophosphate detection for DNA sequencing. *Science, 281*, 363–364.

Ronaghi, M., Karamohamed, S., Pettersson, B., Uhlén, M., & Nyrén, P. (1996). Real-time DNA sequencing using detection of pyrophosphate release. *Analytical Biochemistry, 242*(1), 84–89.

Ronaghi, M., Uhlén, M., & Nyrén, P. (1998). A sequencing method based on real-time pyrophosphate. *Science, 281*(5375), 363–365.

Rothberg, J. M., Hinz, W., Rearick, T. M., Schultz, J., Mileski, W., Davey, M., . . . & Bustillo, J. (2011). An integrated semiconductor device enabling non-optical genome sequencing. *Nature, 475*(7356), 348–352.

Sanger, F. (1949). The terminal peptides of insulin. *Biochemical Journal, 45*(5), 563.

Sanger, F., & Coulson, A. R. (1975). A rapid method for determining sequences in DNA by primed synthesis with DNA polymerase. *Journal of Molecular Biology, 94*(3), 441–448.

Sanger, F., Nicklen, S., & Coulson, A. R. (1977). DNA sequencing with chain-terminating inhibitors. *Proceedings of the National Academy of Sciences, 74*(12), 5463–5467.

Sanger, F., & Tuppy, H. (1951). The amino-acid sequence in the phenylalanyl chain of insulin. 2. The investigation of peptides from enzymic hydrolysates. *Biochemical Journal, 49*(4), 481–490.

Schloss, P. D., & Handelsman, J. (2005). Metagenomics for studying unculturable microorganisms: Cutting the Gordian knot. *Genome Biology, 6*(8), 1–4.

Schloss, P. D., Westcott, S. L., Ryabin, T., Hall, J. R., Hartmann, M., Hollister, E. B., . . . & Weber, C. F. (2009). Introducing mothur: open-source, platform-independent, community-supported software for describing and comparing microbial communities. *Applied and Environmental Microbiology, 75*(23), 7537–7541.

Schuster, S. C., Chi, K. R., Rusk, N., Kiermer, V., Wold, B., & Myers, R. M. (2008). Method of the year, next-generation DNA sequencing. *Functional Genomics and Medical Applications. Nat. Methods, 5*, 11–21.

Senapathy, P., Bhasi, A., Mattox, J., Dhandapany, P. S., & Sadayappan, S. (2010). Targeted genome-wide enrichment of functional regions. *PLoS One, 5*(6), e11138.

Shapiro, B., & Hofreiter, M. (2010). Analysis of ancient human genomes: using next generation sequencing, 20-fold coverage of the genome of a 4,000-year-old human from Greenland has been obtained. *Bioessays, 32*(5), 388–391.

Shendure, J., & Ji, H. (2008). Next-generation DNA sequencing. *Nature Biotechnology, 26*(10), 1135–1145.

Singleton, A. B., Hardy, J., Traynor, B. J., & Houlden, H. (2010). Towards a complete resolution of the genetic architecture of disease. *Trends in Genetics, 26*(10), 438–442.

Soroka, M., & Burzyński, A. (2010). Complete sequences of maternally inherited mitochondrial genomes in mussels Unio pictorum (Bivalvia, Unionidae). *Journal of Applied Genetics, 51*(4), 469–476.

Tawfik, D. S., & Griffiths, A. D. (1998). Man-made cell-like compartments for molecular evolution. *Nature Biotechnology, 16*(7), 652–656.

Thompson, J. F., & Steinmann, K. E. (2010). Single molecule sequencing with a HeliScope genetic analysis system. *Current Protocols in Molecular Biology, 92*(1), 7–10.

Thudi, M., Li, Y., Jackson, S. A., May, G. D., & Varshney, R. K. (2012). Current state-of-art of sequencing technologies for plant genomics research. *Briefings in Functional Genomics, 11*(1), 3–11.

Travers, K. J., Chin, C. S., Rank, D. R., Eid, J. S., & Turner, S. W. (2010). A flexible and efficient template format for circular consensus sequencing and SNP detection. *Nucleic Acids Research, 38*(15), e159–e159.

Valen, E., Pascarella, G., Chalk, A., Maeda, N., Kojima, M., Kawazu, C., . . . & Carninci, P. (2009). Genome-wide detection and analysis of hippocampus core promoters using DeepCAGE. *Genome Research, 19*(2), 255–265.

Valouev, A., Ichikawa, J., Tonthat, T., Stuart, J., Ranade, S., Peckham, H., . . . & Johnson, S. M. (2008). A high-resolution, nucleosome position map of C. Elegans reveals a lack of universal sequence-dictated positioning. *Genome Research, 18*(7), 1051–1063.

Venter, J. C., Adams, M. D., Myers, E. W., Li, P. W., Mural, R. J., Sutton, G. G., . . . & Kalush, F. (2001). The sequence of the human genome. *Science, 291*(5507), 1304–1351.

Venter, J. C., Remington, K., Heidelberg, J. F., Halpern, A. L., Rusch, D., Eisen, J. A., . . . & Smith, H. O. (2004). Environmental genome shotgun sequencing of the Sargasso Sea. *Science, 304*(5667), 66–74.

Vermeer, S., Hoischen, A., Meijer, R. P., Gilissen, C., Neveling, K., Wieskamp, N., . . . & Knoers, N. (2010). Targeted next-generation sequencing of a 12.5 Mb homozygous region reveals ANO10 mutations in patients with autosomal-recessive cerebellar ataxia. *The American Journal of Human Genetics, 87*(6), 813–819.

Vezzi, F. (2012). *Next generation sequencing revolution challenges: Search, assemble, and validate genomes* (Thesis, Università degli Studi di Udine).

Vissers, L. E., de Ligt, J., Gilissen, C., Janssen, I., Steehouwer, M., de Vries, P., . . . & Veltman, J. A. (2010). A de novo paradigm for mental retardation. *Nature Genetics, 42*(12), 1109–1112.

Voelkerding, K. V., Dames, S., & Durtschi, J. D. (2010). Next generation sequencing for clinical diagnostics-principles and application to targeted resequencing for hypertrophic cardiomyopathy: a paper from the 2009 William Beaumont Hospital Symposium on Molecular Pathology. *The Journal of Molecular Diagnostics, 12*(5), 539–551.

Walsh, T., Lee, M. K., Casadei, S., Thornton, A. M., Stray, S. M., Pennil, C., . . . & King, M. C. (2010). Detection of inherited mutations for breast and ovarian cancer using genomic capture and massively parallel sequencing. *Proceedings of the National Academy of Sciences, 107*(28), 12629–12633.

Wang, W. C., Lin, F. M., Chang, W. C., Lin, K. Y., Huang, H. D., & Lin, N. S. (2009). miRExpress: analyzing high-throughput sequencing data for profiling microRNA expression. *BMC Bioinformatics, 10*(1), 1–13.

Wang, Y., Yang, Q., & Wang, Z. (2015). The evolution of nanopore sequencing. *Frontiers in Genetics, 5*, 449.

Wang, Z., Gerstein, M., & Snyder, M. (2009). RNA-Seq: A revolutionary tool for transcriptomics. *Nature Reviews Genetics, 10*(1), 57–63.

Watson, J. D., & Crick, F. H. (1953). Molecular structure of nucleic acids: A structure for deoxyribose nucleic acid. *Nature, 171*(4356), 737–738.

Werner, T. (2010). Next generation sequencing in functional genomics. *Briefings in Bioinformatics, 11*(5), 499–511.

Westra, H. J., Peters, M. J., Esko, T., Yaghootkar, H., Schurmann, C., Kettunen, J., . . . & Franke, L. (2013). Systematic identification of trans eQTLs as putative drivers of known disease associations. *Nature Genetics, 45*(10), 1238–1243.

Wheeler, D. A., Srinivasan, M., Egholm, M., Shen, Y., Chen, L., McGuire, A., . . . & Rothberg, J. M. (2008). The complete genome of an individual by massively parallel DNA sequencing. *Nature, 452*(7189), 872–876.

Xi, R., Kim, T. M., & Park, P. J. (2010). Detecting structural variations in the human genome using next generation sequencing. *Briefings in Functional Genomics, 9*(5–6), 405–415.

Yang, J. H., Shao, P., Zhou, H., Chen, Y. Q., & Qu, L. H. (2010). deepBase: A database for deeply annotating and mining deep sequencing data. *Nucleic Acids Research, 38*(suppl_1), D123–D130.

Yu, L., Guo, J., Xu, N., Li, Z. M., & Ju, J. Y. (2010). DNA sequencing-by-synthesis using novel nucleotide analogs. *The Handbook of Plant Mutation Screening: Mining of Natural and Induced Alleles*, 319–338.

Zallen, D. T. (2003). Despite Franklin's work, Wilkins earned his nobel. *Nature, 425*(6953), 15.

Zhou, X., Ren, L., Li, Y., Zhang, M., Yu, Y., & Yu, J. (2010). The next-generation sequencing technology: a technology review and future perspective. *Science China Life Sciences, 53*(1), 44–57.

Zhou, X., Ren, L., Meng, Q., Li, Y., Yu, Y., & Yu, J. (2010). The next-generation sequencing technology and application. *Protein Cell, 1*, 520–536.

Zoghbi, H. Y., & Warren, S. T. (2010). Neurogenetics: Advancing the "next-generation" of brain research. *Neuron, 68*(2), 165–173.

3 Role of Omics Technologies in Animal Science

Monika Sodhi, Nampher Masharing, Vishal Sharma, RC Sobti and Manishi Mukesh

CONTENTS

3.1 INTRODUCTION

World population increased from three billion in 1960 to seven billion in 2011, and it is expected to reach 11 billion by 2100. To feed this population, it is required to double the world agriculture production. Upsurge in demand for food is also expected due to expanding population, urbanization, and high living standards (FAO, 2011). An FAO (2018) survey showed that domestic animals contribute about 18% of the total calories and 39% of protein requirements for humans. Besides food, animal by-products have found many applications in the fields of pharmaceuticals,

DOI: 10.1201/9781003220831-3

cosmetics, and household and industrial products. To meet the ever-increasing demand of animal products for the growing population, selective breeding has been practiced to propagate superior germplasm that is capable of higher production, less susceptibility to diseases, high fertility, etc. Traditional breeding was done without the molecular knowledge of the genes present in quantitative traits loci (QTL). Quantitative traits are controlled by multiple genes and are also regulated by multiple external factors, but traditional breeding was not able to take into account the effect of all these factors on the production of animals, as it was mostly based on the breeding value of the animals. Though the traditional breeding techniques have increased the milk production in cattle and feed conversion efficiency in pigs and other animals, their use becomes limited for the traits that are difficult to measure, have low heritability, are expressed at particular or later stages of life cycle, and are under the control of multiple genes or external factors. Similarly, transcripts, and protein markers may also help in determining the physiological stage and health status of the animals. Such markers help in designing the appropriate strategy to counter any adverse effect on the animal welfare, production, and health. Thus to ameliorate the productive, reproductive, growth performance and disease resistance in animals, an interest has been generated in researchers across the globe for deeper understanding of animal biology, genomics, proteomics and biotechnology, as domestic animals have always played a crucial role in catering to human nutritional requirements and other needs.

Omics approaches are the high-throughput holistic molecular technologies which analyze "everything," including all genes (genomics), all transcripts (transcriptomics), all proteins (proteomics), all metabolites (metabolomics), accessibility of DNA for transcription (epigenomics), and characterizing microbial groups (metagenomics) in one go. Use of these state-of-the-art omics technologies can help us to understand the complexity of the relationship between genotype and phenotype, and translate these complex traits into heritable and easily measurable proxy traits that help breeders to better define the complex traits and also go for decoupling correlated traits.

Indeed, advanced omics approaches can be used towards improving breeder selection programs, the quality of meat and dairy products and animal health to contribute towards greater sustainability and productivity. Also, in the era of global warming, understanding of adaptation mechanisms of resilient breeds and/or species can facilitate the sustainability of animal production. Herein, we elaborate on the general characteristics and benefits of different omics approaches.

3.2 GENOMICS IN ANIMAL SCIENCE

Use of genomics in animal science has traversed a long way through discoveries of cytogenetic maps, synteny analysis, somatic cell genetics, in situ hybridization, radiation-hybrid maps, and constructing high-resolution comparative mapping. In the 1970s, molecular genetics was used in the animal breeding programs by using DNA markers for identifying the genes controlling the preferred traits. Genetic tests were developed for detecting gene defects, identification of QTLs to make early decisions for the selection of animals through marker-assisted selection for the breeding programs (Soller, 1978; Lande and Thompson, 1990). The applications of gene identification, mapping, and QTL discoveries produced a number of DNA markers that were incorporated in selection programs across many species. Use of these gene marker assays allowed breeders to make improvements in the animals and gave birth to a new discipline, marker-assisted selection (MAS), in the 1980s. MAS resulted in a global pursuit for quantitative trait loci in all animal species as part of their overall improvement programs.

During the last few decades, advances in the techniques of the genome studies like microarray and next-generation sequences has started a new "genomics era" in animal science. The genomics era started with the starting of the Human Genome project that involved the sequencing of the complete genome of humans. This led to the development of high-throughput sequencing platforms that became more efficient, less time-consuming, and less expensive with time. The rapid improvements in high-throughput methods for SNPs genotyping technologies and the development of the

SNP arrays have opened the possibility of using genomic information in animal breeding programs through a novel method called genomic selection. Genomic selection evaluates an equation in a reference population having genotype and phenotype data. This prediction equation is then used to predict the breeding values in the animals without phenotype data. In fact, genomic selection can calculate accurate estimated breeding value (EBV) before sexual maturity. In addition, Genomics through sequencing has also offered new avenues to study and identify pathogens and assist in their subsequent vaccine development. Thus, the use of these high end techniques has helped in addressing and understanding the scientific basis of the animal biology, enabling the researchers to make successful prediction about the animals. This has resulted in overall improvement in terms of their production, feed conversion efficiency, and disease resistance.

3.2.1 Characterization and Conservation of Genetic Diversity

During the past three decades, the extinction of animal species has occurred at an alarming rate. According to an estimate, about 26% of the global breeds are at the edge of extinction (FAO, 2018). The extinction of a particular breed or population or species means the loss of its unique genes or gene combinations responsible for specific attributes. However, conservation initiatives to preserve the diversity can only be taken if the individuals are assigned to a specific breed and uniqueness/genetic distinctness of the specific breed is ascertained (Eding et al., 2002). Hence, characterization of the breeds is essential for preserving precious and irreplaceable diversity, which is key to the future breeding programs and long term sustainability of animal production systems under changing food demands and climatic conditions (Boettcher et al., 2015; Phocas et al., 2016).

In the last two decades, a different set of markers (phenotypic & biochemical markers, RFLP, minisatellites, microsatellites, RAPDs, AFLPs, VNTRs, mtDNA, Y-chr) have been implied for valuation of genetic diversity, genetic uniqueness and process of domestication (Joshi et al., 2012). However, these markers target only a few loci in the genome and thus represents only a fraction of the genetic information (Avise, 2010). With advent of new technical developments, the genetic diversity assessment of animal species based on large sets of markers can be carried out efficiently and effectively (Goodwin et al., 2016).

High-throughput sequencing technology allows the accessibility of higher marker density. Many research groups have sequenced the reference genomes of their native species followed by resequencing genomes of individuals and populations. With the availability of huge data sets and statistical advancements, it is possible to answer the basic questions of conservation biology including phylogenetic relationship amongst the different breeds; loci responsible for adaptation and fitness to local, diverse agro climatic conditions, inbreeding depression; loci responsible for speciation and interaction among species (McMahon et al., 2014). Capturing the rare variants holds important for local native breeds, as these breeds with low population size and high inbreeding rates have a higher chance of losing rare alleles. This allowed the assessment of neutral and adaptive variation in the populations for more logical conservation decisions (Allendorf et al., 2010; Frankham, 2010; Angeloni et al., 2012; Shafer et al., 2015; Benestan et al., 2016).

Further, using the high throughput sequencing data, SNP chips have been developed which has become one of the most cost effective technologies for animal characterization. In SNP array/chip single nucleotide polymorphism information is based on the whole genome information covering coding, noncoding, regulatory regions and thus covers both neutral and functional genetic variations that are widely spread along their entire genomes. Being abundant and widespread throughout the animal genome, SNPs are rapidly becoming the marker of choice for genomic analysis of animals. Taking the advantage of the availability of sequence draft and use of recent techniques for SNP discovery, SNP arrays have already been developed for most of the species, including chicken (Illumina 60K/18K iSelect chip), cattle (Bovine High density770K/50K Illumina™ iSelect chip), pig (Illuminaiselect pig DNA chip, 60K), horse (Equine illumina 60k SNP chip; Illumina Equine SNP70 Bead Chip), buffalo (Axiom® Buffalo Genotyping array), sheep (Illumina's OvineSNP50;

Ovine/Infinium® HD SNP BeadChip), and so on and so forth. For many other animal species/breeds, the chip development is in progress.

Overall, genomics approaches open new possibilities for programs on characterization and conservation of genetic diversity.

3.2.2 Animal Production

As earth has limited space, it is essential to focus on increasing animal productivity rather than increasing the number of animals to meet ever-increasing food demands. The prerequisite is understanding of structure and function of genes influencing productive traits followed by appropriate selection and breeding decisions. Identification of markers and chromosomal regions containing quantitative trait loci affecting important traits has helped in increasing animal production.

Genomic selection has been applied in many production traits such as milk production in cattle, feed efficiency in chickens, cattle, and pigs, and also functional trait complexes linked to animal health and welfare. With the availability of DNA marker data as well as phenotypes and pedigrees on potential selection candidates, information can be combined to estimate improved EBV (estimated breeding value). EBV obtained from "genome-wide" selection (genomic-estimated breeding values; GEBV) is of comparable robustness to a conventional EBV based on analysis of pedigree and phenotype data. The reliability of GEBV has been observed to be significantly greater than the parental average breeding values (Harris et al., 2008; Hayes et al., 2009; VanRaden et al., 2009). The immediate use of genomic selection implies to superior males, which could be selected on the basis of genomic information alone and used for mating long before their progeny test results are available. This will help in cost savings, reducing generation interval and speeding up the breeding programmes for the improvement of animals for the desired traits.

Genome-wide association (GWA) studies facilitate testing associations between SNPs across the genomes and a trait in related or unrelated individuals to identify genotype–phenotype associations. GWAS for traits such milk production, health, reproduction, fertility, calving, faster and leaner growth, high litter size, etc., has been carried out across different species and different countries (Jiang et al., 2010; Pryce et al., 2010; Ibtisham et al., 2017; Nayeri et al., 2017; Raphael et al., 2019), with considerable rates of genetic improvement. The reported accuracy of genomic prediction applied in dairy cattle was >0.8 for production traits and >0.7 for fertility, longevity, somatic cell count, and other traits (Wiggans et al., 2011). SNP assays for genomic selection (GS) use all available genome-wide dense SNP markers simultaneously and thus overcome the disadvantage of MAS which could program trace limited numbers of QTLs using a small number of DNA markers. Further, the strategy of genome-wide association analysis based on the assumption that specific SNP markers could be in genetic linkage disequilibrium with the causative mutation is also being implied across animal species to study complex traits.

The information from genomics about the genes involved in the preferred traits in combination with the advanced reproductive technologies like artificial insemination, embryo transfer, and cloning have greatly improved the quality and efficiency of production within a short span of time.

3.2.3 Adaptation to Climate Change

Climate change is expected to affect animal production significantly. These effects will likely affect feed availability, increased drought and heat stress, and changes in disease prevalence. In coming times, climate change might be the major force determining resilience of food production systems. In animal species, with increasing global warming, heat stress is the major factor responsible for reduced production efficiency, decrease in milk production, and increased susceptibility to diseases. The negative impact of heat stress on dairy industry is huge (Hansen, 2007; Marai and Habeeb, 2010) and for long-term solutions a "landscape approach," that involves selection of animals that are more adaptable to the climate changes is to be followed. A key component of adaptation, in general, is the genetic ability of an organism to survive under stressful conditions.

Discovering markers and screening polymorphism with genetic markers forms the basic need for genetic measures required for selection of animals to counter the effects of climate change. Genomics can play a major role in such transformations including tapping of available genetic variations. Characterization with high-throughput SNP assays or genome sequencing is necessary for assessment of genetic variation responsible for thermotolerance/adaptability and unraveling the physiological basis for adaptation. Species-wide HapMap studies across the diverse geographical regions and production systems can help in generating valuable information related to genomic regions playing a role in adaptation.

Recently, genomic research in Bactrian Camels has revealed that genes involved in the insulin pathway may allow these camels to go long periods of time without water in hot dry environments. Such studies attempted in cattle, sheep, goats, and chickens have identified potential candidate genes (heat shock protein families, heat shock factor) and SNPs associated with thermo-tolerant genes, revealed significant genomic differences as well possible signatures of selection. Overall, all the information on novel resilience traits along with genomic selection can accelerate selective breeding programs for adaptation.

3.2.4 Animal Health

Diseases of animals result in a huge loss to the industry and excess use of antimicrobials affects the welfare of animals as well as consumers. There is a need to understand the biological determinants of disease in order to develop disease-resistant animals. Since genetic gain through genomic selection in case of low heritable traits is far more than highly heritable traits, research work has been extended for health traits. This has resulted in animal welfare as well economic gain for the animal industry as health issues like hypocalcemia, displaced abomasum, ketosis, mastitis, retained placenta, and metritis could be taken care of (Rexroad et al., 2019).

Through whole genome sequencing approaches, it has become possible to identify the strain of a pathogen accurately that could lead to isolation of specific antigens for development of new and more effective vaccines, e.g. tracing of members of the Mycoplasma mycoides cluster, causative agent of contagious pleuropneumonia in ruminants (Fischer et al., 2012). Similarly, high-throughput sequencing has allowed to obtain more detailed maps of genetic recombination like comparative genome analysis of *T. orientalis* relative to their highly pathogenic Theileria species, a protozoan parasite of cattle (Hayashida et al., 2012), and then identification of their antigens. Genomic analysis of parasites have also helped in understanding their resistance to drug treatments and consequently in developing new alternative treatments and their eradication. However, large numbers of pathogens that affect animal production are yet to be identified. Genomics along with computational and immunological technologies can help to identify different pathogens and understand comprehensive interactions of pre- and probiotics, vectors, vaccines, and healthy microbiomes to have a better understanding of diseases and their curative measures.

3.2.5 Traits Related to Environmental Issues

Genomics can help to address environmental issues on several levels. Development of breeds that have feed conversion efficiency can result in decreased waste output and less production of methane. Ruminants produce large amount of methane gas causing environmental pollution. Application of omics approaches to identify the genomic regions or genes involved in the reduction of methane production, or high feed conversion efficiency leading to less waste output and lower production of methane and combining the same with the reproduction technologies can solve many of the environmental issues. Pigs also pollute the environment by producing manure with high concentration of phosphorus. Genomic approaches can be applied in towards identification of SNPs associated with fast growth and no adverse effects under lower amount of phosphorus in the diet. Considering the public acceptance, solution to change genes affecting environment is genomic approaches but not the genetic modification especially for meat producing animal species.

3.3 PROTEOMICS IN ANIMAL SCIENCE

Proteomics refers to the large-scale study of protein structure, function, composition, and interaction, while proteome is the set of proteins expressed by an organism under certain environmental conditions. It has the potential for simultaneous identification and quantification of thousands of proteins in one go. It can identify differential expression, structural modifications, or interactions of different proteins under case control conditions, development stages, growth conditions, or physiological processes and yields the wholesome picture. The importance of proteins as the primary effector molecules of biology is well understood, and thus interest and advances in proteome studies is growing rapidly. It can very well be implied in all areas of animal science like animal physiology, production, health and welfare, reproduction, muscle, lactation biology, and post-harvest modifications of animal products (Martins et al., 2012; Almeida et al., 2015; Bick et al., 2018).

The common tools employed in proteomics are western blotting, 2D gel electrophoresis, X-ray crystallography, NMR, MALDI-TOF/MS, protein microarrays, and two-hybrid screening. For protein identification and quantification, three different strategies involve either measurement of the intact proteins, known as top-down approach, or analysis of incomplete protein digests and protein fragments, called middle-down approach, or analysis of peptides (digested proteins) named bottom-up approach. Amongst these, bottom-up approach is the most preferred one for both gel-based as well as LC-based proteomic approaches.

Recent advances in mass spectrometry technologies have led to the development of novel strategies aimed at identification of biomarkers across different tissues and body fluids such as serum, urine, milk, saliva, cerebrospinal fluid, and semen. These markers can facilitate bio-monitoring on animal health and provide valuable information, on production, feeding status, as well as animal–environment interaction. Some of the applications of proteomics in animal sciences are discussed hereunder.

3.3.1 Biomarkers for Quality and Quantity Traits

Using proteomics approach, accessibility of battery of biomarkers for different traits like higher production (mainly milk and meat), better product quality, and animal welfare can be addressed simultaneously. Milk proteome maps provide in-depth knowledge related to milk quality processes as differential protein profiles not only represent the genetic variations but also epigenetic and environmental factors including post-translational modifications (glycosylation, phosphorylation, and proteolysis) (Smolenski et al., 2007). Information on colostrum bioactive proteins/biomolecules has helped to understand their role in host defense, gastrointestinal growth and development of the neonates. Further, proteomic analysis has also facilitated the development of rapid and sensitive methods for detection of adulterants in milk (Calvano et al., 2013). Thus proteomics studies may help in increasing the animals production and quality of the product.

3.3.2 Animal Reproduction

Reproduction is a very complex phenomenon involving interactions of different proteins. Though genomic information has helped to understand the complex phenomenon of reproduction, still, prediction of the reproductive phenotypes and behavior based on molecular data is time-consuming and tedious. Proteome profiles of male and female reproductive proteins have eased the understanding of this mechanism. Understanding the mechanism and factors responsible for reproductive performance can impact the animal production positively and increase farmer portability. In dairy bulls, differential protein profiles of seminal fluid clearly revealed the role of type and quantity of proteins on fertility indexes (Kelly et al., 2006; Moura et al., 2006). In pigs, proteins aminopeptidase and retinol-binding protein 4 were associated with large litter size as deduced from sow placental protein profile (Lee et al., 2015). In sheep, a proteomic-based approach has been used to understand the prolificacy (Miao et al., 2016) and also in vitro selection of oocytes for IVF by proteomic profile of follicular fluid (Paula Junior et al., 2017).

In dairy cattle, proteomic studies on the follicular fluid of preovulatory follicle revealed the set of proteins differentially expressed in low-fertile dairy cows compared to controls (Zachut et al., 2016). Similar studies have been conducted by different researchers in buffalo (Fu et al., 2016), mare (Fahiminiya et al., 2011), and bitch (Fahiminiya et al., 2010). In boars, the proteins in the spermatozoa have been associated with the cryopreservation tolerance and thus may serve as biomarkers for the cryopreservation of the semen (Guimarães et al., 2017).

Overall, the protein molecules may help in selection of animals having better reproductive performance, diagnosis of reduced reproductive performance, and thus possible treatment for reproduction.

3.3.3 Animal Health and Diseases

Use of proteomics in veterinary medicine is a comparatively new and evolving field. Proteomics can be easily applied in animal health sciences, as animals offer availability of diverse biological samples viz., different tissues, milk, semen, saliva, urine, plasma, which allows investigation of experimental as well as natural disease processes. Proteomics can add new dimensions in understanding disease pathogenesis (biology and pathology) in animal and poultry species. Using the approach of proteomics in conjunction with peptidomics and metabolomics, differentially up/down regulated proteins have helped in the identification of diseased animals. The differentially expressed proteins can suggest the mechanism of host response under diseased conditions. Using a proteomics approach, a number of studies have been conducted in dairy cattle with major focus on mastitis, the most common diseases in bovines, to understand the disease pathogenesis in order to find biomarkers of early diagnosis. Increase in the level of serum albumin and transferrin with concurrent decreases in all caseins, β-lactoglobulin, and α-lactoglobulin, was detected in cows suffering from mastitis (Hogarth et al., 2004), suggesting that failure of the milk-blood barrier might be responsible for transport of serum proteins into blood. Similarly, Smolenski et al. (2007) detected increase in the levels of apolipoprotein A-I, cathelicidin-I, heat shock 70kD protein, and acute-phase protein serum amyloid A in the milk of cows affected with mastitis. Using proteomics tools, a total of 300 milk proteins linked to host defense were identified from milk fat globule membrane, exosome and whey proteins, of which 94 were differentially expressed in the animals affected with mastitis (Reinhardt et al., 2013). Proteomics studies have helped to better understand other animal diseases like brucellosis, hemorhaggicsepticaemia, foot and mouth disease, blue tongue, parasitic diseases, etc. Similar to livestock species, proteomics has also been applied to poultry and aqua species like understanding pathogenesis, new biomarkers of vaccines, etc. Further information generated through comparative proteome across animal species and human can support new advances in veterinary medicine. Use of large animals as model can facilitate better predictions and thus address the health problems (Kues and Niemann, 2004). Animals with better disease resistance can be identified and detailed information on diagnostic tools with better accuracy, new therapies, or biomarkers can be developed. Overall changes in the proteins or pathways or both can be identified through proteomics to explicate the important pathogen or host proteins involved in disease progression and recovery.

3.3.4 Stress Biomarkers in Animals

Stress in animals has a negative impact on growth, production, and disease resistance, as it reduces the innate as well as acquired immune response. In dairy cows, stress has been observed to be associated with increased susceptibility to infectious diseases such as mastitis, salmonellosis, and bovine respiratory complex, infertility, uterine diseases ketosis and milk fever, especially during the periparturient period (Nir, 2003). Proteome analysis has also helped to elucidate the stress-related markers to monitor animal welfare. It also holds importance as it is known the consumption of products obtained from stressed animals might affect human health. Different proteomic studies have been conducted across animal species from different tissues to understand the protein profile alterations under different types of stress, e.g. differentially expressed proteins from mammary tissue and liver in heat-stressed cows (Ma et al., 2019); serum proteins showing changes in response to different housing (individual vs. mixed stalls) in pigs. Overall, the proteomics may help in early identification of animals under stress and thus possible steps may be taken to eliminate the stressors, which may help in increasing the animal production, and improve health status and animal welfare.

3.4 TRANSCRIPTOMICS IN ANIMAL SCIENCE

Transcriptomics is the study of transcripts present in cell, tissue, or organ of an organism under a specific set of physiological conditions or developmental stages. As mRNA acts as intermediate between the genome and proteome of an individual, transcriptome of an organisms gives an important clue about the changes taking place at the molecular level in an organism. Animal development, physiology, and pathology involve a complex interplay of different mechanisms, which are mediated by the transcripts. Gene expression analysis is very important for animal research as it helps in understanding the expressed genes pattern and provides insights into complex regulatory pathways and identification of genes relevant to biological processes. The techniques for the evaluation of transcripts have progressed from simple methods of reverse transcriptase PCR, real-time PCR, Northern blotting to the techniques aimed at identifying differential expression of genes between or among experimental samples like, subtractive hybridization, expressed sequence tags (ESTs), serial analysis of gene expression (SAGE), microarrays, next generation sequencing, etc.

The two most common high-throughput techniques used for the study of whole transcriptome are microarray and RNA sequencing. Both these techniques study large number of transcripts present in the animal tissue, cells, or organs in specific conditions but there are certain differences between these techniques. Microarray relies on the availability of the prior information of transcripts that exist (or supposed to exist) in an animal and provides only limited information about absolute expression levels of known genes (Held et al., 2006). RNA sequencing does not need any prior information about the transcripts and thus can help in detecting novel transcripts. Next generation sequencing (NGS) technology has revolutionized the transcriptomics research during the last two decades and these methods have replaced the microarrays almost completely as have advantage over microarray that no prior knowledge about the transcripts is required. RNA sequencing (RNA-seq) based on NGS platforms, has improved the efficiency and quality of transcriptome sequencing (Buermans and den Dunnen, 2014; Goodwin et al., 2016) and has also reduced the cost and time required for the identification and quantification of transcripts under different conditions. RNA-seq has clear advantages due to high coverage and it captures a wider range of expression values. It can quantify even very low transcripts (in microarrays background noise interference is a problem), and can identify novel transcripts that have not been previously annotated. However, a major limitation of transcriptomics study is that it does not account for post-transcriptional and post-translational regulation of protein expression (Rodrigues et al., 2012). Therefore, combined analysis of the transcriptomic and proteomic analyses can provide deeper insights in the physiology, development, and patho-physiology of the animals.

3.4.1 Animal Product Safety

Residual drugs or growth promoters/hormones used in the animals to increase the production may have an adverse effect on animal welfare and on the health of consumers. Use of illegal growth promoters or drugs to increase animal production brings changes in the level of transcripts. Transcriptomic studies may help in identifying novel biomarkers that can serve as tools to detect the animals that have been doped and devise suitable strategy to monitor the use of illegal drugs in animals. RNA sequencing technology was used to identify the candidate genes to understand the effects of dimethyl ether acetic acid and estradiol on bovine livers. These candidate genes were verified on other animals and may serve as biomarker involved in detecting illegal use of synthesized metabolic agents to increase the animal production (Riedmaier et al., 2012, 2014). Similarly, transcriptomic biomarkers have been discovered to monitor the illegal use of trenbolone acetate (TBA) and 17 β-estradiol in bovine (Elgendy et al., 2015), β2-receptor drugs in lamb breeding (Zhao et al., 2016a, 2017), and amantadine drugs in broiler breeding (You et al., 2019).

3.4.2 Animal Behavior

Transcriptomics can help in better understanding unique animal behaviors like hibernation, nest building, reproduction, swarming, migration, dominance, territory marking, nest building, camouflage, etc. Animals share an important relation with their environment and their behaviors are influenced by the conditions prevailing in the environment. Any shift from the balance may lead to changes in the behavior of animals. Transcriptomics helps in identifying the genes involved in the behavioral change in the animals. Transcriptomic studies in different species of animals have been conducted by different researchers under the conditions of heat stress, hypoxia, drought, and environmental pollution, and have helped in the deep understandings of the underlying mechanisms involved in the adaptation to the micro- and macroenvironments. In pigs, transcriptomic studies helped in identifying the genes and physiological pathways that have a role in tail biting. Differential transcriptomic analysis of brain (prefrontral cortex and hypothalamus) helped in identifying the candidate genes involved in tail-biting behavior in pigs. Some of the genes related to the production traits in the pig, like PDK4-pyruvate dehydrogenase kinase, isozyme 4 (Lan et al., 2010), EGF (epidermal growth factor), GHRL gene coding for ghrelin, and the COMP (Cartilage Oligomeric Matrix Protein) gene (Brunberg et al., 2013) were also associated with the tail-biting behavior.

3.4.3 Understanding of Animal Physiology

Transcriptomics helps in deep understanding of different pathways involved in animal physiology. Researchers are now carrying out research on the cell or tissue of their interest via cost-effective sequencing technologies and are able to gauge deeper in the regulatory pathways involved in the animal physiology. Several microarray-based transcriptomic studies have helped in better understanding of the immune functions and regulation of genes controlling the immunity trait during host-pathogen interaction (Jiang et al., 2008). Differential gene expression analysis by RNA-seq in the resistant and susceptible rainbow trout exposed to F. psychrophilum infection provided insights into the up-regulation of immune-related factors not previously identified, including leptin, C-type lectin, TLR5, and haptoglobin (Langevin et al., 2012; Marancik et al., 2015).

Transcriptomic studies have also provided key insights about the reproductive biology of different animals. Caetano et al. (2004) identified differentially expressed genes in ovaries and ovarian follicles of pigs selected for increased ovulation rate to seek new insights into ovarian physiology and the quantitative genetic control of reproduction in swine. Transcriptomic analysis of pre-implantation embryos helped in identifying genes involved in embryonic development of bovine (Chitwood et al., 2013; Graf et al., 2014; Zhao et al., 2016b; Zolini et al., 2020), pigs (Bick et al., 2018; Zeng et al., 2019), mice (Fan et al., 2015), buffalo (Sood et al., 2019), and monkey (Ma et al., 2019). Transcriptome analysis led to the identification of the biomarkers of freezability and fertility of semen in different animal species like boar (Fraser et al., 2020), Buffalo (Zhang et al., 2021), and bovine (Keles et al., 2021).

To understand the complexity that underlies mammary gland development and function, several gene expression analysis by RNAseq or microarray provided insight into the mechanisms that ultimately allow mammary glands to function in a coordinated fashion throughout puberty, pregnancy, lactation, and involution. These days, initiatives like elucidating the signaling mechanisms underlying the functional development of mammary gland and regulation of milk fat/protein synthesis throughout the lactation cycle by generating whole genome expression pattern coupled with metabolic/hormonal pathways has become a high priority area of research in animal genomics that can yield a wealth of information on as yet unknown molecular adaptations in response to physiological stage of the animal. Such inputs can relate functional development of mammary gland of animals with coordinated changes in the global expression pattern to understand the basic biology of mammary gland development that is far from complete.

Overall transcriptomics has helped in better understanding of the various physiological changes in the animals in different developmental stages which can help in finding solutions for sustainable animal production, health and reproduction.

3.4.4 Understanding Pathology of Diseases

Gene expression analysis has also resulted in better understanding of the mechanisms involved in the development of cancer (Chen et al., 2020), bacterial and viral infection (Langevin et al., 2012; Marancik et al., 2015; Huang et al., 2011; Pereiro et al., 2012), and immune response to the diseases. This has further helped in developing the targets for the treatments of the disease, developing vaccines, biomarkers for the early diagnosis of the diseases.

3.5 APPLICATION OF BIOINFORMATICS IN OMICS TECHNOLOGIES

Using the omics approaches huge data is generated, which most of the time has artifacts as well. Bioinformatics help in separation of these artifacts from good and usable information. The sequence information generated for nucleotides or transcripts or proteins/peptides is aligned to reference sequence using bioinformatics tools and different metrics that determine confidence in the gene or protein identification are calculated. Genes or transcripts or proteins/peptides are then analyzed for association with different biological processes, pathways, and networks. All these steps can be accomplished with various software packages. The sequential running of these computer programs is referred to as a pipeline because the output of one step was used as the input for the second step and so forth. However, as the number of sequenced genomes has considerably increased, inadequate computational power might become a bottleneck for research in evolutionary bioinformatics. Finding all common genes between any two different species that come from a single gene of the last common ancestor, referred to as orthologs, might take more than 60 years of computation with a modern personal computer. So we need to have cluster computing services that enable users to execute multiple computing jobs in parallel.

Analysis of Next Generation Sequencing data is one of the most demanding applications in bioinformatics (Pérez-Sánchez et al., 2015). A major limitation associated with NGS data analyses is the requirement of large data storage and high-performance computing facilities (Kadarmideen, 2014). After generation of omics data that may be nucleotide sequencing data or proteomics or transcriptome or metablome data, the first step in data processing is quality control. In nucleotide or transcriptome data, linker sequences used in the process of nucleotide sequencing are trimmed from the reads. Similarly, for the proteomics data, exclusions can be made to eliminate spectra which are either too small or lack good signal-to-noise ratio. This followed by the mapping the nucleotide reads, transcripts or peptide spectra to a reference database. For proteomics data, mapping peptide spectra to protein database sequence is a prediction based on acceptable tolerances and variation resulting from incomplete processing reactions, or biological modification.

Starting from procedures like alignments and variant calling to more complex challenges like genome-wide annotations and biomarkers correlation to diseases, NGS analyses are time-consuming. High performance computing provides advantages in this field of genomics applied to medicine and healthcare.

3.6 WEB RESOURCES

3.6.1 Databases Used in Genomic and Proteomics Research

NCBI	National Center for Biotechnology Information (NCBI). www.ncbi.nlm.nih.gov
GenBank	GenBank is the NIH genetic sequence database, an annotated collection of all publicly available DNA sequences. www.ncbi.nlm.nih.gov/genbank/
RefSeq	NCBI Reference Sequence Database is a collection of sequences, which provides a comprehensive, integrated, non-redundant, well-annotated set of sequences, including genomic DNA, transcripts, and proteins. www.ncbi.nlm.nih.gov/refseq
PubMed	PubMed comprises more than 23 million citations for biomedical literature from MEDLINE, life science journals, and online books. Citations may include links to full-text content from PubMed Central and publisher web sites. www.ncbi.nlm.nih.gov/pubmed/
OMIM	OMIM is a comprehensive compendium of human genes and genetic phenotypes. Its official home is omim.org.www.ncbi.nlm.nih.gov/omim.
dbSNPs	Database of single nucleotide polymorphisms (SNPs) and multiple small-scale variations that include insertions/deletions, microsatellites, and non-polymorphic variants. www.ncbi.nlm.nih.gov/snp/
EST	The EST database is a collection of short single-read transcript sequences from GenBank. These sequences provide a resource to evaluate gene expression, find potential variation, and annotate genes. www.ncbi.nlm.nih.gov/est
dbEST	dbEST is a division of GenBank that contains sequence data and other information on "single-pass" cDNA sequences, or "Expressed Sequence Tags," from a number of organisms. www.ncbi.nlm.nih.gov/genbank/dbest
WGS	Whole Genome Shotgun (WGS) sequencing projects are incomplete genomes or incomplete chromosomes that are being sequenced by a whole genome shotgun strategy. www.ncbi.nlm.nih.gov/genbank/wgs
HTG	The High-Throughput Genomic (HTG) sequences division contains unfinished DNA sequences generated by the high-throughput sequencing centers. Sequence data in this division are available for BLAST homology searches against either the "htgs" database or the "month" database, which includes all new submissions for the prior month. It was done in a coordinated effort among the International Nucleotide Sequence databases, DDBJ, EMBL, and GenBank. www.ncbi.nlm.nih.gov/genbank/htgs
EMBL	European Bioinformatics Institute, www.embl.org/1000 Genomes. It is a catalog of shared human genetic variation in population groups worldwide. www.1000genomes.org/
ArrayExpress	This is a database of functional genomics experiments that can be queried and the data downloaded. It includes gene expression data from microarray and high-throughput sequencing studies. www.ebi.ac.uk/arrayexpress/
	Database of Genomic Variants archive The Database of Genomic Variants archive (DGVa) is a repository that provides archiving, accessioning and distribution of publicly available genomic structural variants, in all species. www.ebi.ac.uk/dgva/
	PromoterWise It compares two DNA sequences allowing for inversions and translocations, ideal for promoters.

EBI	Metagenomics A resource for the analysis and archiving of metagenomic data. www.ebi.ac.uk/metagenomics/EMBOSS Tools-Selected EMBOSS tools for sequence analysis, providing pairwise sequence alignment, sequence format conversion, sequence translation and back-translation, and sequence statistics. www.ebi.ac.uk/Tools/emboss/
Ensemble	The Ensemble project produces genome databases for vertebrates and other eukaryotic species. www.ensembl.org/index.html
European	Nucleotide Archive www.ebi.ac.uk/ena/home. The European Nucleotide Archive (ENA) provides a comprehensive record of the world's nucleotide sequencing information, covering raw sequencing data, sequence assembly information, and functional annotation.
Immuno	Polymorphism Database www.ebi.ac.uk/ipd/. The Immuno Polymorphism Database (IPD) was developed in 2003 to provide a centralized system for the study of polymorphism in genes of the immune system. The IPD project was established by the HLA Informatics Group of the Anthony Nolan Research Institute in close collaboration with the European Bioinformatics Institute.
Pfam	http://pfam.sanger.ac.uk/. The Pfam database is a large collection of protein families, each represented by multiple sequence alignments and hidden Markov models (HMMs).
Rfam	http://rfam.sanger.ac.uk/. This database is a collection of RNA families, each represented by multiple sequence alignments, consensus secondary structures, and covariance models (CMs).
GenomeNet	www.genome.jp. GenomeNet is a Japanese network of database and computational services for genome research and related research areas in biomedical sciences, operated by the Kyoto University Bioinformatics Center.

3.6.1.1 GenomeNet Database Resources

DBGET: Integrated Database Retrieval System
KEGG: Kyoto Encyclopedia of Genes and Genomes, www.genome.jp/kegg/pathway.html
KEGG PATHWAY – Systems information: pathways
KEGG BRITE – Systems information: ontologies
KEGG Organisms – Organism-specific entry points
KEGG GENES – Genomic information
KEGG LIGAND – Chemical information

3.6.1.2 GenomeNet Bioinformatics Tools

KEGG	Database resource for understanding high-level functions and utilities of the biological system, such as the cell, the organism, and the ecosystem, from molecular-level information used for mapping of molecular data sets in genomics, transcriptomics, proteomics, and metabolomics for biological interpretation.
KEGG	PATHWAY Collection of manually drawn pathway maps for various prokaryotes and eukaryotes representing molecular interaction and reaction networks for metabolism (carbohydrate, energy, lipid, nucleotide, amino acid, other amino glycan cofactor/vitamin, other secondary metabolite, xenobiotics chemical structure), genetic information processing, environmental information processing, cellular processes, organismal systems, and human diseases.
DDBJ	(DNA Data Bank of Japan) www.ddbj.nig.ac.jp/ DNA Data Bank of Japan (DDBJ) is the sole nucleotide sequence data bank in Asia, which is officially certified to collect nucleotide sequences from researchers and to issue the internationally recognized accession number to data submitters. Since we

	exchange the collected data with ENA/EBI; European Bioinformatics Institute and NCBI; National Center for Biotechnology Information on a daily basis, the three data banks share virtually the same data at any given time. The virtually unified database is called "INSD; International Nucleotide Sequence Database." DDBJ collects sequence data mainly from Japanese researchers, but of course accepts data and issues the accession number to researchers in any other countries.
GenomeReviews	European Bioinformatics Institute, www.ebi.ac.uk/GenomeReviews
UniProt	European Bioinformatics Institute, www.uniprot.org/. The mission of UniProt is to provide the scientific community with a comprehensive, high-quality and freely accessible resource of protein sequence and functional information.
UNIProtKB	Protein knowledgebase consists of two sections: Swiss-Prot, which is manually annotated and reviewed, and TrEMBL, which is automatically annotated and is not reviewed. It includes complete and reference proteome sets.
UniProt/SwissProt	(Swiss Institute of Bioinformatics) www.ebi.ac.uk/swissprot/
Protein	Data Bank in Europe (PDBe) www.ebi.ac.uk/pdbe. PDBe is the European resource for the collection, organization, and dissemination of data on biological macromolecular structures. In collaboration with the other worldwide Protein Data Bank (wwPDB) partners – the Research Collaboratory for Structural Bioinformatics (RCSB) and BioMagResBank (BMRB) in the USA and the Protein Data Bank of Japan (PDBj) – we work to collate, maintain, and provide access to the global repository of macromolecular structure data.
PDBj	Protein Data Bank Japan, http://pdbj.org/. It maintains a centralized PDB archive of macromolecular structures and provides integrated tools, in collaboration with the RCSB, the BMRB in USA, and the PDBe in EU. PDBj is supported by JST-NBDC and Osaka University.
The	Worldwide Protein Data Bank (wwPDB), www.PDB http://www.wwpdb.org/, consists of organizations that act as deposition, data processing, and distribution centers for PDB data.1 Members are RCSB PDB (USA), PDBe (Europe) and PDBj (Japan), and BMRB (USA). The wwPDB's mission is to maintain a single PDB archive of macromolecular structural data that is freely and publicly available to the global community.
PDB-Protein	Data Bank, www.rcsb.org/. An information portal to biological macromolecular structures.
PROSITE	(Swiss Institute of Bioinformatics), prosite.expasy.org/
UniProt/PIR	(National Biomedical Research Foundation), http://pir.georgetown.edu/

3.6.2 ALIGNMENT

Clustal	Omega is a new multiple sequence alignment program that uses seeded guide trees and HMM profile-profile techniques to generate alignments. www.ebi.ac.uk/Tools/msa/clustalo/
ClustalW2-Phylogeny	Commonly used phylogenetic tree generation methods provided by the ClustalW2 program. www.ebi.ac.uk/Tools/phylogeny/clustalw2_phylogeny/
DaliLite	Pairwise alignment of protein structures. DaliLite computes optimal and suboptimal structural alignments between two protein structures. It compares all chains in the first structure against all chains in the second (unless specific chain IDs are given). The resulting superimposed coordinate files can be downloaded or viewed interactively in Jmol. www.ebi.ac.uk/Tools/structure/dalilite/
Multiple	Sequence Alignment (MSA), www.ebi.ac.uk/Tools/msa/

ClustalOmegaNew	MSA tool that uses seeded guide trees and HMM profile-profile techniques to generate alignments. Suitable for medium-large alignments.
ClustalW2	Popular MSA tool that uses tree-based progressive alignments. Suitable for medium alignments.
DbClustal	Create a Multiple Sequence Alignment from a protein BLAST result using the DbClustal program.
Kalign	Very fast MSA tool that concentrates on local regions. Suitable for large alignments.
MAFFT	MSA tool that uses Fast Fourier Transforms. Suitable for medium-large alignments.
MUSCLE	Accurate MSA tool, especially good with proteins. Suitable for medium alignments.
MView	Transform a Sequence Similarity Search result into a Multiple Sequence Alignment or reformat a Multiple Sequence Alignment using the MView program.
T-Coffee	Consistency-based MSA tool that attempts to mitigate the pitfalls of progressive alignment methods. It is suitable for small alignments.
WebPRANK	The EBI has a new phylogeny-aware multiple sequence alignment program which makes use of evolutionary information to help place insertions and deletions.
Pairwise	Sequence Alignment, www.ebi.ac.uk/Tools/psa/
Global	Alignment: Global alignment tools create an end-to-end alignment of the sequences to be aligned. There are separate forms for protein or nucleotide sequences.
Needle	EMBOSS Needle creates an optimal global alignment of two sequences using the Needleman-Wunsch algorithm.
Stretcher	Stretcher uses a modification of the Needleman-Wunsch algorithm that allows larger sequences to be globally aligned.
Local	Alignment Local alignment tools find one, or more, alignments describing the most similar region(s) within the sequences to be aligned. There are separate forms for protein or nucleotide sequences.
Water	Water uses the Smith-Waterman algorithm (modified for speed enhancements) to calculate the local alignment of two sequences.
Matcher	Matcher identifies local similarities between two sequences using a rigorous algorithm based on the LALIGN application.
LALIGN	LALIGN finds internal duplications by calculating non-intersecting local alignments of protein or DNA sequences.
Genomic	alignment tools concentrate on DNA (or to DNA) alignments while accounting for characteristics present in genomic data.
Wise2DBA	Wise2DBA (DNA Block Aligner) aligns two sequences under the assumption that the sequences share a number of colinear blocks of conservation separated by potentially large and varied lengths of DNA in the two sequences.
GeneWise	GeneWise compares a protein sequence to a genomic DNA sequence, allowing for introns and frameshifting errors.

3.6.3 Protein Analysis Web Resources

The Protein Functional Analysis tools described on this page are provided using our new bioinformatics analysis tools framework. At present a subset of the protein functional analysis tools available at EBI are available in the new framework. www.ebi.ac.uk/Tools/pfa/

CENSOR	Identify and/or mask repeat sequences in protein sequence data.
FingerPRINT ScanIdentify	the closest matching PRINTS sequence motif fingerprints in a protein sequence.
InterProScan	Protein Functional Analysis using the InterProScan program.
Phobius	Prediction of transmembrane topology and signal peptides using the Phobius program.
Pratt	Search patterns conserved in sets of unaligned protein sequences.
PROSITE	Scan Compare a protein sequence against the signatures in PROSITE.
RADAR	Detection and alignment of repeats in protein sequences.
InterPro	Protein sequence analysis and classification www.ebi.ac.uk/interpro/
InterPro	Provides functional analysis of proteins by classifying them into families and predicting domains and important sites. We combine protein signatures from a number of member databases into a single searchable resource, capitalizing on their individual strengths to produce a powerful integrated database and diagnostic tool.
PRF	SEQDB Protein Research Foundation, Japan, www.prf.or.jp/seqdb-e.html
ExplorEnz	Trinity College Dublin, www.enzyme-database.org/
EPD	Swiss Institute for Experimental Cancer Center, http://epd.vital-it.ch/
ProDom	Rhone-Alpes Bioinformatics Cente, prodom.prabi.fr
AAindex	Kyoto University Bioinformatics Center, www.genome.jp/aaindex/
LinkDB	Kyoto University Bioinformatics Center, www.genome.jp/linkdb/
BLASTp	Protein database searching tool
NRDB	Database containing non-redundant protein sequences
DEG	Database of essential genes required for survival of the organism
COBALT	Multiple protein sequence alignment tool
Pfam	The Pfam database is a large collection of protein families, each represented by multiple sequence alignments and hidden Markov models. It can be used for prediction of Pfam families, motifs, repeats, and clans.
Motif	search For prediction of protein motifs.
ProtParam	For prediction of physio-chemcal attributes of the protein.
GlobPlot	Protein globularity analysis.
PDB	Structural database for 3D structuring of biomolecules.
ClustalW	Multiple sequence alignment tool.
HHpred	For protein homology determination and structure prediction of the protein.
MODELLER	For 3D structure prediction of protein.
PROCHECK	Evaluates the steriochemical properties of the protein structure based on Ramchandran plot.
ProSA	For recognition of errors in 3D structure of protein.
MetaPocket	2.0 For predicting the active site in protein sequence.
Autodockvina	Virtual screening of ligands to protein receptors.
ZINC	Ligand database.
Ligplot	For protein-ligand interaction analysis.
MATLAB	To calculate properties of peptide sequences, amino acid composition, ORF finding.
GROMACS	To perform molecular dynamics of protein specifically bonding, interaction.
Discovery	studio Comprehensive life science modeling and simulation suite of applications for drug discovery process, protein ligand docking, protein homology modeling, protein–protein docking, antibody modeling

3.7 METABOLOMICS IN ANIMAL SCIENCE

Metabolomics is the study of a comprehensive set of metabolites that are the end products of complex interactions that occur both inside the cell (the genome) and outside the cell or organism (the environment). Analytical chemistry, data mining, and data interpretation are all part of this multidisciplinary approach. Metabolomics gives molecular information about a biological system's biochemical state, which can be represented by metabolic species and their levels, which are obtained from a specific combination of gene expression activity and environmental influences. In other words, metabolomics allows investigators to obtain a sensitive and more complete description of the phenotype. With this power to noninvasively identify phenotypic changes, innate phenotypic proclivities and dietary responses makes it an ideal tool for livestock research, breeding, and assessment. Further, it can easily be applied to predict feed efficiency and residual feed intake (RFI) (Karisa et al., 2014), disease diagnosis (Hailemariam et al., 2014; LeBlanc et al., 2005; Sundekilde et al., 2013), assess carcass merit (Weikard et al., 2010), and production (Chapinal et al., 2012; Melzer et al., 2012). The advantage of metabolomics over many traditional phenotypic measurements such as those related to animal feed consumption and RFI are, metabolomics is inexpensive and time saving. In addition, it allows reproductive trait evaluation, before the animals reach sexual maturity. Overall, metabolomics allows many trait measurements to be conducted earlier, more routinely, and is cost effective than other conventional techniques. The commonly used technologies applied in metabolomics includes bioanalytical approaches based on mass spectrometry (MS), high-performance liquid-phase chromatography (HPLC), and nuclear magnetic resonance (NMR) spectroscopy that can measure a large number of metabolites in many different samples. Besides, the analysis and identification of analytes is further enhanced with the availability of different tools, software, and databases at the public domain. Some of the applications of metabolomics in animal sciences are discussed later.

3.7.1 Biomarkers for Quality and Quantity Traits

Metabolomics can be used in the evaluation and/or prediction of production traits such as residual feed intake (RFI), carcass merit, and milk quality. It offers a unique opportunity for indirect, non-invasive, and inexpensive marker-assisted measurement of these economically relevant traits. Information of metabolites related to feed efficiency and RFI has helped to understand the pathways involved and to select animals with better RFI, carcass and milk quality. Metabolomics in addition, can serve as a tool for identifying metabolites which can be used as biomarkers.

3.7.2 Animal Health and Diseases

Metabolomics is used to identify novel disease biomarkers and provide new information on disease pathogenesis. The identification of metabolites facilitates a better understanding of the complex changes that occur in metabolic and biochemical pathways. Detecting complex changes in metabolite levels can aid in disease diagnosis. By using metabolomics different volatile metabolites have been identified in diseased animals. The composition, relative abundance, and interactions among metabolites in a given organism can suggests the mechanism involved in the host in response to different disease conditions. By applying metabolomics number of studies have been conducted in bovine with major focus on mastitis. A total of 1,105 metabolites were detected with an upsurge in di- and tri-peptides in the milk samples of cows infected by Escherichia coli and Staphylococcus aureus. Besides, pathway analyses showed an upregulated level of arachidonic in mastitic milk samples (Mansor et al., 2012). Similarly, when compared with healthy milk samples, decrease in glucose, d-glycerol-1-phosphate, sn-glycero-3-phosphocholine, 4-hydroxyphenyllactate, lcarnitine citrate, and hippurate were detected in the clinical mastitic milk samples, and less d-glycerol-1-phosphate, l-carnitine, benzoic acid, and cis-aconitate were found in the subclinical mastitic milk samples (Xi et al., 2017).

Metabolomics represents a powerful platform for the identification of metabolites as biomarkers associated with physiological alterations resulting from environmental influences such as heat stress. It is well established that stress in dairy cows has a negative impact in production, reproduction, growth, immunity, physiology, and overall well-beingness (Chase, 2006; Salak-Johnson and McGlone, 2007; Marai and Habeeb, 2010; Hansen, 2013; Wohlgemuth et al., 2016; Tao et al., 2018; Liu et al., 2019). Different metabolomic studies have been conducted in dairy animals to understand the changes in metabolomic profile under different types of stress e.g. differentially expressed metabolites from blood plasma (Tian et al., 2015), milk (Yue et al., 2020), mammary tissues and liver in (Ma et al., 2019) heat stressed cows. These metabolites identified under stressed conditions can serve as biomarkers which can facilitate disease diagnosis and health problems in dairy animals.

3.8 EPIGENETICS IN ANIMAL SCIENCE

Epigenetics focuses on the analysis of factors that control or changes the genetic information of DNA independent of the changes in DNA sequences. Epigenetics can alter the gene expression in individual by DNA methylation, histone modifications by methylation, acetylation or phosphorylation, and noncoding RNAs like siRNA, miRNAs, long noncoding RNAs, etc. Different studies in humans and animals have revealed that epigenetic marks in addition to the DNA sequence, may be inherited from one generation to next and influence the phenotype of offspring (Jablonka and Raz, 2009; Burggren and Crews, 2014; Burggren, 2014). Animal nutrition, environment, age, mutagens, toxic agents, radiation, infectious agents, and immunological factors can bring the epigenetic changes in an animal. Different tools to study DNA methylation are digestion assays (involving restriction enzymes that are sensitive or resistant to the methylated DNA sequences) and bisulfite sequencing. Digestion assays identify only the global methylation and do not provide information about the methylation at single nucleotide level. Bisulfite sequencing can provide information about the methylation at single nucleotide level. Bisulfite treatment converts cytosine to uracil, while the methylated cytosine remains unaffected. The resulting DNA can be analyzed by sequencing (most common), microarray, and MALDI-TOF mass spectrometry. DNA methylation can also be studied by other methods like immunoprecipitation and affinity enrichment-based assays. Histone modifications are mostly studied by chromatin immunoprecipitation, and MALDI-TOF mass spectrometry, while noncoding RNAs are mostly studied by next generation sequencing.

Epigenetic marks in the cell may determine the fate of the cell. In vertebrates, during the development of sexually dimorphic organisms, formation of bipotential gonad is the first and critical step. Sex determination during the development requires a unidirectional commitment, of an initially bipotential gonad, toward either the male or female gonad. Different epigenetic mechanisms decide the fate of the bipotential gonad by fine tuning the timing and expression level of sex determining factors (Garcia-Moreno et al., 2018). ScRNA-seq in the Zebrafish and frog also revealed how the epigenetic regulators participate in the differentiation of the germ layers during early embryogenesis and organogenesis (Wagner et al., 2018).

Various environmental and nutritional factors change the phenotype of the individual by the epigenetic changes. A classic example of the effect of the environmental factor on an organism phenotype is the development of honey bees. The magic formula responsible for the development of a queen is royal jelly, produced by nurse workers by their head glands, which is consumed by the queen larvae. The phenotypic difference in the honey bees shows the link between epigenetic changes and environmental factors in deciding the phenotype of an individual (Maleszka, 2008; Hunt BG et al., 2013). Milk production in dairy cows is also influenced by different factors like nutrition, photoperiod, environmental temperature, hormones, local factors in the mammary gland, diseases, etc. Adverse environmental effects (suboptimal nutritional and health status), reduce milk yield, and body energy reserves of cows which in turn may have an impact on the phenotype of the offspring of the cows, which is partially because of epigenetic modifications during the in utero development. Experimental evidence suggest a positive association of dam's milking performance and nutritional status with the daughter's first lactation performance (Singh et al., 2012). Similarly in sheep, maternal OB and increased nutrient intake before and during gestation can lead to altered growth, adiposity, and glucose tolerance in adult offspring (Long et al., 2010; Zhang et al., 2011).

DNA methylation study in the cattle sperm and somatic cells revealed that in sperm cells partially methylated regions were more in number than in somatic cells. This helped in the identification of meiosis-related genes such as KIF2B and REPIN1, which are hypomethylated in sperm but hypermethylated in somatic cells (Zhou et al., 2018). The nutritional effects on carcass traits, gene expression, and DNA methylation were also investigated down the male line in a three-generation pig pedigree and it was revealed that the diet of the mother effects these traits of the offspring and are effectively transmitted to the F2 offspring (Braunschweig et al., 2012).

BIBLIOGRAPHY

Allendorf FW, Hohenlohe PA, Luikart G. Genomics and the future of conservation genetics. *Nat Rev Genet.* 2010. 11: 697–709.

Almeida AMD, Bassols A, Bendixen E, Bhide M, Ceciliani F, Cristobal S, Eckersall PD, Hollung K, Lisacek F, Mazzucchelli G, McLaughlin M. Animal board invited review: advances in proteomics for animal and food sciences. *Animal.* 2015. 9(1):1–17.

Angeloni F, Wagemaker N, Vergeer P, Ouborg J. Genomic toolboxes for conservation biologists. *Evol Appl.* 2012. 5(2): 130–143.

Avise JC. Perspective: conservation genetics enters the genomics era. *Conserv Genet.* 2010. 11(2): 665–669.

Baguisi A, Behboodi E, Melican DT, Pollock JS, Destrempes MM, Cammuso C. Production of goats by somatic cell nuclear transfer. *Nat Biotechnol.* 1999. 17: 456–461.

Benestan LM, Ferchaud AL, Hohenlohe PA, Garner BA, Naylor GJP, BaumsIB, . . . Luikart G. Conservation genomics of natural and managed populations: building a conceptual and practical framework. *Mol Ecology.* 2016. 25(13): 2967–2977. doi: 10.1111/mec.13647

Bick JT, Flöter VL, Robinson MD, Bauersachs S, Ulbrich SE. Small RNA-seq analysis of single porcine blastocysts revealed that maternal estradiol-17beta exposure does not affect miRNA isoform (isomiR) expression. *BMC Genomics.* 2018 Aug 6. 19(1): 590. doi: 10.1186/s12864-018-4954-9

Blanchard P, Mahé D, Cariolet R, Keranflec'h A, Baudouard MA, Cordioli P, Albina E, Jestin A. Protection of swine against post-weaning multisystemic wasting syndrome (PMWS) by porcine circovirus type 2 (PCV2) proteins. *Vaccine.* 2003. 21(31): 4565–4575.

Bodzsar N, Eding H, Revay T, Hidas A., Weigend S. Genetic diversity of Hungarian indigenous chicken breeds based on microsatellite markers. *Anim Genet.* 2009. 40: 516–523.

Boettcher PJ, Hoffmann I, Baumung R, Drucker AG, McManus C, Berg P, Stella A, Nilsen LB, Moran D, Naves M, Thompson MC. Genetic resources and genomics for adaptation of livestock to climate change. *Front Genet.* 2015. 5: 461. doi: 10.3389/fgene.2014.00461

Braunschweig M, et al. Investigations on transgenerational epigenetic response down the male line in F2 pigs. *PLoS One.* 2012. 7(2): e30583.

Brunberg E, Jensen P, Isaksson A, Keeling LJ. Brain gene expression differences are associated with abnormal tail biting behavior in pigs: Gene expression and abnormal behavior in pigs. *Genes Brain Behav.* 2013. 12: 275–281.

Buermans HPJ, den Dunnen, JT. Next generation sequencing technology: Advances and applications Biochim. *Biophys Acta.* 2014. 1842: 1932–1941. doi: 10.1016/j.bbadis.2014.06.015

Burggren WW, Crews D. Epigenetics in comparative biology: Why we should pay attention. *Integr Comp Biol.* 2014. 54(1): 7e20.

Burggren WW. Epigenetics as a source of variation in comparative animal physiology – or – Lamarck is lookin' pretty good these days. *J Exp Biol.* 2014. 217(Pt 5): 682e9.

Caetano AR, Johnson RK, Ford JJ, Pomp D. Microarray profiling for differential gene expression in ovaries and ovarian follicles of pigs selected for increased ovulation rate. *Genetics.* 2004. 168: 1529–1537

Chapinal N, Carson ME, LeBlanc SJ, Leslie KE, Godden S, Capel M, et al. The association of serum metabolites in the transition period with milk production and early-lactation reproductive performance. *J Dairy Sci.* 2012. 95: 1301–1309. doi: 10.3168/jds.2011-4724 PMID: 22365212

Chase LE. Climate change impacts on dairy cattle, fact sheets. Climate Change and Agriculture: Promoting Practical and Profitable Responses, Department of Animal Sciences 2006. http://www.climateand farming.org/pdfs/FactSheets/III.3Cattle.pdf

Chitwood JL, Rincon G, Kaiser GG, Medrano JF, Ross PJ. RNA-seq analysis of single bovine blastocysts. *BMC Genomics*. 2013 May 25.14: 350. doi: 10.1186/1471-2164-14-350. PMID: 23705625; PMCID: PMC3668197.

Cibelli JB, Stice SL, Golueke PJ, Kane JJ, Jerry J, Blackwell C, Ponce De Leon, FA Robl, JM. Cloned transgenic calves produced from non quiescent fetal fibroblasts. *Science*. 1998. 280: 1256.

Cole JB, Wiggans GR, Ma L, Sonstegard TS, Lawlor TJ Jr, Crooker BA, Van Tassell CP, Yang J, Wang S, Matukumalli LK, Da Y. Genome-wide association analysis of thirty one production, health, reproduction and body conformation traits in contemporary U.S. Holstein cows. *BMC Genomics*. 2011. 11; 12: 408. doi: 10.1186/1471-2164-12-408.

Eding H, Crooijmans PMA, Groenne, MAM and Meuwissen THE. Assessing the contribution of breeds to genetic diversity in conservation schemes. *Genet Sel Evol*. 2002. 34: 613–633.

Elgendy R, Giantin M, Montesissa C, Dacasto M. The transcriptome of muscle and liver is responding differently to a combined trenbolone acetate and estradiol implant in cattle. *Steroids*. 2015. 106: 1–8.

Fahiminiya S, Labas V, Dacheux JL, Gérard N. Improvement of 2D-PAGE resolution of human, porcine and equine follicular fluid by means of hexapeptide ligand library. *Reprod Domest Anim*. 2011. 46(3): 561–563.

Fahiminiya S, Reynaud K, Labas V, Batard S, Chastant-Maillard S, Gérard N. Steroid hormones content and proteomic analysis of canine follicular fluid during the preovulatory period. *Reprod Biol Endocrinol*. 2010. 8: 132.

Fan X, Zhang X, Wu X, Guo H, Hu Y, Tang F, Huang Y. Single-cell RNA-seq transcriptome analysis of linear and circular RNAs in mouse preimplantation embryos. *Genome Biol*. 2015 Jul 23. 16(1): 148. doi: 10.1186/s13059-015-0706-1

Fischer A, Shapiro B, Muriuki C, Heller M, Schnee C, Bongcam-Rudloff E, Vilei EM, Frey J, Jores J. The origin of the Mycoplasma cluster coincides with the domestication of ruminants. *PLoS One*. 2012: e36150.

Food and Agriculture Organization of the United Nations [FAO]. Mapping Supply and demand for animal-source foods to 2030. *Animal Production and Health Working Paper by Robinson TP*, Pozzi T. 2011.

Food and Agriculture Organization (FAO). Shaping the Future of Livestock. *Berlin: The 10th Global Forum for Food and Agriculture*. 2018.

Frankham R. Challenges and opportunities of genetic approaches to biological conservation. *Biol Conserv*. 2010.143: 1919–1927.

Fraser L, Brym P, Pareek CS, Mogielnicka-Brzozowska M, Paukszto Ł, Jastrzębski JP, Wasilewska-Sakowska K, Mańkowska A, Sobiech P, Żukowski K. Transcriptome analysis of boar spermatozoa with different freezability using RNA-Seq. *Theriogenology*. 2020 Jan 15. 142: 400–413.

Fu Q, Huang Y, Wang Z, Chen F, Huang D, Lu Y, Liang X, Zhang M. Proteome profile and quantitative proteomic analysis of buffalo (Bubalusbubalis) follicular fluid during follicle development. *Int J Mol Sci*. 2016. 17(5): 618.

Garcia-Moreno SA, Plebanek MP, Capel B. Epigenetic regulation of male fate commitment from an initially bipotential system. *Mol Cell Endocrinol*. 2018. 468: 19e30.

Golovan SP, Meidinger RG, Ajakaiye A, Cottrill M, Wiederkehr MZ, Barney DJ, Plante C, Pollard JW, Fan MZ, Hayes MA, Laursen J, Hjorth JP, Hacker RR, Phillips JP, Forsberg CW. Pigs expressing salivary phytase produce low-phosphorus manure. *Nat Biotechnol*. 2001. 19: 741–745. doi: 10.1038/90788

Goodwin S, McPherson JD, McCombie WR. Coming of age: Ten years of next-generation sequencing technologies. *Nat Rev Genet*. 2016. 17(6): 333–351. doi: 10.1038/nrg.2016.49

Graf A, Krebs S, Zakhartchenko V, Schwalb B, Blum H, Wolf E. Fine mapping of genome activation in bovine embryos by RNA sequencing. *Proc Natl Acad Sci U S A*. 2014 Mar 18. 111(11): 4139–4144.

Guimarães DB, Barros TB, van Tilburg MF, Martins JAM, Moura AA, Moreno FB, Monteiro-Moreira AC, Moreira RA, Toniolli R. Sperm membrane proteins associated with the boar semen cryopreservation. *Anim Reprod Sci*. 2017. 183: 27–38. doi: 10.1016/j.anireprosci.2017.06.005

Hailemariam D, Mandal R, Saleem F, Dunn SM, Wishart DS, Ametaj BN. Identification of predictive biomarkers of disease state in transition dairy cows. *J Dairy Sci*. 2014. 97: 2680–2693. https://doi.org/10.3168/jds.2013-6803. PMID: 24630653

Hansen PJ. Exploitation of genetic and physiological determinants of embryonic resistance to elevated temperature to improve embryonic survival in dairy cattle during heat stress. *Theriogenology*. 2007. 68S: S242–S249. doi: 10.1016/j.theriogenology.2007.04.008

Hansen PJ. Antecedents of mammalian fertility: Lessons from the heat-stressed cow regarding the importance of oocyte competence for fertilization and embryonic development. *Anim. Front*. 2013. 3(4):34–39.

Harris BL, Johnson DL, Spelman RJ. *Genomic selection in New Zealand and the implications for national genetic evaluation*. ProcInterbull Meeting, Niagara Falls, Canada. 2008.

Hayashida K, Hara Y, Abe T, Yamasaki C, Toyoda A, Kosuge T, Suzuki Y, Sato Y, Kawashima S, Katayama T, Wakaguri H, Inoue N, Homma K, Tada-Umezaki M, Yagi Y, Fujii Y, Habara T, Kanehisa M, Watanabe H, Ito K, Gojobori T, Sugawara H, Imanishi T, Weir W, Gardner M, Pain A, Shiels B, Hattori M, Nene V, Sugimoto C. Comparative genome analysis of three eukaryotic parasites with differing abilities to transform leukocytes reveals key mediators of Theileria-induced leukocyte transformation. *mBio.* 2012. 3(5): e00204–12. doi: 10.1128/mBio.00204-12

Hayes BJ, Bowman PJ, Chamberlain AJ, Goddard ME. Invited review: Genomic selection in dairy cattle: progress and challenges. *J Dairy Sci.* 2009 Feb. 92(2): 433–443. doi: 10.3168/jds.2008-1646

Held GA, Grinstein G, Tu Y. Relationship between gene expression and observed intensities in DNA microarrays – a modeling study. *Nucleic Acids Res.* 2006. 34. doi: 10.1093/nar/gkl122e70.

Hogarth CJ, Fitzpatrick JL, Nolan AM, Young FJ, Pitt A, Eckersall PD. Differential protein composition of bovine whey: A comparison of whey from healthy animals and from those with clinical mastitis. *Proteomics.* 2004. 4(7): 2094–2100.

Huang XN, Wang ZY, Yao CL. Characterization of Toll-like receptor 3 gene in large yellow croaker. *Pseudosciaenacrocea Fish Shellfish Immunol.* 2011. 31: 98–106. doi: 10.1016/j.fsi.2011.04.009

Hunt BG, et al. The function of intragenic DNA methylation: Insights from insect epigenomes. *Integr Comp Biol.* 2013. 53(2): 319e28.

Ibtisham F, Zhang L, Xiao M, Lilong A, Muhammad B, et al. Genomic selection and its application in animal breeding. *Thai J Vet Med.* 2017. 47(3): 301–310.

Jablonka E, Raz G. Transgenerational epigenetic inheritance: Prevalence, mechanisms, and implications for the study of heredity and evolution. *Q Rev Biol.* 2009. 84(2): 131e76.

Jiang N, Leach LJ, Hu X, Potokina E, Jia T, Druka A, Waugh R, Kearsey MJ. Luo Z. Methods for evaluating gene expression from Affymetrix microarray datasets. *BMC Bioinformatics* 2008. 9: 284.

Jiang L, Liu J, Sun D, Ma P, Ding X, Yu Y, Zhang Q. Genome wide association studies for milk production traits in Chinese Holstein population. *PLoS One.* 2010. 5(10):e13661. doi: 10.1371/journal.pone.0013661

Joshi BK, Sodhi M, Mukesh M, Mishra BP. Genetic characterization of farm animal genetic resources of India: a review. *Indian J Anim Sci.* 2012. 82(3):1259–75.

Kadarmideen HN. Genomics to systems biology in animal and veterinary sciences: progress, lessons and opportunities. *Livest. Sci.* 2014. 166:232–248.

Karisa BK, Thomson J, Wang Z, Li C, Montanholi YR, Miller SP, et al. Plasma metabolites associated with residual feed intake and other productivity performance traits in beef cattle. *Livest Sci.* 2014. 165: 200–211

Keles E, Malama E, Bozukova S, Siuda M, Wyck S, Witschi U, Bauersachs S, Bollwein H. The micro-RNA content of unsorted cryopreserved bovine sperm and its relation to the fertility of sperm after sex-sorting. *BMC Genomics.* 2021 Jan 7. 22(1): 30. doi: 10.1186/s12864-020-07280-9

Kelly VC, Kuy S, Palmer DJ, Xu Z, Davis SR, Cooper GJ. Characterization of bovine seminal plasma by proteomics. *Proteomics.* 2006. 6(21): 5826–5833. doi: 10.1002/pmic.200500830

Khandelwal A, Renukaradhya GJ, Rajasekhar M, Sita GL, Shaila, MS. Systemic and oral immunogenicity of hemagglutinin protein of rinderpest virus expressed by transgenic peanut plants in a mouse model. *Virology.* 2004. 323(2): 284–291.

Kues WA, Niemann H. The contribution of farm animals to human health. *Trends Biotechnol.* 2004. 22 (6): 286–294.

Kumar D, Chauhan TK, Agarwal RK, Dhama K, Goswami PP, Mariappan AK. A double-stranded probe coupled with isothermal amplification for qualitative and quantitative detection of avian reovirus. *Arch Virol.* 2017. 162: 979–985.

Lan J, Lei MG, Zhang YB, Wang JH, Feng XT, Xu DQ, Gui JF, Xiong YZ. Characterization of the porcine differentially expressed PDK4 gene and association with meat quality. *Mol Biol Rep.* 2009. 36: 2003–2010.

Lande R, Thompson R. Efficiency of marker-assisted selection in the improvement of quantitative traits. *Genetics.* 1990. 124(3): 743–756.

Langevin C, Blanco M, Martin SAM, Jouneau L, Bernardet JF, Houel A, Lunazzi A, Duchaud E, Michel C, Quillet E, Boudinot P. Transcriptional responses of resistant and susceptible fish clones to the bacterial pathogen Flavobacteriumpsychrophilum. *PLoS One.* 2012. 7. doi: 10.1371/journal.pone.0039126 e39126

LeBlanc SJ, Leslie KE, Duffield TF. Metabolic predictors of displaced abomasum in dairy cattle. *J Dairy Sci.* 2005. 88: 159–170. doi: 10.3168/jds.S0022-0302 (05)72674-6. PMID: 15591379.

Lee DG, Nam J, Kim SW, Kang YM, An HJ, Kim CW, Choi JS. Proteomic analysis of reproduction proteins involved in litter size from porcine placenta. *Biosci Biotechnol Biochem.* 2015. 79(9): 1414–1421. doi: 10.1080/09168451.2015.1039478

Liu J, Li L, Chen X, Lu Y, Wang D. Effects of heat stress on body temperature, milk production, and reproduction in dairy cows: a novel idea for monitoring and evaluation of heat stress — A review. *Asian-Australas J Anim Sci*. 2019 Jan 4;32(9):1332–1339. doi: 10.5713/ajas.18.0743. PMID: 30744345; PMCID: PMC6722315.

Long NM, et al. Maternal obesity and increased nutrient intake before and during gestation in the ewe results in altered growth, adiposity, and glucose tolerance in adult offspring. *J Anim Sci*. 2010. 88(11): 3546e53.

Ma Lu Yang Y, Zhao X, Wang F, Gao S, Bu D. Heat stress induces proteomic changes in the liver and mammary tissue of dairy cows independent of feed intake: An iTRAQ study. *PLoS One*. 2019. 14.1: e0209182.

Madan ML. Animal biotechnology: Applications and economic implications in developing countries. *Rev Sci Tech*. 2005. 24(1): 127–139.

Maleszka R. Epigenetic integration of environmental and genomic signals in honey bees: the critical interplay of nutritional, brain and reproductive networks. *Epigenetics*. 2008. 3(4): 188e92.

Mansor R. *Proteomic and metabolomic studies on milk during bovine mastitis* (Doctoral dissertation, University of Glasgow). 2012.

Marai IFM, Habeeb AAM. Buffalo's biological functions as affected by heat stress. A review. *Livest Sci*. 2010. 127: 89–109. doi: 10.1016/j.livsci.2009. 08.001

Marancik D, Gao G, Paneru B, Ma H, Hernandez AG, Salem M, Yao J, Palti Y, Wiens GD. Whole-body transcriptome of selectively bred, resistant-, control-, and susceptible-line rainbow trout following experimental challenge with Flavobacteriumpsychrophilum *Front Genet*. 2015: 5. doi: 10.3389/fgene.2014.00453

Martins RP, Collado-Romero M, Martinez-Gomariz M, Carvajal A, Gil C, Lucena C, Moreno A, Garrido JJ. Proteomic analysis of porcine mesenteric lymph-nodes after Salmonella typhimurium infection. *J Proteomics*. 2012. 75: 4457–4470.

Mashooq M, Kumar D, Niranjan AK, Agarwal RK, Rathore R. Development and evaluation of probe based real time loop mediated isothermal amplification for Salmonella: A new tool for DNA quantification. *J Microbiol Methods*. 2016. 126: 24–29.

McMahon BJ, Teeling EC, Höglund J. How and why should we implement genomics into conservation? *Evol Appl*. 2014. 7(9): 999–1007. doi: 10.1111/eva.12193

Melzer N, Wittenburg D, Hartwig S, Jakubowski S, Kesting U, Willmitzer L, et al. Investigating associations between milk metabolite profiles and milk traits of Holstein cows. *J Dairy Sci*. 2012. 96: 1521–1534.

Meuwissen THE, Hayes BJ, Goddard ME. Prediction of total genetic value using genome-wide dense marker maps. *Genetics*. 2001. 157: 1819–1829.

Miao X, Luo Q, Zhao H, Qin X. Ovarian proteomic study reveals the possible molecular mechanism for hyperprolificacy of Small Tail Han sheep. *Sci Rep*. 2016. 6: 27606. doi: 10.1038/srep27606

Moura AA, Chapman DA, Koc H, Killian GJ. Proteins of the caudaepididymal fluid associated with fertility of mature dairy bulls. *J Androl*. 2006. 27(4): 534–541. doi: 10.2164/jandrol.05201

Nayeri S, Sargolzaei M, Abo-Ismail MK, Miller S, Schenkel F, Moore SS, Stothard P. Genome-wide association study for lactation persistency, female fertility, longevity, and lifetime profit index traits in Holstein dairy cattle. *J Dairy Sci*. 2017. 100(2): 1246–1258.

Niemann H, Kues W, Carnwath JW. Transgenic farm animals: Present and future. *Revue Scientifiqueet Technique (International Office of Epizootics)* 2005. 24: 285–298.

Nir O. What are production diseases, and how do we manage them? *Acta Vet Scand*. 2003. 44(1): 1–12.

Nwanta JA, Umoh JU, Abdu PA, Ajogi I, Egege SC, Adeiza AA. Field trial of Malaysian thermostable Newcastle disease vaccine in village chickens in Kaduna State, Nigeria. *Sokoto J Vet Sci*. 2008. 7(2).

Oura CAL, Bishop R, Wampande EM, Lubega GW, Tait A. The persistence of component Theileriaparva stocks in cattle immunized with the 'Mugugacocktail'live vaccine against East Coast fever in Uganda. *Parasitology*. 2004. 129(1): 27–42.

Palya V, Kiss I, Tatár-Kis T, Mató T, Felföldi B, Gardin Y. Advancement in vaccination against Newcastle disease: Recombinant HVT NDV provides high clinical protection and reduces challenge virus shedding with the absence of vaccine reactions. *Avian Dis*. 2012. 56(2): 282–287.

Pantawane PB, Dhanze H, Ravi KG, Dudhe NC, Bhilegaonkar KN. TaqMan real-time RT-PCR assay for detecting Japanese encephalitis virus in swine blood samples and mosquitoes. *Anim Biotechnol*. 2018. 23. 1–6

Paula Junior AR, van Tilburg MF, Lobo MDP, Monteiro-Moreira ACO, Moreira RA, Melo CHS, Souza-Fabjan JMG, Araújo AA, Melo LM, Teixeira DIA, Moura AA, Freitas VJF. Proteomic analysis of follicular fluid from tropically-adapted goats. *Anim Reprod Sci*. 2018. 188: 5–44. doi: 10.1016/j.anireprosci.2017.11.005

Pereiro P, P. Balseiro, A. Romero, S. Dios, G. FornCuni, B. Fuste, J.V. Planas, S. Beltran, B. Novoa, A. Figueras High-throughput sequence analysis of turbot (Scophthalmusmaximus) transcriptome using

454-pyrosequencing for the discovery of antiviral immune genes. *PLoS One*. 2012. 7. doi: 10.1371/journal.pone.0035369

Peréz-Sánchez H, Fassihi A, Cecilia J, Ali H, Cannataro M. Applications of high-performance computing in bioinformatics, computational biology and computational chemistry. In: Ortuño F, Rojas I, editors. *Bioinformatics and biomedical engineering*. Springer, Heidelberg; 2015. p 527–41.

Phocas F, Belloc C, Bidanel J, Delaby L, Dourmad JY, Dumont B, Ezanno P, Fortun-Lamothe L, Foucras G, Frappat B, González-García E, Hazard D, Larzul C, Lubac S, Mignon-Grasteau S, Moreno CR, Tixier-Boichard M, and Brochard M. Review: Towards the agroecological management of ruminants, pigs and poultry through the development of sustainable breeding programmes. II. Breeding strategies. *Animal*. 2016. 10(11): 1760–1769. doi: 10.1017/S1751731116001051

Pryce J, Bolormaa S, Chamberlain A, Bowman P, Savin K, Goddard M, et al. A validated genome-wide association study in 2 dairy cattle breeds for milk production and fertility traits using variable length haplotypes. *J Dairy Sci*. 2010. 93: 3331–3345. doi: 10.3168/jds.2009-2893

Raphael M, Julie M K O, Okeyo, A M and Joram M M. Genomic selection and use of molecular tools in breeding programs for indigenous and crossbred cattle in developing countries: current status and future prospects. *Front Genet*. 2019. 9: 694.

Reinhardt TA, Sacco RE, Nonnecke BJ, Lippolis JD. Bovine milk proteome: quantitative changes in normal milk exosomes, milk fat globule membranes and whey proteomes resulting from Staphylococcus aureus mastitis. *J Proteomics*. 2013. 82: 141–154. doi: 10.1016/j.jprot.2013.02.013

Rexroad C, Vallet J, Matukumalli LK, Reecy J, Bickhart D, Blackburn H, Boggess M, Cheng H, Clutter A, Cockett N, Ernst C, Fulton JE, Liu J, Lunney J, Neibergs H, Purcell C, Smith TPL, Sonstegard T, Taylor J, Telugu B, Eenennaam AV, Tassell CPV, Wells K. Genome to Phenome: Improving animal health, production, and well-being: A new USDA blueprint for animal genome research 2018–2027. *Front Genet*. 2019. 10: 327. doi: 10.3389/fgene.2019.00327

Riedmaier I, Benes V, Blake J, Bretschneider N, Zinser C, et al. RNA-Sequencing as useful screening tool in the combat against the misuse of anabolic agents. *Anal Chem*. 2012. 84(15): 6863–6868.

Riedmaier I, Spornraft M, Pfaffl MW. Identification of a potential gene expression biomarker signature in bovine liver to detect the abuse of growth promoters. *Food Addit Contam A*. 2014. 31(4): 641–649.

Rodrigues PM, Silva TS, Dias J, Jessen F. PROTEOMICS in aquaculture: Applications and trends J. *Proteomics*. 2012. 75: 4325–4345. doi: 10.1016/j.jprot.2012.03.042

Salak-Johnson JL, McGlone JJ. Making sense of apparently conflicting data: Stress and immunity in swine and cattle. *J. Anim. Sci*. 2007. 85(suppl13): E81-E88.

Seidel GE Jr. Sexing mammalian sperm – intertwining of commerce, technology, and biology. *Anim Reprod Sci*. 2003. 79(3–4): 145–156. doi: 10.1016/s0378-4320(03)00162-3. PMID: 14643102.

Shafer ABA, Wolf JBW, Alves PC, Bergström L, Bruford MW, Brännström I, Zieliński P. Genomics and the challenging translation into conservation practice. *Trends Ecol Evol*. 2015. 30(2): 78–87. doi: 10.1016/j.tree.2014.11.009

Shkap V, de Vos AJ, Zweygarth E. Jongejan F. Attenuated vaccines for tropical theileriosis, babesiosis and heartwater: the continuing necessity. *Trends Parasitology*. 2007. 23(9): 420–426.

Simon JP, Wilmut I, Clark AJ, Archibald AL, Bishop JO, Lathe K. Gene transfer into sheep. *Biotechnology*. 1998. 6: 179–183.

Singh K, et al. Epigenetics: A possible role in acute and transgenerational regulation of dairy cow milk production. *Animal*. 2012. 6(3): 375e81.

Smolenski G, Haines S, Kwan FYS, Bond J, Farr V, Davis SR, et al. Characterization of host defense proteins in milk using a proteomic approach. *J Prot Res*. 2007. 6(1): 207–215.

Soller M. The use of loci associated with quantitative effects in dairy cattle improvement. *Anim Production*. 1978. 27: 133–139.

Sood TJ, Lagah SV, Mukesh M, Singla SK, Chauhan MS, Manik RS, Palta P. RNA sequencing and transcriptome analysis of buffalo (Bubalusbubalis) blastocysts produced by somatic cell nuclear transfer and in vitro fertilization. *Mol Reprod Dev*. 2019 Sep. 86(9): 1149–1167. doi: 10.1002/mrd.23233

Sundekilde UK, Poulsen NA, Larsen LB, Bertram HC. Nuclear magnetic resonance metabonomics reveals strong association between milk metabolites and somatic cell count in bovine milk. *J Dairy Sci*. 2013. 96: 290–299. doi: 10.3168/jds.2012-5819 PMID: 23182357.

Tao S, Orellana RM, Weng X, Marins TN, Dahl GE, Bernard JK. Symposium review: The influences of heat stress on bovine mammary gland function. *J. Dairy Sci*. 2018. 101(6):5642–5654.

Tian H, Wang W, Zheng N, Cheng J, Li S, Zhang Y, Wang J. Identification of diagnostic biomarkers and metabolic pathway shifts of heat-stressed lactating dairy cows. *J Proteom.* 2015. 125: 17–28.

VanRaden PM, Van Tassell CP, Wiggans GR, Sonstegard TS, Schnabel RD, Taylor JF, Schenkel FS. Invited review: Reliability of genomic predictions for North American Holstein bulls. *J Dairy Sci.* 2009. 92(1): 16–24. doi: 10.3168/jds.2008-1514

Wagner DE, et al. Single-cell mapping of gene expression landscapes and lineage in the zebrafish embryo. *Science.* 2018. 360(6392): 981e7.

Weikard R, Altmaier E, Suhre K, Weinberger KM, Hammon HM, Albrecht E, et al. Metabolomic profiles indicate distinct physiological pathways affected by two loci with major divergent effect on Bos taurus growth and lipid deposition. *Physiol Genomics.* 2010. 42A: 79–88. doi: 10.1152/physiolgenomics.00120.2010. PMID: 20647382.

Wiggans GR, Vanraden PM, Cooper TA. The genomic evaluation system in the United States: Past, present, future. *J. Dairy Sci.* 2011. 94(6):3202–3211. doi:10.3168/jds.2010-3866.

Wilmut I, Schnieke AE, McWhir J, Kind AJ, Campbell KH. Viable offspring derived from fetal and adult mammalian cells. *Nature.* 1997. 385(6619): 810–813.

Wohlgemuth SE, Ramirez-Lee Y, Tao S, Monteiro APA, Ahmed BM, Dahl GE. Effect of heat stress on markers of autophagy in the mammary gland during the dry period. *J. Dairy Sci.* 2016. 99(6):4875–4880.

Xi, Xiaomin, Lai-Yu Kwok, Yuenan Wang, Chen Ma, ZhihuiMi, and Heping Zhang. Ultra-performance liquid chromatography-quadrupole-time of flight mass spectrometry MSE-based untargeted milk metabolomics in dairy cows with subclinical or clinical mastitis. *J Dairy Sci.* 2017. 100(6): 4884–4896.

You X, Xu M, Li Q, Zhang K, Hao G, et al. Discovery of potential transcriptional biomarkers in broiler chicken for detection of amantadine abuse based on RNA sequencing technology. *Food Addit Contam.* 2019. 36(2): 254–269.

Yue S, Ding S, Zhou J, Yang C, Hu X, Zhao X, Wang Z, Wang L, Peng Q, Xue B. Metabolomics approach explore diagnostic biomarkers and metabolic changes in heat-stressed dairy cows. *Animals.* 2020. 10(10): 1741.

Zachut M, Sood P, Levin Y, Moallem U. Proteomic analysis of preovulatory follicular fluid reveals differentially abundant proteins in less fertile dairy cows. *J Proteomics.* 2016. 139: 122–129. doi: 10.1016/j.jprot.2016.03.027

Zeng S, Bick J, Kradolfer D, Knubben J, Flöter VL, Bauersachs S, Ulbrich SE. Differential transcriptome dynamics during the onset of conceptus elongation and between female and male porcine embryos. *BMC Genomics.* 2019 Aug 28. 20(1): 679. doi: 10.1186/s12864-019-6044-z

Zhang L, et al. Maternal obesity in ewes results in reduced fetal pancreatic beta-cell numbers in late gestation and decreased circulating insulin concentration at term. *Domest Anim Endocrinol.* 2011. 40(1): 30e9.

Zhang XY, Li TT, Liu YR, Geng SS, Luo AL, Jiang MS, Liang XW, Shang JH, Lu KH, Yang XG. Transcriptome analysis revealed differences in the microenvironment of spermatogonial stem cells in seminiferous tubules between pre-pubertal and adult buffaloes. *Reprod Domest Anim.* 2021 Jan 25. doi: 10.1111/rda.13900

Zhao L, Yang S, Cheng Y, Hou C, You X, et al. Identification of transcriptional biomarkers by RNA-sequencing for improved detection of β2-agonists abuse in goat skeletal muscle. *PLoS One.* 2017. 12(7): e0181695.

Zhao L, Yang S, Zhang Y, Zhang Y, Hou C, et al. New analytical tool for the detection of ractopamine abuse in goat skeletal muscle by potential gene expression biomarkers. *J Agr Food Chem.* 2016a. 64(8): 1861–1867.

Zhao XM, Cui LS, Hao HS, Wang HY, Zhao SJ, Du WH, Wang D, Liu Y, Zhu HB. Transcriptome analyses of inner cell mass and trophectoderm cells isolated by magnetic-activated cell sorting from bovine blastocysts using single cell RNA-seq. *Reprod Domest Anim.* 2016b Oct. 51(5): 726–735. doi: 10.1111/rda.12737

Zhou Y, et al. Comparative whole genome DNA methylation profiling of cattle sperm and somatic tissues reveals striking hypomethylated patterns in sperm. *Gigascience.* 2018. 7(5).

Zolini AM, Block J, Rabaglino MB, Tríbulo P, Hoelker M, Rincon G, Bromfield JJ, Hansen PJ. Molecular fingerprint of female bovine embryos produced in vitro with high competence to establish and maintain pregnancy. *Biol Reprod.* 2020 Feb 14. 102(2): 292–305. doi: 10.1093/biolre/ioz190

4 Analysis of Gene Expression at mRNA Level

Technological Evolution

Anju Kumari, Harjot Kaur, Ravinder Kumar and Vijay Kumar

CONTENTS

4.1 GENE AND ITS EXPRESSION

A gene is a basic physical and functional unit of heredity. The term 'gene' was introduced by Danish botanist, plant physiologist, and geneticist Wilhelm Johannsen in 1909 (Johannsen, 1909). It is made up from deoxyribonucleic acid (DNA) which encodes the synthesis of a gene product, i.e. ribonucleic acid (RNA), which further helps in formation of protein. Basically a gene has a fixed position on chromosomal DNA and each chromosome consists of many genes (Hirsch, 2002). Gene expression is the process in which DNA directs the synthesis of functional products, such as proteins through RNA intermediate. In other words, when any gene is turned 'ON' for the synthesis of RNA which further are responsible for the production of protein then this process is known as gene expression. Gene expression was first discovered by Francis Crick in 1958, which is included

DOI: 10.1201/9781003220831-4

in central dogma of molecular biology (Crick, 1970). According to this theory, DNA basically contains information for making protein which is copied by RNA.

The process of gene expression is utilized by all of biological life, such as eukaryotes, prokaryotes, and viruses to generate the macromolecules. Gene expression is the most elemental or basic level in which genotype (DNA) gives rise to phenotype (RNA or protein). Cells can regulate gene expression at various stages of development. Mainly genetic information passes from DNA to RNA and finally to protein. This whole process requires mediator molecule for passing out information steadily. That key mediator is messenger RNA (mRNA) (Alberts et al., 2002).

The levels of gene expression can be analyzed by measuring mRNA or protein. Gene expression is a complex process operating at multiple levels. It involves association of potent events that are regulated at the transcriptional level, the posttranscriptional level, the translational level, and at the posttranslational level. These processes involve structural and chemical changes to DNA, in which proteins bind to DNA binding domain elements and regulate transcription that further leads to translation of mRNA. However most of the genes are regulated at the transcriptional level only (Phillips, 2008). Transcription is the first part of central dogma, i.e. DNA to RNA. It involves transferring of genetic instructions from DNA to mRNA, which is further utilized by a cell.

4.2 MESSENGER RNA IN GENE EXPRESSION

Messenger RNA (mRNA) is the single stranded intermediate molecule. Its structure consists of adenine, cytosine, guanine and uracil, exons, 5'-cap and 3'-poly-tail which links genes to proteins. They carry the coding sequences for protein synthesis, and therefore are called transcripts. The mRNA was first described by Elliot Volkin and Lazarus Astrachan in 1956. So the smooth and regular flow of molecular information (i.e. DNA to RNA to protein) helps organisms to coordinate and function well. For performing each and every part of body function, proteins are needed, which is directed by mRNA. The proteins which are produced by cells play major roles in the form of enzymes, structural molecules or transport machinery, hormones, and glands inside the body and influence the entire metabolism of an organism (Xu et al., 2020).

By analyzing gene expression at transcriptional as well as translational levels we can better understand normal cellular processes such as differentiation and also abnormal processes to study biological systems. This will also help to understand gene function, cell behavior, cell signaling, gene regulation development, and disease progression processes. In this chapter, mainly we are focusing on the study and analyzing of the expression of mRNA. So the tools and techniques which are used to study mRNA expression will be discussed here.

4.3 METHODS FOR mRNA QUANTIFICATION

Gene expression studies rely upon techniques for the identification and quantitation of mRNA species coding for specific proteins. There are various methods which are available for detection and quantification of messenger RNA; each one of them contributes to distinct advantages and disadvantages. Using modern molecular methods and future advancements in diagnostic technology allows for the detection of any disease at earlier stages of its development (Orakpoghenor and Markus, 2020). Before the invention of PCR, single gene transcripts of mRNA were analyzed by northern blotting, RNAse protection assay, and in situ hybridization techniques. After the invention of PCR, various other methods were developed for the study of gene expression, such as polymerase chain reaction (PCR) itself, competitive PCR, Real-Time PCR, DNA microarray, etc. Here we are discussing the general principle, applications, merits, and demerits of these techniques in brief. Northern blot is the most widely used method; it is also known as RNA blot. This technique is used in molecular biology research to study gene expression by analyzing RNA (isolated mRNA) in a sample (Gilbert, 2000; Alberts et al., 2008; Kevil et al., 1997). The northern blot technique was developed in 1977 by James Alwine, David Kemp, and George Stark at Stanford University, with contributions from Gerhard Heinrich (Alwine et al., 1977). The term 'northern blot' actually refers specifically to the capillary transfer of RNA from the electrophoresis gel to the blotting membrane (Trayhurn, 1996). In this, electrophoresis is used to separate RNA samples by size, and its detection is done with a hybridization probe complementary to part of or the entire target sequence. A cDNA radioactively labeled with 32P is the most commonly used hybridization probe (Liang and Pardee, 1995; Engler-Blum et al., 1993). With the help of this technique one can observe a particular gene's expression pattern between tissues, organs, developmental stages, environmental stress levels, and pathogen infection (Mori et al., 1991; Liang and Pardee, 1995; Baldwin et al., 1999).

Northern blot is helpful in detection of RNA size, the observation of alternate splice products, and the use of probes with partial homology. The quality and quantity of RNA can be measured on the gel prior to blotting, and the membranes can be stored for a longer period of time and later on re-probed for years after blotting (Streit et al., 2009). This is the only technique that provides information about the transcript size. Because of mRNA size measurement it is helpful in detecting mutations that causes abnormal size of mRNA. The main problem in northern blotting is sample degradation by RNases, which can be avoided by proper sterilization of glassware and the use of RNase inhibitors such as DEPC (Diethylpyrocarbonate) (Trayhurn, 1996).

RNAse protection assay (RPA) technique is useful for the detection and quantification of an mRNA molecule in a varied RNA sample taken out from cells. From this, one can recognize one or more RNA molecules of known sequence even at very low concentration. The principle of RPA is based on the hybridization in solution of a labeled single-stranded antisense RNA probe with a target mRNA molecule. After hybridization, single-strand specific RNases are then used to digest away un-hybridized RNA. The hybrid can be resolved by a denaturing gel. Subsequent detection will reveal the appropriate-sized gel band corresponding to the target mRNA, which further corresponds to level of transcription of the gene in a cell (Qu and Boutjdir, 2007). This method provides higher sensitivity than northern blot analysis, allowing the detection of low abundance mRNAs (Sambrook et al., 1989). They are also used to detect the presence of double-stranded RNA, which detects the presence of RNA interference. The major advantages of RPA are the high sensitivity and the simultaneous detection and quantification of multiple mRNA targets in a single RNA sample. They are also commonly used to precisely map mRNA termini and intron/exon junctions. The primary limitation of RPA is the lack of information on transcript size. Another drawback to RPAs is the lack of probe flexibility.

In Situ Hybridization (ISH) is a powerful and versatile tool for the localization of specific mRNAs in cells or tissues. This technique was invented by French biologist Mary-Lou Pardue and Joseph G. Gall (O'Connor, 2008; Gall and Pardue 1969; McKusick, 1997). ISH does not require the isolation or electrophoretic separation of RNA. Hybridization of the probe takes place within the cell or tissue. The cellular structure is maintained throughout the procedure, so it provides information about the location of mRNA within the tissue sample. For the preservation of target mRNA within tissues, crosslinking fixatives (such as formaldehyde) should be used (Lehmann et al., 1994). It is a laboratory technique in which a single-stranded DNA or RNA sequence probe is allowed to form complementary base pairs with DNA or RNA present in a tissue or chromosome sample. A radioactive label is attached to the probe so that its binding can be observed (Jensen, 2014). This is a powerful technique for identifying specific mRNA species within individual cells in tissue sections, so it provides understanding of physiological processes and disease pathogenesis. It is helpful in physical mapping of chromosomes. In developmental biology, it is used for gene expression profiling in embryonic tissues. A major advantage of in situ hybridization is that it permits maximum use of tissue that is difficult to obtain (e.g., embryos and clinical biopsies). Hundreds of different hybridizations can be performed on the same tissue. Libraries of tissues can be formed and stored in the freezer for future use. The main procedure or protocol is very time-consuming, laborious and may require specialized equipment for preparing samples and visualizing results of the experiment. There is difficulty in identifying targets that have low DNA and RNA copies number. Moreover, quantitation of gene expression is not as straightforward as with the other techniques.

By an old survey, well-established methods (e.g., northern blotting, RNase protection assays, and dot/slot blot) disclose major deficiencies such as the requirement for large samples size, their relative insensitivity, and the lack of sufficient productivity to make these techniques routinely suitable for detection or viewing of large number of individuals (Farr and Dunn, 1999). So to overcome these shortcomings, several modern molecular techniques have been used for the analysis of gene expression at mRNA level such as PCR, Competitive PCR, RT-PCR, and DNA microarray, which will be discussed in this chapter.

Polymerase Chain Reaction (PCR): PCR is a revolutionary invention in the molecular biology field, which is used nowadays in research. PCR was invented by American biochemist Kary Mullis in 1983 at Cetus Corporation, for which he was awarded the Nobel Prize in 1993. PCR is a very simple and economical technology to amplify a specific segment of DNA. This technique has transformed the study of DNA to such an area that it has now become the choice of tool for current research in the molecular biology field (Zhu et al., 2020). The whole procedure of PCR depends upon thermal cycling. It basically amplifies a specific region of a DNA strand. With the thermal cycling process the reactants undergo repetitive cycles of heating and cooling for melting of DNA and enzyme-linked replication of DNA. For this process, it consists of various components, i.e. DNA template, dNTPs, primer, and a TaqDNA polymerase. PCR involves the initial denaturation of the target DNA, followed by the annealing of target-specific oligonucleotide primers to each strand of DNA and finally the extension of the primers by polymerase enzyme to produce a copy of the selected region. By repeating this cycle of denaturation, primer annealing, and extension, the exponential amplification of the targeted sequence occurs (NCBI, 2017).

The essential component of PCR is heat stable enzyme (Taq DNA polymerase) and it was discovered in 1976 from the thermophilic bacterium Thermusaquaticus. In 1971, Kjell Kleppe and his co-workers first described a method of using an enzymatic assay to replicate a short DNA template with primers (in vitro) in the laboratory of Har Gobind Khorana. But they did not receive much attention at that time and the main acknowledgment regarding the invention of PCR was given to Kary Mullis in 1983 (Rabinow, 1996). In *Scientific American Journal* Mullis summarized the procedure:

> Beginning with a single molecule of the genetic material DNA, the PCR can generate 100 billion similar molecules in an afternoon. The reaction is easy to execute. It requires no more than a test tube, a few simple reagents, and a source of heat.
>
> **(Mullis, 1990)**

PCR has many diverse applications in the molecular research field. It allows selective DNA isolation by amplification of a specific DNA segment. It is helpful in DNA sequencing to know amplified PCR sequences. This technique is helpful in amplification and quantification of DNA to study gene expression. For this purpose quantitative PCR is an important tool that measures the accumulation of DNA product after each round of PCR amplification. PCR has a valuable role in medical and diagnostic applications. It has infectious disease applications, forensic analysis and research applications.

It is a very simple and easy to handle technique that produces results within a short period of time. Meanwhile, this is a highly sensitive tool, too, to produce millions to billions copies of a specific product for sequencing, cloning and analysis. Prior information of targeted sequence is necessary for generating primers that will be later on used for amplification process (Garibyan and Avashiya, 2013). Error rate of DNA polymerases are high so mutations are common and they cause mutations in PCR products (Zhou et al., 1991). Several environmental conditions also affect PCR amplification, which leads to irrelevant results. However, we know that PCR is a highly sensitive method for the detection of a small amount of DNA so the exact quantification of the desired product is not possible. Also the amplified final product of PCR reaction does not correspond to the initial template, which was present at that time. This is mainly due to 'plateau phase' of the PCR cycle, which is caused by deactivation of Taq DNA polymerase, having nucleotide substrate in small amounts, shortage of primer, inhibition by pyrophosphate, and re-annealing of amplified PCR products. So, to overcome these problems, normal PCR, competitive PCR, real-time (RT PCR), and DNA microarray techniques have been developed for analyzing gene expression at mRNA level.

Competitive PCR is the most widely used method for the quantification of cellular mRNA. It is a highly rapid, versatile, and most sensitive technique (McCulloch et al., 1995). It involves the co-amplification of a target DNA sample with known amounts of a competitor DNA that shares most of the nucleotide sequence with the target (Zentilin and Giacca, 2007). In this PCR, a known amount of a DNA fragment (competitor) is added to the sample. The competitor has sequences for the same primers used to amplify the target. When correct primers are available then the whole protocol of quantification is done. When the target DNA and competitor are amplified together, both templates will compete for the same set of primers. Due to this competition, the ratio of the amounts of the two amplified products reflects the ratio of the amounts of the target DNA and competitor. Since the initial amount of the competitor is known, the amount of the target DNA can then be estimated according to the T:C ratio. T corresponds to amount of amplified product from target DNA or RNA and C is equal to the amount of amplified product from competitor. When the T:C ratio = 1, the initial amount of target DNA or RNA will correspond to the amount of competitor. The final result is measured by digital analysis of gel image or generation of a regression line (Raeymaekers, 1993).

Competitive RT-PCR is a type of competitive PCR. It is used for absolute quantification. It is basically performed similar to competitive PCR. There are two methods for its performance. The first method is competition between complementary DNA (cDNA) and DNA competitor. This method uses a DNA competitor when cDNA is formed. The relative amounts of cDNA among different samples can be determined. Since the yield and efficiency of the RT reaction is not reflected in this method, the absolute quantity of the target mRNA cannot be determined. The second method is competition between mRNA and RNA competitor. This method is based on competition between mRNA and a RNA competitor during the RT reaction. The relative amounts of target mRNA among different samples can be determined. As the yield of the RT reaction is reflected in this method, the absolute quantity of the target mRNA can be determined (Tang et al., 2013; Tsai and Wiltbank, 1996). There are two types of competitors which are used: **Homologous Competitor** – a competitor that has the same nucleotide sequences as the target DNA (RNA) but contains a deletion or an insertion, or has a different restriction site introduced by site-specific mutagenesis. When designing this type of competitor, care should be taken to avoid the formation of a hetero-duplex, which would interfere with subsequent quantitation. The second type is **Heterologous Competitor** – a competitor that has nucleotide sequences different from the target DNA (RNA) except for the sequences of the primer annealing sites. A hetero-duplex will not be formed. Since the primer sequences are the same, the difference in amplification efficiency between the target and heterologous competitor is usually minor.

Practical applications of competitive PCR involve easy quantification of mRNA transcript and side by side estimation of multiple targets of gene (Zentilin and Giacca, 2010). This type of method can be done without any expensive equipment but it requires parallel running of several PCR tubes. The main advantage of this technique is that results are not affected by tube to tube variations in amplification efficiency. Also, it is insensitive to the effects of PCR saturation. While doing competitive PCR there is a chance of contamination in the handling of tubes so this method requires proper stringent precautions to avoid contamination of samples and tool. For quantification of each sequence, proper control measures should be done. Applying all these safety measures will be helpful in obtaining accurate quantification of mRNA as this is a highly sensitive method too after all (Raeymaekers, 1993).

4.4 REAL-TIME PCR

Differential gene expression at the time of development, growth, reproduction is crucial and significant and requires assays for quantification of mRNA levels. Real-time reverse transcriptase PCR (real-time RT PCR) is emphatically established in the core research technology because of its large dynamic range, high sensitivity, and absolute quantification of the RNA targets (Bustin, 2000; Huggett et al., 2005). Though theoretically, real-time RT PCR differs from traditional PCR only by the addition of an initial step, RNA-dependent DNA polymerase (reverse transcriptase) based conversion of RNA into a DNA template but practically, this additional preliminary step results in a much more fragile and flexible assay (Bustin and Nolan, 2004).

The first ever demonstration of real-time PCR was achieved by Higuchi and co-workers at Roche Molecular Systems and Chiron by video graphing the accumulation of ethidium bromide with DNA under ultraviolet light, which causes EtBr to fluoresce (Higuchi et al., 1992; Higuchi et al., 1993). It was only because of this fluorescent chemistry of ethidium bromide with PCR and real-time videography that real-time PCR was born in the early 1990s. The technology gradually expanded into a competitive market, dominating the commercial world and becoming scientifically powerful. Commercially, in 1996 Applied Biosystems made the real-time PCR instrumentation available for the first time; subsequently, many other companies supplemented new machines into the market. Bio-Rad, Roche Applied Science, Applied Biosystems, MJ Research, Bioneer, Idaho Technology, Cepheid, Stratagene, Corbett Research and BioGene, are the companies that offer real-time PCR instrumentation (Valasek and Repa, 2005).

For reverse transcription step some of the commonly used reverse transcriptase's are Moloney Murine Leukemia Virus (M-MuLV), Avian Myeloblastosis Virus (AMV), or Enhanced-AMV reverse transcription enzymes, each engineered to best create cDNA from RNA that is difficult to amplify. The cDNA is then used as raw material for a real-time PCR reaction (Valasek and Repa, 2005).

4.4.1 Detection Methods

The detection system can be either intercalator-based or probe-based real-time PCR, also known as TaqMan PCR.

4.4.1.1 Probe-Based Real-Time PCR

TaqMan is a non-extendable fluorogenic probe having a reporter florescent dye at the 5' end and a quencher molecule at the 3' end. In an intact probe, based on fluorescence resonance energy transfer, quencher dye in close proximity absorbs the fluorescence emitted by reporter dye (Cardullo et al., 1988). Primer and TaqMan probe anneal to the template strand, after its denaturation. TaqMan with fluorogenic probe attach downstream to the primer site and during the extension phase of PCR it is cleaved by the 5' nuclease activity of the polymerase; unraveling the fluorescent dye from the quencher and therefore enhancing fluorescence from the reporter probe. Also, removal of fluorescent TaqMan probe by cleavage permits the extension of target template by primer and subsequent release of fluorescent probes with each cycle results in increased fluorescence intensity directly proportional to the amount of target template present (Arya et al., 2005).

4.4.1.2 Intercalator-Based System

An intercalator-based system requires a dye that would bind to the newly generated double-stranded DNA and produce a fluorogenic signal. SYBR green I is the most commonly used intercalating fluorescent dye among other uncommonly used dyes, e.g., 4-[(3-Methyl-6-(benzothiazol-2-yl)-2,3-dihydro-(benzo-1,3-thiazole)-2-methylidene)]-1-methylpyridinium iodide (BEBO), YOYO-1, TOTO-1, etc. SYBR green I binds to the minor groove of DNA and absorbs light of wavelength 480-nm and emits light of 520-nm wavelength. The fluorogenic dye emits 1,000-fold greater fluorescence than when it is free in solution (Valasek and Repa, 2005). This results in an increase in the fluorescence signal as the polymerization proceeds, when monitored in real time. The PCR product can be verified by plotting fluorescence as a function of temperature in order to create a melting curve of the amplicon (Ririe et al., 1997).

4.4.2 Quantification of Real-Time PCR

The two basic ways to quantify mRNA in real-time PCR are either a one-step reaction, whereby whole of the reaction from reverse transcription to PCR amplification is performed in a single tube, or as a two-step reaction, i.e., synthesizing cDNA first by a reverse transcription reaction separately and then amplifying the cDNA in a PCR reaction.

4.4.2.1 One-Step Real-Time PCR Quantification

In one-step real-time PCR, reactions occur in a single tube, hence it is believed to minimize experimental variation. Certain pros associated with this method are that it involves a lesser amount of sample handling, thus narrowing chances of contamination, pipetting errors, as well as other sources of error (Battaglia et al., 1998). Also, it is less expensive yet quicker to set up. However, gene-specific primers are used with the one-step method, and both reverse transcription and amplification occur in one tube; thereby, other genes of interest cannot be amplified for later analysis (Wong and Medrano, 2005). Also, RNA template is used as a starting point, which if not handled efficiently is susceptible to quick degradation. Hence, this reaction may not prove to be a feasible one where the same sample is tested on several occasions over a period of time and certainly requires aliquots that will be archived for future testing. One-step real-time PCR reactions have been reported to be less sensitive than two-step real-time PCR reactions (Wacker and Godard, 2005).

4.4.2.2 Two-Step Real-Time PCR Quantification

Two-step real-time PCR separates the reverse transcription reaction from the real-time PCR reaction, allowing different real-time PCR assays on a same cDNA. As reverse transcription process is infamous for its vastly variable reaction efficiency (Mannhalter et al., 2000), using aliquots of same cDNA template safeguards those reactions from subsequent assays that have the same amount of template as those assayed earlier. Data from the two-step reaction method is relatively reproducible with Pearson correlation coefficients ranging from 0.974 to 0.988. When using a DNA binding dye like SYBR Green I, a two-step real-time PCR reaction may be favored because it is easier to remove primer-dimers through the manipulation of melting temperatures (Tms) (Mannhalter et al., 2000). The disadvantage involved with the reaction is increased opportunities of DNA contamination.

4.4.3 Methods of Real-Time Quantification

The method of quantification should be manifested beforehand before setting up a real-time RT-PCR reaction. Absolute quantification and relative quantification are the two methods for quantification of real-time RT-PCR results. Both these methods employ the cycle time (C_T) to determine the level of gene expression. C_T value signifies that fluorescence signal has reached at a threshold level and the fluorescence intensity is higher than the background fluorescence and it is displayed by the software in the form of amplification curves. Greater amount of mRNA at the start of reaction will reach the threshold in lesser time and thus yielding a lower C_T value (Heid et al., 1996).

4.4.3.1 Relative Quantification

Relative quantification compares the variability in steady-state levels of mRNA among different samples and indicates them relative to the levels of an internal control RNA. Relative quantification does not require standards with known concentrations but its sequence must be known (Bustin,

2002). The internal control or reference gene needs to be carefully chosen based on the study that one is conducting and is the one which is indeed invariant in its expression across the treatment regime. Housekeeping genes such as β-actin, cyclophilin, glyceraldehyde-3-phosphate dehydrogenase, cyclophilin, a glycolytic enzyme, ribosomal protein 36B4, 18S rRNA, transferrin receptor, predominantly serve as internal control or reference gene (Bustin and Nolan, 2004; Valasek and Repa, 2005).

Relative quantification utilizes comparison in expression levels of the gene to be tested versus an internal control gene and, in many experiments, it is adequate and may not be of concern as the fold change may be informative irrespective of the absolute value (Valasek and Repa, 2005). Quantitation is done using just mathematical equations, and the need for standard curves is not required in this method. The expression level of target gene is normalized to the reference gene by first subtracting the C_T of the reference gene from the C_T of the target gene and the resulting difference in cycle number (ΔC_T) is then expressed as an exponent of the base 2 (due to the doubling function of PCR), thereby indicating the fold difference of template for these two genes (Arya et al., 2005). The method of relative quantitation has both positive and negative connotations. With careful consideration, the internal control must be selected and the assumption must be made that the control gene chosen does not differ in expression level or copy number amongst the samples of study. If this assumption does not hold true, then the samples run on any given plate or multiple plates that will won't be comparable.

4.4.3.2 Absolute Quantification

Absolute quantification determines the number of copies of a specific RNA per cell or unit mass of tissue (Wang and Spadoro, 1998). The method requires a sample of copy number of the gene of interest or the target gene that can be diluted in order to generate a standard curve; it is an external 'absolute standard' (Kuhne and Oschmann, 2002). A linear relationship is generated between C_T and amounts of overall RNA or cDNA present at the start of the experiment by the standard curve, thus on the basis of their C_T values concentration of target gene is determined and unknown samples are thus compared with the standard curve for absolute quantitation (Heid et al., 1996). DNA standard molecules, like genomic DNA, recombinant plasmid DNA (recDNA), RT-PCR product, and commercially synthesized bigoligonucleotide, cRNA bearing the target sequence can be used to generate standard curves. The method of absolute quantitation is more accurate but is labor-intensive (Bustin and Nolan, 2004). This method is majorly applied in circumstances when absolute quantitation is analytic such as in quantitation of viral load (Borg et al., 2003; Lin et al., 2004; Castelain et al., 2004) or when estimating a few numbers of genes in either small or large sized samples (Dumur et al., 2002; Jurado et al., 2003).

4.4.4 Real-Time PCR Applications

4.4.4.1 Detection of Alleles or Mutation Detection

Differences in genes due to alterations lead to variation in gene expression and are associated with individuality, individual response to medicine, diversity in the population, and susceptibility to diseases. Real-time RT-PCR is of great efficiency to screen large numbers of samples for allelic variation.

TaqMan probe assay making use of probes labelled withdifferent fluorescent reporter dyes has been used for allelicdiscrimination (Lee et al., 1993; Livak, 1999) under competitive conditions where the probes are present in the same tube. Hybridization of the specific probe corresponding to exact sequence amplifies the target sequence and also prevents the binding of mismatchedprobe. Homozygosity is indicated by a surge in either fluorescentsignal whereas rise in both fluorescent signals point towardsheterozygosity (Bustin, 2000). A modification involving use of probes having incorporation of 5-propyne-2'-deoxyuridine instead of thymidine is convenient when amplifying high A/T content sequences, as the melting temperature of these probes is higher than that of standard Taqman probes. However, this improvement is of less use where PCR reaction needs to be set between 60°C and 62°C as it is tough to achieve perfect discrimination between exact-match probes and mismatch probes. It has certain relevance only when a small number of the templates will carry the chosen mutation such as detection of mutant K-ras colorectal cancer patient (Bustin, 2000).

Other method to detect alleles is by designing FRET hybridization probes which would span the SNP of interest and would perform melting curve analysis after the amplification phase (Uhl et al., 2002; Sloan et al., 2004; Uhl and Cockerill, 2004). In case of presence of a variation in the nucleotide sequence under one of the probes, a change will be observed in the melting temperature, signaling the possible presence of a SNP. The speed and efficiency of the method is improved as only amplified products from those particular samples with different melting temperatures are sequenced to detect the variation (Houghton and Cockerill, 2006).

4.4.4.2 mRNA Splice Variants Detection

mRNA splice variants are a product of the same promoter but after reaching the tissue the RNA transcript is spliced in a particular way in order to generate a tissue-specific mRNA and protein, thereby, promoting a tissue-specific level regulation. Thus, alternative splicing of RNAs by tissue-specific gene regulation contributes to functional diversification of proteins as well as refinement of gene expression (Lopez, 1998).

Real-time RT-PCR assays have been shown to be ideal for the quantification as well as identification of mRNA splice variants, as they can be unfailingly distinguished by a combination of exon-specific primers and an apt probe (Kafert et al., 1999). Quantification of each mRNA splice variant is done by probes that are designed to bind with the particular exon splice junctions. On the other hand, probes designed to hybridize with whole of the exon will result in overall detection of all mRNAs defined by the exon-specific upstream and downstream primer sequences. Consequently, an amalgam of probes permits maximum information about the presence, any number of and the levels of mRNA splice variants to be collected together with a minimum of trouble and complexity. Predominantly, normalization becomes irrelevant as the results attained are ratios of one mRNA splice variant to another (Bustin, 2000).

4.4.4.3 Gene Expression Quantitation

Quantitating gene expression at RNA level refers to the abundance of transcript copy numbers of specific RNA sequences, i.e., measuring DNA amplifications/deletions or messenger RNA (mRNA) molecules expressed by gene activation. Determination of relative number of RNA transcripts for a particular gene has too many potential clinical applications (Bustin, 2002).

Having access to an accurate method to quantify gene expression is vital (Bustin, 2002). Other techniques for quantitation of RNA include northern blot and traditional PCR analysis but there are certain limitations associated with these techniques such as simple PCR, though it is of great use when determining the presence of a specific transcript, but is not precise for quantification. Similarly, northern blot can accurately determine presence of specific RNA sequences but it is unable to distinguish between small differences in amounts of specific RNA transcripts (Mocellin et al., 2003). Real-time PCR method has been shown to have an upper hand in quantitation of gene expression (Schmittgen et al., 2000; Wilhelm and Pingoud, 2003). RNA quantitation can be done efficiently by either of the detection methods such as SYBR Green, FRET hybridization probes, or TaqMan probes. Because of the cost efficacy, SYBR Green is the most common of the detection methods and also it was among the first method to be utilized for real-time PCR testing, but two other formats are gaining popularity due to the decreasing cost as well as increased specificity (Houghton and Cockerill, 2006).

For real-time RT-PCR analysis, first RNA isolation from a sample is done and cDNA is made using reverse transcriptase. The choice of housekeeping genes is very crucial for reliability of the obtained data. Not all housekeeping genes always have the same level of expression, therefore, the choice for housekeeping gene for the experimental set-up must be tested and demonstrated before any reliable data can be obtained (Klein, 2002). Normalization of the amount of cDNA transcripts present for a specific gene is done with the housekeeping gene (Bustin, 2002; Mocellin et al., 2003). For some gene expression studies, like the respective influences of several factors on gene expression levels, relative method of quantification is satisfactory, as relative or comparative quantification would solve the purpose of the study. But other applications require absolute numbers, which are calculated relative to the standard (RNA, cDNA, plasmid DNA, genomic DNA) and largely depend on the accuracy of the used standard (Klein, 2002).

4.4.4.4 Analysis of Cytokine Levels

Cytokines are small molecules secreted majorly by multiple types of immune cells like lymphocytes, antigen-presenting cells, monocytes, endothelial cells, and other cells like fibroblasts. They play a central role in the immune network by modulating lymphocyte activation, proliferation, differentiation, and apoptosis (Giulietti et al., 2001). Reliable and accurate quantitation of cytokine mRNA expression profiles is of utmost importance when analyzing and measuring specific immunological responses (Overbergh et al., 1999). Analysis of levels of cytokines by Q-PCR methods has provided with a better understanding of action of the broad spectrum of cytokines and also has refined the role of cytokines in a number of biological contexts (Patel et al., 2001; Ginzinger, 2002).

Extraction of RNA from the tissue of interest is followed by reverse transcription leading to synthesis of cDNA. Primers and fluorogenic probes are designed after which amplification is performed in a real-time manner and quantification is done using standard curve analysis or absolute quantification method shows quantitation in absolute copy numbers per cell, total RNA concentration, or total tissue (Overbergh et al., 2003).

4.4.5 Advantages of Quantitative Real-Time RT-PCR

Quantitative real-time RT-PCR has many advantages over traditional PCR. qRT-PCR measures the data simultaneously in the exponential growth phase while in traditional PCR it is measured at the End Point or the plateau phase. Enhanced intensity of fluorescent signal of reporter gene is directly proportional to the number of amplicons produced in the qRT-PCR reaction (Schmittgen and Livak, 2008). The cleaved probe provides a permanent record of amplification of an amplicon. The range of amplicon detection is increased dynamically in a qRT-PCR reaction. Moreover, traditional PCR requires post-processing steps which are not required in qRT-PCR (Klein, 2002). Risk of cross contamination is minimized in qRT-PCR and it offers a wide dynamic range of quantification, high technical sensitivity, precision, throughput.

However, qRT-PCR, to its full capacity, in the current times, can only detect hundreds of transcripts in a single assay and is not capable of genome-wide analysis. The emergence of microarrays in the recent times has led to comparative analysis of gene expression of thousands of genes simultaneously in a single experiment.

4.5 DNA MICROARRAYS

DNA microarray is an influential, highly specific, rapid technology to analyze comparative mRNA expression levels of thousands of genes under study in a snapshot of time. The technology has not only substantially accelerated molecular biological research but also provides a technological intersection between biology and computers enabling expression analysis of potentially all of the genes in an organism with much specificity (King and Sinha, 2001; Karakach et al., 2010).

The technique has become the core of expression genomics and is based on the basic principle of hybridization of complementary nucleic acids to immobilized target sequences. Thousands of DNA probes are usually immobilized on a solid surface with the capacity to recognize matching molecules explicitly (Pirrung, 2002). Sample DNAs are labeled with fluorophores having different emission spectra, thus, when allowed to interact with target probes, the samples could be interrogated simultaneously. After subsequent hybridization of sample DNA to complementary target in the array, intensity of fluorescence at each spot on the array provides a quantitative measure corresponding to the expression level of the particular gene in the sample, thus determining their respective concentrations.

4.5.1 Types of DNA Microarrays

DNA microarrays are categorized into two basic categories:

4.5.1.1 In Situ Hybridization

In in situ hybridization, also known as high-density oligonucleotide arrays or oligo chips, sets of target DNA probes are synthesized in situ on glass surfaces with the help of a photolithography or inkjet process (Harrington et al., 2000). Photolithography method involves usage of a photolithographic mask in order to determine the position for deposition of a new probe on the array by photo-induced deprotection of a previously deposited functionalized nucleotide (Karakach et al., 2010). Due to possible failure of photo-induced deprotection at each step of the probe synthesis, this method produces short probes of 25 nucleotides. Probes of a longer length, up to 60 nucleotides, are generated by inkjet system, where the DNA probe is synthesized by changing the base at each location via the inkjet system (Arcellana-Panlilio and Robbins, 2002; Karakach et al., 2010).

4.5.1.2 Spotted Arrays

Spotted arrays are produced with mechanical deposition of 50–70-nucleotide-long synthesized oligomer or cDNA amplicons on a functionalized substrate. For the generation of cDNA amplicons, reverse transcription of all mRNAs from an organism under study is done to generate so-called expressed sequence tags (ESTs), followed by amplification by polymerase chain reaction, purification, and spotting on microarray glass slides.

4.5.2 Applications of DNA Microarray

Microarray has a great amount of applicability in tumor progression as well as in distinguishing between cancers of similar pathology helping in classification of number of different cancers into further subcategories (Khan et al., 2001). Whole genome sequencing with the advent of microarray technology has opened new doors in the field of biomedicine. In examining the virtual response of a drug on the organism, toxicological research, drug resistance, microarrays have proved to be extremely useful (Chicurel and Dalma-Weiszhausz, 2002; Waring and Halbert, 2002). Another application of microarray is in the immune system where gene expression profiles inquiring effects of different bacterial endotoxins are showing potential differences in the cellular response to various bacteria (Aderem and Ulevitch, 2000). Various studies have proved that lipopolysaccharides can initiate expression of a number of genes, including proinflammatory cytokines, chemokines, and adhesion molecules (Arcellana-Panlilio and Robbins, 2002).

BIBLIOGRAPHY

Aderem, A., and Ulevitch, R. J. 2000. Toll-like receptors in the induction of the innate immune response. *Nature* 406(6797):782–787.

Alberts, B., Johnson A., Lewis J., Raff M., Roberts K. and Walter P. 2002. "The shape and structure of proteins." In *Molecular Biology of the Cell 4th edition*. Garland Science.

Alberts, B., Johnson A., Lewis J., Raff M., Roberts K. and Walter P. 2008. "*Molecular biology of the cell 5th edition*. Garland Science.":921–944.

Alwine, J. C., Kemp D. J. and George Stark R. 1977. "Method for detection of specific RNAs in agarose gels by transfer to diazobenzyloxymethyl-paper and hybridization with DNA probes." *Proceedings of the National Academy of Sciences* 74(12):5350–5354.

Arcellana-Panlilio, M., and Robbins, S. M. 2002. Global gene expression profiling using DNA microarrays. *American Journal of Physiology-Gastrointestinal and Liver Physiology* 282(3):G397–G402.

Arya, M., Shergill, I. S., Williamson, M., Gommersall, L., Arya, N., and Patel, H, R. 2005. Basic principles of real-time quantitative PCR. *Expert review of Molecular Diagnostics* 5(2):209–219.

Baldwin, D., Crane V. and Rice D. 1999. "A comparison of gel-based, nylon filter and microarray techniques to detect differential RNA expression in plants." *Current opinion in plant biology* 2, (2):96–103.

Battaglia, M., Pedrazzoli, P., Palermo, B., Lanza, A., Bertolini, F., Gibelli, N., Da Prada, G.A., Zambelli, A., Perotti, C. and dellaCuna, G. R. 1998. Epithelial tumour cell detection and theunsolved problems of nested RT-PCR: a new sensitive one step method without false positive results. *Bone Marrow Transplantation* 22(7):693–698.

Borg, I., Rohde, G., Löseke, S., Bittscheidt, J., Schultze-Werninghaus, G., Stephan, V., and Bufe, A. 2003. Evaluation of a quantitative real-time PCR for the detection of respiratory syncytial virus in pulmonary diseases. *European Respiratory Journal* 21(6):944–951.

Bustin, S. A. 2000. Absolute quantification of mRNA using real-time reverse transcription polymerase chain reaction assays. *Journal of Molecular Endocrinology* 25(2):169–193.

Bustin, S. A. 2002. Quantification of mRNA using real-time RT-PCR. Trends and problems. *Journal of Molecular Endocrinology* 29(1):23–39.

Bustin, S. A., and Nolan, T. 2004. Pitfalls of quantitative reverse transcription polymerase chain reaction. *Journal of Biomolecular Techniques: JBT* 15(3):155–166.

Cardullo, R. A., Agrawal, S., Flores, C., Zamecnik, P. C., and Wolf, D. E. 1988. Detection of nucleic acid hybridization by nonradiative fluorescence resonance energy transfer. *Proceedings of the National Academy of Sciences* 85(23):8790–8794.

Castelain, S., Descamps, V., Thibault, V., François, C., Bonte, D., Morel, V., Izopet, J., Capron, D., Zawadzki, P., and Duverlie, G. 2004. TaqMan amplification system with an internal positive control for HCV RNA quantitation. *Journal of Clinical Virology* 31(3):227–234.

Chicurel, M. E., and Dalma-Weiszhausz, D. D. 2002. Microarrays in pharmacogenomics – advances and future promise. *Pharmacogenomic* 3(5):589–601.

Crick, F. 1970. "Central dogma of molecular biology." *Nature* 227, (5258):561–563.

Dumur, C. I., Dechsukhum, C., Wilkinson, D. S., Garrett, C. T., Ware, J. L., and Ferreira-Gonzalez, A. 2002. Analytical validation of a real-time reverse transcription – polymerase chain reaction quantitation of different transcripts of the Wilms' tumor suppressor gene (WT1). *Analytical Biochemistry* 309(1):127–136.

Englerblum, G., Meier, M., Frank, J. and Muller, G.A. 1993. "Reduction of background problems in nonradioactive Northern and Southern blot analyses enables higher sensitivity than 32P-based hybridizations." *Analytical biochemistry* 210(2):235–244.

Farr, S. and Dunn, R. T. 1999. 2nd. "Concise review: gene expression applied to toxicology." *Toxicological sciences: an official journal of the Society of Toxicology* 50(1):1–9.

Gall, J. G. and Pardue, M. L. 1969. "Formation and detection of RNA-DNA hybrid molecules in cytological preparations." *Proceedings of the National Academy of Sciences* 63(2):378–383.

Garibyan, L. and Avashia, N. 2013. "Research techniques made simple: polymerase chain reaction (PCR)." *The Journal of investigative dermatology* 133(3):e6.

Gilbert, S. F. 2000. "Early development of the nematode *Caenorhabditis elegans*." In *Developmental Biology 6th edition*. Sinauer Associates:428–431.

Ginzinger, D. G. 2002. Gene quantification using real-time quantitative PCR: An emerging technology hits the mainstream. *Experimental Hematology* 30(6):503–512.

Giulietti, A., Overbergh, L., Valckx, D., Decallonne, B., Bouillon, R., and Mathieu, C. 2001. An overview of real-time quantitative PCR: applications to quantify cytokine gene expression. *Methods* 25(4):386–401.

Harrington, C. A., Rosenow, C., and Retief, J. 2000. Monitoring gene expression using DNA microarrays. *Current opinion in Microbiology* 3(3):285–291.

Heid, C. A., Stevens, J., Livak, K. J., and Williams, P. M. 1996. Real time quantitative PCR. *Genome Research* 6(10):986–994.

Higuchi, R., Dollinger, G., Walsh, P.S., and Griffith, R. 1992. Simultaneous amplification and detection of specific DNA sequences. *Biotechnology* 10(4):413–417.

Higuchi, R., Fockler, C., Dollinger, G., and Watson, R. 1993. Kinetic PCR analysis: Real-time monitoring of DNA amplification reactions. *Biotechnology* 11(9):1026–1030.

Hirsch, E. D., Joseph F. K. and James S. 2002. Trefil. *The new dictionary of cultural literacy*. Houghton Mifflin Harcourt.

Houghton, S. G., and Cockerill, F. R. 2006. Real-time PCR: overview and applications. *Surgery* 139(1):1–5.

Huggett, J., Dheda, K., Bustin, S., and Zumla, A. 2005. Real-time RT-PCR normalisation; Strategies and considerations. *Genes & Immunity* 6(4):279–284.

Jensen, E. 2014. "Technical review: In situ hybridization." *The Anatomical Record* 297(8): 1349–1353.

Johannsen, W. 1909. *Elemente der exakten Erblichkeitslehre*. Fischer.

Jurado, J., Prieto-Álamo, M. J., Madrid-Rísquez, J., and Pueyo, C. 2003. Absolute gene expression patterns of thioredoxin and glutaredoxin redox systems in mouse. *Journal of Biological Chemistry* 278(46):45546–45554.
Kafert, S., Krauter, J., Ganser, A., and Eder, M. 1999. Differential quantitation of alternatively spliced messenger RNAs using isoform-specific real-time RT-PCR. *Analytical Biochemistry* 269(1):210–213.
Karakach, T. K., Flight, R. M., Douglas, S. E., and Wentzell, P. D. 2010. An introduction to DNA microarrays for gene expression analysis. *Chemometrics and Intelligent Laboratory Systems* 104(1):28–52.
Kevil, C. G., Walsh, L., Laroux, F. S., Kalogeris, T., Grisham, M. B. and Alexander, J. S. 1997. "An improved, rapid Northern protocol." *Biochemical and biophysical research communications* 238(2):277–279.
Khan, J., Wei, J. S., Ringner, M., Saal, L. H., Ladanyi, M., Westermann, F., Berthold, F., Schwab, M., Antonescu, C. R., Peterson, C., and Meltzer, P. S. 2001. Classification and diagnostic prediction of cancers using gene expression profiling and artificial neural networks. *Nature Medicine* 7(6):673–679.
King, H. C., and Sinha, A. A. 2001. Gene expression profile analysis by DNA microarrays: Promise and pitfalls. *Jama* 286(18):2280–2288.
Klein, D. 2002. Quantification using real-time PCR technology: Applications and limitations. *Trends in Molecular Medicine* 8(6):257–260.
Kuhne, B. S., and Oschmann, P. 2002. Quantitative real-time PCR using hybridization probes and imported standard curves for cytokine gene expression analysis. *Biotechniques* 33(5):1078–1084.
Lee, L. G., Connell, C. R., and Bloch, W. 1993. Allelic discrimination by nick-translation PCR with fluorogenic probes. *Nucleic Acids Research* 21(16):3761–3766.
Lehmann, R. and Tautz, D. 1994. "In situ hybridization to RNA." *Methods in cell biology* 44:575–598.
Liang, P. and Pardee, A. P. 1995. "Recent advances in differential display." *Current opinion in immunology* 7(2):274–280.
Lin, J. C., Wang, W. Y., Chen, K. Y., Wei, Y. H., Liang, W. M., Jan, J. S., and Jiang, R. S. 2004 Quantification of plasma Epstein – Barr virus DNA in patients with advanced nasopharyngeal carcinoma. *New England Journal of Medicine* 350(24):2461–2470.
Livak, K. J. 1999. Allelic discrimination using fluorogenic probes and the 5' nuclease assay. *Genetic Analysis* 14(5–6):143–149.
Lopez, A. J. 1998. Alternative splicing of pre-mRNA: Developmental consequences and mechanisms of regulation. *Annual Review of Genetics* 32(1):279–305.
Mannhalter, C., Koizar, D., and Mitterbauer, G. 2000. Evaluation of RNA isolation methods and reference genes for RT-PCR analyses of rare target RNA. *Clinical Chemistry and Laboratory Medicine* 38(2):171–177.
McCulloch, R. K., Choong, C. S. and Hurley, D. M. 1995. "An evaluation of competitor type and size for use in the determination of mRNA by competitive PCR." *Genome Research* 4(4):219–226.
McKusick, V. A. 1997. "Mapping the human genome: retrospective, perspective and prospective." *Proceedings of the American Philosophical Society* 141(4): 417–424.
Mocellin, S., Rossi, C. R., Pilati, P., Nitti, D., and Marincola, F. M. 2003. Quantitative real-time PCR: A powerful ally in cancer research. *Trends in Molecular Medicine* 9(5):189–195.
https://www.ncbi.nlm.nih.gov/pubmed/
Mori, H., Takeda-Yoshikawa, Y., Hara-Nishimura, I. and Nishimura, M. 1991. "Pumpkin malate synthase: cloning and sequencing of the cDNA and Northern blot analysis." *European journal of biochemistry* 197(2):331–336.
Mullis, K. B. 1990. "The unusual origin of the polymerase chain reaction Scientific American.":56–65.
O'Connor, C. 2008. "Fluorescence in situ hybridization (FISH)." *Nature Education* 1(1): 171.
Orakpoghenor, O. and Markus, T. P. 2020. "Diagnostic Techniques in Molecular Biology–An Overview." *MedRead Journal of Family Medicine* 1(2):1008.
Overbergh, L., Giulietti, A., Valckx, D., Decallonne, B., Bouillon, R., and Mathieu, C. 2003. The use of real-time reverse transcriptase PCR for the quantification of cytokine gene expression. *Journal of Biomolecular Techniques: JBT* 14(1):33.
Overbergh, L., Valckx, D., Waer, M., and Mathieu, C. 1999. Quantification of murine cytokine mRNAs using real time quantitative reverse transcriptase PCR. *Cytokine* 11(4):305–312.
Patel, L., Charlton, S. J., Chambers, J. K., and Macphee, C. H. 2001. Expression and functional analysis of chemokine receptors in human peripheral blood leukocyte populations. *Cytokine* 14(1):27–36.
Phillips, S. J. and Dudík, M. 2008. "Modeling of species distributions with Maxent: new extensions and a comprehensive evaluation." *Ecography* 31(2):161–175.

Pirrung, M. C. 2002. How to make a DNA chip. *Angewandte Chemie International Edition* 41(8):1276–1289.
Qu, Y. and Boutjdir, M. 2007. "RNase protection assay for quantifying gene expression levels." In *Cardiac Gene Expression* 145–158. Humana Press.
Rabinow, P. 2011. *Making PCR: A story of biotechnology*. University of Chicago Press.
Raeymaekers, L. 1993. "Quantitative PCR: theoretical considerations with practical implications." *Analytical biochemistry* 214(2):582–585.
Ririe, K. M., Rasmussen, R. P., and Wittwer, C. T. 1997. Product differentiation by analysis of DNA melting curves during the polymerase chain reaction. *Analytical Biochemistry* 245:154–160.
Sambrook, J., Fritsch, E. F. and Maniatis, T. 1989. *Molecular cloning: a laboratory manual*. No. Ed. 2. Cold spring harbor laboratory press.
Schmittgen, T. D., and Livak, K. J. 2008. Analyzing real-time PCR data by the comparative CT method. *Nature Protocols* 3(6):1101–1108.
Schmittgen, T. D., Zakrajsek, B. A., Mills, A. G., Gorn, V., Singer, M. J., and Reed, M. W. 2000. Quantitative reverse transcription-polymerase chain reaction to study mRNA decay: Comparison of endpoint and real-time methods. *Analytical Biochemistry* 285(2):194–204.
Sloan, L. M., Uhl, J. R., Vetter, E. A., Schleck, C. D., Harmsen, W. S., Manahan, J., Thompson, R. L., Rosenblatt, J. E., and Cockerill III, F. R. 2004. Comparison of the Roche LightCyclervanA/vanB detection assay and culture for detection of vancomycin-resistant enterococci from perianal swabs. *Journal of Clinical Microbiology* 42(6):2636–2643.
Streit, S., Michalski, C. W., Erkan, M., Kleef, J. and Friess, H. 2009. "Seabuckthorn: A Fruit for Future." *Nature Protocols* 4(1):37–43.
Tang, Y. W., Stratton, C. W. and Tang, Y. W. 2013. *Advanced techniques in diagnostic microbiology*. New York: Springer.
Trayhurn, P. 1996. "Northern blotting." *Proceedings of the Nutrition Society* 55(1B):583–589.
Tsai, S. J. and Wiltbank, M. C. 1996. "Quantification of mRNA using competitive RTPCR with standard-curve methodology." *Biotechniques* 21(5):862–866.
Uhl, J. R., Bell, C. A., Sloan, L. M., Espy, M. J., Smith, T. F., Rosenblatt, J. E., and Cockerill III, F. R. 2002. Application of rapid-cycle real-time polymerase chain reaction for the detection of microbial pathogens: The Mayo-Roche rapid anthrax test. *Mayo Clinic Proceedings* 77(7):673–680.
Uhl, J. R., and Cockerill, F. R. 2004. The fluorescence resonance energy transfer system. In *Molecular Microbiology: Diagnostic Principles and Practice*, ed. D. H. Persing, 295–306. Washington, DC: ASM Press.
Valasek, M. A., and Repa, J. J. 2005. The power of real-time PCR. *Advances in Physiology Education* 29(3):151–159.
Wacker, M. J., and Godard, M. P. 2005. Analysis of one-step and two-step real-time RT-PCR using SuperScript III. *Journal of Biomolecular Techniques: JBT* 16(3):266.
Wang, Z., and Spadoro, J. 1998. Determination of target copy number of quantitative standards used in PCR-based diagnostic assays. In *Gene Quantification*, ed. F. Ferré, 31–43. Boston, MA: Birkhäuser.
Waring, J. F., and Halbert, D. N. 2002. The promise of toxicogenomics. *Current Opinion in Molecular Therapeutics* 4(3):229–235.
Wilhelm, J., and Pingoud, A. 2003. Real-time polymerase chain reaction. *Chembiochem* 4(11):1120–1128.
Wong, M. L., and Medrano, J. F. 2005. Real-time PCR for mRNA quantitation. *Biotechniques* 39(1):75–85.
Xu, L., & Seki, M. 2020. Recent advances in the detection of base modifications using the Nanopore sequencer. *Journal of human genetics*, 65(1):25–33.
Zentilin, L. and Giacca, M. 2007. "Competitive PCR for precise nucleic acid quantification." *Nature protocols* 2(9):2092–2104.
Zentilin, L. and Giacca, M. 2010. "The renaissance of competitive PCR as an accurate tool for precise nucleic acid quantification." In *RT-PCR Protocols* 233–248. Humana Press, Totowa, NJ.
Zhou, Y. H., Zhang, X. P. and Ebright, R.H. 1991. "Random mutagenesis of gene-sized DNA molecules by use of PCR with Taq DNA polymerase." *Nucleic acids research* 19(21):6052.
Zhu, H., Zhang, H., Xu, Y., Laššáková, S., Korabečná, M. and Neužil, P. 2020. "PCR past, present and future." *BioTechniques* 69(4):317–325.

5 Concept of Cre-Lox Recombination System

Ravneet Kaur

CONTENTS

5.1 INTRODUCTION

Cre-lox system is an invaluable recombination technique that is site-specific to modify DNA (in vivo) for genetic manipulation. This allows the regulation of gene expression where the gene knock

DOI: 10.1201/9781003220831-5

system may turn lethal or if gene expression gets unchecked. The Cre-lox system has two functional components:

- A Cre recombinase enzyme
- LoxP site, the recognition site for Cre recombinase enzyme

5.2 A CRE RECOMBINASE

This enzyme identifies and recombines a pair of short-target sequences called the lox sequences. Both the components, i.e. Cre enzyme and the loxP sequence have been derived from bacteriophage P1.Cre recombinase, 38 kDa in size, which belongs to the integrase family that catalyses the site-specific recombination event between two loxP sites (DNA recognition sites). It is a tyrosine recombinase that utilizes a topoisomerase I-like mechanism to carry out the reaction and requires no additional proteins or cofactors (such as ATP) to function.

Structurally, Cre recombinase consists of 343 amino acids arranged into two distinct domains: the amino terminal domain and the carboxy terminal domain.

The amino terminal domain includes 20–129 residues and contains 5 alpha helical segments (A, B, C, D, E) linked by a series of short loops. Of these, A and E helices help in the formation of the recombinase tetramer with the C terminal region of helix E that forms contacts with the C terminal domain of adjacent subunits. B and D helices form direct contacts with the major groove of the loxP DNA.

The carboxy terminal domain of the enzyme is composed of 132–341 amino acids and it occupies the active site of the enzyme. The structure of this domain resembles the catalytic domain of other enzymes of the integrase family. This domain has nine distinct helical structures (F–N). The terminal N helix projects from the main body of the carboxy domain. It plays a key role in mediating interactions with other subunits. Studies on crystal structures have demonstrated that terminal N helix buries its hydrophobic surface into the acceptor pocket of an adjacent Cre subunit.

5.2.1 The Active Site of Cre Recombinase Enzyme Contains Two Types of Residues

1. A conserved catalytic residue that contains a triad of three amino acids: Arginine 173, Histidine 289, Arginine 292. All three residues are found in a single subunit of the enzyme.
2. Nucleophilic residues contain two residues: Tyrosine 324 and Tryptophan 315. Tyrosine 324 forms a covalent 3′-phosphotyrosine linkage to the DNA substrate that is coordinated by the side chains of the three amino acid residues of the catalytic triplet (Arginine 173, Histidine 289 & Tryptophan 315).

5.2.2 Methods for Introducing Cre Recombinase

There are two main ways to get Cre recombinase to be expressed in a cell. The first is to use a virus to introduce the gene for re-recombinase into the cell. The other method is to generate a transgenic mouse that inserts the gene for Cre recombinase after a specific promoter. Because this promoter can be expressed in certain cell types, the expression of Cre recombinase can also be limited to these cells.

5.3 A RECOGNITION SITE, LOXP

LoxP sites are present in P1 bacteriophage, formed of 34 bp sequences with two recognition sites of 13 bp separated by an asymmetric spacer region of 8 bp. The spacer region is variable except for the middle two bases in between two sets of symmetric 13 bp sequences. The loxP sequence is directional because of 13 bp sequences being palindromic and not 8 bp spacer region. LoxP sites are always used in pairs, often flanking a gene of interest or a reporter gene (Table 5.1).

5.4 MECHANISM OF CRE-LOX RECOMBINATION (TABLE 5.2)

The outcome of recombination is determined by the orientation of loxP sites. There are three outcomes that can be achieved via this technology.

The binding of 2 Cre molecules with a single loxP site contains two active sites (Figure 5.1). A synapse tetramer is formed when two Cre-loxP complexes join with four distinct active sites. In this tetramer, the indole nitrogen of tryptophan 315 forms a hydrogen bond with the phosphate that cleaves the DNA, thus freeing 5′ hydroxyl group. The reaction is catalyzed in the active site of two out of the four recombinase subunits present at the synapse tetramer. A holiday junction

TABLE 5.1
LoxP Site Sequence, Made Up of Two Flanking Palindromic Recognition Regions of 13 bp and a Spacer Region of 8 bp Giving It Directionality (*Source*: Internet)

13 bp Recognition Region	8 bp Spacer	13 bp Recognition Region
ATAACTTCGTATA	ATGTATGC	TATACGAAGTTAT

TABLE 5.2
Showing Steps for Cre-loxP Recombination

Step 1	The 2 Cre proteins first recognize a loxP site and bind to it, thus forming a dimer.
Step 2	The two Cre-loxP dimers join to form a tetramer. The two loxP sites come together with opposite directionality.
Step 3	Finally, cleavage of double-stranded DNA occurs in the center of the loxP site and crossing over occurs.

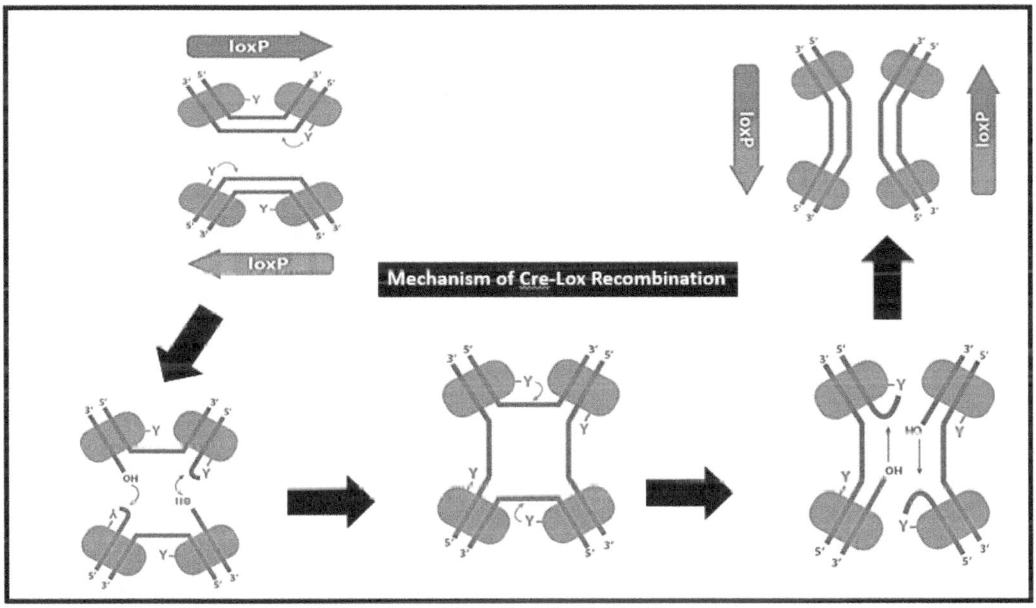

FIGURE 5.1 Mechanism of Cre-lox recombination.

intermediate is formed by the reaction of 5′ hydroxyl groups that attack the 3′-phosphotyrosine linkage one pair of the DNA strands.

1. **Excision:** If the two loxP sites are in the same orientation, the sequence flanked by two loxP sites is excised or removed from its original locus. This outcome is essentially irreversible (Figure 5.2).

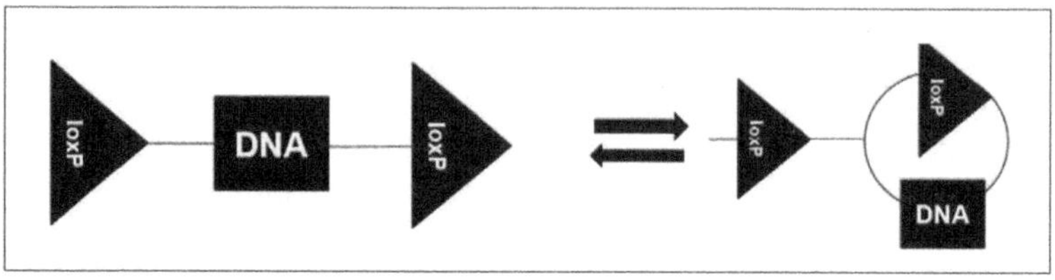

FIGURE 5.2 Excision of DNA sequence.

2. **Inversion:** If the two loxP sites are in the opposite orientation, the sequence flanked by two loxP sites is inverted or reversed. This result is reversible as the lox sites remain unchanged. Therefore, the gene can flip back and forth between both orientations indeterminately (Figure 5.3).

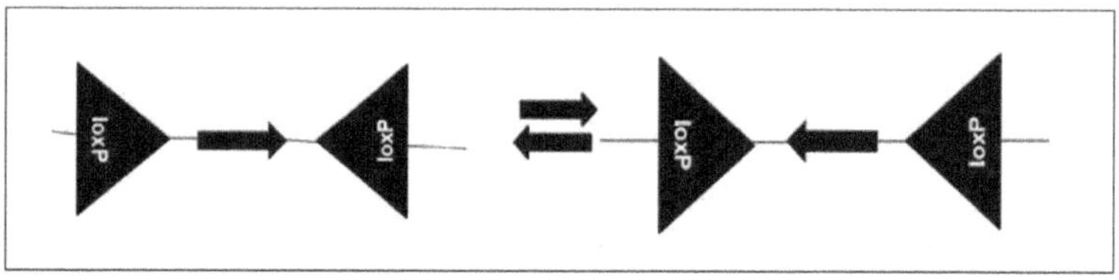

FIGURE 5.3 Inversion of DNA sequence.

3. **Translocation:** If the two lox sites are present on separate segments of DNA, the recombination results in the translocation that is reversible. This phenomenon is called recombinase-mediated cassette exchange, which is a time-saving method for genetic manipulation (Figure 5.4).

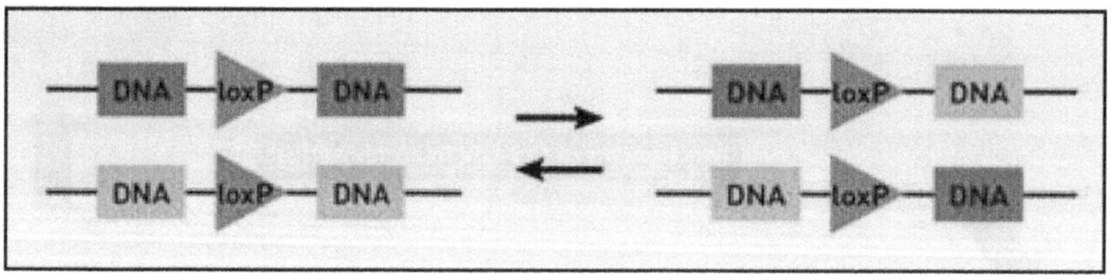

FIGURE 5.4 Translocation of DNA sequence.

5.5 APPLICATIONS OF THE TECHNOLOGY

5.5.1 Gene Alterations

The system has been created to carry out various gene alterations like deletions, insertions, inversions and translocation at specific sites in the DNA. The modification is site specific and triggered by external stimulus. It has been implemented in prokaryotes and eukaryotes (Figure 5.5).

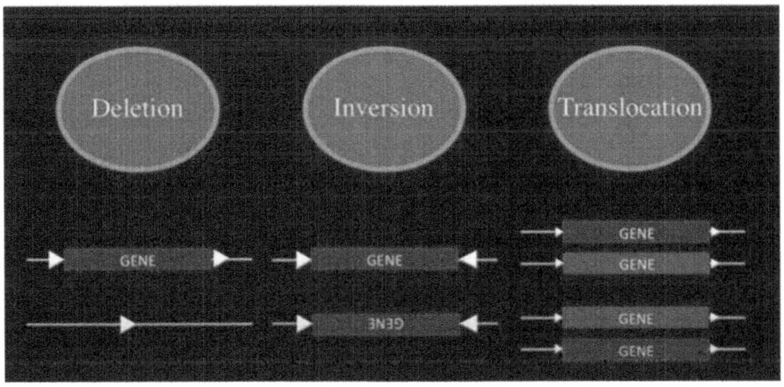

FIGURE 5.5 Outcomes of recombination by Cre-loxP technology (deletion, inversion and translocation by Cre-lox system).

5.5.2 Lineage Analysis

One example of lineage analysis was used by Melton lab in determining the origin of adult pancreatic beta cells (Ma et al. 2018). They used an inducible version of Cre to determine whether new beta cells were derived from stem cells or preexisting beta cells. In their experiment, transgenic mice were generated in which the insulin promoter drove the expression of tamoxifen-dependent cre recombinase. The enzyme then existed in the cytosol as a complex with the estrogen receptor and het shock protein. When tamoxifen was injected into the cell, the CreER complex separated, and the Cre recombinase entered the nucleus to perform recombination. Once in the nucleus, the Cre recombinase cut out the lacZ stop sequence, and thus allowed the expression of HPAP label, short for human placental alkaline phosphatase. Through this process, cells that normally produce insulin, namely the pancreatic beta cells, were tagged with the HPAP label. Using these mice, the Melton lab performed a pulse chase experiment, the pulse being the set of cells labeled by HPAP using cre recombinase in the pancreas, and the 'Chase' being the set of cells that had the label after a given amount of time. Starting with a population of fully labeled beta cells, they hypothesized that if new beta cells arose from preexisting ones, the percentage of stained cells wouldn't change. However, if they developed from unlabeled stem cells, the new population would have a much smaller percentage of labeled cells. The data showed that the percentage of labeled cells remained constant, which suggested that new beta cells were derived from preexisting ones rather than from stem cells.

5.5.3 Genome Transformation in *B. Subtilis*

This method inactivated a specific gene. The genome of *Bacillus subtilis* was engineered by fusion using PCR of the two flanking homologous regions with an antibiotic resistance gene, bordered by two mutant lox sites, i.e. lox71 and lox66. The results of PCR were used directly to transform *Bacillus subtilis*. The expression of transient enzyme (Cre recombinase) in these transformants recombined lox71 and lox66 sites into a double-mutant lox72 site, thereby excising the marker gene (Yan et al. 2008).

5.5.4 Recycling of Antibiotic Marker

This system has proved to be very important for many bacterial species that lack sophisticated genetic tools. The experiment was conducted on two different bacterial strains, i.e. Methylobacterium extorquens AM1 and *Burkholderia fungorum* LB400 to generate unmarked deletions of genes. The unmarked mutants were created and this marker recycling allowed the generation of such bacterial strains that showed multiple genome manipulations in those organisms for which few antibiotic markers are currently available in the market (Marx & Lidstrom 2002).

5.5.5 Expression of Prokaryotic Cre-Lox Site-Specific Recombination System in a Eukaryote

The prokaryotic Cre-lox site-specific recombination system of coliphage P1 was shown to function in an efficient manner in a eucaryote, the yeast Saccharomyces cerevisiae. The Cre gene (site-specific recombinase) was placed under control of the yeast GALI promoter. Lox sites flanking the LEU2 gene were integrated into two different chromosomes in both orientations. The loss of LEU2 gene was measured by excisive recombination at the lox sites by the Cre protein that was dependent upon induction by galactose. The results demonstrated that a prokaryotic recombinase can enter a eukaryotic nucleus and, moreover, that the ability of the Cre recombinase to perform precise recombination events on the chromosomes of S. cerevisiae is unimpaired by chromatin structure (Sauer 1987).

5.5.6 Recombination Footprint in Transgenic Rice

The Cre gene under the control of soybean heat-shock promoter proved to be an effective tool for conditional removal of the marker gene in rice. A Cre gene was inserted into the rice genome with the recombination target (lox) construct. Only one line out of six transgenic lines contained a single copy of the Cre gene that maintained the excision of marker gene in the regenerated plants. Also, the production of marker-free progeny indicated the inheritance of the recombination footprint (Khattri et al. 2011).

5.5.7 Multiple Gene Deletions in Lactobacillus Plantarum

Cre-lox-based system was used for multiple gene deletions in Lactobacillus plantarum that has been adapted to be used in Gram-positive bacteria. An effective mutagenesis vector (pNZ5319) allowed selection of double-crossover mutants based on their antibiotic resistance phenotype. The resulting mutants showed the replacement of target gene a lox66-P32-cat-lox71 cassette, where lox66 and lox71 are mutant variants of loxP and P32-cat represents chloramphenicol resistance cassette. In double-crossover mutants, the expression of transient Cre recombinase leads to recombination of the lox66-P32-cat-lox71 cassette into a double-mutant loxP site, named lox72. This method strongly reduced recognition by Cre protein (Lambert et al. 2007).

5.5.8 Construction of Modular Tandem Expression Vectors for the Green Alga Chlamydomonas Reinhardtii

The method construct tandem vectors expressing Renilla luciferase and a metabolic selection marker (ARG7) by in vitro fusion of two plasmids. Linearized tandem vectors increased the co-expression frequency by 77% as compared with standard cotransformation procedures. An arginine auxotrophic Chlamydomonas strain was transformed using linear and supercoiled plasmids. The co-expression rates and levels of luciferase activity were monitored using different promoters (HSP70A, LHCBl, PSAD, chimeric HSP70A/RBCS2). The results indicated that the artificial HSP70A/RRBCS2 promoter was the most active for luciferase expressions. The expression was enhanced by heatshock protein in about 69% of all luminescent transformants carrying the HSP70A promoter luciferase, indicating physiological promoter function in a transgenic construct (Heitzer & Zschoernig 2007).

5.5.9 Gene Excision in Transgenic Maize

The Cre-lox system was tested from bacteriophage P1 for its ability to precisely excise stable integrated marker genes from chromosomes in transgenic maize plants by using two strategies, i.e. crossing and autoexcision (Zhang et al. 2003).

1. **Crossing strategy:** In this technique, plants that expressed the Cre recombinase are crossed with plants bearing a transgene construct in which the selectable marker gene is flanked by directly repeated lox sites. The experiments showed complete excision of somatic and germline marker gene in the F1 plants from most crosses with multiple independent Cre and lox lines.
2. **Autoexcision strategy:** In this strategy, the Cre gene is worked under the control of a heat shock-inducible promoter, which is then excised along with the nptII marker gene. The results indicated that if transient heat shock treatment is provided to primary transgenic callus, it is sufficient to induce Cre and for excising the Cre and nptII genes. The precise, stable and complete removal of marker gene was confirmed by genetic segregation and molecular analysis.

5.5.10 Study of Muscle Cell Dedifferentiation after Skeletal Muscle Injury of Mice with a Cre-Lox System

Based on the activation of the marker gene, i.e. β-galactosidase, a Cre-lox-β-galactosidase system was developed that specifically tagged differentiated multinuclear myotubes and myotube-generated mononuclear cells in an adult mouse line. It was observed that on muscle injury, the system generated β-galactosidase positive mononuclear cells from β-galactosidase positive multinuclear myofibers. It was also seen that these mononuclear cells diverged into different types of muscle cell lineages, i.e. myoblasts, muscle derived stem cells, satellite cells etc. (Mu et al. 2011).

5.5.11 Conferring Resistance against Pests

Two binary vectors were created – one containing the insecticidal gene Allium sativum leaf agglutinin (ASAL), and a selectable nptII marker gene cassette flanked by lox sites; second with the chimeric Cre gene construct. These vectors were used independently to transform tobacco plants. Each of the T0 lox plants was crossed with T0 Cre plants. PCR analyses and the sequencing of the target T-DNA part of the hybrid T1 plants showed the deletion of the nptII gene in a stable and precised manner in some T1 hybrid lines that corresponded to about 19.2% in the hybrids. The crossing strategy using Cre-lox system to excise marker genes was proved to be very effective, as marker-free plants expressed ASAL efficiently and also there was decrease in the survivability of Myzus persiceae, which is known to be the deadly pest of tobacco as compared to the control tobacco plants. The results of PCR and southern blotting also revealed the excision of Cre and nptII genes from some of the T2 plants (Chakraborti et al. 2008).

5.5.12 Site-Specific Integration of Agrobacterium T-DNA

The Agrobacterium-mediated transformation have been used as a tool to obtain site-specific integration using Cre-lox system at a lox site in the genome of *Arabidopsis thaliana*. Site-specific recombinants were preferentially selected by activation of a target gene nptII (silent lox-neomycin phosphotransferase). The T-DNA vectors harboring a single lox sequence or two lox sequences were used to analyze the effectiveness of Agrobacterium-mediated transfer. The cotransformation approach was used to control the reversibility of the integration reaction for Cre recombinase (Vergunst & Hooykaas 1998).

5.5.13 As Therapeutics

The Cre-lox technique is a powerful tool for targeting therapeutic effectors to control various human disorders. The system has shown promise against Wnt-positive colorectal cancer cell lines (Bordonaro 2009).

5.5.14 Co-Expression of Multiple Reporters

Double-floxed inverse ORF (DIO) vectors are used for Cre-inducible expression for the gene of interest. The DIO vector system allows the expression of the target gene only in cells or tissues that also express the Cre recombinase protein. This allows tissue-specific gene expression. This combined system is also called Cre-ON or Cre-Switch. Also, DIO vectors enable to control co-expression of multiple reporters in which reporter vector is co-transfected with high molar ratio, with addition of a limiting amount of cre-expressing factor (Villa et al. 2016) (Figure 5.6).

5.5.15 Studying Functional Mutations

The lox sites can be placed at different positions to activate one gene and simultaneous excision of the other gene. The system easily replaces the wild gene with a mutated version. Since the traditional methods of gene knock out have proved to be quite lethal, Cre-lox system is an excellent method to study important genes and their functionality.

5.5.16 Genetic Fate Mapping Experiments

The site-specific Cre recombinases adapted from the bacteriophage P have also been used in mice for genetic fate mapping (Austin et al. 1981). The method provides a direct comparison between wild-type and mutant cells behaviors. But the system has some limitation, i.e. cell division is needed to create the recombination; secondly, labelin and mutagenesis can occur only in dividing cells expressing active Cre protein. However, the gene of interest that is mutated must be on the same chromosome that carries the marker (Legué & Joyner 2010).

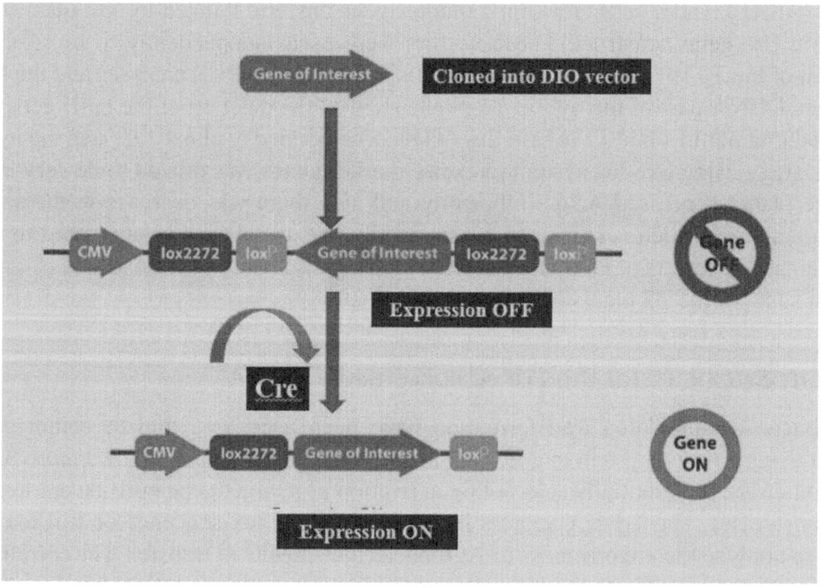

FIGURE 5.6 DIO vectors are also extremely useful for controlling co-expression of multiple reporters.

5.5.17 Cre-Induced Auto-Excision in Transgenic Rice

Excision has been achieved in transgenic rice using a Cre-loxP system under floral specific promoter (OsMADS45). The binary vector in combination with Cre recombinase was constructed and introduced into rice via Agrobacterium-mediated transformation. The transgenic rice plants were generated and analyzed by Southern blotting for auto-excision of a selectable marker gene (Bai et al. 2008).

5.6 LIMITATIONS OF THE CRE-LOXP TECHNIQUE

1. The recombination reaction can reverse back, making cassette exchange inefficient.
2. Sequence excision/deletion can happen in transform instead of in cis cassette exchange event. Therefore, loxP mutants are formed to avoid these problems.
3. The Cre-lox system is known to produce numerous problems that affect various physiological parameters. These changes are unexploited when proper control strains are not used.
4. Since the promoter is the most critical component of this system, promoter-driven Cre transgenic animals must be analyzed several times and tested critically.
5. The expression of a transgene is determined by integration site into the chromosomal locus. But this integration varies between the lines, and even between offspring produced from the same line.
6. All the cells targeted may not express Cre recombinase. This irregular phenomenon is known as variegation.
7. The right location for insertion of loxP into target gene sequence is a strong concern.
8. It should be made sure that insertion of loxP sites should not affect the normal expression of the target gene. This must be checked prior to Cre-mediated recombination.
9. LoxP sites should be inserted between the coding genes (exons) that are critical for the biological activity of the protein.
10. Sometimes, Cre-mediated recombination may generate biologically active truncated protein that makes the data inconclusive and needs careful interpretation.
11. The interpretation may be misleading if target gene is inactivated during embryogenesis and early phenotypic changes are still observed.

5.7 CONCLUSION

There are many applications of Cre recombinases, one of which is lineage analysis, which is the determination of the origin or fate of cells. Thus the Cre-lox system can be used for numerous types of modifications. Because it can be designed in many ways, it has been used in a variety of different research studies that involve many different tissue types and organ systems. Perhaps in the future, this technology can also be used for cures and treatments in humans. It has already been demonstrated that Cre-lox functions in human embryonic stem cells. There is definitely great potential for this system, not only in research but also in clinical settings.

BIBLIOGRAPHY

Austin S et al. (1981). A novel role for site-specific recombination in maintenance of bacterial replicons. *Cell* 25: 729–736.

Bai X, Wang Q and Chu C (2008). Excision of a selective marker in transgenic rice using a novel Cre/loxP system controlled by a floral specific promoter. *Transgenic Res* 17: 1035–1043. https://doi.org/10.1007/s11248-008-9182-7

Bordonaro M (2009). Modular Cre/lox system and genetic therapeutics for colorectal cancer. *Journal of Biomedicine & Biotechnology*: 358230. https://doi.org/10.1155/2009/358230

Chakraborti D, Sarkar A, Mondal HA, Schuermann D, Hohn B, Sarmah BK and Das S (2008). Cre/lox system to develop selectable marker free transgenic tobacco plants conferring resistance against sap sucking homopteran insect. *Plant Cell Reports* 27(10): 1623–1633. https://doi.org/10.1007/s00299-008-0585-y

Heitzer M and Zschoernig B (2007). Construction of modular tandem expression vectors for the green alga Chlamydomonas reinhardtii using the Cre/lox-system. *BioTechniques* 43(3): 324–326. https://doi.org/10.2144/000112556.

Khattri A, Nandy S and Srivastava V (2011). Heat-inducible Cre-lox system for marker excision in transgenic rice. *Journal of Biosciences* 36: 37–42. https://doi.org/10.1007/s12038-011-9010-8

Lambert JM, Bongers RS and Kleerebezem M (2007). Cre-lox-based system for multiple gene deletions and selectable-marker removal in Lactobacillus plantarum. *Applied and Environmental Microbiology* 73(4): 1126–1135. https://doi.org/10.1128/AEM.01473-06

Legué E and Joyner AL (2010). Genetic fate mapping using site-specific recombinases. *Methods in Enzymology* 477: 153–181. https://doi.org/10.1016/S0076-6879(10)77010-5

Ma H, Wert KJ, Shvartsman D, Melton DA and Jaenisch R (2018). Establishment of human pluripotent stem cell-derived pancreatic β-like cells in the mouse pancreas. *Proceedings of the National Academy of Sciences of the United States of America* 115(15): 3924–3929. PUBMED: 29599125. https://doi.org/10.1073/PNAS.1702059115

Marx CJ and Lidstrom ME (2002). Broad-host-range cre-lox system for antibiotic marker recycling in gram-negative bacteria. *BioTechniques* 33(5): 1062–1067. https://doi.org/10.2144/02335rr01

Mu X, Peng H, Pan H, Huard J and Li Y (2011). Study of muscle cell dedifferentiation after skeletal muscle injury of mice with a Cre-Lox system. *PloS One* 6(2): e16699. https://doi.org/10.1371/journal.pone.0016699

Sauer B (1987). Functional expression of the cre-lox site-specific recombination system in the yeast Saccharomyces cerevisiae. *Molecular and Cellular Biology* 7(6): 2087–2096. https://doi.org/10.1128/mcb.7.6.2087-2096.1987

Vergunst AC and Hooykaas PJ (1998). Cre/lox-mediated site-specific integration of Agrobacterium T-DNA in Arabidopsis thaliana by transient expression of cre. *Plant Molecular Biology* 38(3): 393–406. https://doi.org/10.1023/a:1006024500008

Villa KL, Berry KP, Subramanian J, Cha JW, Oh WC, Kwon HB, Kubota Y, So PT and Nedivi E (2016). Inhibitory synapses are repeatedly assembled and removed at persistent sites in vivo. *Neuron* 89(4): 756–769. https://doi.org/10.1016/j.neuron.2016.01.010

Yan X, Yu HJ, Hong Q and Li SP (2008). Cre/lox system and PCR-based genome engineering in Bacillus subtilis. *Applied and Environmental Microbiology* 74(17): 5556–5562. https://doi.org/10.1128/AEM.01156-08

Zhang W, Subbarao S, Addae P, Shen A, Armstrong C, Peschke V and Gilbertson L (2003). Cre/lox-mediated marker gene excision in transgenic maize (Zea mays L.) plants. TAG. Theoretical and applied genetics. *Theoretische und Angewandte Genetik* 107(7): 1157–1168. https://doi.org/10.1007/s00122-003-1368-z

6 Applications of Tag-SNPs in Quantitative Trait Loci (QTL) Identification

CS Mukhopadhyay and Bhawanpreet Kaur

CONTENTS

6.1 INTRODUCTION

Single nucleotide polymorphism (SNP) is the most common form of polymorphism in human and other eukaryotic genomes, although other variations like insertions and deletions, (INDELS) microsatellites, copy number variants, and epigenetic markers are vital to consider and can impact disease. SNP is the single nucleotide polymorphism occurring between DNA samples concerning a single base. Therefore, even if there is an insertion/deletion of a single base, those changes cannot be recognised as SNPs. This polymorphism is identified when two homologous DNA fragments have the same sequence except for a specific one point where it differs. For example, in Figure 6.1, two individuals have the following different sequences of nucleotide for an exon:

DOI: 10.1201/9781003220831-6

```
Seq_1 AGGGTTGGGGGCCTGGTCAGCCCCTGGCTGACCTTGCTCGTG 42
Seq_2 AGGGTTGGGAGCCTGGTCAGCCCCTGGCTGACCTTGCTCGTG 42
      ********* ********************************
```

FIGURE 6.1 Exon contains different nucleotide sequences.

In the example in Figure 6.1, position 10 (bold) has an SNP, showing the transition between Guanine (G) to Adenine (A). Figure 6.2 depicts the electropherogram representing the SNP ("G" to "A" transition at nucleotide position 10 of variants A and B) in the fourth intron of the TSPY gene of Indian water buffalo (Mukhopadhyay et al., 2011):

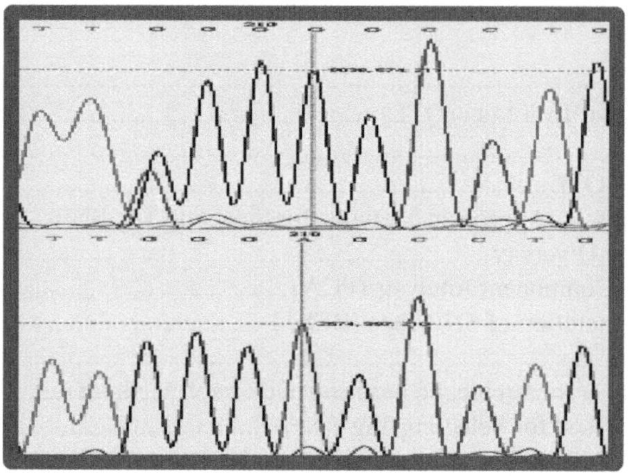

FIGURE 6.2 Electropherogram represents the SNP ("G" to "A" transition at nucleotide position 10 of variants A and B) detected by SSCP.

Some of the fundamental features of SNP are:

- Usually all common SNPs are bi-allelic (only two alleles).
- In genomic regions, SNPs are not homogenous but rather inconsistently heterogeneous.
- Besides natural selection, point mutation is the major contributing factor to the creation of SNP.
- SNPs are more frequently found within the noncoding regions as compared to the exons, since within the genome, survival or natural selection cannot operate in the noncoding region (Barreiro et al., 2008).

Throughout the genome, SNPs occur at a frequency of 1 in 1,000 bp. SNP markers allow to automatize and maximize tenfold the efficiency of genotype analysis. For studying complex genetic traits and understanding genome evolution, SNPs act as excellent markers. SNPs are usually bi-allelic, even though in principle any of the four (A, T, C, G) nucleotides can be present at any location in a stretch of DNA strand. This is due to the low mutation frequency that leads to new SNPs. According to their location in the genome, SNPs can be classified as synonymous (silent) SNP and non-synonymous (missense) SNP (nsSNPs). SNPs are the most ubiquitous genetic variations in the human genome. SNPs occur in the coding region and noncoding regions of the genome. Approximately 50% of SNPs lie in the noncoding regions named ncSNPs, 25% lie in the coding region called cSNPs, and 25% are the silent mutations, also named synonymous SNPs (sSNP; they do not change encoded amino acids), while, on the other hand, nonsynonymous SNPs (nSNPs, or nsSNP change-encoded amino acids) (Halushka et al., 1999) may produce pathology and are subject to natural selection. The low mutation rate and simplicity of SNPs also make them excellent markers for studying complex genetic traits and a tool for the understanding of genome evolution, analyzing genetic variations responsible for differential expression of several economically important traits and susceptibility to complex diseases. Due to lower mutation rates, SNPs are highly abundant and more stable than short tandem repeats (STRs). The basic genetic variations that occur at the genic level between two individuals can be tracked through single nucleotide polymorphisms (Collins et al., 1997). However, genome-wide scanning of SNPs and genotyping all the existing SNPs in experimental samples is time- and resource-intensive.

Genome-wide association studies (GWAS) have identified the quantitative trait loci (QTL) affecting many common complex diseases, and for dissecting the genetic basis of polygenic traits (McCarthy et al., 2008). The first ever GWAS was conducted on patients with age-related macular degeneration, which was published in 2005 and reported two SNPs between case and control individuals (Klein et al., 2005). Thereafter, several studies on GWAS (and using co-expression network analysis) in mice, humans, and other species have been reported (Luo, 2015; Laber et al., 2017; Zhang-James et al., 2019; Zhonggang et al., 2020).

6.1.1 Quantitative Trait Loci (QTL)

"QTL", aterm coined by Geldermann (1975), is a "genomic region that is strongly related to particular phenotypic trait effect, which varies in degree and which may be attributed to polygenic effects". Quantitative trait loci, or QTLs, are the regions in the genome in which genes link with a quantitative trait, and the process of developing linkage maps and operating QTL analysis is referred to as QTL mapping (Paterson et al., 1996). QTL analysis is based on the principle of identifying a connection between phenotype and genotype of markers. QTLs are usually found on different chromosomes. The genetic architecture of a trait is estimated by the number of QTLs, which explains the phenotypic trait variation. It may indicate that the height of a plant is controlled by many genes of small effect, or large effect by few genes. In addition, a single phenotypic trait is usually determined by many genes. Several QTL are associated with a single trait. It is also used to identify the candidate genes underlying a specific trait. Conceptually, a QTL can be a single gene, or it may be a cluster of linked genes that affect the trait. Different software are used for the QTL mapping. Table 6.1 shows some useful software for QTL mapping.

6.2 WHY QTL

QTL offers a direct channel to investigate the number of genes that influence a particular trait, to find out the exact location of the gene, and to know the effect of dosage of these genes on a variation of the trait. An urgent challenge of plant breeding is to integrate the recent innovations

TABLE 6.1
Different Types of SNPs and Their Functions

Types	Functions
cSNP (coding SNP)	SNP present within the coding region of a gene
ncSNP (noncoding SNP)	SNP present within a noncoding region
eSNP (expression SNP)	SNP that affects the gene expression
rSNP (regulatory SNP)	SNP present in a gene regulatory region
sSNP (synonymous or "silent" SNP)	SNP lies within a protein, codon that does not result in an amino acid change
nSNP, nsSNP (nonsynonymous SNP)	SNP present inside a protein, codon that results in an amino acid change

TABLE 6.1
Some Useful Software for QTL Mapping

Name	Functions
QGENE	Single-marker analysis (SMA), and multiple-trait analysis
MQTL	Simple Interval Mapping, QE interactions, and can perform permutation tests

in biology and genetics to enhance crop improvement. Most cloned QTL elements involved in gene regulation, candidate approach, comparing the position of plant phenotype QTLs, expression of genes QTLs (eQTLs), and network eQTLs may enhance chances to clone QTLs of agronomic traits.

6.3 FACTORS AFFECTING QTL

- Number of genes that control the target traits
- Heritability and position of the traits
- Type and size of mapping population in QTL mapping
- Type and number of markers in linkage maps

The way to reduce the genotyping effort and simplify the analysis is to select the representative SNPs, which maximizes the information; called tag-SNPs or tag-SNP-set (abbreviated as tSNPs).

A tag-SNP is defined as single nucleotide polymorphism (SNP) with high frequency of linkage disequilibrium in the particular region of the genome. This specific type of SNPs helps in depiction of a group of single nucleotide polymorphics known as a Haplotype. The establishment of genetic variation and its resemblance with phenotypes can be possible even without any genotyping of each and every SNP in the chromosome. So, it may helpful in saving time as well as money used in the mapping of those genomic regions which belong to the particular types of disease, and also, there may be no requirement to study SNPs individually. Table 6.2 shows the different tools for SNP detection. When large numbers of SNPs are genotyped in the genome, these tag-SNPs play an important role in the study of SNPs linkage in the whole genome.

Linkage disequilibrium (LD) plays an important role in genome-wide association studies (GWAS) in identifying and fine mapping of complex disease genes.

Through the indirect manipulation of QTLs improvement of crop yield has been possible. It helps to understand the genetic basis of crop performance under environmental conditions. Tuberosa et al., 2002 investigated QTLs for root architecture and their effects on yield under varying moisture regimes in rice and maize. In maize, a major QTL was originally reported for leaf ABA concentration (Tuberosa et al., 1998). In model organisms, disease-specific QTL are often obtained by intensive genetic crossing and analysis and are expected to contain genetic elements that contribute to the phenotype trait. QTL regions are a platform to enhance the efficiency of selective breeding by targeting the only regions that are responsible for the variability of the target trait.

Numerous loci at which frequent variations effect disease risk or quantitative features have been effectively identified by GWAS.

6.3.1 Bioinformatics Approach: Methods for Selecting Tag-SNPs

Different methods were used by the researchers for tagging the SNPs to select the minimum set of tag-SNPs (tSNPs) that might be enough to capture most of the variation in defined regions of the genome. Different methods have been described to select tSNPs.

TABLE 6.2
Bioinformatics Tools for SNPs Detection

Name	URL
NovelSNPer	http://www2.hu-berlin.de/wikizbnutztier/software/NovelSNPer/
SNiPlay	http://sniplay.cirad.fr/
QualitySNPng	www.bioinformatics.nl/QualitySNPng
JSNP database	http://snp.ims.u-tokyo.ac.jp/
HaploSNPer	www.bioinformatics.nl/tools/haplosnper/
Pupa SNP	http://pupasnp.bioinfo.cnio.es/
SNPper	http://snpper.chip.org.
SNPsnap	www.broadinstitute.org/mpg/snpsnap/
PolyPhred	http://droog.gs.washington.edu/polyphred/
FASTSNP	http://fastsnp.ibms.sinica.edu.tw
SNP detector	http://lpg.nci.nih.gov

6.3.2 Haplotype Diversity

A dynamic algorithm for haplotype block partitioning is used to reduce the number of SNPs. In each block, the quality by a function of haplotypes is characterized by the SNPs in that vary block. "*D*" is the haplotype diversity that represents the subsets of SNPs in the block (Clayton 2001). Patil et al. (2001) define the common haplotypes using the concept of ambiguous and unambiguous haplotypes. He developed a greedy algorithm that gives a solution to the problem.

6.3.3 Principal Component Analysis (PCA)

In PCA, selection of the foremost set of SNPs, often known as "haplotype tagging SNPs" (htSNPs), is a major challenge. The haplotype tagging (htSNPs) is by the use of principal components analysis (PCA). In order to understand how each SNP adds information to data sets, first identify the eigenSNPs using the correlation matrix and then mapping principal SNP components to characteristic SNPs – that is, mapping htSNPs. It is applied to access multivariate SNP correlations to infer groups of SNPs in LD-groups and to establish an optimal set of group-tagging SNPs (gtSNPs) that provide the most comprehensive coverage of intra-genic diversity and reduce the resources necessary to perform informative association analysis (Lin et al., 2004).

6.3.4 Best Enumeration of SNP Tags (BEST)

Identify haplotype tagging SNP (htSNPs). In contrast with other equivalent methods, the identification of a minimum set of SNPs tagging depicts this part as a perfect analytical method. The BEST method helps to identify the minimum set of SNPs tagging. Selection of a few SNPs is enough for haplotypic variation representation in a population, and when the haplotype length will be enlarged, the haplotypic variation will be reduced at a drastic level.

6.3.5 ldSelect

The number of common polymorphisms is finite within candidate genes but direct assay of all existing polymorphisms is not sufficient due to the strong association between maximum number of sites. The selection of a descriptive subset of tag-SNPs in candidate gene association studies is based on the algorithm based on r^2 linkage disequilibrium (LD) statistical method. Although common variation tends to be shared between populations with different ancestries, tag-SNP (tagging the SNPs) should be selected separately for populations (Carlson et al., 2004).

6.3.6 CLUSTAG (Hierarchical Clustering and Graph Methods)

In the CLUSTAG or hierarchical clustering and graph methods, those small parts of tag-SNPs that recognized all the identical SNPs in a particular chromosomal region can be found by developing cluster and set-cover algorithms and must have a squared correlation, R2>C, with at least one tSNP, where C is specified by the user.

6.3.7 Software Used for Selecting Tag-SNPs

6.3.7.1 SNP Tagger

The most recently used program for tagging the SNPs is called SNP Tagger. It was developed by Paul de Bakker. The server is online at www.broad.mit.edu/mpg/tagger/; it incorporates a novel combination of both genotype and haplotype information.

6.3.7.2 FastTagger

r^2 LD (linkage disequilibrium) statistics have gained popularity as r^2 is directly related to statistical power in detection of association of diseases. Most of the existing r^2-based algorithms use pairwise linkage disequilibrium (LD). FastTagger uses two different models for selecting the tag-SNPs. The one-vs-the-rest model in which two ways $2^{2^k}-2$ possible ways to group the $2k$ haplotypes into two non-empty groups. There are $2k$ potential haplotypes over the k loci, given k SNPs, S = SNP1, SNP2, SNP3, SNPk. One group contains only one haplotype, while another group contains all the other haplotypes. It calculates the r^2 statistics for all the $2k$ groupings and then selects the one with the highest r^2 statistic. Advance studies show that multi-marker LD can help further reduce the number of tag-SNPs. Tag-SNP selection algorithms based on multi-marker LD are both time- as well as memory-consuming. It works on chromosomes containing more than 100 k SNPs using length-3 tagging rules. The tool is several times faster than existing multi-marker-based tag-SNP selection algorithms. It calculates the r^2 statistic between S and SNP x and, by merging equivalent SNPs, it not only helps to save computation cost but also reducing storage overhead.

6.3.7.3 TAGster

Used for visualization, selection, and evaluation of linkage equilibrium tag-SNPs for both single and multiple populations. The method proposed by Carlson et al. (2004) helps in improvement of selection efficiency of tag-SNPs in multiple populations. An additional repetitive step is used in elimination of SNPs to boost the tag-SNP search-based algorithm. The TAGster software is an open source-based software that is freely available at at www.niehs.nih.gov/research/resources/software/tagster/.

6.3.7.4 Genetic Tagger (Gtagger)

Gtagger is a genetic algorithm for solving the htSNP selection problem. The input of the algorithm is full haplotype patterns, each with nSNPs, which are presented as a matrix H. The output will be a set of htSNPs. A genetic algorithm begins with an initial population of many individuals or solutions. The algorithm then adds some better individuals by letting the best individuals do crossover and mutating the results. Then, in the population, all individuals compete with each other to survive for the next generation. The genetic algorithm uses two evolutionary operators, mutation and crossover, and a mechanism for selecting candidate individuals to do these operations. Thus, a genetic algorithm is efficient if the evolutionary operators provide a good exploration of the problem-solution space.

6.3.7.5 Quick SNP

A webserver called QuickSNP helps to provide a cost-effective selection of SNPs, and to fill in some of the gaps in existing SNP selection tools. The advanced feature of QuickSNP is the option to select only gene-centric SNPs from a chromosomal region in an automated fashion. Other useful features of the server include an automated selection of coding non-synonymous SNPs, filtering of SNP based on inter-SNP distances, and information regarding the availability of genotyping assays for SNPs and whether they are present on whole-genome chips. The server also produces summary tables, results, and a link to a UCSC Genome Browser track illustrating the position of the selected tagSNPs concerning genes and other genomic features. QuickSNP is a user-friendly server and freely available. It can be accessed at http://bioinformoodics.jhmi.edu/quickSNP.pl.

6.3.8 SNPs and TagSNPs

Among people, single nucleotide polymorphisms (SNP) are the genetic differences that only affect one of the DNA nucleotides – adenine (A), guanine (G) called purines and thymine (T), or cytosine (C) called pyrimidines – in a segment of a DNA that occurs in more than 1% of a population.

An example of an SNP is the substitution of T for A in the nucleotide sequence “ATCCAT”, accordingly producing the sequence “AACCAT”.

To recognize the regions of DNA, SNPs can function as chromosomal tags and they explore those regions for identifying those variations that are associated with any disease or disorder in human beings. Those SNPs that may be associated with any disease may be helpful in terms of diagnostic work.

In contrast to this is identification of differences that are involved in altering responses to drugs and could facilitate the formation of personalized medicine.

SNPs are helpful in recognition and confirmation of location of genes on the chromosome. The scrutinization of genome region to find the position of SNPs helps the researcher to construct chromosome maps for the identification of genes associated to particular traits.

A set of all of the SNPs in the genome, tagging the SNPs (tag-SNP) can be extremely useful for testing the association of a marker locus with a qualitative or quantitative trait locus, since it is not necessary to genotype all of the SNPs. The tag-SNPs can be selected based on the haplotypes present in the genome. Most of the sequencing technologies supply the information solely based on genotypic characteristics and these sequences do not provide any details about haplotypes. Haplotypes determination can be done through molecular methods like Allele-Specific PCR and many more. Therefore, using such methods are expensive, laborious, and time-consuming.

6.4 APPLICATIONS OF LINKAGE DISEQUILIBRIUM IN MOLECULAR BREEDING

Mapping the complex traits: Association mapping or linkage disequilibrium (LD) mapping is a new powerful platform to map complex traits. The method identifies genetic loci which are associated with phenotypic trait variation in a collection of individuals. LD uses natural diversity, which represents recombination events in the history of the population, providing better resolution.

Crop improvement: LD is used in the study of genetic diversity in natural populations and germplasm collections, where it can be employed for the evaluation of population genetics and crop improvement programs.

Background genetic markers: In association studies, background markers are scaled to achieve genome-wide coverage to broadly characterize the genetic composition of individuals. These are useful in assigning individuals to populations, preventing fake associations if population structure and relatedness exist and estimating kinship and inbreeding. Due to lower mutation rates, higher genome density, and better amenability to high-throughput detection systems, SNPs are rapidly becoming the marker of choice for complex trait dissection studies.

6.5 LINKAGE DISEQUILIBRIUM (LD)

Linkage equilibrium (LE) states that alleles at different loci are randomly associated. Linkage disequilibrium (LD) states the association of alleles non-randomly belonging to two or more loci located in the same or different chromosomes of a species, representing a population (Lewontin, 1964). Other names for LD are “allelic association” or “gametic phase disequilibrium” (GPD). The concept was given by Jennings in 1917 and its quantification (*D*) was developed by Lewontin and Kojima (1960). Mendel’s law of random assortment states that genes (alleles) are independently transferred across generations (random assortment–linkage equilibrium); but not in the case when two genetic loci are physically close to each other.

LD gives information about past events. LD is strongly affected by natural selection, genetic drift, recombination, population structure, and mutation. D is the measure of linkage disequilibrium. *D*-value ranges between 0 and 1. If $D = 0$, two loci are in linkage equilibrium (LE). If $D \neq 0$, two loci are in linkage disequilibrium (LD). The recombination rate (lower the recombination rate, higher the LD) and distance between the SNPs (larger the distance, lower the LD) are the two parameters that greatly affected linkage disequilibrium. Table 6.3 shows various tools and their web addresses for LD estimation. Genome-wide association studies (GWAS) methods based on the linkage disequilibrium offer to detect genetic variations. LD/GPD is remarkable to GWAS because it allows identifying genetic markers for tagging the particular causal variants to complex human diseases and understanding disease mechanisms. Analysis of big data for genomic selection (GS) of livestock also applies the concept of linkage disequilibrium (LD) between the specific loci of markers and SNPs. The strength of LD is of crucial importance for genome-based analysis. The widely used measure of LD in breeding and genome-wide association mapping is $r2$., LD extends to a much longer distance in self-pollinated crops such as wheat than in cross-pollinated species such as maize. If LD decays within a short distance, mapping resolution is expected to be high (a large number of markers are required) and if LD decay extends a long distance, then mapping resolution will be low (a relatively small number of markers are required). A graphical view of LD can be presented either as LD decay plot of D' or $r2$ over physical or genetic distance or as in a linear arrangement of LD between polymorphic sites within a gene or loci along a chromosome.

6.6 MEASURES OF LINKAGE DISEQUILIBRIUM (LD)

LD calculation is divided into five steps.

1. Calculate allele frequencies of SNP1 and SNP2.

If $p1$ and $p2$ are the frequency of the alleles at SNP1 and $q1$ and $q2$ are the frequency of the alleles at SNP2.

The C and G nucleotides are SNPs to each other as there is substitution of a single nucleotide at a particular position. Likewise T and A are referred to as SNPs.

AC	C	GTAT . . . . . . GATC	T	CCAG
AC	G	GTAT . . . . . . GATC	T	CCAG
AC	G	GTAT . . . . . . GATC	A	CCAG

Allele	SNP 1	SNP 2
Allele 1	C	T
Allele 2	G	A

SNP 1		SNP2	
Allele	**Frequency**	**Allele**	**Frequency**
C	$p1$	T	$q1$
G	$p2$	A	$q2$

2. Calculate the haplotype frequencies.

From the two alleles of each SNP, the calculated haplotypes are

		SNP 2	
	Allele	T	A
SNP 1	C	CT	CA
	G	GT	GA

Suppose if the haplotype frequency of CT = p11

$$CA = p12$$
$$GT = p21$$
$$GA = p22$$

3. Calculation of linkage equilibrium (LE).

When haplotype frequencies are equal to the product of their corresponding allele frequencies, which depicts both, the loci are in linkage equilibrium.

$$p11 = p1q1$$
$$p12 = p1q2$$
$$p21 = p2q1$$

4. Linkage disequilibrium (LD).

Conclude linkage disequilibrium for each haplotype as the deviation of observed haplotype frequency from its corresponding allelic frequencies expected under equilibrium.

$$D11 = p11 - p1q1$$
$$D12 = p12 - \mathrm{p}1q2$$
$$p22 - p2q2$$
$$D21 = p21 - p2q1$$
$$D22 = p22 - p2q2$$

5. Calculation of linkage disequilibrium measure *D*

The most commonly used measure of linkage disequilibrium, D equals to $p11$ $p22$ $p12$ $p21$

$$D = (p11)(p22) - (p12)(p21)$$

- If two loci are in linkage equilibrium (LE), then $D = 0$
- If the two loci are in linkage disequilibrium (LD), then $D \neq 0$

In Standardization method

Sometimes, depending on the allele frequency of two loci, the value of D can be negative, but actual gametic frequencies cannot be negative. To overcome this issue, standardization methods have been proposed.

A relative measure of disequilibrium D′

$$\text{When } D \text{ is positive, } D_{max} = \min\ [\ (p1q2) \text{ or } (p2q1)\ ]$$
$$\text{When } D \text{ is negative, } D_{max} = \min\ [\ (p1q1) \text{ or } (p2q2)\ ]$$

Standardization makes D-values range between 0 and 1

r^2 is the Pearson's (product-moment) coefficient of the correlation that describes the predictive value of the allelic state at one polymorphic locus on the allelic state at another polymorphic locus.

Squared correlation coefficient varies 0-1

$r^2 = 0$, Loci are in complete equilibrium (LE)
$r^2 = 1$, Loci are in complete disequilibrium (LD)

Used extensively in disease gene or phenotype mapping through association testing. D' statistic was developed to partially normalize the D value concerning the maximum value possible for the allele frequencies and has a range between 0 and 1. The $r2$ statistic is the same because the squared value of the Pearson's (product-moment) correlation coefficient and has an expectation of 1/(1 + 4Nc), where N is the effective population size and c is the recombination rate in morgan.

6.7 TESTS OF ASSOCIATION: QUANTITATIVE TRAIT LOCUS (QTL) BY LD

For QTL, linkage disequilibrium mapping studies typically use either family samples or unrelated individuals. When family data are used, the transmission/disequilibrium test (TDT) for quantitative traits can be used to test for linkage or association (Sun et al., 2000). Different methods have been employed for mapping QTL using unrelated population individuals (Fisher, 1918). The regression method is also used to test whether there is an association between a marker locus and a QTL.

The possibility to increase crop yields through the indirect manipulation of QTLs, which regulate the physiological processes that affect biomass production and its partitioning as well as heritable variability of traits.

The various QTL-based approaches help toward the better recognition of the genetic basis of crop performance under environmentally constrained conditions.

TABLE 6.3
Some Tools for Estimation of Linkage Disequilibrium (LD)

Name	URL
LDMAP	http://cedar.genetics.soton.ac.uk/public_html/helpld.htmlhttp://
lds (calculator)	www.bioinf.mdc-berlin.de//span
HapGraph	http://bioinformatics.med.utah.edu//span
Haploview	www.broad.mit.edu/personal/jcbarret/haplo/index.ph
LDtrait	https://ldlink.nci.nih.gov/?tab=ldtrait

6.8 SIGNIFICANCE OF TAG-SNPS

The aim of GWAS is to govern the condition of genetic risk factors, which may be helpful in assessment of disease risk, to make new strategies for their treatment and prevention. Single nucleotide polymorphisms (SNPs) can be used for the identification of associations of genetic characteristics with phenotypic characters and clinical conditions. The significant relationships between large numbers of SNPs with a broad range of phenotypes can be featured by using different GWAS databases. GWAS has dominated genetic analysis looking for variants associated with a trait of interest. GWAS utilizes genome-wide genotyping arrays that contain between a hundred thousand and several million SNPs. The SNPs in LD are referred to as "tag-SNPs". GWAS arrays are designed to detect common variants with small to modest effect sizes.

6.9 GENETIC MARKERS IN PREDICTION OF DISEASES

Tag-SNPs serve as genetic markers for the detection of diseases that represent the genotype status of other SNPs. GWAS plays an important role in improving our clinical prediction and disease treatment, it is undisputed that these studies satisfy the expectation of discovery of new molecular mechanisms and contribute to our knowledge of pathophysiological pathways of disease. GWAS established the genotype–phenotype association and also helps to detect the disease-causing variants (often not identified by GWAS tag-SNPs) or to narrow the risk interval regions, and improve the power to detect low-frequency variants.

BIBLIOGRAPHY

Barreiro, L. B, Laval, G., Quach, H., Patin, E., Quintana-Murci, L. (2008). Natural selection has driven population differentiation in modern humans. *Nature Genetics* 40(3), 340–345. https://doi.org/10.1038/ng.78

Carlson, C. S., Eberle, M. A., Rieder, M. J., Yi, Q., Kruglyak, L., Nickerson, D. A. (2004). Selecting a maximally informative set of single-nucleotide polymorphisms for association analyses using linkage disequilibrium. *American Journal of Human Genetics* 4, 106–120.

Clayton, D. (2001). www.nature.com/ng/journal/v29/n2/extref/ng1001-233-S10.pdf

Collins, S., Guyer, M. S., Charkravarti, A. (1997). Variations on a theme: cataloguing human DNA sequence variation. *Science* 2(8), 1580–1581. https://doi.org/10.1126/science.278.5343.1580

Fisher, R. A. (1918). The correlation between relatives on the supposition of Mendelian inheritance. *Transactions of the Royal Society of Edinburgh* 52, 399–433.

Geldermann, H. (1975). Investigations on inheritance of quantitative characters in animals by gene markers. I. Methods. *Theoretical and Applied Genetics* 46, 319–330.

Halushka, M. K., Fan, J. B, Bentley, K., Hsie, L., Shen, N. P., Weder, A., Cooper, R., Lipshutz, R., Chakravarti, A. (1999). Patterns of single nucleotide polymorphisms in candidate genes for blood pressure homeostasis. *Nature Genetics* 22, 239–247.

Klein, R. J., Zeiss, C., Chew, E. Y., et al. (2005). Complement factor H polymorphism in age-related macular degeneration. *Science* 308, 385–389.

Laber, S., Cox, R. D. (2017). Mouse models of human GWAS hits for obesity and diabetes in the post genomic era: time for reevaluation. *Frontiers in Endocrinology* 8(11). https://doi.org/10.3389/fendo.2017.00011

Lewontin, R. C. (1964). The interaction of selection and linkage. I. General considerations; heterotic models. *Genetics* 49, 49–67. https://doi.org/10.1093%2Fgenetics%2F49.1.49

Lewontin, R. C., Kojima, K. (1960). The evolutionary dynamics of complex polymorphisms. *Evolution* 14, 458–472. https://doi.org/10.1111/j.1558-5646.1960.tb03113

Lin, Z., Altman, R. B. (2004). Finding haplotype tagging SNPs by use of principal components analysis. *American Journal of Human Genetics* 75, 850–861.

Luo, J. (2015). Metabolite-based genome-wide association studies in plants. *Current Opinion in Plant Biology* 24, 31–38. https://doi.org/10.1016/j.pbi.2015.01.006

McCarthy, M., Abecasis, G., Cardon, L. et al. (2008). Genome-wide association studies for complex traits: consensus, uncertainty and challenges. *Nature Reviews Genetics* 9, 356–369. https://doi.org/10.1038/nrg2344

Mukhopadhyay, C. S., Gupta, A. K., Yadav, B. R., Mohanty, T. K. (2011). Exploration of Y-chromosome-specific markers to discover SNP associated with subfertility traits in dairy bulls. *Indian Journal of Biotechnology* 10, 178–182.

Paterson, A. H., Lan, T. H., Reischmann, K. P., Chang, C., Lin, Y. R., Liu, S. C., Burrow, M. D., Kowalski, S. P., Katsar, C. S., DelMonte, T. A., Feldmann, K. A, et al. (1996). Toward a unified genetic map of higher plants, transcending the monocot-dicot divergence. *Nature Genetics* 14, 380–382.

Patil, N., Berno, A. J., Hinds, D. A., et al. (2001). Blocks of limited haplotype diversity revealed by high-resolution scanning of human chromosome 21. *Science* 294, 1719–1723.

Sun, F. Z., Flanders, W. D., Yang, Q. H., Zhao, H. Y. (2000). Transmission/disequilibrium tests for quantitative traits. *Annals of Human Genetics* 64, 555–565. https://doi.org/10.1017/S000348000000840X

Tuberosa, R., Salvi, S., Sanguineti, M. C., Landi, P., Maccaferri, M., Conti, S. (2002). Mapping QTLs regulating morpho-physiological traits and yield: case studies, shortcomings and perspectives in drought-stressed maize. *Annals of Botany (Lond)* 89, 941–963.

Tuberosa, R., Sanguineti, M. C, Landi, P., Salvi, S., Casarini, E., Conti, S. (1998). RFLP mapping of quantitative trait loci controlling abscisic acid concentration in leaves of drought-stressed maize (*Zea mays* L.). *Theoretical and Applied Genetics* 97, 744–755.

Zhang-James, Y., Fernàndez-Castillo, N., Hess, J. L. et al. (2019). An integrated analysis of genes and functional pathways for aggression in human and rodent models. *Mol Psychiatry* 24, 1655–1667. https://doi.org/10.1038/s41380-018-00687

Zhonggang, Li, James, A., Votava, Gregory, J. M., Zajac, Jenny, N., Nguyen, Fernanda, B., Leyva Jaimes, et al. (2020). Integrating mouse and human genetic data to move beyond GWAS and identify causal genes in cholesterol metabolism. *Cell Metabolism* 31(4), 741–754, e5. https://doi.org/10.1016/j.cmet.2020.02.015

7 Flow Cytometry and Its Applications

Dr. Himanshu Tillu, Dr. Rekha R. Gour, Dr. Hemant Agrawal

7.1 INTRODUCTION

Flow cytometry has become an essential requirement for biological studies owing to its attributes of speed, objectivity, precision, accuracy, and high throughput; its extensive multiplexing capabilities; and its ability to analyze a large number of cells individually. As the name suggests, flow cytometry refers to the measurement of physical, chemical, and biological properties of cells and cell-like particles while the cells are flowing through a stream at a rapid pace. A beam of light (usually a laser) illuminates the stream of cells, which are passing one by one through the flow chamber/flow cell due to hydrodynamic focusing (and in some cases both hydrodynamic and acoustic focusing). This makes the cells scatter or fluoresce (internal or external fluorescence) light in all directions. Detectors are placed in such a way that they can capture either scattered or fluorescent light. The scattered light is collected at two angles: in the line of the laser beam (forward scatter, or FSC) and perpendicular to the laser beam path (side scatter, or SSC). FSC is said to be roughly proportional to the size of the cell. SSC is primarily a measurement of refracted and reflected light, collected at approximately 90 degrees from the laser beam path, and proportional to cell granularity or internal complexity. Flow cytometers are also equipped with detectors that are capable of detecting emitted fluorescent light from cells that are labeled with fluorophore-conjugated antibodies or a fluorescent dye. It's the combination of light scatter (FSC, SSC) and fluorescence signals that give us information about the cells and allow researchers to study multiple subpopulations of cells in a heterogeneous sample. The photodetectors that are used in flow cytometers to capture light are mainly photodiodes (PDs), photomultiplier tubes (PMTs), and avalanche photodiodes (APDs). The captured light is further filtered spectrally and is converted to electrical signals (voltage). The voltage data is then quantified by signal processors (analog-to-digital converter, ADC) into numerical values. This numerical data is then transferred to the cytometer attached to the computer for storage and further analysis.

DOI: 10.1201/9781003220831-7

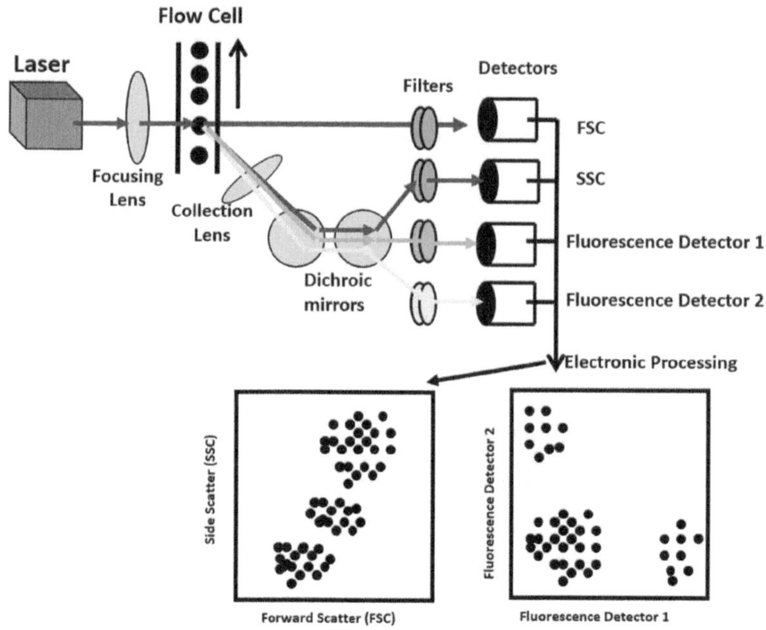

FIGURE 7.1 Outline of a flow cytometer.

Furthermore, with time, flow cytometers have become more compact, cost-effective, more user-friendly, and well suited for higher throughput than ever before. All these aspects make flow cytometry a powerful tool for detailed and rapid analysis of complex cell populations.

7.1.1 History of Flow Cytometry

Many early scientific developments in the field of cytometry built the foundation for the current field of flow cytometry (**1, 2**): notably, the very early work done by Antonie van Leeuwenhoek in the field of microscopy to study microbes by using single-lensed microscopes of his design (late 1600); the description of the staining properties of Fluorescein by Paul Ehrlich (1880s); the photodetection of cells and particles by Moldavan (1934); the detection of *Streptococcus pneumonia* by UV-excitable fluorochrome-conjugated antibodies by Coons (1942); the sizing of microscopic particles by Wallace H. Coulter using an impedance-based system known as the Coulter principle (1950s); the droplet deflection sorter by Fulwyler (1965), which led to the design for modern cell sorters; and the development of the spectrophotometer by Loius A. Kamentsky et al. (1965) to measure and display multiple properties of biological cells. In the late 1960s, Bonner, Sweet, Hulett, and Herzenberg (Stanford University) designed and later patented (1972) the first Fluorescence Activated Cell Sorter (FACS) instrument. In 1974, using the Stanford University patent and expertise of Herzenberg et al., the first commercial Fluorescent Activated Cell Sorter (FACS-II) instrument was produced by Becton Dickinson. During the same time, Wolfgang Göhde, from the University of Münster, Germany, developed a fluorescence-based flow cytometry device (1968) and named it Impulszytophotometer (ICP-11) (**2**). It was then commercialized in 1968–69 by German developer and manufacturer Partec through Phywe AG in Göttingen. This was followed by the development of PAS-8000 (1973) and ICP-22 (1975) from Partec. In 1977–78, Coulter came up with the Epics system. Up to 1978, flow cytometry was known as pulse cytophotometry (German: *Impulszytophotometrie*). In 1978, the term 'pulse cytophotometry' was changed to 'flow cytometry' at the Conference of the American Engineering Foundation in Pensacola, Florida; the term 'pulse cytophotometer' quickly become popular and is used to date.

In 1975, George Köhler and Cesar Milstein developed a hybridoma technology to make monoclonal antibodies, which brought a revolution to the field of flow cytometry (**21**). In 1984, Köhler and Milstein were awarded a share of the Nobel Prize in physiology or medicine for their breakthrough research. By the late 1970s and 1980s, biomedical researchers began making monoclonal antibodies against antigens of white blood cells and labeling them with fluorescent dyes. As the number of antibodies and fluorescent dyes increased, the numbers of lasers on flow cytometers increased. By the 1990s, flow cytometers were capable of measuring four or more fluorescence parameters simultaneously, and, in 2000, BD Biosciences introduced the LSRII™, with 14-color capability. Today some flow cytometers can characterize more than 40 fluorescent parameters simultaneously, which allows us to study the complexity of the immune system. As researchers, we owe a lot to all the contributions made by physicists, biophysicists, biologists, and computer engineers toward the advancement of this amazing technology.

7.1.2 How Does a Flow Cytometer Work?

Successful operation of any flow cytometer requires three systems – fluidics, optics, and electronics – to work efficiently in tandem (Figure 7.2).

Flow cytometry-based analysis of cells requires individual cells to be in suspension. The cells can be previously stained by fluorescent dyes directed against specific cellular biomolecules or by fluorescently tagged antibodies that are generated against specific surface or intracellular (IC) antigens. The cells in the suspension are passed in a stream through a flow cytometer where individual cells are illuminated by single or multiple lasers at the same point (collinear lasers) or different points (spatially separated lasers). The point at which a laser illuminates the cells is termed the point of interrogation. Lasers perform two roles in a flow cytometer – to illuminate the cells to enable light scattering and to excite the cell-associated fluorochromes (35).

Individual cells in the suspension are focused onto the point of interrogation by a surrounding stream of sheath fluid, allowing the cells to pass individually through the interrogation point. This process is critical to the functioning of a flow cytometer and is termed 'hydrodynamic focusing' (Figure 7.3). Inside the flow cytometer, the sample stream is introduced at the center of the sheath stream at a slightly higher velocity than the sheath fluid. Sample and sheath fluid form a system of laminar flow where they are in contact with each other but do not mix. The stability of flow can be characterized by Reynolds Number (Re), which is a dimensionless constant (**36**). If d is the

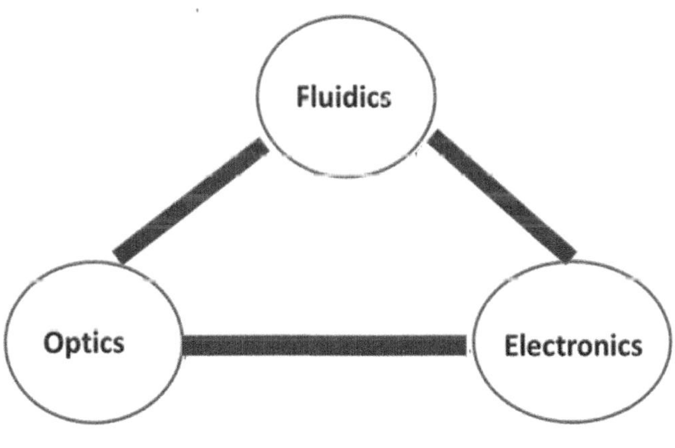

FIGURE 7.2 The three cornerstones of flow cytometry systems. Proper working of flow cytometers requires the efficient co-operation of fluidics, optics, and electronics systems.

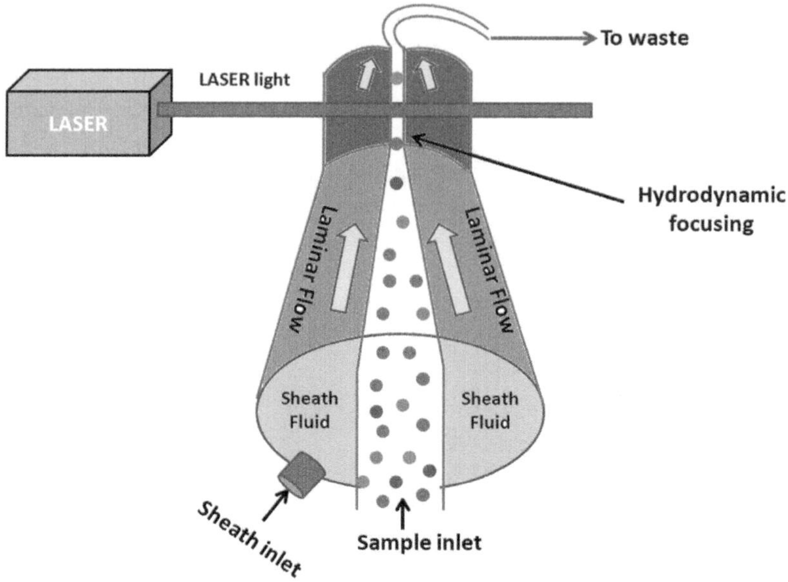

FIGURE 7.3 Hydrodynamic focusing in a typical analyzer. Sheath fluid and sample are introduced into the flow cell inside a flow cytometer. Inside a flow cell, the cells are hydrodynamically focused by sheath fluid at the point of interrogation so that the cells pass the laser one cell at a time.

diameter of the stream, in cm, p is the density of the fluid, in gm/cm^3; ή is the fluid viscosity, in poise (1 poise = 1 g/cm-s); and v_{av} is the average fluid flow velocity, in cm/s:

$$Re = dpv_{av}/ή$$

Laminar flow only occurs at values Re < 2300. At higher values than that, turbulent flow is more likely, which would be undesirable for flow cytometry measurements. The fluidic systems of flow cytometers are designed to maintain laminar flow conditions. Inside the flow cell of a flow cytometer, the flow runs in a parabolic profile, where the flow at the center travels at twice the velocity as that of the flow near the walls of the flow cell. This flow profile creates an inverse pressure gradient, with the maximum pressure at the walls and the lowest pressure at the center (Bernoulli's Principle). Consequently, the particles relocate to the faster central zone.

The sheath fluid and cell suspension are introduced into the flow cytometers employing various pumps, giving rise to the classification of flow cytometers into positive or negative pressure systems, systems involving peristaltic pumps. The use of precision syringe/injection pumps allows precise volumetric counting of absolute cell numbers. Although hydrodynamic focusing is shown to be highly effective at particle positioning inside the flow cell, there have been reports about development of portable flow cytometers, which work without the use of sheath fluid and analyze particles with high sensitivity and speed (**10**).

Once the cells pass the point of interrogation, the generated scatter and fluorescence signals are captured by detectors, which can either be photodiodes, photomultiplier tubes (PMTs), or avalanche photodiodes (APDs). Photodiodes are usually used as the detectors for bright FSC light. The obscuration bar is used to block the undeflected (direct) laser light from reaching the detector. The acquired light signals are converted into electrical signals, and following analog-to-digital conversion are displayed to the users in the form of single parameter histograms or a variety of two-dimensional plots. The design of a typical flow cytometer is represented in Figure 7.1.

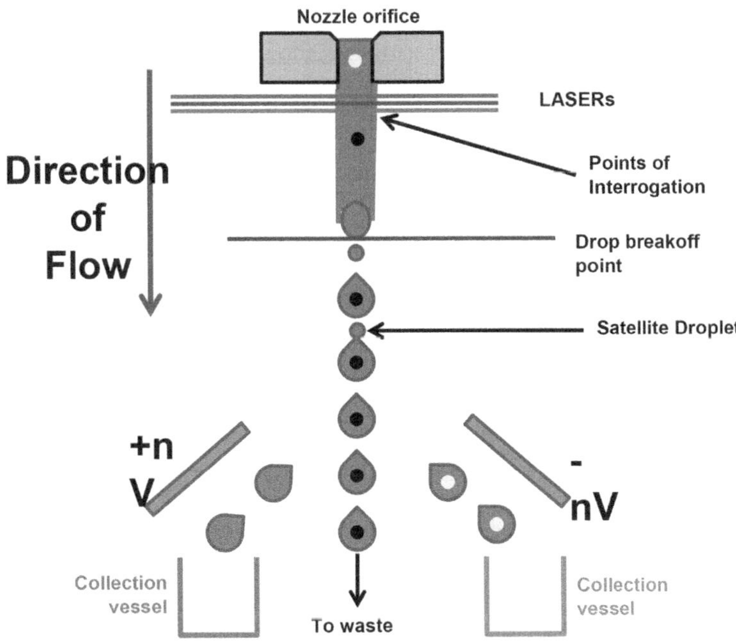

FIGURE 7.4 Design of a typical cell sorter. The cells are detected at the point of interrogation. The sample stream is broken off into droplets once the cells pass the point of interrogation. Once the drop containing the cell of interest arrives at the droplet break off point, a charge pulse is applied to the stream resulting in the droplet of interest getting charged. The charged droplet is then deflected using potential difference that is applied to metallic plates, and the deflected drops are collected into a collection vessel. The undeflected drops continue into the waste trough.

Flow cytometry systems can also be used for sorting a pure population (purity > 99%) of cells from a mixture of various cell types (**11**). A cell of interest that needs to be sorted is measured at the point of interrogation and moves down the stream to the break off point. During the separation of the droplet with the cell in it from the intact liquid jet, a charge is given to the liquid stream so that droplets containing the cells of interest get charged and can be deflected in an electric field between two deflection plates for sorting. The sorting instrumentation contains metallic plates to which are applied high potential difference. The charged droplets carrying the cells of interest are deflected by the metallic plates and are collected into appropriate collection tubes or vessels, which are placed below the deflection plates (Figure 7.4). The flow cytometers that are capable of sorting cells are termed cell sorters. Modern cell sorters can allow the sorting of up to six different cell types simultaneously. Single cell sorting can be accomplished in a sterile environment, and some of the modern cell sorters can also be accommodated inside a laminar air-flow hood. The sorted-out cells can be used further to set up cultures or are subjected to genomics, transcriptomics, or proteomics profiling, as per the requirement.

7.1.3 What Does a Flow Cytometer Measure?

Flow cytometry measures optical and fluorescence characteristics of single cells (or any other particles, such as nuclei, polystyrene, silica, and latex beads). In flow cytometry, differences in physical properties are reflected by differences in the intensity of scattered light and are used to resolve different populations from each other and from debris. Although light is scattered in all directions, scattered light is collected at two different locations, yielding information about different physical properties. One of the detectors (often a photodiode or, in some cases, a PMT) is placed to collect light that is scattered in the forward direction, that is, in the direction of the incident laser light. This is the so-called forward scatter, or FSC. The forward scatter measurements in flow cytometry are based on Mie's analysis, named after Gustav Mie. The concept that FSC measurements are said to represent the relative size of the particles scattering the light is based on the work carried out by Mullaney et al. (**29**), showing that the intensity of light scattered at small angles (0.5° to 2°) from an incident laser beam was roughly proportional to particle volume. Here, it is important to note that FSC is not a monotonic function of particle size, and several other factors apart from particle size influence forward scatter (**18, 19, 24, 25, 26, 39, 40, 41, 42, 43, 45, 46, 47, 48, 53, 54,** and **55**). A few of the parameters that influence the forward scatter measurements are the wavelength of the incident light, differences in the refractive indices of the cells and the surrounding medium, the presence of any biomolecules displaying strong absorptive properties at the incident wavelength, and the range of angles used to collect scattered light. The intensity of scattering is inversely proportional to the wavelength of incident light. Hence, to analyze small particles, violet laser light (405 nm) is preferred over the standard blue laser (488 nm). If the refractive index of the cells is the same as that of the surrounding medium, the cells will not scatter any light. The effectiveness of absorptive biomolecules can be illustrated by the fact that although granulocytes in the blood have a higher volume (350 fL) compared to lymphocytes (200 fL), granulocytes generate smaller FSC compared to lymphocytes in some flow cytometers.

The range of angles used to collect scattered light depends on the focal length and numerical aperture of the collecting lenses and the size, shape, and position of irises, slits, and obscuration bar used in the optical system. The obscuration bar is used to prevent the strong, non-scattered incident light from hitting the FSC detector.

It is not possible to reliably estimate the exact sizes of cells of different types from FSC signals (**13**). However, FSC signals have their use as an approximate estimate of relative cell sizes, allowing the users to resolve populations of cell types differing in size or cell volume.

In 1975, Salzman et al. demonstrated the resolution of unlabeled lymphocyte, monocyte, and granulocyte populations from each other by using the measurements of forward and orthogonal (90°) scatter, now called side scatter (SSC) (**41**). However, it is not mandatory to place an SSC detector at a 90° angle. Steen et al. (**50**) have observed that scatter signals collected at angles as small as 18° relative to the beam axis represent identical cellular features, as the signals collected at a 90° angle to the axis. Most of the information required for the resolution of these populations is derived from SSC. Granulocytes that have a granule-rich cytoplasm, as the name suggests, afford significant opportunity for the light to scatter, and consequently have the highest SSC among the leukocytes. Lymphocytes, with relatively little and very uniform cytoplasm, have the lowest SSC, while monocytes display an intermediate SSC profile. Apart from analyzing leukocytes, SSC signals have also been used in other kinds of applications. Nusse et al. (**31**) and Zucker et al. (**58**) have demonstrated mitotic nuclei identification on a bivariate plot of DNA content (propidium iodide fluorescence) versus side scatter. Dubelaar et al. (**8**) have used FSC and SSC signals to discriminate between cyanobacteria with and without gas vacuoles. Nordstrom et al. (**30**) report that increased side scatter signal is a good indicator of whether or not insect cells have been infected with recombinant baculoviruses, eliminating the need for costly fluorescent markers.

Fluorescence is used in various ways in flow cytometry. Perhaps the most common application of fluorescence in flow cytometry is the identification of specific cell types. Since the advances in multicolor flow cytometry over the past decades, fluorescence flow cytometry has become the mainstay of the diagnosis of hematological malignancies and other kinds of disorders. In addition, fluorescence is also used to gauge the metabolic states of cells in functional assays. Fluorochrome-conjugated antibodies targeted against specific molecules expressed either on the cell surface or intracellular, termed markers, are used as probes to identify the cells of interest. It is possible to use modern flow cytometers to analyze more than 50 markers simultaneously. Additionally, fluorescent dyes may also be used to tag specific biomolecules to get more information about the metabolic state of the cells. For instance, DNA binding dyes like Propidium Iodide, DAPI, Hoechst, and so on, can resolve different phases of the cell cycle, changes in ploidy levels in plant and animal tissues, and various generations of cells can be resolved from each other by analyzing the distribution and dilution of dye in dividing cells. Fluorescence is also widely used to detect DNA damage and define various stages of cell death, to identify the changes in mitochondrial membrane potential changes. In addition, it is also possible to recognize certain kinds of stem cells by analyzing the efflux of dyes by cellular transporters pumps. The multiplexing capabilities of the flow cytometers, the availability of fluorochromes with unique properties and which can be excited by lasers across the spectrum, and clever assay design have enabled flow cytometers to solve diverse kinds of biological problems (**3, 4, 5, 34**).

7.1.4 Controls for Flow Cytometry

Controls are essential to maximizing the probability of success in an experiment. In case the experiment fails, controls allow the users to pinpoint the causes of failure. Controls ensure that an instrument is performing well, populations are resolved properly and accurately, and proper interpretation of data is possible. A wide variety of controls have been described in the literature. The type of control that is used depends on the assay that is being performed and the question the assay is aimed at to answer. The following types of controls are usually required in a flow cytometry assay.

1. Instrument Quality Controls (QC)
2. Staining Controls
3. Gating Controls
4. Compensation Controls

7.1.4.1 Instrument QC Controls

Instrument QC controls consist of hard-dyed beads, which are uniform in physical properties and the extent of fluorescent staining. They are stained with different fluorochromes, which can fluoresce in all the detectors of a given flow cytometer. All manufacturers of flow cytometry equipment recommend certain commercially available beads to be used for Instrument QC. Instrument QC beads serve a variety of purposes. They check laser alignment, power, time delay, Q (sensitivity) and B (background) values, and the integrity of fluidics. Certain ranges are recommended by the manufacturer for each of these parameters. If the QC beads give values that are within these ranges, the flow cytometer is deemed to be passing the QC criteria and is considered to be functionally optimal.

It is recommended to run instrument QC daily or before acquiring any data on the flow cytometer. It is also important to run QC if any problem is suspected with the flow cytometer.

7.1.4.2 Staining Controls

Antibodies would be expected to bind to their targets in an extremely specific manner. However, in reality, nonspecific signals can occur as a result of the binding of the antibodies to Fc receptors (FcR) on target cells, nonspecific antibody interactions with cellular proteins, carbohydrates, lipids, and antibodies nonspecifically trapped in the cells during intracellular staining. The main function of staining controls is to check for nonspecific binding of antibodies to cells of interest resulting in false positives. There are multiple staining controls (**17**) – isoclonic, isotype, viability, blocking reagents, secondary only.

Isoclonic Controls: Isoclonic controls involve the addition of excess unlabeled antibodies of the same clone and specificity to cells that are previously stained with fluorescence-tagged antibodies of that clone. If antibody binding of the fluorescence-tagged antibodies is specific, adding an excess of unlabeled antibodies will result in a decrease in fluorescence intensity. No reduction in fluorescence indicates nonspecific binding.

Isotype Controls: Isotype control is an antibody that lacks specificity to the target antibody but is also of the same clone and originates from the same species as that of the fluorescence-tagged antibody. In addition, isotype controls also match the heavy and light chain class and the fluorescence: protein ratio.

Isotype controls are very useful to differentiate between specific and nonspecific binding. They are especially useful when testing for panels of antibodies and antibodies that are newly acquired in the laboratory for specificity. However, isotype controls cannot address cross-reactivity of antibodies with different epitopes on the same or different antigens. In addition, in contrast to the popular myth, isotype controls are not good gating controls. In a landmark study, Maecker and Trotter demonstrated that three different isotype controls, each used at the recommended concentration gave different levels of background staining when compared with unstained cells in intracellular staining (**28**). Isotype control may be used as a gating control only if it is of the same isotype as that of the test antibody, is used at the same concentration, has the same fluorophore: protein ratio, and exhibits the same degree of aggregation as that of the test antibody.

Blocking Reagents: Antibodies can bind to cells expressing Fc Receptors such as monocytes, B cells, macrophages, NK cells, and dendritic cells using their Fc portion. FcR blocking reagents (anti CD16/32) or proteins like BSA are commonly used to block unwanted binding of antibodies. False-positive results caused by Fc receptor binding are also reduced by adding IgG from the animal species in which the secondary antibody is raised.

Viability Controls: Dead cells can lead to significant problems in flow cytometry assays due to increased autofluorescence (**57**), altered antigen expression (**32**) and DNA content (**3, 6, 7, 23, 32, 51**), and their ability to bind the fluorescence-tagged antibodies nonspecifically (**9, 32, 37, 44, 52**), leading to false-positive signals.

7.1.4.3 Gating Controls

Unstained: Unstained controls are very commonly used in single-color assays involving fluorescent dyes or direct labeling of cellular antigens by fluorescently tagged antibodies. The requirement of unstained controls stems from the fact that every cell has an inherent fluorescence associated, which is termed autofluorescence, which shows up in the various 2D and 3D plots in the absence of fluorescence labeling. Specific fluorescence will then show up beyond the outer boundary of the unstained population.

If the stained and unstained cells give rise to two discrete easily resolvable populations, then using unstained cells as a gating control is not strictly required. However, in cases where the markers that are stained by fluorescence-conjugated antibodies or fluorescent dyes are weakly expressed, the resultant population is not resolvable from the unstained population and instead appears as an extension of the unstained population. In such cases, using unstained cells is invaluable for use as a gating control.

Fluorescence Minus One (FMO): When multiple fluorochromes are simultaneously used to stain cells, it results in spectral spillover into different channels. The measurement of light signals by detectors is always associated with photon-counting error (**38**). These two factors lead to the spreading of the data in different fluorescence channels. The extent of spread is not equal in all the channels and the FMO or **F**luorescence **M**inus **O**ne control affords a way to measure the spread. FMO controls contain all the antibodies in the panel except one of them. FMO controls are used to set the upper boundary for the background signal on the missing label, and thus help to identify and gate positive populations in multicolor experiments. FMO control can be used to define the limits of the gate for a positive population when the positive population is not clearly separated from the negative population due to dim or variable expression levels. When using more than one fluorochrome in the assays simultaneously, the limits of the gate should not be set based on negative populations; rather, FMO control should be used in such a case.

Biological Controls: In certain experiments, biological controls can be better gating controls than even FMO controls. For example, unstimulated or mock-stimulated cells can serve as appropriate gating controls for stimulated cells. Similar to FMO control, the mock-stimulated control can account for spillover spread in the channel of interest since it includes all of the fluorescently tagged antibodies used in the panel. In addition, similar to an isotype control, it can be used to detect any background staining in the channel of interest. And since the same antibody is used to stain stimulated and mock-stimulated cells, there is no need to match the control antibody to that of the test. In some situations, favoring the FMO control instead of biological control may lead to inappropriate placement of gates, as is seen in (**28**).

The use of biological control as a gating control should be avoided in cases when the mock-stimulated cells show considerable background signals in the channel of interest.

Internal Control: Another useful gating control is internal control. An internal control is a population of cells within the stained tube that does not express the antigen of interest and hence remains unlabeled after staining. The main benefit of using an internal negative control is that the unlabeled population of cells is exposed to identical experimental conditions (like fixation and permeabilization, washing and even exposure to the specific fluorescence-conjugated antibody directed to the antigen of interest) as that of the target population under study. Ideally, the fluorescence intensity of internal control should be equivalent to the autofluorescence level of the unstained control. However, practically speaking, some level of nonspecific antibody binding always occurs, the level of which depends on the total absolute amount of antibody rather than the antibody concentration (**28**).

7.1.4.4 Compensation Controls

Compensation is a process that deals with the correction of artifacts resulting from the spectral overlap of different fluorophores with one another when used simultaneously (**14**, **38**). Compensation is calculated by generating what is called the compensation matrix, where the percentage of spillover in every other detector (2° signal) is subtracted from the 1° signal in every detector. Accurate generation of the compensation matrix requires recording data from single-color compensation controls. Single-color compensation controls consist of a negative and positive population and are prepared using cells or antibody capture beads labeled with one of the fluorescent-tagged antibodies in the panel. The number of single-color controls is equal to the number of fluorochromes used in the panel. The single-color controls must be generated according to the following rules (**16**) for accurate calculation of the compensation matrix.

1. Compensation controls must be as bright as or brighter than the experimental samples they will be applied to.
2. The compensation fluorophore must be the same as the experimental fluorophore. For example, Alexa Fluor 488 or GFP cannot be used to compensate for FITC.
3. Autofluorescence of the positive cells/beads and negative/dim cells/beads in the same tube must be exactly the same. Some of the compensation controls for a single panel can be antibody capture beads and some others can be singly stained cells. However, the use of a mixture of beads and cells in a single tube is not recommended to be used as compensation controls.
4. The compensation matrix needs to be reset if the lot number of dyes is changed (very important for tandem dyes) or if the detector voltage for fluorescence parameters is changed (e.g., after preventive maintenance).
5. There should be enough events (around 2000–5000 events) in the negative and positive populations of the compensation controls.
6. Compensation controls should be treated in the same way as that of a sample, and this includes the fixation/permeabilization treatments that are given to the sample cells.

7.2 DATA ANALYSIS AND PRESENTATION

There has been enormous progress in flow cytometry in the last few years in terms of instruments, the number of parameters that can be analyzed, availability of reagents, and data analysis software, thus allowing detailed analysis of cells or cell-like particles. Data analysis and presentation are very important aspects of any successful flow cytometry experiment. Data analysis involves correct scaling, display, gating, statistics, and subsequently accurate interpretation and presentation (**15, 20, 22, 27, 49, 56**).

The data generated on flow cytometers are usually stored in a standardized file format known as .fcs file (flow cytometry standard). This format of file extension was developed by the International Society for Advancement of Cytometry (ISAC). The .fcs file has three portions, namely, header, text, and data. It all started with fcs 1.0, and now the current version of the fcs file format is fcs 3.1. Once the data is acquired on a flow cytometer, it can further be analyzed either using the software available on the flow cytometer, such as Cell Quest Pro, FACSDiVa, CytExpert, Attune NxT Software, and so on, or using a third-party data analysis software like FCS Express, FlowJo, GemStone, and so on. These third-party software help to analyze data acquired on any flow cytometer and thus the user is not restricted with the choice of the flow cytometer to be used for the acquisition of data.

The flow cytometry data can be plotted as univariate or 1-D plots known as histograms, where the X-axis represents the parameter's signal and the Y-axis represents the number of events per channel number. In 2-D or bivariate plots, such as dot plots, density plots, zebra plots, contour plots, and color-dot plots, the X-axis represents one parameter, and the Y-axis is the second parameter (Figure 7.5). With advancement of the number of parameters that can be analyzed simultaneously, there is a growing demand of high-dimensional data analysis computational tools, which can help the user to look beyond manual methods for analyzing high parametric data (**33**).

There are mainly two types of data scaling used for plotting flow cytometry data, namely, linear (Lin) and logarithmic (Log). Linear scaling is applied for data where differences in signals obtained

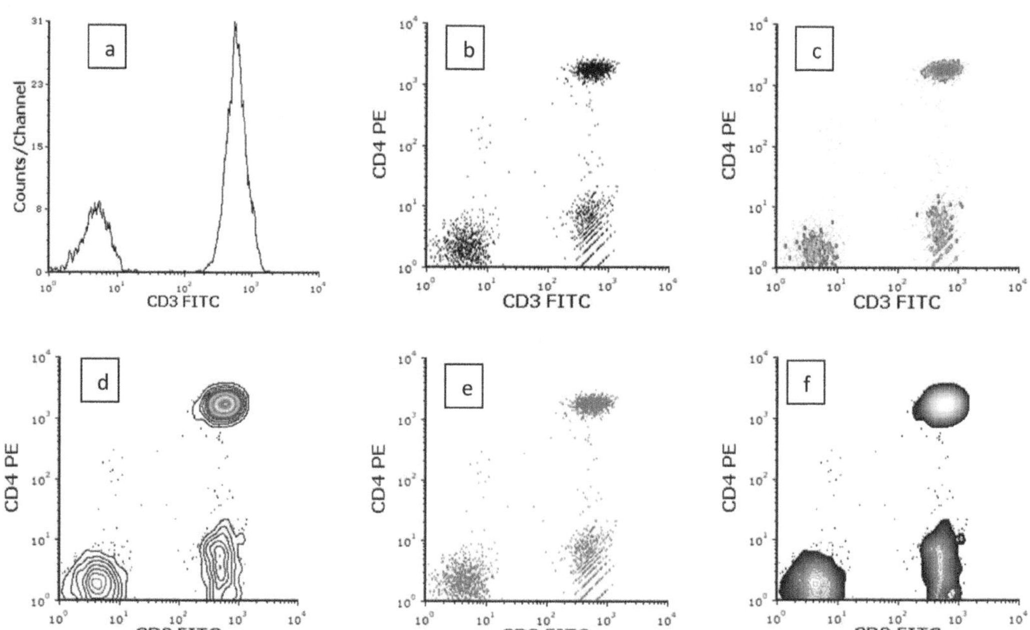

FIGURE 7.5 Different types of 1-D and 2-D plots used in flow cytometry: (a) histogram, (b) dot plot, (c) density plot, (d) contour plot, (e) color-dot plot, (f) zebra plot.

from various populations are in the range of twofold to tenfold, such as DNA Cell Cycle or FSC versus SSC scatter plot to represent white blood cell populations in RBC lysed blood (Figure 7.6). The linear scale is displayed as channel numbers that depend on the bits of the analog-to-digital converter (ADC) utilized in the machine, for example, machines like BD FACSCalibur have an ADC of 10 bit, so the values of linear scales lie between 0 and 1024 (2^{10}). The logarithmic scale is utilized when signal differences are greater than 100-fold or more, for example, data generated using various fluorescence-labeled antibodies. It can also be utilized to represent small cell populations on FSC/SSC scatter plots such as red blood cells, platelets, microparticles, bacteria, and so on. Generally, the logarithmic scale used in the flow cytometry data is 4- to 7-decade logs, depending on which flow cytometer one uses to generate the data.

Other than the linear and the logarithmic scale, there is another kind of scaling known as logicle or bi-exponential scaling, which allows improved representation of cells with minimal fluorescence signal. Cells with minimal fluorescence signal can be visualized properly with logicle or bi-exponential displays, in contrast to logarithmic displays, where the cells are compressed on the axis. This generally happens due to a photon counting error and gets revealed as data spread in the compensated data, and often leads to visual misrepresentations of the flow cytometry data. To avoid this data misinterpretation, this 'logicle or bi-exponential' scaling display is better than the classical logarithmic displays (**12, 38**) (Figure 7.7).

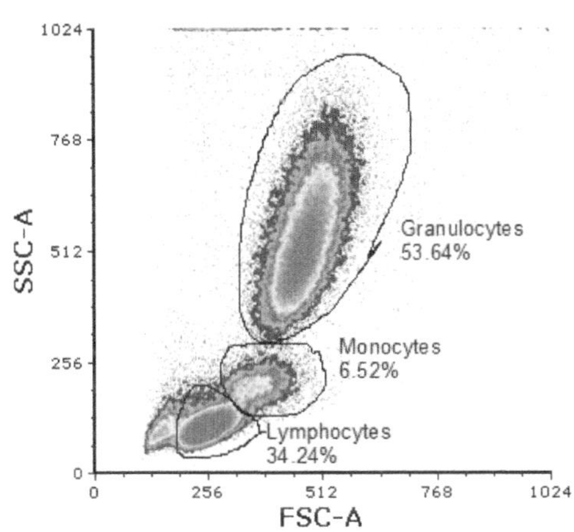

FIGURE 7.6 FSC versus SSC plot of RBC lysed whole blood.

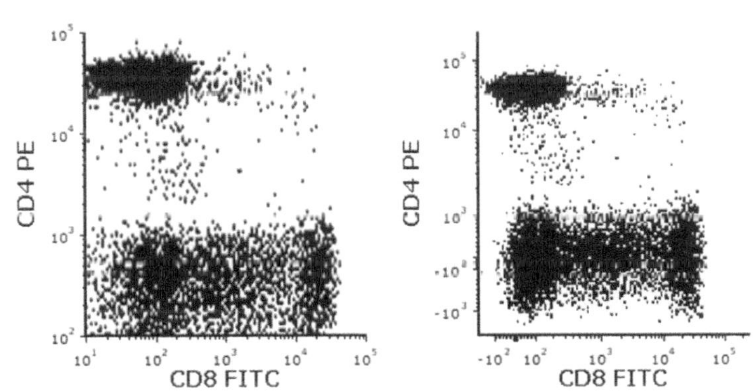

FIGURE 7.7 Logarithmic (a) versus bi-exponential (b) scaling display.

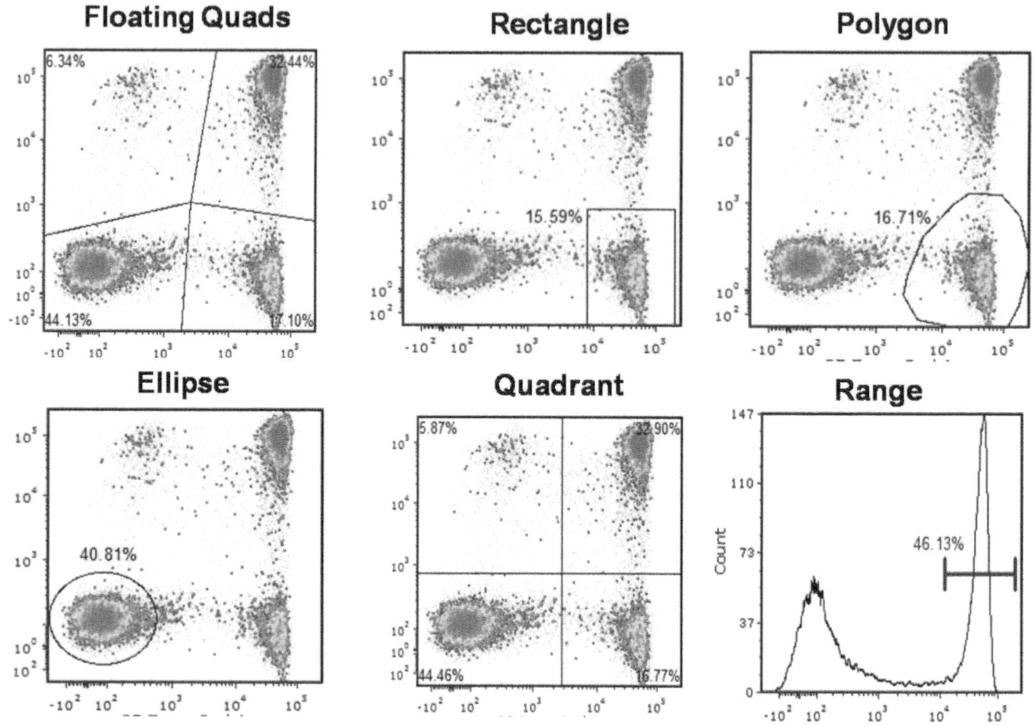

FIGURE 7.8 Different types of gating.

After data display and data scaling, another important thing in data analysis is data gating. Gating is defined as a selection of a population of interest by drawing a region or a gate around the population for further characterization. Gating in flow cytometry is somewhat subjective, so it should always be applied with reasoning and with proper gating controls (**17**, **28**), such as FMO, biological control, internal control, positive control, and so on, to get accurate gating. Various gating options can be utilized while gating the cells, such as ellipse, rectangle, polygon, quadrants, and so on (Figure 7.8). The main things to remember while gating are information about the size/complexity of the cell, change in the size of the cell on treatment, marker expression, and so on. It has been established that to analyze and interpret flow cytometry data properly, gating plays a major role, so we need to be very careful while placing gates over a population of interest.

7.3 FLOW CYTOMETRY APPLICATIONS

Flow cytometry is routinely used in basic research, clinical practice, and clinical trials to study multiple biological aspects of cells (Figure 7.9). The applications of flow cytometry include the following:

1. **Determining cell characteristics and function through immunophenotyping:** Using fluorescence-conjugated antibodies specific for a protein of interest, specific cell types expressing that protein may be distinguished from a mixed population.
2. **Cell cycle analysis based on DNA content:** Cellular DNA can be stained by a variety of DNA binding dyes, such as propidium iodide and 4'6-diamidino-2-phenylindole (DAPI). The premise of these dyes is that they are stoichiometric, that is, they bind in proportion to the amount of DNA present in the cell. Thus, the fluorescent signal generated by stained

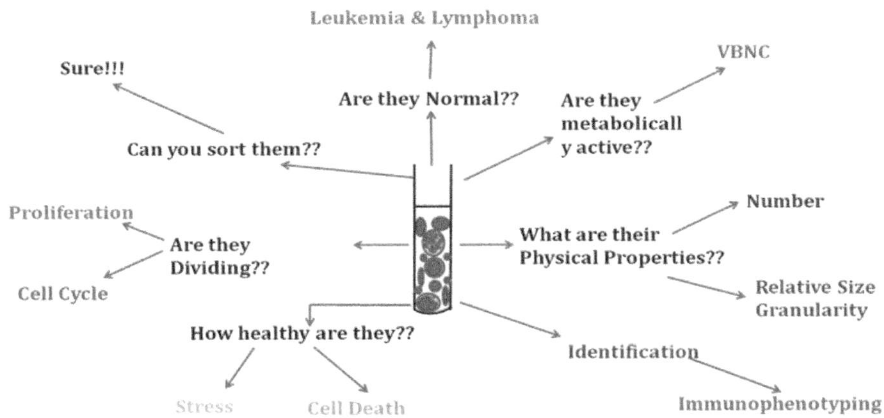

FIGURE 7.9 Flow cytometry and different applications. Flow cytometry can be used to answer several questions in biology.

cells is directly proportional to the amount of DNA in the nucleus. This approach reveals the distribution of cells in three major phases of the cycle (G1 vs. S vs. G2/M). This is because the cells that are in the S phase have more DNA than cells in G1. They take up proportionally more dye and fluoresce more brightly until they have doubled their DNA content. The cells in G2 are approximately twice as bright as the cells in G1. Along with determining cell cycle replication states, the assay can measure cell aneuploidy associated with chromosomal abnormalities.

3. **Transfection efficiency:** It is possible to determine transfection efficiency if a fluorescent protein (e.g., GFP) is used as a marker for transfection.
4. **Apoptosis measurement:** Live cells have intact membranes and are impermeable to dyes such as propidium iodide, which only leaks into cells with compromised membranes. Cells can be stained with propidium iodide and fluorescence-conjugated Annexin V, which is known to possess a high affinity for phosphatidylserine that is present on the surface of early apoptotic cells. Thus, dual parameter flow cytometry analysis using propidium iodide in conjunction with labeled Annexin V allows the discrimination between viable and apoptotic cells.
5. **Cell proliferation:** Cell proliferation assays are widely used in cell biology to measure cellular mitotic activity in response to growth factors, cytokines, and other regulators. A flow cytometer can measure proliferation by labeling resting cells with fluorescent dye, carboxyfluorescein succinimidyl ester (CFSE). CFSE is passed from parent to daughter cells so that the cells of each generation have half the fluorescent intensity of their parent cells. By measuring the reduction of the fluorescence signal, cellular proliferation can be determined.
6. **Cell sorting:** A particular subset of cells may be sorted from a mixture of cells based on particular properties. The flow cytometer cell sorter instrument analyzes the emission signals from each cell in the sample to sort the cells into one of two sample collection tubes. All the remaining cells are diverted to the waste reservoir. Modern cell sorters can sort up to six cell types simultaneously.
7. **Diagnosis of health disorders such as blood cancers:** In clinical labs, immunophenotyping is useful in diagnosing hematological malignancies such as lymphomas and leukemia.
8. **Quantification of soluble molecules:** Soluble/circulating antigens or antibodies can be quantified by flow cytometry if standard cells or beads are used.

REFERENCES

1. A Brief History of Flow Cytometry – Purdue Cytomation website (http://www.cyto.purdue.edu/cdroms/cyto5/sponsors/cytomate/flowhist.htm).
2. A History of Fluorescence-based Cytometry, Becton-Coulter Life Science (https://www.mybeckman.in/resources/technologies/flow-cytometry/history).
3. Afanasyev, V.N., Korol, B.A., Matylevich, N.P., Pechatnikov, V.A. and Umansky, S.R., 1993. The use of flow cytometry for the investigation of cell death. *Cytometry: The Journal of the International Society for Analytical Cytology*, 14(6), pp. 603–609.
4. Cossarizza, A., Chang, H.D., Radbruch, A., Acs, A., Adam, D., Adam-Klages, S., Agace, W.W., Aghaeepour, N., Akdis, M., Allez, M. and Almeida, L.N., 2019. Guidelines for the use of flow cytometry and cell sorting in immunological studies. *European Journal of Immunology*, 49(10), pp. 1457–1973.
5. Cossarizza, A., Chang, H.D., Radbruch, A., Akdis, M., Andrä, I., Annunziato, F., Bacher, P., Barnaba, V., Battistini, L., Bauer, W.M. and Baumgart, S., 2017. Guidelines for the use of flow cytometry and cell sorting in immunological studies. *European Journal of Immunology*, 47(10), pp. 1584–1797.
6. Darzynkiewicz, Z., Bruno, S., Del Bino, G., Gorczyca, W., Hotz, M.A., Lassota, P. and Traganos, F., 1992. Features of apoptotic cells measured by flow cytometry. *Cytometry*, 13(8), pp. 795–808.
7. Dive, C., Gregory, C.D., Phipps, D.J., Evans, D.L., Milner, A.E. and Wyllie, A.H., 1992. Analysis and discrimination of necrosis and apoptosis (programmed cell death) by multiparameter flow cytometry. *Biochimica et Biophysica Acta (BBA)-Molecular Cell Research*, 1133(3), pp. 275–285.
8. Dubelaar, G.B., Visser, J.W. and Donze, M., 1987. Anomalous behaviour of forward and perpendicular light scattering of a cyanobacterium owing to intracellular gas vacuoles. *Cytometry: The Journal of the International Society for Analytical Cytology*, 8(4), pp. 405–412.
9. Frankfurt, O.S., 1983. Assessment of cell viability by flow cytometric analysis using DNase exclusion. *Experimental Cell Research*, 144(2), pp. 478–482.
10. Goddard, G., Martin, J.C., Graves, S.W. and Kaduchak, G., 2006. Ultrasonic particle-concentration for sheathless focusing of particles for analysis in a flow cytometer. *Cytometry Part A: The Journal of the International Society for Analytical Cytology*, 69(2), pp. 66–74.
11. Herzenberg, L.A., Parks, D., Sahaf, B., Perez, O., Roederer, M. and Herzenberg, L.A., 2002. The history and future of the fluorescence activated cell sorter and flow cytometry: a view from Stanford. *Clinical Chemistry*, 48(10), pp. 1819–1827.
12. Herzenberg, L.A., Tung, J., Moore, W.A., Herzenberg, L.A. and Parks, D.R., 2006. Interpreting flow cytometry data: a guide for the perplexed. *Nature Immunology*, 7(7), pp. 681–685.
13. Howard, M. *Shapiro Practical Flow Cytometry*, 4th ed., July 2003. Wiley Publication ISBN: 978-0-471-41125-3 (https://www.wiley.com/en-us/Practical+Flow+Cytometry%2C+4th+Edition-p-9780471411253).
14. http://www.drmr.com/compensation/
15. https://expert.cheekyscientist.com/analyze-facs-data-prepare-cytometry-figures-scientific-papers/
16. https://flowjo.typepad.com/the_daily_dongle/2011/09/three-rules-for-compensation-controls.html
17. Hulspas, R., O'Gorman, M.R., Wood, B.L., Gratama, J.W. and Sutherland, D.R., 2009. Considerations for the control of background fluorescence in clinical flow cytometry. *Cytometry Part B: Clinical Cytometry: The Journal of the International Society for Analytical Cytology*, 76(6), pp. 355–364.
18. Kerker, M., 1983. Elastic and inelastic light scattering in flow cytometry. *Cytometry: The Journal of the International Society for Analytical Cytology*, 4(1), pp. 1–10.
19. Kerker, M., Chew, H., McNulty, P.J., Kratohvil, J.P., Cooke, D.D., Sculley, M. and Lee, M.P., 1979. Light scattering and fluorescence by small particles having internal structure. *Journal of Histochemistry & Cytochemistry*, 27(1), pp. 250–263.
20. Khowawisetsut, L., Sukapirom, K. and Pattanapanyasat, K., 2018. Data analysis and presentation in flow cytometry. *ScienceAsia*, 44, p. 19.
21. Köhler, G. and Milstein, C., 1975. Continuous cultures of fused cells secreting antibody of predefined specificity. *Nature*, 256(5517), pp. 495–497.
22. Lee, J.A., Spidlen, J., Boyce, K., Cai, J., Crosbie, N., Dalphin, M., Furlong, J., Gasparetto, M., Goldberg, M., Goralczyk, E.M. and Hyun, B., 2008. International society for advancement of cytometry data standards task F. Scheuermann RH, Brinkman RR. MIFlowCyt: the minimum information about a Flow Cytometry experiment. *Cytometry Part A: The Journal of the International Society for Analytical Cytology*, 73, pp. 926–930.

23. Lockshin, R.A. and Zakeri, Z.F., 1992. Physiology and protein synthesis in programmed cell death: Early synthesis and DNA degradation A. *Annals of the New York Academy of Sciences*, 663(1), pp. 234–249.
24. Loken, M.R. and Herzenberg, L.A., 1975. Analysis of cell populations with a fluorescence-activated cell sorter. *Annals of the New York Academy of Sciences*, 254(1), pp. 163–171.
25. Loken, M.R. and Houck, D.W., 1981. Light scattered at two wavelengths can discriminate viable lymphoid cell populations on a fluorescence-activated cell sorter. *Journal of Histochemistry & Cytochemistry*, 29(5), pp. 609–615.
26. Loken, M.R., Sweet, R.G. and Herzenberg, L.A., 1976. Cell discrimination by multiangle light scattering. *Journal of Histochemistry & Cytochemistry*, 24(1), pp. 284–291.
27. Maecker, H.T. and McCoy, J.P., 2010. A model for harmonizing flow cytometry in clinical trials. *Nature Immunology*, 11(11), pp. 975–978.
28. Maecker, H.T. and Trotter, J., 2006. Flow cytometry controls, instrument setup, and the determination of positivity. *Cytometry Part A: The Journal of the International Society for Analytical Cytology*, 69(9), pp. 1037–1042.
29. Mullaney, P.F., Van Dilla, M.A., Coulter, J.R. and Dean, P.N., 1969. Cell sizing: a light scattering photometer for rapid volume determination. *Review of Scientific Instruments*, 40(8), pp. 1029–1032.
30. Nordström, T., Willamo, P., Arvela, M., Stenroos, K. and Lindqvist, C., 1999. Detection of baculovirus-infected insect cells by flow cytometric side-scatter analyses. *Cytometry: The Journal of the International Society for Analytical Cytology*, 37(3), pp. 238–242.
31. Nüsse, M., Jülch, M., Geido, E., Bruno, S., Vinci, A.D., Giaretti, W. and Ruoss, K., 1989. Flow cytometric detection of mitotic cells using the bromodeoxyuridine/DNA technique in combination with 90 and forward scatter measurements. *Cytometry: The Journal of the International Society for Analytical Cytology*, 10(3), pp. 312–319.
32. Ormerod, M.G., Sun, X.M., Snowden, R.T., Davies, R., Fearnhead, H. and Cohen, G.M., 1993. Increased membrane permeability of apoptotic thymocytes: a flow cytometric study. *Cytometry: The Journal of the International Society for Analytical Cytology*, 14(6), pp. 595–602.
33. Palit, S., Heuser, C., de Almeida, G.P., Theis, F.J. and Zielinski, C.E., 2019. Meeting the challenges of high-dimensional single-cell data analysis in immunology. *Frontiers in Immunology*, 10, p. 1515.
34. Perez Gonzalez and Jameson Introduction to Flow Cytometry Melbourne Cytometry Platform (https://biomedicalsciences.unimelb.edu.au/__data/assets/pdf_file/0012/3549837/MCP-introduction-to-flow-cytometry-final.pdf).
35. Picot, J., Guerin, C.L., Le Van Kim, C. and Boulanger, C.M., 2012. Flow cytometry: Retrospective, fundamentals and recent instrumentation. *Cytotechnology*, 64(2), pp. 109–130.
36. Pinkel, D. and Stovel, R., 1985. Flow chambers and sample handling. In M.A. Van Dilla, P.N. Dean, O.D. Laerum, M.R. Melame (Eds.), *Flow cytometry: Instrumentation and data analysis*. London: Academic Press, pp. 77–128, ISBN: 0-12-712150-1
37. Riedy, M.C., Muirhead, K.A., Jensen, C.P. and Stewart, C.C., 1991. Use of a photolabeling technique to identify nonviable cells in fixed homologous or heterologous cell populations. *Cytometry: The Journal of the International Society for Analytical Cytology*, 12(2), pp. 133–139.
38. Roederer, M., 2001. Spectral compensation for flow cytometry: visualization artifacts, limitations, and caveats. *Cytometry: The Journal of the International Society for Analytical Cytology*, 45(3), pp. 194–205
39. Salzman, G.C., Singham, S.B., Johnston, R.G. and Bohren, C.F., 1990. Light scattering and cytometry. In M.R. Melamed, T. Lindmo, M.L. Mendelsohn (Eds.), *Flow cytometry and sorting*. 2nd ed. New York: Wiley-Liss, pp. 81–107
40. Salzman, G.C., Crowell, J.M., Goad, C.A., Hansen, K.M., Hiebert, R.D., LaBauve, P.M., Martin, J.C., Ingram, M.L. and Mullaney, P.F., 1975. A flow-system multiangle light-scattering instrument for cell characterization. *Clinical Chemistry*, 21(9), pp. 1297–1304.
41. Salzman, G.C., Crowell, J.M., Martin, J.C., Trujillo, T.T., Romero, A., Mullaney, P.F. and LaBauve, P.M., 1975. Cell classification by laser light scattering: identification and separation of unstained leukocytes. *Acta Cytologica*, 19(4), pp. 374–377.
42. Salzman, G.C., Mullaney, P.F. and Price, B.J., 1979. Light-scattering approaches to cell characterization. In M.R. Melamed, P.F. Mullaney and M.L. Mendelson (Eds.), *Flow Cytometry and Cell Sorting*. New York: John Wiley and Sons, pp. 105–124.

43. Salzman, G.C., Wilder, M.E. and Jett, J.H., 1979. Light scattering with stream-in-air flow systems. *Journal of Histochemistry & Cytochemistry*, 27(1), pp. 264–267.
44. Schmid, I., Krall, W.J., Uittenbogaart, C.H., Braun, J. and Giorgi, J.V., 1992. Dead cell discrimination with 7-amino-actinomcin D in combination with dual color immunofluorescence in single laser flow cytometry. *Cytometry: The Journal of the International Society for Analytical Cytology*, 13(2), pp. 204–208.
45. Shapiro, H.M., 1977. Fluorescent dyes for differential counts by flow cytometry: does histochemistry tell us much more than cell geometry? *Journal of Histochemistry & Cytochemistry*, 25(8), pp. 976–989.
46. Sharpless, T.K., Traganos, F., Darzynkiewicz, Z. and Melamed, M.R., 1975. Flow cytofluorimeuy: discrimination between single cells and cell aggregates by direct size measurements. *Acta Cytol* 19(6), pp. 577–581.
47. Sharpless, T., Bartholdi, M. and Melamed, M.R., 1977. Size and refractive index dependence of simple forward angle scattering measurements in a flow system using sharply-focused illumination. *Journal of Histochemistry & Cytochemistry*, 25(7), pp. 845–856.
48. Sharpless, T.K. and Melamed, M.R., 1976. Estimation of cell size from pulse shape in flow cytofluorometry. *Journal of Histochemistry & Cytochemistry*, 24(1), pp. 257–264.
49. Spidlen, J., Breuer, K. and Brinkman, R., 2012. Preparing a minimum information about a flow cytometry experiment (MIFlowCyt) compliant manuscript using the International Society for Advancement of Cytometry (ISAC) fcs file repository (FlowRepository. org). *Current Protocols in Cytometry*, 61(1), pp. 10–18.
50. Steen, H.B., 1990. Light scattering measurement in an arc lamp-based flow cytometer. *Cytometry: The Journal of the International Society for Analytical Cytology*, 11(2), pp. 223–230.
51. Telford, W.G., King, L.E. and Fraker, P.J., 1992. Comparative evaluation of several DNA binding dyes in the detection of apoptosis-associated chromatin degradation by flow cytometry. *Cytometry: The Journal of the International Society for Analytical Cytology*, 13(2), pp. 137–143.
52. Terstappen, L.W.M.M., Shah, V.O., Conrad, M.P., Recktenwald, D. and Loken, M.R., 1988. Discriminating between damaged and intact cells in fixed flow cytometric samples. *Cytometry: The Journal of the International Society for Analytical Cytology*, 9(5), pp. 477–484.
53. Uličný, J., 1992. Lorenz-Mie light scattering in cellular biology. *General Physiology and Biophysics*, 11, pp. 133–151.
54. Watt, S.M., Burgess, A.W., Metcalf, D. and Battye, F.L., 1980. Isolation of mouse bone marrow neutrophils by light scatter and autofluorescence. *Journal of Histochemistry & Cytochemistry*, 28(9), pp. 934–946.
55. Weil, G.J. and Chused, T.M., 1981. Eosinophil autofluorescence and its use in isolation and analysis of human eosinophils using flow microfluorometry. *Blood*, 57, pp. 1099–1104.
56. Welsh, J.A., Van Der Pol, E., Arkesteijn, G.J., Bremer, M., Brisson, A., Coumans, F., Dignat-George, F., Duggan, E., Ghiran, I., Giebel, B. and Görgens, A., 2020. MIFlowCyt-EV: A framework for standardized reporting of extracellular vesicle flow cytometry experiments. *Journal of Extracellular Vesicles*, 9(1), p. 1713526.
57. Zamai, L., Bareggi, R., Santavenere, E. and Vitale, M., 1993. Subtraction of autofluorescent dead cells from the lymphocyte flow cytometric binding assay. *Cytometry: The Journal of the International Society for Analytical Cytology*, 14(8), pp. 951–954.
58. Zucker, R.M., Elstein, K.H., Easterling, R.E. and Massaro, E.J., 1988. Flow cytometric discrimination of mitotic nuclei by right-angle light scatter. *Cytometry: The Journal of the International Society for Analytical Cytology*, 9(3), pp. 226–231.

8 CRISPR/Cas Gene Editing System in the Silkworm, *Bombyx mori*

Venkatesh R Kumar, Devika Srivastava and Rajendra Mundkar

CONTENTS

DOI: 10.1201/9781003220831-8

8.1 INTRODUCTION

Currently the novel combination of technology and biology has undoubtedly made spectacular changes in the genome editing of living organisms. Three types of nucleases are used for targeted DNA Double Strand Break (DSB) formation, namely, transcription activator-like effectors nuclease (TALEN), zinc finger nuclease (ZFN) (protein-driven genome editors) and clustered regularly interspaced short palindromic repeat-associated (CRISPR-Cas) system that is an RNA-guided genome editing system (Li and Peng 2019).

CRISPR/Cas system is considered as a significant part of bacterial adaptive immunity; it is composed of Cas nuclease and two individual RNA components, a fixed tracrRNA (trans-activating RNA) and a programmable crRNA (CRISPR RNA). Cas protein is competent to cut the entering phage DNA into small pieces, and they are incorporated into the CRISPR array as a spacer; afterward, this array is transcribed to produce crRNA and a complementary tracrRNA, which form a double-stranded RNA structure that employs Cas proteins for cleavage (Mei et al. 2016). Thus, the defensive CRISPR/Cas system represents an impressive system for explicit genome editing, as in vivo double-stranded breaks (DSB) are repaired chiefly through homology-directed repair (HDR) or non-homologous end joining (NHEJ) pathways introducing insertions or deletions (Taning et al. 2017). Since ZFNs and TALENs were showed a serious deficiency in the gene-editing system, in 2012, CRISPR/Cas9 technology has been emerged as a powerful tool for genome editing that overcomes the shortcomings of earlier techniques due to its accuracy (Rodríguez–Rodríguez et al. 2019). The CRISPR/Cas9 system has *Streptococcus pyogenes* Cas 9 nuclease that is directed to introduce a site-directed double-strand DNA break using a small guide RNA molecule. Cas9 nuclease, which can identify the protospacer adjacent motifs (PAM) sequence NGG (N is A, T, C or G) (Jiang and Doudna 2017; Rodríguez-Rodríguezet al. 2019; Taning et al. 2017). CRISPR/Cas9 has become ubiquitous in biological sciences including insect sciences. Silkworm (*Bombyx mori*) is a lepidopteran insect with the immense economically important and finest model organism for a variety of scientific studies. *B. mori* has been one of the earliest insect models for which a genomic editing tool was applied to elucidate the function of genes. Silkworm was the first model organism of agriculture insect and second arthropods after *Drosophila* where the CRISPR system was used for the first time. The efficacy of three different genomic engineering techniques, namely ZFNs, TALENs and CRISPR/Cas9 was compared by researchers with reference to the silkworm research and reported that CRISPR/Cas9 the system was considered as the most convenient and effective method (Sun et al. 2017). Diversified studies have been carried out on silkworms by applying CRISPR/Cas 9 gene-editing tool to understand the inhibition of microbial infections (Chenet al. 2017; Dong et al. 2019a), various developmental processes of silkworm (Zhanget al. 2015; Liu et al. 2019b), gene targeting by knocking out NHEJ pathway (Zhu et al. 2016) and behavioural studies (Zhang et al. 2019a; Liu et al. 2017). Further, transgenic silkworms were also produced with the combination of spider and mulberry silk (Zhang et al. 2019b). This chapter covers the application of CRISPR/Cas 9 technology with diversified research on silkworms with updated information and provides a potential outlook for this most promising field of research.

8.2 UNDERSTANDING CRISPR/CAS TECHNOLOGY

The CRISPR/Cas9 system consists of two key molecules that cut DNA strands: (i) an enzyme called Cas9, which acts as a pair of 'molecular scissors' that can cut the two strands of DNA at a specific location in the genome so that bits of DNA can then be added or removed; and (ii) a piece of RNA called guide RNA (gRNA). This consists of a small piece of pre-designed RNA sequence (about 20 bases long) located within a longer RNA scaffold. The scaffold part binds to DNA at CRISPR site and the pre-designed sequence 'guides' Cas9 to the right part of the genome. This makes sure that the Cas9 enzyme cuts at the right point in the genome.

The guide RNA is designed to find and bind to a specific sequence in the DNA. The sequence of gRNA pairs with the complementary sequence of the target DNA. The Cas9 follows the guide RNA to the same location in the DNA sequence and makes a cut across both strands of the DNA. DNA has an innate repair mechanism. The cut end joins with the DNA segments available at the site. Scientists introduce DNA segments with known and desired sequences. These sequences join the cut end thus new sequence (gene) is added to the original DNA strand.

8.3 IMPORTANT TECHNICAL TERMINOLOGIES TO UNDERSTAND CRISPR/CAS

Anti-CRISPR: Anti-CRISPR, or Acr, are small proteins found in phages that have evolved ways to overcome the endogenous CRISPR systems to protect against inactivation of their own genomes. Some Acr proteins interfere with DNA binding at the PAM site, whereas others interfere with the Cas9 HNH endonuclease domain. Niopek (2016) developed a way to opto-genetically control genome editing by fusing the anti-CRISPR protein AcrIIA4 with the photo-sensitive LOV2 domain. This allows researchers to switch genome editing on and off using light.

Cas9 (CRISPR-associated protein 9) is an endonuclease, or enzyme, that acts as 'molecular scissors' to cut DNA at a location specified by a guide RNA. It is an enzyme that uses CRISPR sequences as a guide to recognize and cleave specific strands of DNA that are complementary to the CRISPR sequence. Cas9 enzymes together with CRISPR sequences form the basis of a technology known as CRISPR/Cas9. Emmanuelle Charpentier and Jennifer Doudna were awarded the Nobel Prize in chemistry in 2020 for the development of the CRISPR/Cas9 genome editing technique.

CRISPR: Clustered Regularly Interspaced Short Palindromic Repeats.

crRNA: CRISPR RNA (crRNA).

dsDNA cleavage: Double-strand DNA cleavage. Cleavage of both the strand of DNA at the spacer complementary site, called protospacer.

Endonuclease: Endonucleases are enzymes that cleave the phosphodiester bond within a polynucleotide chain. Some, such as deoxyribonuclease I, cut DNA randomly, while restriction endonucleases, cleave only at very specific nucleotide sequences.

Guide RNA (gRNA): It is a type of RNA molecule that binds to Cas9 and specifies, based on the sequence of the gRNA, the location at which Cas9 will cut DNA.

Palindromic sequence: A palindrome reads the same backwards as forwards, e.g. madam. A palindromic nucleic acid sequence reading in one direction (e.g. 5' to 3') on one strand matches the sequence reading in the opposite direction (e.g. 3' to 5') on the complementary strand. Thus complementary strands become palindromic of each other. E.g., 5'- G A A T T C -3' and 3'- C T T A A G -5' are palindromic complementary strands.

PAM: PAM is Protospacer Adjacent Motif. It is a specific nucleotide sequence that is compulsorily required for dsDNA cleavage. The most commonly used Cas9 nuclease, derived from S. pyogenes (spCas9), recognizes a PAM sequence of NGG that is found directly downstream of the target sequence in the genomic DNA, on the non-target strand. N stands for any nucleotide (A, T, G, C). Cas14 or Prime editing system does not need PAM.

Phosphodiester bonds: Phosphodiester bonds make up the backbones of DNA and RNA. The phosphate is attached to the 5' carbon. The 3' carbon of one sugar is bonded to the 5' phosphate of the adjacent sugar. Specifically, the phosphodiester bond links the 3' carbon atom of one sugar molecule and the 5' carbon atom of another (hence the name, 3', 5' phosphodiester linkage).

Repeats: Repeats are repeated nucleotide sequences that occur in multiple copies throughout the genome. In many organisms, a significant fraction of the genomic DNA is highly repetitive. In humans, over two-thirds of the sequences are repetitive sequences.

Spacer: The part of the crRNA sequence that is complementary to the target sequence is known as a spacer.

ssDNA cleavage: Single strand DNA cleavage. Cleavage of only the complementary strand of DNA at the spacer site, called protospacer.

tracrRNA: Trans-activating crispr RNA (tracrRNA) is a small trans-encoded RNA. In an RNA-mediated defence system of bacteria, which protects against viruses and plasmids, the defensive pathway has three steps. First a copy of the invading nucleic acid of virus is integrated into the CRISPR locus. Next, CRISPR RNAs (crRNAs) are transcribed from this CRISPR locus. The crRNAs are then incorporated into effector complexes, where the crRNA guides the complex to the invading nucleic acid and the Cas proteins degrade this nucleic acid. Type II CRISPR/Cas systems require a tracrRNA which plays a role in the maturation of crRNA. The tracrRNA is partially complementary to and base pairs with a pre-crRNA forming an RNA duplex. This is cleaved by RNase III, an RNA-specific ribonuclease, to form a crRNA/tracrRNA hybrid. This hybrid acts as a guide for the endonuclease Cas9, which cleaves the invading nucleic acid of virus.

8.4 EVOLUTION OF CRISPR/CAS TECHNOLOGY

A number of CRISPR/Cas systems have been developed. Collectively, the 93 Cas genes are grouped into 35 families based on sequence similarity of the encoded proteins. Eleven of the 35 families form the Cas core, which includes the protein families Cas1 through Cas13. A complete CRISPR/Cas locus has at least one gene belonging to the Cas core (Makarova et al. 2015).

CRISPR/Cas systems grouped into two classes.

Class 1 systems use a complex of multiple Cas proteins to degrade foreign nucleic acids. They use several Cas proteins together with the CRISPR RNAs (crRNA) to build a functional endonuclease. Class 1 is further divided into types I, III, and IV. Most CRISPR-Cas systems have a Cas1 protein.

Class 2 systems use a single large Cas protein for the same purpose. Class 2 is divided into types II, V, and VI. The six system types are divided into 19 subtypes. Each type and most subtypes are characterized by a 'signature gene' found almost exclusively in the category. Classification is also based on the complement of Cas genes that are present (Makarova et al. 2015).

Many organisms contain multiple CRISPR/Cas systems, suggesting that they are compatible and may share components. The sporadic distribution of the CRISPR/Cas subtypes suggests that the CRISPR/Cas system is subject to horizontal gene transfer during microbial evolution (Makarova et al. 2015) (Table 8.1).

TABLE 8.1
Classification of Cas

Classification of Cas (CRISPR-associated system)											
Class	**Type**	**Cas name**									
Class 1	Type I	Cas 1	Cas 2	Cas 4	Cas 5	Cas 6	Cas 7	SS	Cas 8	Cas 3'	Cas 3"
	Type III	Cas 1	Cas 2	RT	Cas 5	Cas 6	Cas 7	SS	Cas 10/LS		
	Type IV	Cas 1	Cas 2		Cas 5		Cas7	SS	Cas 12/Csf1		
Class 2	Type II	Cas 1	Cas 2							Cas 9	
	Type V	Cas 1	Cas 2	Cas 12	Cas 12a (cpf1)	Cas 12b (C2C1)	Cas12c (C2c3)	Cas12d (CasY)	Cas12e (CasX)	Cas12f (Cas14)	Cas 12
								Cas12g	Cas12h	Cas12i	Cas12k (C2c5)
	Type VI	Cas 1	Cas 2				Cas 13	Cas 13a (C2c2)	Cas 13b	Cas 13c	Cas 13d

8.5 APPLICATION OF CRISPR/CAS TOOLS IN SILKWORM

Zeng et al. (2016) describe that in most cases of CRISPR/Cas9 system application, single guide RNA (sgRNA) targeting sites have been designed as GN19NGG or GGN18NGG, because of restriction of the initiation nucleotide for RNA Pol III promoters. However, authors demonstrated that the U6 promoter from a lepidopteran model insect, *Bombyx mori*, effectively expressed the sgRNA initiated with any nucleotide bases (adenine, thymine, guanine or cytosine), which further expands the CRISPR targeting space, with the highest increase occurring on the Z sex chromosome.

Brady et al. (2021) report that CRISPR/Cas9 technology is being rapidly adopted as the most efficient molecular tool for generating silkworm lines carrying mutations in target genes. They have illustrated a complete and efficient workflow to screen, characterize rapidly and follow mutations through generations, allowing the generation of *B. mori* lines, stably inheriting single CRISPR/Cas9-induced mutations. This approach relies on the use of different molecular methods, the heteroduplex assay, cloning followed by Sanger sequencing and the amplification refractory mutation system PCR. They use these methodologies in a sequential combination, which allows the identification of CRISPR/Cas9-induced mutations in genes mapping on both autosomes and sex chromosomes, and the selection of appropriate individuals to found stable mutant *B. mori* lines. They say that this protocol could be further applied to screen CRISPR/Cas9 mutations in haploid insects.

8.5.1 NPV Resistance

Bombyx mori nuclear polyhedrosis virus (BmNPV) infection in silkworms has often been lethal and is challenging to prevent when compared to other silkworm diseases. Its lethality is associated with the nature of the viral particle and silkworm strains (Xu et al. 2015). Among major silkworm diseases, BmNPV infection in silkworms exerts maximum economic losses to the silk industry. In India, BmNPV infection is accountable for >50% of silk cocoon crop losses (Gani et al. 2017). However, available gene-editing tools, particularly CRISPR/Cas9, can curb viral infection in silkworms. To boost virus clearance in silkworms, Chen et al. adopted the combination of transposon-based transgenesis and (CRISPR)/CRISPR-associated 9 (Cas9) systems for direct cleaving of the *Bombyx mori* nuclear polyhedrosis virus (BmNPV) genome DNA. Transgenic silkworm constitutively expressing Cas9 and guide RNAs (gRNAs) targeted BmNPV immediate early-1 (ie-1) and me53 gene efficiently encouraged target-specific cleavage and successive mutagenesis, chiefly deletion of large segments in BmNPV genomes consequently exerted vigorous repression of BmNPV proliferation (Chen et al. 2017).

Dong et al. successfully developed CRISPR/Cas9 expressing transgenic lines and the double sgRNA targeting *Bombyx mori* nuclear polyhedrosis virus (BmNPV) immediate early-1 (ie-1) gene in the silkworm, respectively, and by G1 generation hybridization obtained four transgenic hybrid lines, namely Cas9(−)/sgRNA(−), Cas9(+)/sgRNA(−), Cas9(−)/sgRNA(+) and Cas9(+)/sgRNA(+). Further, the target site of the BmNPV genome was successfully edited by Cas9(+)/sgRNA(+) transgenic lines. As part of a study, subsequent to BmNPV infection, large fragment deletion was observed. Furthermore, the antiviral study of the Cas9(+)/sgRNA(+) transgenic lines revealed that after inoculation with occlusion bodies, the median lethal dose (LD50) was 1,000-fold higher than the normal lines (Dong et al. 2018).

Dong et al. (2018) report that they had established a highly efficient virus-inducible CRISPR/Cas9 system in transgenic silkworms. That system included the baculovirus-inducible promoter 39K, which directs transcription of the gene encoding, the Cas9 protein and the U6 promoter, which targets the sgATAD3A site of the ATPase family AAA domain-containing protein 3 (ATAD3A) gene. The double-positive transgenic line sgATAD3A × 39K-Cas9 (ATAD3A-KO) was obtained by hybridization; antiviral activity in this hybrid transgenic line was induced only after *Bombyx mori* nucleopolyhedro virus (BmNPV) infection. They also report that the BmNPV-inducible system significantly reduced off-target effects and did not affect the economically important characteristics of the transgenic silkworms. Most importantly, this novel system efficiently and consistently edited target genes, inhibiting BmNPV replication after the transgenic silkworms were inoculated with occlusion bodies (OBs). The suppression of BmNPV by the virus-inducible system was comparable to that of the stably expressed CRISPR/Cas9 system. They claim that they have developed BmNPV-resistant ATAD3A-KO transgenic silkworm line.

In 2019, Dong et al. for the first time, used multiplex CRISPR/Cas9 system to inhibit nuclear polyhedrosis virus replication in silkworms. Researchers had created a one-vector (pSL1180-Cas9-U6-sgRNA) system that was found to express multiple sgRNA and Cas9 protein for the excision of *Bombyx mori* nuclear polyhedrosis virus (BmNPV) in insect cells (Dong et al. 2019a). To control viral infections, many cells undergo apoptosis, with a resulting decline in the release of progeny virus. However, viruses on the other hand have developed multiple distinct mechanisms for modulating host cell apoptosis (Thomson 2001). Similarly, Baculovirus possess the mechanism to modulate the apoptosis in silkworms and the gene involved in this mechanism is the BmNPV inhibitor of apoptosis 2 (iap2). Huang et al. designed a transgenic silkworm using CRISPR/Cas 9 gene tool in order to knock out the BmNPV inhibitor of apoptosis 2 (iap2). The sequencing outcome revealed that all the sequences could edit the target site of the BmNPV iap2 gene. Significant reduction in mortality rate and about 100 times higher of a median lethal dose (LD50) was observed in the BmNPV iap2 knockout strain than the control group. After BmNPV iap2 being edited in transgenic silkworm, the expression levels of apoptotic related gene namely BmDreed, Bmiap2 and BmICE were found to be remarkably affected also the caspase 9 activity was reported to be increased (Huang et al. 2021).

8.5.2 Microsporidian Resistance

Nosema bombycis, a causal organism of pebrine disease in the *Bombyx mori*, is a microsproridian that is responsible for huge economic losses to the sericulture industry (Huang et al. 2018; Ma et al. 2013). Dong et al. created a potent technique that delivers resistance to *N. bombycis* in silkworms. This system was comprised of a target BmATAD3A gene line and an HSP70 promoter-induced expression of the Cas9 protein line. Through hybridization and activation of the CRISPR/Cas9 system under the condition of microsporidia infection, the double-positive HSP70-Cas9 (+) ×sgATAD3A (+) lines were obtained, while it was found to be silent in non-infected individuals. The editing of mATAD3A gene could efficiently be done by the system and induce large deletions. Further, authors also reported that after microsporidian infection, the survival rate of transgenic silkworm could effectively be improved by an HSP70-induced system and it inhibits the key microsporidia expression. In addition to this, the authors observed no noteworthy developmental difference between normal individuals and transgenic silkworms infected with microsporidia (Figure 8.1) (Dong et al. 2019b).

8.5.3 BmWnt1 Gene Inactivation

The embryonic development in insects and genome mining systems has given augment to ever-growing insect developmental studies. The importance of the Wnt1 signaling pathway has always been witnessed in a number of studies. The Wnt signaling pathways are a group of signal transduction pathways that begin with proteins that pass signals into a cell through cell surface receptors. The name Wnt is created from the names Wingless and Int-1. Studies conducted in *Bombyx mori* have reported that the Wnt1 gene has played a significant role in spot patterns formation and normal abdominal segments (Sun et al. 2017). Zhang et al. designed a CRISPR/Cas9 mediated gene mutagenesis and accomplished functional analysis of *Bombyx* Wnt1 during embryogenesis. Direct injection of Cas9 mRNA and BmWnt1-specific sgRNA into *Bombyx* embryos induced a typical Wnt-deficient phenotype: injected embryos could not hatch and exhibited severe defects in body segmentation and pigmentation in a dose-dependent manner. Quantitative real-time PCR (qRT-PCR) analysis revealed that Hox genes were down-regulated after BmWnt1 depletion (Figure 8.2) (Zhang et al. 2015).

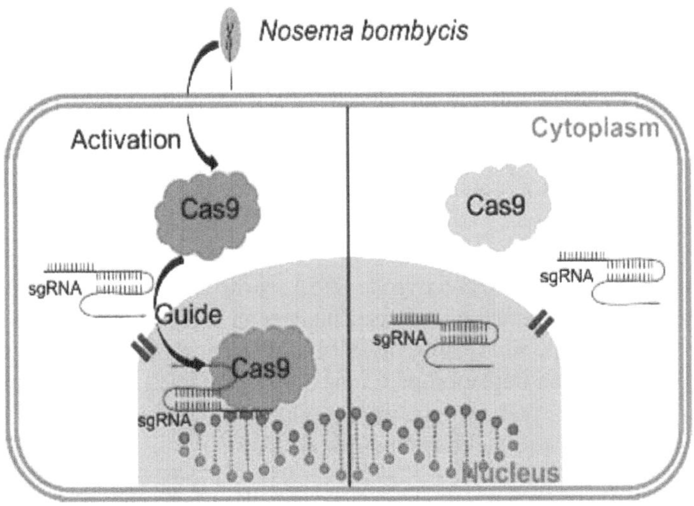

FIGURE 8.1 Schematic diagram illustrating *Nosema bombycis* resistance with high specificity is deliberated by microsporidia-inducible transgenic CRISPR/Cas9 system.

Source: Adapted with permission from *Applied Microbiology and Biotechnology* 2019, 103(23–24), 9583–9592, Springer

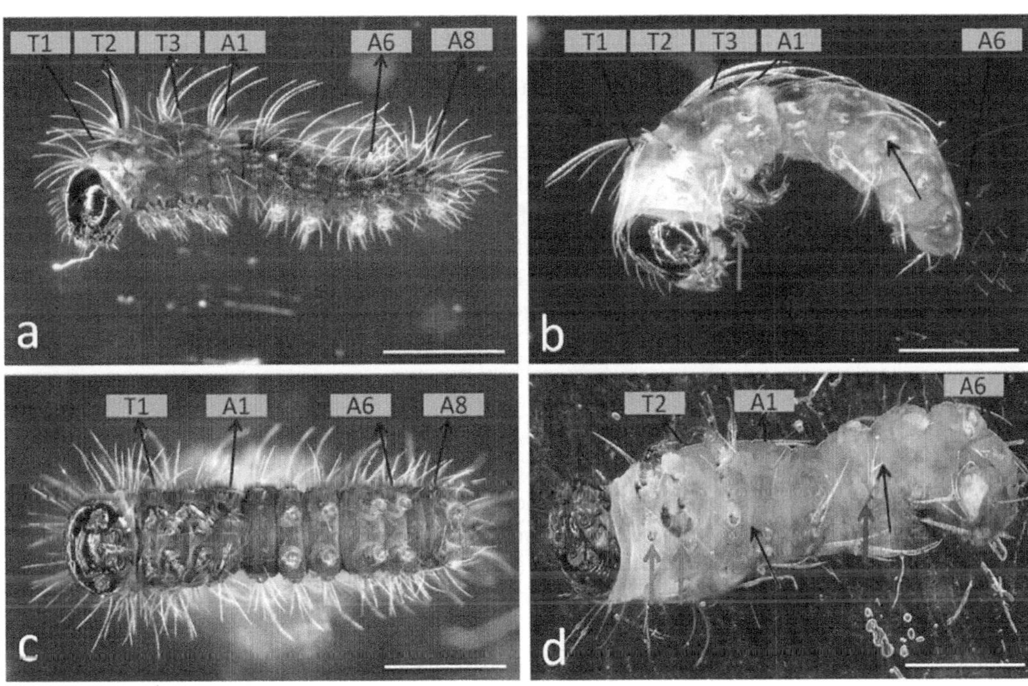

FIGURE 8.2 Pictures showing CRISPR/Cas9 driven abnormal segmentation and pigmentation in Wnt-deficient mutant silkworm larvae. (a) Wild type larva (lateral view). (b) Mutant larvae with segment posterior to A6 are missing. (c) Wild type larva (ventral view). (d) Mutant embryo (ventral view), abnormal thoracic legs and malformed prolegs are indicated by white and thick grey with white dot.

Source: Adapted with permission from the *Journal of Insect Physiology* 2015, 79:73–9, Elsevier

8.5.4 Bm-miRNA Excision

MicroRNAs (miRNA) have crucial role in the development of insects. Liu et al. (2020a) opined that miR-34 directly targeted BmE74 and BmCPG4 by using a dual luciferase reporter assay in HEK293T cells. The loss-of-function of BmCPG4 caused wing defects that was similar to the phenotype of miR-34 overexpression in animals (Liu et al. 2020a).

8.5.5 JH gene Inactivation

Juvenile hormone (JH) and 20-hydroxyecdysone (20E) are the two major hormones that are responsible for the growth and development of insects. Zhang et al. performed a loss-of-function analysis of JH esterase (JHE) in *Bombyx mori*, using CRISPR/Cas9 transgenic system. As a result, an extension of larval stages due to the depletion of *B. mori JHE* (*BmJHE*) was reported by the authors. Further, a larger size of larvae and cocoons were produced by prolonging larval stages, which are significantly beneficial to silk production (Zhang et al. 2017).

8.5.6 Fibroin Gene Inactivation

Cui et al. efficaciously conducted genome editing of Bmfib-H gene (largest silk protein) in *B. mori* with the aid of CRISPR/Cas9 system. The loss-of-function mutations resulted in naked pupa and thin cocoon mutant phenotypes. Further their results revealed posterior gland of knockout individuals exhibited degradation into fragments in advance of programmed cell death of silk gland cells when compared with wild type (Cui et al. 2018).

8.5.7 BmTCTP Gene Functional Analysis

In insects, translational controlled tumor protein (TCTP), which is a highly conserved protein, is known to be essential for cell growth and proliferation. To understand the role of TCTP in *Bombyx mori*, Liu et al. carried out the functional analysis of BmTCTP with the help of a binary transgenic CRISPR/Cas9 system. Further, the development was arrested and consequent lethality in third instar larvae was observed due to disruption of BmTCTP. The analysis showed impairment of growth initiated from reduced cell size, and proliferation and differentiation of intestinal epithelial cells were found to be affected. Depletion of BmTCTP significantly affected the genes involved in the metabolism of carbohydrate and lipid and digestive system pathways (Liu et al. 2019b).

8.5.8 BIGFLP Gene Functional Analysis

In insects and other vertebrates, insulin family peptides are known to be significant regulators of metabolism and growth. Insects are known for two types of insulin family peptides, namely insulin-like peptides and insulin-like growth factor (IGF)-like peptides (IGFLPs). Fujinaga studied the role of BIGFLP in *B. mori* using CRISPR and CRISPR/Cas9 systems. When compared with wild type moths, *BIGFLP* knockout moths showed reduced body weights and size of the appendages. Knockout females exhibited significantly lower ovary weights and the number of eggs than wild type females. However, when mutant ovaries were transplanted into wild-type hosts, pupae acquired similar weight to wild-type ovaries that were transplanted into wild-type hosts, indicating that in the haemolymph IGFLP stimulates ovarian development (Fujinaga et al. 2019).

8.5.9 BmShadow Gene Knockout

Diapause is a period that is characterised by the development and growth thereby, enabling insects to surpass and survive unfavorable environmental conditions. In *Bombyx mori*, the nature of diapause is determined by the expression of genetic characters and endocrinological mechanisms under the impact of environmental stimuli, such as photoperiod and temperature (Singh et al. 2013). Therefore, in order to fully understand the mechanism and nature of diapause, study at the gene stage is crucial for coherent conclusion of any research. Zhu et al. in 2021 performed a study to interpret the role of Shadow gene (BmShadow) and its link with diapause using novel CRISPR/Cas9 technique. The BmShadow mutant had a 42-base deletion related precisely to the regions of the amino acid from positions 155 to 168. The BmShadow knockout larvae could not moult at the pre-moulting stage of the second instar of the silkworm. However, moulting was achieved when BmShadow knockout larvae were fed with 20E analogue at the later stage of the second instar and were rescued and developed into the third instar. Further the moulting of mutant larvae into the first instar could not be achieved when they hatched from eggs after being stored at 5°C for 40 days, which indicates that the 20E transported from the mother was partially consumed in the diapause maintenance stage. Authors report that BmShadow is involved in the synthesis of 20E and a 14-amino acids region from position 155 to 168 is indispensable for its function (Zhu et al. 2021).

8.5.10 BmEL Gene Functional Analysis

BmELAV-like-1 (BmEL-1) and BmELAV-like-2 (BmEL-2) are part of the ELAV/Hu family. These two orthologs share 55–71% and 47–62% identity with that of in *Drosophila* and humans, respectively. BmEL-1 and BmEL-2 are well studied in *Drosophila*. However, this family is less studied *Bombyx mori*. A study was carried out to elucidate the functions of these two genes in *Bombyx mori*Bmel-1/-2 mutants created using CRISPR/Cas9 technology. The expression of four silk protein genes, namely Fib-L, Fib-H, P25 and Ser-1, were found to be increased due to the mutations of both genes. The cocoon parameters such as body weight, pupae, cocoon and cocoon shell were affected differently in Bmel-1/-2 mutants. Further, reduction in the ability of oviposition was observed in Bmel-2 females (Sun et al. 2021).

8.5.11 BmKu70 Gene Functional Analysis

Ma et al. (2014), with the aid of CRISPR/Cas9 system, reported the multiplex genome editing and heritable mutagenesis of BmKu70, which is required for the NHEJ pathway. In Bmku70 knocked out *Bombyx mori*, the frequency of homologous recombination (HR) mediated knock-in was found to be increased (Ma et al. 2014).

8.5.12 BE3 Base Editor Analysis

CRISPR/Cas9-dependent base editor (BE3) efficiently converted cytidine (C) to thymine (T) with a high frequency in the silkworm. Li et al. used BE3 as a knockout tool and inactivated endogenous and exogenous genes by base-editing-induced nonsense mutations. Further, authors reported that 96.5% of *B. mori* genes had one or more targetable sites that could be edited through BE3 for inactivation (Li et al. 2018).

8.5.13 Bm-ok, BmKMO, BmTH, Bmtan Functional Analysis

With four loci, namely, Bm-ok, BmKMO, BmTH and Bmtan as candidates, Wei et al. ascertained that by direct microinjection of specific guide RNA and Cas9-mRNA into silkworm embryos, genome alterations at specific sites could be induced. Authors performed a test for mutant heritability and showed that when two Bm-ok mosaic mutant moths when crossed with each other, in 93.6% of the offspring mutations appeared in both alleles of Bm-ok gene (Wei et al. 2014).

8.5.14 BmOvo Gene Functional Analysis

Bi et al. (2019) used the transgenic-based CRISPR/Cas9 genome editing system to study the function of the Ovo gene in *Bombyx mori*. BmOvo is the homolog of a factor important in germline sex determination in *Drosophila melanogaster*. BmOvo mutants had abnormally shaped eggs that were disordered in the ovarioles, and gonad development was abnormal. Wing discs and wings did not develop properly, and most of the mutants failed to eclose. Gene expression analyses by qRT-PCR showed that BmOvo gene was highly expressed in the wing disc and epidermis. Genes involved in the WNT signaling pathway and wing development genes BmWCP10 and BmE74 were downregulated in the BmOvo mutants when compared with wild-type animals. These results demonstrate that the BmOvo gene product plays an important role in wing metamorphosis. Thus, this study provides new insights into the multiple functions of BmOvo beyond germline sex determination.

8.5.15 Kynurenine 3-Monooxygenase (KMO) Gene Functional Analysis

Hong et al. (2020) conducted genetic and phenotypic analyses to examine the molecular breeding of *Bombyx mori* through CRISPR/Cas9-mediated editing of the kynurenine 3-monooxygenase (KMO) gene. The synthesized guide RNAs (gRNAs) were analyzed using T7 endonuclease I after introduction into the BM-N silkworm cell line. To edit the silkworm gene, K1P gRNA, and Cas9 complexes were microinjected into silkworm embryos. After microinjection, the hatching rate and the incidence of mutation were determined as 18.1% and 60%, respectively. Gene mutation was verified in the heterozygous G0 generation, but no phenotypic change was observed; however, certain embryos and moths produced through sib-mating had significant differences compared to the wild-type. In successive generations, a distinct phenotypic change was also observed by continuous mating. Thus, according to authors, although there were limitations in the phenotypic expression in breeding through the induction of deletion mutations, the process is believed to yield successful results within a shorter period compared to traditional breeding and is safer than transgenic technology.

8.5.16 Maelstrom (Mael) Gene Function Analysis

The Maelstrom (Mael) gene is identified essential for spermatogenesis in both *Drosophila* and mouse, though the mechanisms appear to differ. Chen et al. (2019) obtained *Bombyx mori* Mael mutants by using a binary transgenic CRISPR/Cas9 system and analyzed the function. The results showed that BmMael is not necessary for piRNA pathway in the ovary of silkworm, whereas it might be essential for transposon elements (TEs) repression in testis. The BmMael mutation resulted in male sterility, and further analysis established that BmMael was essential for spermatogenesis. The spermatogenesis defects occurred in the elongation stage and resulted in nuclei concentration arrest. RNA-seq and qRT-PCR analyses demonstrated that spermatogenesis defects were associated with tight junctions and apoptosis. They also found that BmMael was not involved in the silkworm sex determination pathway.

8.5.17 BmBLOS2 Gene Function Analysis

Zeng et al. (2016) reported that an endogenous gene, BmBLOS2, was disrupted by using various types of N20NGG sgRNAs, and the cleavage efficiency of N20NGG sgRNAs with different initial nucleotides and GC contents was evaluated in vitro. Transgenic silkworms expressing Cas9 and sgRNAs targeting the BmBLOS2 gene were generated with many types of mutagenesis. The typical transparent skin phenotype in knockout silkworms was stable and inheritable, suggesting that N20NGG sgRNAs function sufficiently in vivo.

8.5.18 Sterile Insect Technique Using CRISPR/Cas9

The sterile insect technique has become a regular constituent of area-wide integrated pest management programs to tackle numerous major agricultural pests (Marec and Vreysen 2019). Recently, researchers have started exploring the efficacy of CRISPR/Cas 9 tool in sterile insect technique using lepidopteran *Bombyx mori*. Combining the TALEN and the clustered regularly interspaced short palindromic repeats (CRISPR)/CRISPR-associated protein 9 (Cas9) technologies, Zhang et al. developed a binary, female-specific, embryonic lethality system that contains one strain with U6-derived single-guide RNA (sgRNA) expression targeting transformer 2 (tra2), which is a crucial gene for embryonic development of silkworm, and another strain with TALEN-mediated, W-specific Cas9 expression driven by the silkworm germ cell-specific nanos (nos) promoter. Authors reported that during embryonic stages, complete female-specific lethality was exhibited by Filial 1 (F1) hybrids. This study paves the way for the development of sterile insect techniques for lepidopteron pests with WZ/ZZ sex chromosome systems (Zhang et al. 2018).

In 2021, using the CRISPR/Cas9 tool, researchers studied the gene function of the egg-specific protein (Esp), which is one of the yolk proteins found in *Bombyx mori*. Authors found that Esp mutation gives rise to female sterility, however, posed no effect on male sterility. Further, it was reported that eggs laid by female mutants were smaller in size and lighter in color when compared with wild-type females and the most significant was that eggs laid by female mutants did not hatch. Though the authors were silent on how to maintain the breed when the female is sterile, they report that the stable inheritance of female sterility due to ESP mutation was observed through successive generations. This clearly indicates that ESP encodes a yolk protein that is vital for the reproductive success of the female and is a potential target for pest control (Article in press).

8.5.19 Behavioral Gene Function Analysis

With the help of CRISPR/Cas9 system, in 2019, gene GR66 was decoded, which encodes alleged bitter gustatory receptor, is accountable for the mulberry-specific feeding preference of domesticated silkworm (*Bombyx mori*). Using this technique, the GR66 locus was genetically mutated and silkworm strains that were homozygous mutant with truncated gustatory receptor 66 (GR66) proteins were created. In addition to mulberry leaves, GR66 mutant larvae acquired and showed the ability to feed on different plant species (Zhang et al. 2019a).

Olfaction has a crucial role in several important insect behaviors such as reproduction and feeding. An odorant receptor co-receptor (Orco) is required to detect olfactory stimuli. Liu et al., using binary transgene-based (CRISPR)/Cas9 system, deleted Orco gene in *Bombyx mori*. Orco gene knocked-out individuals resulted in impaired olfactory sensitivity. Orco mutant larvae showed defective selection for mulberry leaves. Further, single sensillum recordings showed that the antenna of the homozygous mutant was incapable to respond to either of the two sex pheromones, bombykol or bombykol. Further, an adult mating behavior analysis showed that Orco knockout individuals exhibited notably impaired mating selection behavior in response to a pheromone released by a wild-type female moth, as well as to bombykol/bombykol mixture (Figure 8.3) (Liu et al. 2017).

In 2020, Xu et al. with the help of CRISPR/Cas9 and electrophysiological techniques, supported the existence of two genetic pathways in *Bombyx mori*, namely Bmdsx-BmOR1-bombykol and Bmfru-BmOR3-bombykol, which are responsible for controlling distinct aspects of sexual behavior of males that are known to be modulated by olfaction (Xu et al. 2020).

The circadian clock is an internal regulator in the cells of organisms that are known to coordinate behavioral and physiological activities with daily environmental changes within 24 h cycles (Farhud and Aryan 2018). Endogenously, genetically encoded molecular clocks produce circadian rhythmicity (Patke et al. 2020). In *Drosophila melanogaster*, the timeless (tim) gene is responsible for encoding cardinal components of the fly's endogenous clock (Benna et al.

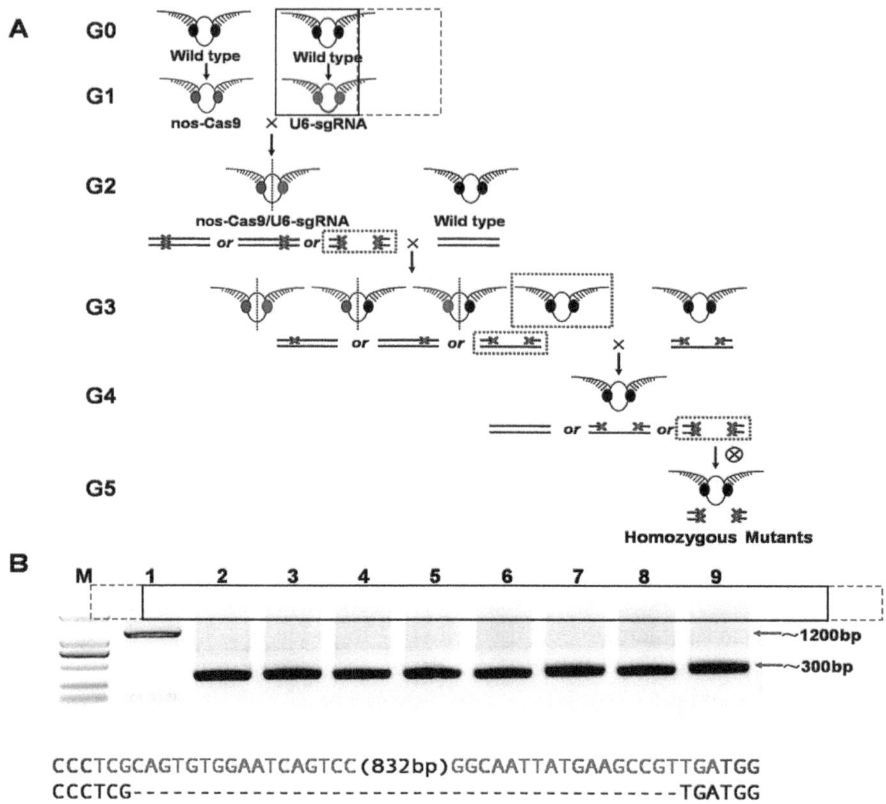

FIGURE 8.3 Schematic diagram illustrating crossing scheme to produce homozygous mutations and diagnosis for mutations (A) Crossing scheme to produce homozygous mutations. The pre-blastderm silkworm embryos (G0) were injected with transgenic plasmids pBac [IE1-EGFP-nos-Cas9-SV40] or pBac [IE1-DsRed2-U6-Orco-sgRNA (2")] to produce two transgenic silkworm lines, which were regarded as parental silkworm strains (G1). Subsequently, the two transgenic lines were hybridized to produce founder animals (G2), which express both Cas9 and Orcosg RNAs. The silkworm with the inter-space fragment between two target sites deleted was backcrossed with wild-type to obtain G3 offspring. The silkworms with none fluorescence and thorough deletion event were in-crossed to generate the homozygous G4 animals and the stable BmOrco mutant line was established in G5. (B) Deletion mutations in BmOrco homozygous mutants. Lane 1 shows the fragment amplified from the genome of wild-type silkworm. Lane 2 to lane 9 represents the fragments amplified from eight randomly selected G5 silkworms. The sequence below shows the mutation event. The sgRNA-targeting sequence is in solid box and the PAM sequence is in dashed box. The number of 832 bp in bracket refers to the 832 bp inter-space fragment between the two target sites.

Source: Adapted with permission from *Insect Biochemistry and Molecular Biology* 2017, 86. 58–67, Elsevier.

2000). However, there is little known regarding the role of tim gene in the negative feedback loop responsible for behavioral activities in *Bombyx mori*. Nartey et al., using the CRISPR/Cas9 gene editing system, studied the role of tim in the circadian clock of *B. mori*. Two circadian behavioral activities that were adult emergence and embryo hatching rhythms were found to be affected in knockouts. Further, the authors reported a reduction in expression levels and oscillation amplitudes of Bumper and Bmclk mRNAs during both the photophase and scotophase were induced in tim knockouts. Furthermore, in a tim knockout, the daily rhythmic expression of Bmdbt was reported to be down-regulated in the scotophase and was up-regulated in the photophase (Nartey et al. 2020).

8.5.20 Pollution-Sensitive Gene Screening

Liu et al. (2021) performed a CRISPR-based functional genomic screen to measure the genetic response of genes to the pollutants. They identified 751 NaF-resistance genes, 753 NaF-sensitive genes, 757 CdC12-resistance genes and 725 CdC12-sensitive genes, the silencing of which conferred resistance to fluoride or cadmium. Functional analysis of the resistant- and sensitive-genes demonstrates enrichment of multiple signaling pathways, among which the MAPK signaling pathway and DNA damage and repair are both required for fluoride- or cadmium-induced cell death.

8.5.21 Host-Pathogen Interaction Gene Screening

Chang et al. (2020) developed a piggyBac strategy to deliver pooled sgRNA libraries stably into cell lines. They used this strategy to conduct a screen-based CRISPR-Cas9 in *Bombyx mori* cells. They constructed a single guide RNA (sgRNA) library containing 94,000 sgRNAs, which targeted 16,571 protein-coding genes. Then generated knockout collections in BmE cells using the piggyBac transposon. They identified 1006 genes that are essential for cell viability under normal growth conditions. Of the identified genes, 82.4% (829 genes) were homologous to essential genes in seven animal species. They also performed context-specific positive screens for resistance to biotic or nonbiotic stresses using temperature and baculovirus separately, which identified several key genes and pathways from each screen.

8.5.22 Fibroin Gene (FibH)

Fibroin heavy chain (FibH) gene, encoding the core component of silk protein, is specifically and highly expressed in silk gland cells but not in the other cells. Although the silkworm FibH gene has been well studied in transcriptional regulation, its biological functions in the development of silk gland cells remain elusive. Hu et al. (2021) constructed a CRISPRa system to activate the endogenous transcription of FibH in *Bombyx mori* embryonic (BmE) cells, and the mRNA expression of FibH was successfully activated maximally at 60 h after transient transfection of sgRNA/dCas9-VPR at a molar ratio of 9:1. The qRT-PCR analysis showed that the expression levels of cellular stress response-related genes were significantly up-regulated along with activated FibH gene.

8.5.23 Spider-Silkworm Hybrid Silk Fiber

Spider silk is one of the strongest known biomaterials and is well-known for its striking mechanical properties. Farming of spiders for their silk is impracticable, which has paved the way for the manufacturing of recombinant spider silk proteins (spidroins) in more acquiescent hosts. For the heterologous production of spider silk, silkworms (*B. mori*) are perfect candidates as a host system (Whittall et al. 2021). For spinning spider silk-like fibers, use of transgenic silkworms with their natural spinning apparatus has reported to be a favorable approach.

However, incorporation of native size spider silk proteins and getting inheritable transgenic silkworm strains are the challenges. Zhang et al. in their study reported successful incorporation of spider silk protein genes into the genome of *Bombyx mori* with the aid of CRISPR/Cas9 technology. The obtained silk fibers were as strong as native spider silks (1.2 GPa tensile strength). The transgenic silkworms, with a normal inheritance of the transgenic silkworms, were reported when tracked for several generations for spinning spider silk-like fibers (Figure 8.4) (Zhang et al. 2019a).

8.5.24 Other CRISPR Tools

Delivery of the CRISPR/Cas9-guide RNA (gRNA) complex to any cell results in modification of the target sequence. There are several methods available to deliver CRISPR/Cas9 and its components to the target cells, including viral, non-viral and physical methods to deliver plasmid or

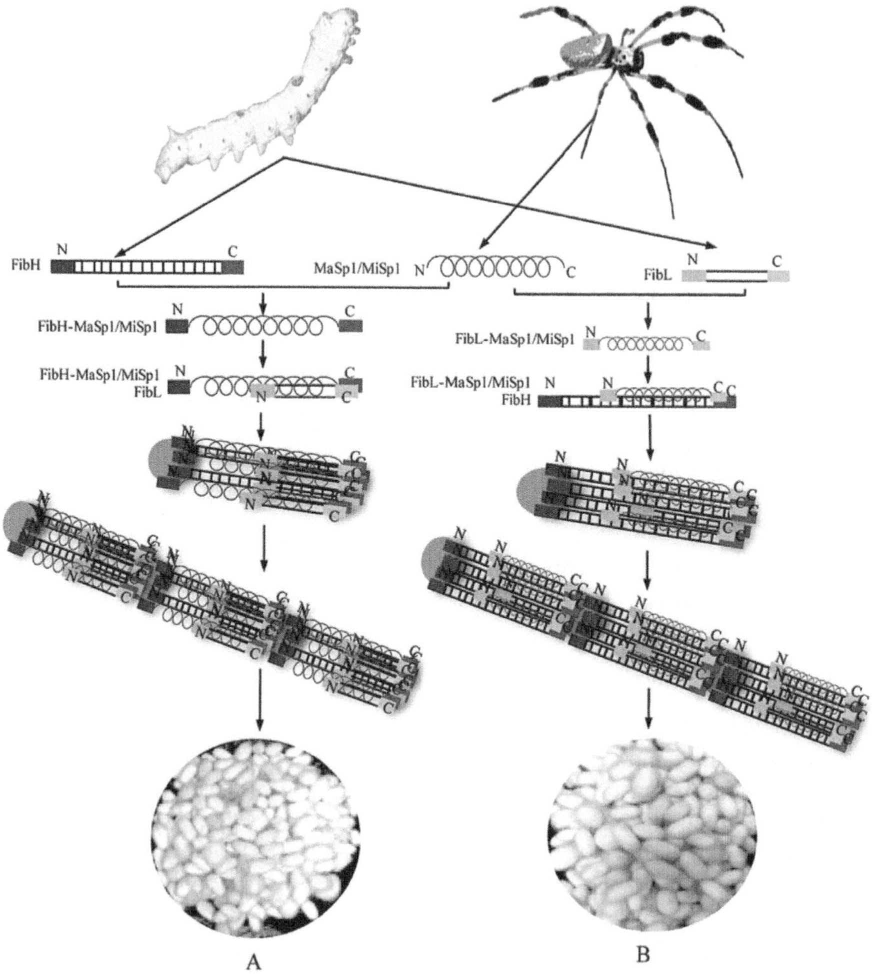

FIGURE 8.4 A schematic diagram illustrating the structural formation of the spider silk-like fibers made by transgenic silkworms.

Source: Adapted with the permission from *Bio Macromolecules* 2019, 20, 2252–2264, American Chemical Society.

ribonucleoprotein (RNP) of CRISPR components. But in vivo CRISPR/Cas9 delivery remains challenging to the researchers due to insertional mutagenesis, targeted delivery, immunogenicity and off-targets. There are several factors that hinder the delivery of the CRISPR/Cas9 complex, such as the large size of the plasmid or high dosage of the chemical agent. Scientists are constantly trying to develop more efficient ways of Cas technology.

8.5.24.1 Off-Target Issues of Cas9 System

Ma et al. (2021) performed in-vivo multiplex editing, deep sequencing, and comprehensive analysis of its editing outcomes in *Bombyx mori*. A total of 31,161 editing events from nine single-guide RNA (sgRNA) sites in 16 individuals were generated and analyzed. They found that Cas9 introduces mutations with some regularity rather than via stochastic indels. The editing efficiency varies with sgRNA sequences, individuals and orientation. Insertions are formed by diverse mechanisms, including direct acquisition of free genomic fragments, duplication of broken ends, replication of adjacent sequences or random addition of free nucleotides.

Liu et al. (2014) showed that CRISPR/cas9 system can induce targeted genome editing of a single gene locus in BmNs cells. They showed that CRISPR/cas9 system can induce high frequently targeted genomic structure variation such as large chromosomal deletion or inversion. Importantly, when BmNs cells were transfected with 10 gRNAs for six genes, targeted mutations were detected for all target loci.

Wang et al. (2013) opined that the Cas9-sgRNA system effectively induced targeted mutagenesis in *B. mori*. However, some potential problems with this system exist. Mutations in many loci do not cause phenotypic effects that are easy to assess, posing a difficulty in mutation screening. Also, the targeting sequences of the CRISPR/Cas system are limited to 23 bp. Considering the tolerance for mismatches at the 5′-ends of the targeting site, this limitation may lead to serious off-target effects. Nonetheless, we report here a proof-of-principle study that exemplifies targeted mutagenesis of BmBLOS2 by direct microinjection of Cas9 mRNA and sgRNAs. The ensuing highly efficient mutagenesis demonstrates a new, facile genome engineering technology in *B. mori*.

The CRISPR technology is fast evolving and has many alternative and case specific editing systems.

8.5.24.2 Base Editing (BE3) System

DNA base editing is conversion of C-to-T or A into G or A-T into G-C. It requires chimeric protein comprised of an RNA-guided endonuclease and an enzyme able to deaminate an adenine or a cytidine base. The fused deaminating enzyme may be cytidine deaminase (C-to-T) or adenine deaminase (A into G or A-T into G-C). APOBEC (apolipoproteinBmRNA-editing enzyme catalytic polypeptide-like) and AID (activation-induced deaminase) have been used as cytidine deaminase frequently.

Liu et al. (2020a) showed that base editing (BE3) was superior to Cas9 in silkworm cultured cells. BE3 introduced accurate termination codons, whereas Cas9 did not. Moreover, Cas9 induced chromosome translocation, chromosome fragment repetition, and chromosome fragment deletion, with the deletion frequency reaching up to 4.29%. Cas9-derived proteins blocked ribosome advance and mRNA transcription for nine days, with a 9.40% repression effect by combining with double-stranded DNA when single guide RNAs were targeted in the coding region in silkworms.

8.5.24.3 CRISPR Interference (CRISPRi) System

Wang et al. (2019) showed that CRISPR interference (CRISPRi) isa precise tool for regulating gene expression, even in silkworms. When CRISPR/dCas9 was fused with transcriptional repressors, the system could robustly silence endogenous gene expression. The CRISPR/Cas9 tool does not respond to traditional gene suppression techniques, such as RNA interference (RNAi). Authors tested CRISPRi usability and efficiency in *Bombyx mori* cells (BmE). They fused dCas9 to five transcriptional repressors, including KRAB, Hairy, SID, SRDX and ERD. They opine that all five constructs were efficient in BmE cells and have fewer off-target effects. They also caution that, sgRNA distance from transcriptional start site (TSS) and the dCas9: sgRNA ratios are the two limiting factors of CRISPRi efficiency.

8.5.24.4 CRISPR-Cas12a (Cpf1) System

Only a few labs have conducted genomic editing using CRISPR in silkworm, *Bombyx mori*. The commonly used tool was SpCas9, which only recognized target sites containing NGG PAM sequence. Ma et al. (2017) worked on Cpf1 tool to improve the efficiency of by 3.5-fold over SpCas9. The improved high efficiency was also observed at several loci in both BmNs cells and *B. mori* embryos. Genome-wide analysis of targetable sites was further conducted by them and showed that the integrated system covers 69,144,399 sites in *B. mori* genome, and one site could be found in every 6.5 bp. They opine that the efficiency and resolution of this CRISPR/Cpf1 platform will probably accelerate both fundamental researches and applicable studies in B. mori, and perhaps other insects.

CRISPR-Cas12a (Cpf1) is a single RNA-guided endonuclease that provides new opportunities for targeted genome engineering through the CRISPR/Cas9 system. Dong et al. (2020) compared three Cas12a orthologs nucleases, AsCas12a, FnCas12a and LbCas12a, for their editing efficiencies and antiviral abilities. The three Cas12a efficiently edited the *Bombyx mori* nucleopolyhedro virus (BmNPV) genome and inhibited BmNPV replication in BmN-SWU1 cells. The antiviral ability of the FnCas12a system was more efficient than that of the SpCas9 system after infection by BmNPV. They created FnCas12a × gIE1 and SpCas9 × sgIE1 transgenic hybrid lines and evaluated the gene-editing efficiency of different systems at the same target site. We improved the antiviral ability using the FnCas12a system in transgenic silkworm.

8.5.24.5 CRISPR- dCas9-TET1 System

Liu et al. (2019a) showed that fusion protein of catalytically inactive Cas9 (dCas9) with TET1 can efficiently demethylate genomic DNA of silkworm, *Bombyx mori*, in a programmable target region specific manner. They developed an all-in-one vector to maximize the targeting efficiency of dCas9-TET1. They selected three endogenous genes that were harboring methylated DNA, and designed gRNAs within the methylated region. Co-transfection of dCas9-TET1 and gRNA successfully erased methylation marks near the targeting region, with efficiencies from about 17.50% to 40.00%. It was found that the methylation inhibitor dCas9-TET1 is more effective and specific, and has no unwanted impact on whole-genome methylation. DCas9-TET1 provides a powerful tool for investigating the functional significance of DNA methylation in a locus-specific manner, and for exploring the unknown links between methylation and development in insects.

8.5.24.6 Prime Editing System

Prime editing, a genome-editing technology, enables all types of base transitions and transversions to be accomplished, as well as customized insertions (up to 44 nucleotides) and deletions (up to 80 nucleotides) in one step. The key components of a prime editor (PE) are a Cas9 nickase fused with reverse transcriptase and a prime editing guide RNA (pegRNA), which brings the PE system to targeted sites through a standard single-guide RNA (sgRNA) sequence. Upon binding, the Cas9 nickase nicks the target strand and the reverse transcriptase uses the primer binding site (PBS) of the pegRNA to initiate the reverse transcription. The genetic information encoded in the reverse-transcription template (RTT) can then be copied into the targeted site. There are four different prime editing systems: PE1, PE2, PE3 and PE3b.

8.6 CONCLUSION

CRISPR/Cas system is considered to be a significant part of bacterial adaptive immunity. It is composed of Cas nuclease and two individual RNA components, a fixed tracrRNA (trans-activating RNA) and a programmable crRNA (CRISPR RNA). Silkworm (*Bombyx mori*) is a lepidopteran insect of immense economical importance and the finest model organism for a variety of scientific studies. *B. mori* has been one of the earliest insect models for which a genomic editing tool was applied to elucidate the function of genes. Transgenic silkworms were also produced with the hybridization of the spider fibroin gene with the mulberry silkworm fibroin gene. The CRISPR/Cas9 system consists of two key molecules that cut DNA strands. (i) An enzyme called Cas9. This acts as a pair of 'molecular scissors' that can cut the two strands of DNA at a specific location in the genome so that bits of DNA can then be added or removed. (ii) A piece of RNA called guide RNA (gRNA). This consists of a small piece of pre-designed RNA sequence (about 20 bases long) located within a longer RNA scaffold. The scaffold part binds to DNA at CRISPR site and the pre-designed sequence 'guides' Cas9 to the right part of the genome. This makes sure that the Cas9 enzyme cuts at the right point in the genome. Anti-CRISPR technology has also been developed to apply controlled break-over gene editing. In a short span of time, a number of CRISPR/Cas systems have been developed. Collectively the 93 cas genes are grouped into 35 families based on sequence similarity of the encoded proteins. Eleven of the 35 families form the cas core, which includes the protein families Casl through Cas13. A number of studies are conducted in silkworm *Bombyx mori*, in the field on pathology, developmental studies, behavioral studies, gene function analysis, fibroin gene analysis, and so on. Different cas tools are also applied for different kinds of studies. With the pace at which the CRISPR/Cas system is evolving, it will soon become an essential and affordable gene-editing tool across the living kingdom.

BIBLIOGRAPHY

Benna C, Scannapieco P, Piccin A, Sandrelli F, Zordan M, Rosato E, Kyriacou CP et al. A second timeless gene in Drosophila shares greater sequence similarity with mammalian tim. Current Biology 10, no. 14(2000):R512–R513.

Bi H, Xu X, Li X, Zhang Y, Huang Y, Li K, Xu J. CRISPR disruption of Bmovo resulted in the failure of emergence and affected the wing and gonad development in the silkworm Bombyx mori. Insects. 10, no. 8(2019 Aug):254.

Brady D, Saviane A, Cappellozza S, Sandrelli F. An efficient workflow for screening and stabilizing CRISPR/Cas9-mediated mutant lines in bombyx mori. Methods and Protocols. 4, no. 1(2021 Mar):4.

Chang J, Wang R, Yu K, Zhang T, Chen X, Liu Y, Shi R, Wang X, Xia Q, Ma S. Genome-wide CRISPR screening reveals genes essential for cell viability and resistance to abiotic and biotic stresses in Bombyx mori. Genome Research. 30, no. 5(2020 May 1):757–767.

Chen K, Chen S, Xu J, Yu Y, Liu Z, Tan A, Huang Y. Maelstrom regulates spermatogenesis of the silkworm, Bombyx mori. Insect Biochemistry and Molecular Biology. 1, no. 109(2019 Jun):43–51.

Chen S, Hou C, Bi H, Wang Y, Xu J, Li M, James AA, Huang Y, Tan A. Transgenic clustered regularly interspaced short palindromic repeat/Cas9-mediated viral gene targeting for antiviral therapy of Bombyx mori nucleopolyhedrovirus. Journal of Virology. 91, no. 8(2017 Apr 15):e02465–16.

Cui Y, Zhu Y, Lin Y, Chen Y, Feng Q, Wang W, Xiang H. New insight into the mechanism underlying the silk gland biological process by knocking out fibroin heavy chain in the silkworm. BMC Genomics. 19, no. 215(2018).

Dong Z, Dong F, Yu X, Huang L, Jiang Y, Hu Z, Chen P et al. Excision of Nucleopolyhedrovirus form transgenic silkworm using the CRISPR/Cas9 system. Frontiers in Microbiology. 9, no. 209(2018):1–10.

Dong Z, Huang L, Dong F, Hu Z, Qin Q, Long J, Cao M, Chen P, Lu C, Pan MH. Establishment of a baculovirus-inducible CRISPR/Cas9 system for antiviral research in transgenic silkworms. Applied Microbiology and Biotechnology. 102, no. 21(2018 Nov):9255–965.

Dong Z, Long J, Huang L, Hu Z, Chen P, Hu N, Zheng N et al. Construction and application of an HSP70 promoter-inducible genome editing system in transgenic silkworm to induce resistance to Nosemabombycis. Applied Microbiology and Biotechnology. 103, no. 23–24(2019a):9583–9592.

Dong Z, Qin Q, Hu Z, Chen P, Huang L, Zhang X, Tian T et al. Construction of a one-vector multiplex CRISPR/Cas9 editing system to inhibit nucleopolyhedro virus replication in silkworms. VirologicaSinica. 34(2019b):444–453.

Dong Z, Qin Q, Hu Z, Zhang X, Miao J, Huang L, Chen P, Lu C, Pan M. CRISPR/Cas12a Mediated genome editing enhances Bombyx mori resistance to BmNPV. Frontiers in Bioengineering and Biotechnology. 8(2020 Jul 15):841.

Farhud D, Aryan Z. Circadian rhythm, lifestyle and health: A narrative review. Iranian Journal of Public Health. 47, no. 8(2018):1068–1076.

Fujinaga D, Shiomi K, Yagi Y, Kataoka H, Mizoguchi A. An insulin-like growth factorlike peptide promotes ovarian development in the silkmoth Bombyx mori. Scientific Reports. 9, no. 18446(2019).

Gani M, Chouhan S, Babulal R, Gupta K, Khan G, Kumar N B, Saini P et al. Bombyxmorinucleopolyhedrovirus (BmBPV): Its impact on silkworm rearing and management strategies. Journal of Biological Control. 31, no. 4(2017):189–193.

Hong JW, Jeong CY, Yu JH, Kim SB, Kang SK, Kim SW, Kim NS, Kim KY, Park JW. Bombyx mori kynurenine 3-monooxygenase gene editing and insect molecular breeding using the clustered regularly interspaced short palindromic repeat/CRISPR associated protein 9 system. Biotechnology Progress. 36, no. 6(2020 Nov):e3054.

Hu W, Wang X, Ma S, Peng Z, Cao Y, Xia Q. CRISPR-mediated endogenous activation of fibroin heavy chain gene triggers cellular stress responses in bombyx mori embryonic cells. Insects. 12, no. 6(2021 Jun):552.

Huang L, Dong Z-Q, Dong F-F, Yu X-B, Hu Z-G, Liao N-C, Chen P et al. Gene editing the BmNPV inhibitor of apoptosis protein 2 (iap2) as an antiviral strategy in transgenic silkworm. International Journal of Biological Macromolecules. 166(2021):529–537.

Huang Y, Chen J, Sun B, Zheng R, Li B, Li Z, Tan Y et al. Engineered resistance to Nosemabombycis by in vitro expression of a single-chain antibody in Sf9-III cells. PLoS One. 13, no. 2(2018):e0193065, 1–6.

Jiang F, Doudna JA. CRISPR-Cas 9 structures and mechanisms. Annual Review of Biophysics. 46(2017):505–529.

Li Y, Ma S, Sun L, Zhang T, Chang J, Lu W, Chen X et al. Programmable single and multiplex base-editing in bombyx mori using RNA-guided cytidinedeaminases. G3: Genes, Genomes, and Genetics. 8, no. 5(2018):1701–1709.

Li Y, Peng N. Endogenous CRISPR-Cas system-based genome editing and antimicrobials: review and prospects. Front Microbiol. 10(2019):1–9.

Liu Q, Liu W, Zeng B, Wang G, Hao D, Huang Y. Deletion of the Bombyx mori odorant receptor co-receptor (BmOrco) impairs olfactory sensitivity in silkworms. Insect Biochemistry and Molecular Biology. 86(2017):58–67.

Liu Y, Chang J, Yang C, Zhang T, Chen X, Shi R, Liang Y, Xia Q, Ma S. Genome-wide CRISPR-Cas9 screening in Bombyx mori reveals the toxicological mechanisms of environmental pollutants, fluoride and cadmium. Journal of Hazardous Materials. 410(2021 May 15):124666.

Liu Y, Li Y, Liang Y, Wang T, Yang C, Ma S, Xia Q. Comparative analysis of genome editing systems, Cas9 and BE3, in silkworms. International Journal of Biological Macromolecules. 158(2020a Sep 1):486–492.

Liu Y, Ma S, Chang J, Zhang T, Chen X, Liang Y, Xia Q. Programmable targeted epigenetic editing using CRISPR system in Bombyx mori. Insect Biochemistry and Molecular Biology. 110(2019a Jul 1):105–11.

Liu Y, Ma S, Wang X, Chang J, Gao J, Shi R, Zhang J, Lu W, Liu Y, Zhao P, Xia Q. Highly efficient multiplex targeted mutagenesis and genomic structure variation in Bombyx mori cells using CRISPR/Cas9. Insect Biochemistry and Molecular Biology. 49(2014 Jun 1):35–42.

Liu Z, Xu J, Ling L, Luo X, Yang D, Yang X, Zhang X, Huang Y. miR-34 regulates larval growth and wing morphogenesis by directly modulating ecdysone signalling and cuticle protein in Bombyx mori, RNA. Biology. 17, no. 9(2020b):1342–1351.

Liu Z-L, Xu J, Ling L, Zhang R, Shang P, Huang Y-P. CRISPR disruption of TCTP gene impaired normal development in the silkworm Bombyx mori. Insect Science. 26(2019b):973–982.

Ma S, Chang J, Wang X, Liu Y, Zhang J, Lu W, Gao J et al. CRISPR/Cas9 mediated multiplex genome editing and heritable mutagenesis of BmKu70 in Bombyx mori. Scientific Reports. 4, no. 4489(2014):1–6

Ma S, Liu Y, Liu Y, Chang J, Zhang T, Wang X, Shi R, Lu W, Xia X, Zhao P, Xia Q. An integrated CRISPR Bombyx mori genome editing system with improved efficiency and expanded target sites. Insect Biochemistry and Molecular Biology. 83(2017 Apr 1):13–20.

Ma S, Wang A, Chen X, Zhang T, Xing W, Xia Q. Deep sequencing reveals the comprehensive CRISPR-Cas9 editing spectrum in Bombyx mori. The CRISPR Journal. 4, no. 3(2021 May 26):371–380.

Ma Z, Li C, Pan G, Li Z, Han B, Xu J, Lanet X et al. Genome-Wide Transcriptional Response of Silkworm (Bombyx mori) to Infection by the Microsporidian Nosemabombycis. PLoS One. 8, no. 12(2013):e84137, 1–18.

Makarova KS, Wolf YI, Alkhnbashi OS, Costa F, Shah SA, Saunders SJ et al. An updated evolutionary classification of CRISPR-Cas systems. Nature Reviews Microbiology. 13 no. 11(2015 Nov):722–736. doi:10.1038/nrmicro3569. PMC 5426118. PMID 26411297.

Marec F, Vreysen MJB. Advances and challenges of using the sterile insect technique for the management of pest lepidoptera. Insects. 10, no. 11(2019):371.

Mei Y, Wang Y, Chen H, Sun ZS, Ju X-D. Recent progress in CRISPR/Cas9 technology. Journal of Genetics and Genomics 43, no. 2(2016):63–75.

Nartey AM, Sun X, Qin S, Hou C-X, Li M-W. CRISPR/Cas9-based knockout reveals that the clock gene timeless is indispensable for regulating circadian behavioural rhythms in Bombyx mori. Insect Science. 28, no. 5(2020):1414–1425.

Niopek D, Wehler P, Roensch J, Eils R, Di Ventura B. Optogenetic control of nuclear protein export. Nature Communications. 7, no. 1(2016 Feb 8):1–9.

Patke A, Young MW, Axelrod S. Molecular mechanisms and physiological importance of circadian rhythms. Nature Reviews Molecular Cell Biology. 21(2020):67–84.

Rodríguez Rodríguez DR, Ramírez Solís R, Garza Elizondo MA, Garza Rodríguez MD, Barrera Saldaña HA. Genome editing: A perspective on the application of CRISPR/Cas9 to study human diseases (Review). International Journal of Molecular Medicine. 43(2019):1559–1574.

Singh T, Singh P K, Sahaf KA. Egg diapause and metabolic modulations during embryonic development in the silkworm, Bombyx mori L. (Lepidoptera: Bombycidae). Annals of Biological Research. 4, no. 1(2013):12–21.

Sun D, Guo Z, Liu Y, Zhang Y. Progress and prospects of CRISPR/Cas systems in insects and other arthropods. Frontiers in Physiology. 8, no. 608(2017):1–22.

Sun X, Zhang X, Gu J, Yang J, Huang Q, Yan R, Qin S et al. The biological characters of Bmelav-like genes in the development of Bombyx mori. Insect Molecular Biology. 30, no. 1(2021):9–17.

Taning CNT, Eynde BV, Yu N, Ma S, Smagghe G. CRISPR/Cas9 in insects: Applications, best practices and bio safety concerns. Journal of Insect Physiology 98(2017):245–257.

Thomson, B J. Viruses and apoptosis. International Journal of Experimental Pathology. 82, no. 2(2001):65–76.

Wang X, Ma S, Liu Y, Lu W, Sun L, Zhao P, Xia Q. Transcriptional repression of endogenous genes in BmE cells using CRISPRi system. Insect Biochemistry and Molecular Biology. 111(2019 Aug 1):103172.

Wang Y, Li Z, Xu J, Zeng B, Ling L, You L, Chen Y, Huang Y, Tan A. The CRISPR/Cas system mediates efficient genome engineering in Bombyx mori. Cell Research. 23, no. 12(2013 Dec):1414–1416.

Wei W, Xin H, Roy B, Dai J, Miao Y, Gao G. Heritable genome editing with CRISPR/Cas9 in the silkworm. Bombyx mori. PLoS One. 9, no. 7(2014):e101210, 1–6.

Whittall R D, Baker KV, Breitling R, Takano E. Host systems for the production of recombinant spider silk. Trends in Biotechnology. 39, no. 6(2021):560–573.

Xu J, Liu W, Yang D, Chen S, Chen K, Liu Z, et al. (2020) Regulation of olfactory-based sexbehaviours in the silkworm by genes in the sex-determination cascade. PLoS Genetics 16, no. 6(2020):e1008622, 1–20.

Xu K, Li F, Ma L, Wang B, Zhang H, Ni M, Hong F et al. Mechanism of enhanced Bombyx mori Nucleopolyhedrovirus-resistance by titanium dioxide nanoparticles in silkworm. PLoS One. 10, no. 2(2015):e0118222, 1–18.

Zeng B, Zhan S, Wang Y, Huang Y, Xu J, Liu Q, Li Z, Huang Y, Tan A. Expansion of CRISPR targeting sites in Bombyx mori. Insect Biochemistry and Molecular Biology. 72(2016 May 1):31–40.

Zhang X, Xia L, Day BA, Harris TI, Oliveira P, Knittel C, Licon AL et al. CRISPR/Cas9 initiated transgenic silkworms as a natural spinner of spider silk. Biomacromolecules. 20, no. 6(2019a):2252–2264.

Zhang Z, Aslam AFM, Liu X, Li M, Huang Y, Tan A. Functional analysis of Bombyx Wnt1 during embryogenesis using the CRISPR/Cas9 system. Journal of Insect Physiology. 79(2015):73–79. ISSN 0022-1910. https://doi.org/10.1016/j.jinsphys.2015.06.004

Zhang Z, Liu X, Shiotsuki T, Wang Z, Xu X, Huang Y, Li M et al. Depletion of juvenile hormone esterase extends larval growth in Bombyx mori. Insect Biochemistry and Molecular Biology. 81(2017):72–79.

Zhang Z, Niu B, Ji D, Li M, Li K, James AA, Tana A et al. Silkworm genetic sexing through W chromosome-linked, targeted gene integration. PNAS. 115(2018):8752–8756.

Zhang Z-J, Zhang S-S, Niu B-L, Ji D-F, Liu X-J, Li M-W et al. A determining factor for insect feeding preference in the silkworm, Bombyx mori. PLoS Biology. 17, no. 2(2019b):e3000162, 1–17.

Zhu L, Chen YR, Geng T, Tang SM, Zhao QL, Shen XJ. A 14-amino acids deletion in BmShadow results to non-moult on the 2nd instar in the bivoltine silkworm, Bombyxmori. Gene. 777(2021):145450.

Zhu L H, Mon J, Xu J M, Lee T. Kusakabe. CRISPR/Cas9 mediated knockout of factors in non-homologous end joining pathway enhances gene targeting in silkworm cells. Scientific Reports. 5, no. 18103(2016):1–13.

9 Metagenomic Approaches for Harnessing Microbial Enzymes and Bioactive Molecules

Ankita Srivastava, Akanksha Vishwakarma, Vinay Kumar and Digvijay Verma

CONTENTS

9.1 INTRODUCTION

Traditional cultivation approaches have been explored for a long time to capture the hidden microorganisms in an environmental sample. The 19th century was dedicated to exploring the microorganisms using traditional cultivation approaches (Giovannoni and Stingl, 2005). These techniques involve tedious steps, from sterilization, isolation, and purification of pure cultures on a suitable medium. Though the technique is still irreplaceable, this way one loses the major chunk of 99.99% of microorganisms from an environmental sample (Amann et al., 1995). It can surprise anyone that currently running microbial product-based industries survive on less than 1% of microbes that are accessible in the pure form under laboratory conditions. Several physiochemical limitations to simulate the environmental conditions into the laboratory make these microorganisms out of reach. These parameters include suitable nutrient medium, pH, temperature, pressure, and host-microbe interactions. To meet these conditions in an artificial environment of laboratories is almost impossible; therefore, to access the entire microbial community remains an irrational dream of microbiologists. We can understand the microbial load in an environmental sample by estimating the count of microorganisms of that sample. For example, one gram of garden soil 10^{10} number of bacterial cells that are rich in the innumerous count of useful genes (Torsvik et al., 1990). A handful of efforts

DOI: 10.1201/9781003220831-9

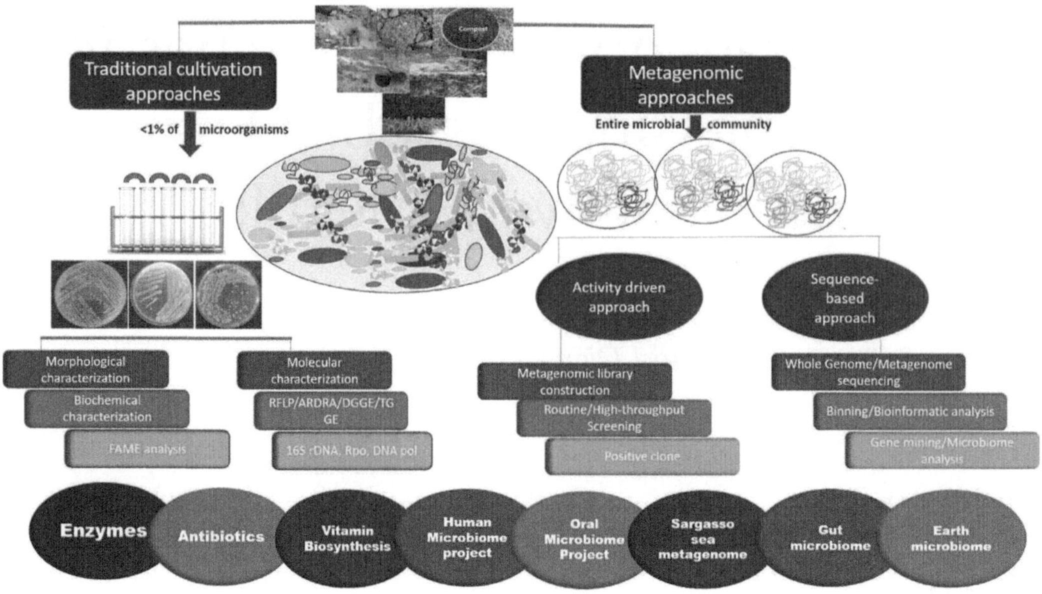

FIGURE 9.1 The picture depicting the overall strategy of cultivation as well as metagenomic approaches to harness the microbial bioactive molecules from different environments as well as for microbiome analysis.

have been made to isolate the microorganisms that were earlier unculturable (Platomaculum thermospropionicum gen.nov.sp.nov., Thermoacetogenium phaeum gen.nov.sp.nov., Sporotamuculum syntrophicum gen.nov.sp.nov). However, such approaches have not contributed much to enhance the number of accessible prokaryotes of that sample. The microorganisms, either prokaryotes or eukaryotes, represent massive genetic elements in their biological pool and must be dug out for the retrieval of novel genes, entire metabolic pathways, and their products (Cowan, 2000). Although the significance of the traditional cultivable approach is remarkable, where one can obtain pure isolates, the reality of the big count of inaccessible microorganisms cannot be ignored. With the best of the cultivation approaches, it is impossible to characterize the entire microbial community for their useful products. Therefore, an alternate way has been adopted to at least access the useful genes of hidden microorganisms of a habitat (Figure 9.1). This nontraditional way to explore the microorganisms came into existence during the late 20th century and was based on direct cloning of the community DNA from an environmental sample. Direct cloning circumvents these obstacles, as methods of sequencing can grab blueprints of the hidden microorganisms either from living or dead (Pace NR, 1997). It was later developed as 'metagenomics' and is now being used for harnessing desirable genes from unculturable associates of prokaryotes. The technique has come out with huge success, where several bioactive molecules have been produced by retrieving their respective genes of interest from various environmental samples. Metagenomic approaches also find their suitability to explore those environments that exhibit low microbial biomass due to extreme conditions. This chapter discusses metagenomics and its approaches extensively along with the development in metagenomics for harnessing microbial bioactive molecules from various environments.

9.2 MAJOR BREAKTHROUGHS OF METAGENOMICS

Before the use of molecular marker or ribosomal RNA gene for taxonomical and classification of microbes proposed by Carl Woese in 1977, for almost 300 years the study was based on various biochemical, morphological, and growth features to identify the microorganisms (Roszak et al.,

1984; Oliver et al., 1991), which limits microbial resolution in several ways. Norman R. Pace and colleagues in 1986 used direct analysis of 5S and 16S rRNA gene for exploring the diversity and complexity of microorganisms by cloning bulk DNA in an environmental sample without using a culturable approach, which undoubtedly changed the researcher's perspective of looking at the microbial population of the specific niche. The picoplankton metagenome from marine picoplankton metagenome and enriched consortium samples from cellulose digesters to mine genes encoding cellulases were used to generate the first few metagenomic libraries (Schmidt et al., 1991; Healy et al., 1995). However, the term "metagenomic library" was coined by Rondon and colleagues during 2000 by making BAC libraries (bacterial artificial chromosome) using DNA directly extracted from soil samples. Moreover, the first attempt of metagenomic sequencing was done by Stein et al. (1996) by using Hawaiian ocean water and sequenced it via a fosmid cloning approach to explore novel metabolism in archaea, although till then metagenomics term was not introduced. Later in 1998, Jo Handelsman recognized that entire sample DNA can be used for exploring novel BCG loci and described the term 'metagenomics'. The whole-genome shotgun sequencing method was applied by Venter et al. (2004) in their study of the Sargasso Sea water sample in which a total of 1.045 billion base pairs of the sequence were analyzed. With the advancement in the DNA sequencing method, i.e., next-generation sequencing (NGS), in which millions of DNA sequences can be dealt with at one time with low cost and in less time, gained much interest, which is comparatively more advanced than Sanger-based sequencing (Kircher and Kelso, 2010). There are several sequencing platforms. Ion Torrent technology, Illumina Hi/MiSeq platforms, and MinIon dominate (Kchouk et al., 2017). However, further advancement led to third- and fourth-generation platforms as single-molecule real-time (SMRT) sequencing by PacBio, nanopore sequencing having advantages over the second generation, such as lower sequencing costs, no PCR processing, and a faster process (Escobar-zepeda et al., 2015). Due to the tremendous advancement in Illumina-based sequencing platforms in their sequencing outputs and inexpensive rates, several microbiome-based projects were successfully executed on this efficient platform (Yasir et al., 2015; Huttenhower et al., 2012). Hence, a significant increase in the microbiome and other metagenome-based projects have been published since 2012, and many studies showed positive impact with these NGS technology in different areas (Logares et al., 2012; Fichot and Norman, 2013; Sanchez-Flores and Abreu-Goodger, 2014). Besides whole genome sequencing (WGS), high-throughput sequencing (HTS) such as 'short-read' (second-generation sequencing/NGS) and 'long-read' (third-generation) sequencing methods have also emerged for whole transcriptome shotgun sequencing (WTSS)/RNA sequencing (Wang et al., 2009), methylation sequencing (Pelizzola et al., 2011), whole-exome sequencing (Rabbani et al., 2014), and targeted or candidate gene sequencing (CGS) (Leo et al., 2015). This development in technology directs researchers to explore novel enzymes with an unculturable approach, which is useful for industrial processes (Bashir et al., 2014). A unique ecological niche was explored for novel enzymes such as sugar cane bagasse for cellulolytic bacteria, oil environment for hydrocarbon-degrading enzymes, and hot spring sample (Mhuantong et al., 2015; Staley et al., 2018; Nie et al., 2016; Saxena et al., 2017), which also provides evidence of the dependence of enzyme characterization on the environment. Metagenomic-based diagnosis and treatment are also being explored and various studies were made for various rare diseases (Pereira et al.,2019; Oliveira et al., 2017). Thus, the NGS is evolving and rapidly gaining broad applicability other than just dealing with traditional genome approaches (Tebani et al., 2016; Ohashi et al., 2015).

Numerous large-scale metagenomic sampling projects have generated large (MAGs) data sets during 2015 onwards (Almeida et al., 2019; Parks et al., 2017; Pasolli et al., 2019; Delmont et al., 2018). Delmont et al. (2018) recovered nearly 1,000 non-redundant microbial population genomes from the TARA Oceans Project, in which they reported the first genomic evidence of the presence of non-cyanobacterial diazotrophs in the surface waters of the ocean. Similarly, Almeida et al. (2019) and Pasolli et al. (2019) studied the human gut microbiome and its relation with human health and diseases. There was a gradual shift from phylogenetics to phylogenomics base on approaches of looking at the microbial tree of life, which is facilitating scientists to elucidate evolutionary science as well (Parks et al., 2018). Metapangenomics is emerging as a new branch where methods of clustering of the genes from a set of metagenomes is carried out to make gene clusters based on single-copy core genes. This is being further explored for several interesting biological insights (Maistrenko et al., 2020; Van Rossum et al., 2020). Besides, analysis of horizontal gene transfer and genomic islands further enhance our understanding of their evolutionary relationship (Bazin et al., 2020).

9.3 METAGENOMICS FOR HARNESSING MICROBIAL BIOACTIVE MOLECULES

Microorganisms are considered as the reservoir for several useful bioactive molecules. During the 19th century, the market of microbially derived compounds emerged and developed exponentially. The foremost reasons to adopt these microbial products were their suitability in several of the industrial bioprocesses in an eco-friendly and cost-effective manner. The extreme conditions of the industries always look for an alternative that can replace the high energy-consuming steps of several downstream processes. For example, enzymes can perform such tasks ideally; especially in industries where several chemical reactions occur in highly extreme conditions for the conversion of substrates into a suitable product. Extremozymes can withstand such extreme conditions to achieve a product of interest such as thermo-alkali-stable and acid-stable enzymes. Similarly, the box of potential antibiotics is also getting emptied and needs to be refilled with novel antibiotics. Harboring 99.99% of microbiota through metagenomics opened the box to fulfill the demand for novel biocatalysts, antibiotics, and other useful microbial products from non-accessible microorganisms (Figure 9.1).

9.3.1 Biocatalysts

The enzymes find their role in several industries that include lignocellulose degradation in the paper/pulp industry, starch liquefaction, detergent, dairy, textile, food/feed brewing, and beverage industries. The processing of the biomolecules is carried out in extreme conditions in industries that are energy-intensive and pollution-generating processes. Therefore, extremozymes are required that act optimally under extreme conditions of the industries (Ferrer et al., 2005; Verma and Satyanarayana, 2020). A plethora of enzymes have been reported, from bacteria to fungi, using cultivation approaches but most of these enzymes do not find suitability under the extreme environment of the industries and get deactivated. Extremophilic microorganisms can solve such problems up to a level because of the inherent properties of the extreme environments that are adapted by these enzymes/proteins (Pace, 1997; Whitman et al., 1998). A handful of reports suggest the employability of using extremophilic microorganisms for harnessing extremozymes, for example, the huge success of Taq-polymerase for their application in the DNA amplification process at high very high temperatures of 72 to 95°C. However, the cultivation of extremophilic bacteria or archaea in itself is a tedious task because of the uncultivability of their major chunk; therefore, metagenomic approaches have been employed where several novel candidates of enzymes have been uncovered from extreme as well as moderate environments.

9.3.1.1 Xylanase

Xylanases are a class of glycosyl hydrolases that are required for the complete degradation of heterogeneous polysaccharide xylan (Kumar et al., 2013a, b). It is one of the highly demanded enzymes for the paper and pulp industry that replace the harmful chlorine used during the bleaching of the pulp. Industries demand thermo-alkali-stable xylanases that can tolerate the alkaline pH of 8 to 10 and temperature of 80–100°C. The pulp obtained through the kraft-cooking process exhibit such conditions theretofore, thermo-alkali-stable xylanases are required that loosen the cell wall of plants for releasing the cellulosic fibers. Several metagenomic xylanases have been reported from various environments with desirable properties (Brennan et al., 2004; Shi et al., 2011; DeCastro et al., 2016; Ellila et al., 2019; Ariaeenejad et al., 2019; Verma and Satyanarayana, 2020). Approximately 200 metagenomic xylanolytic enzymes have been uncovered to date by using metagenomic approaches (Verma et al., 2019). Interestingly, the majority of the metagenomic xylanases belong to the glycosyl hydrolase (GH-10) family, while the number of GH-11 xylanases is comparatively low. Sunna and Bergquist (2003) were the pioneers to report the first metagenomic xylanase that was highly thermophilic xylanase having an optimum temperature of 100°C. A metagenomic xylanase (Mxyl) obtained through a plasmid-based metagenomic library has shown an optimum temperature of 80°C under alkaline pH of 9.0 (Verma et al., 2013). This enzyme was successfully employed for bleaching of pulp samples that reduce the chlorine consumption by 29%, which makes the process highly eco-friendly (Verma et al., 2013). Insect gut was heavily explored for harnessing several acidophilic xylanases that belong to GH-8 and GH-11 type. Compost is another highly investigated environment for uncovering xylanase candidates (Sun et al., 2015; Sae-Lee and Boonmee, 2014; Jeong et al., 2012) that revealed several unusual xylanases. Five acidic xylanases have been reported from compost soil metagenome where all showed optimum activity at and above 50°C (Kwon et al., 2010). Wang et al. (2010) reported several GH-10 and GH-11 xylanases from Alpine tundra soil metagenome that exhibited optimum activity at a very low temperature of 4 to 20°C. Such enzymes find applicability for the dissolution of pulp in the juice extraction industry. Unusual xylanase has been uncovered from camel rumen metagenome that showed activity stimulation by 132% in presence of 5M NaCl (Ghadikolaei et al., 2019). Bifunctional properties of xylanases with endoglucanases find applicability in the bioethanol industry (Verma et al., 2019); such unusual xylanase has been recovered from cow rumen metagenome that exhibits synergistic effect with xylosidases and glucosidase (Chang et al., 2011). Several unusual genes xylanases have been retrieved from community DNA for their applications in various industries. However, the majority of them are still uncharacterized and need extensive characterization for their suitable applicability.

9.3.1.2 β-Xylosidase

β-xylosidase (Exo-1,4-β-D-xylosidase EC 3.2.1.37) represents another important class of enzyme that cleaves the glycosidic bonds at the non-reducing end of xylo-oligosaccharides and liberate xylose sugars. The majority of the β-xylosidases exhibit several limitations that include low thermostability, poor efficiency, and salt sensitivity, and therefore encourage to find industrially relevant β-xylosidases. Several metagenomic β-xylosidases have been recovered from various environmental samples (Zhou et al., 2012; Bastien et al., 2013; Sato et al., 2017; Liu et al., 2018; Rohman et al., 2019; Verma and Satyanarayana, 2020). The metagenome of ruminal fluid of yak (Zhou et al., 2012), buffalo (Singh et al., 2014), and cow (Jordan et al., 2016) revealed several β-xylosidases; this may be due to the hemicellulose degrading microorganisms that dominate in the gut of ruminants. Soil (Matsuzawa et al., 2016), compost (Wang et al., 2015), and hot spring (Sato et al., 2017) have also been explored for identifying several β-xylosidase genes. Highly thermostable β-xylosidases were recovered from metagenome of compost soil that showed optimum activity in the range of 60–75°C (Wang et al., 2015). Similarly, the three β-xylosidases (AR19M-311–2, AR19M-311–11, and AR19M311–21) derived from hot spring metagenome revealed their optimum activity at 90°C along with fair thermo-stability of one hour at 70°C. It might be possible that the source of these enzymes may be from some thermophilic archaea such as Thermoanaerobacter ethanolicus, or Thermotoga maritima, whose β-xylosidases share similar characteristics (Shao and Wiegel, 1992; Shi W. et al., 2013). Thermostable β-xylosidases improve the overall productivity of the industry by reducing the other microbial contamination as well as enhancing the viscosity of the reactants. Matsuzawa et al. (2016) report the first unique metagenomic α-xylosidase (MeXyl31) that belongs to glycosyl hydrolase family GH-31 and exhibit an optimum temperature at 45°C. Interestingly, several metagenomic β-xylosidases share bi-functional activity (DeCastro et al., 2016; Rohman et al., 2019). For example, three β-xylosidases (Rubg3A, Rubg3B, and RubGX1) also showed β-glucosidase activity (Bao et al., 2012). Similarly, several metagenomic β-xylosidases are also characterized for arabinofuranosidase activity (Wagschal et al., 2009; Jordan et al., 2016). Liu et al. (2018) also identified four clones having bifunctional β-xylosidases activity, either of β-arabinosidases or β-glucosidase type. Therefore, metagenomic has uncovered several β-xylosidases with very similarity/identity with the existing homolog in the database.

9.3.1.3 Acetyl Xylan Esterases (AXEs)

Similar to α-arabinosidases activity, this class of enzymes also acts on the side chain of heterogenous xylan substrates for their complete degradation, therefore, it also acts synergistically in a cocktail of xylanolytic enzymes (Hettiarachchi et al., 2019). A handful of acetyl xylan esterases have been reported using metagenomic approaches (Adesioye et al., 2018; Brito et al., 2018). A meso-alkaliphilic acetyl xylan esterase (Axe1NaM1) has been recovered from a hot desert hypolith metagenome that showed an optimum temperature of 30°C under alkaline pH of 8.5 (Adesioye et al., 2018). Brito et al. (2018) also retrieved several acetyl xylan esterases from shipworm gut metagenome that were novel protein sequences having very little similarity of 50–75%.

9.3.1.4 α-Arabinosidases

It is another important class of xylanolytic enzyme for the complete degradation of heterogenous xylan substrates by cleaving α-1,2 and α-1,3 glycosidic bonds (Verma et al., 2019). The enzyme finds immense applications in a cocktail of other xylanolytic enzymes such as environmental waste management and lignocellulosic biomass conversion. Efforts have also been made to retrieve the metagenomic α-arabinosidases from various environmental sources (Matsuzawa et al., 2015; Jordan et al., 2016; Verma and Satyanarayana, 2020). The majority of the metagenomic α-arabinosidases exhibit bi-functional activity, chiefly with β-xylosidases (Wang et al., 2015; Jordan et al., 2016). Three acidic metagenomic α-arabinosidases have been uncovered from the compost soil metagenome that showed their optimum temperature in the mesophilic range of 40°C (Matsuzawa et al., 2015).

9.3.1.5 Cellulase

Cellulase is another group of class that finds applications in lignocellulose biomass processing chiefly in biofuel production, waste management, paper/pulp industry, textile, and food processing industries (Kuhad et al., 2011). The enzyme is composed of three major types of hydrolases (endoglucanases, exoglucanases, and β-glucosidases) to achieve cello-oligosaccharides, cellobiose, and finally the monomeric glucose. The metagenomic approaches have been successfully employed for harnessing several cellulases (Duan et al., 2009; Nguyen et al., 2012; Lee et al., 2018; Wu et al., 2020). The gut of the ruminants is one of the highly explored environments for retrieving cellulase genes due to the presence of abundant cellulosic anaerobes in their gut (Duan et al., 2009; Shah et al., 2017). A total of 61 clones having cellulase activity were recovered from a ruminal metagenomic library (Duan et al., 2009). Of them, one novel cellulase (C67–1) showed its stability in a broad pH range from 3.5 to 10.5 with an optimal activity at a pH of 4.5. Acidic cellulases find applications in plant biomass conversion and waste management. Several such acidic cellulases have been uncovered from the metagenome of ruminal fluid-based metagenomic libraries (Lee et al., 2018; Shah et al., 2017). It may be due to the acidic environment of the ruminal fluid. The majority of the cellulase showed their optimum temperature in the mesophilic range of 30 to 40°C. However, metagenomic also reports the presence of several thermophilic and hyper-thermophilic cellulases that showed optimum temperature beyond 45°C (Escuder-Rodríguez et al., 2018; Pabbathi et al., 2021; Wu et al., 2020). Soil (Garg et al., 2016; Chai et al., 2020) and mangrove (Gonçalves et al., 2015), and coral (Sousa et al., 2016), and compost soil (Pang et al., 2009) environments have also been explored for accessing cellulosic genes. Duan et al. (2009) elaborate the hurdles to obtain thermophilic cellulases using functional metagenomic approaches, as most of the metagenomic libraries have been constructed in mesophilic host *E. coli*, which is not considered as a suitable host for expressing thermophilic proteins. Metagenomics has opened the pandora's box to access the cellulase of unique properties (Garg et al., 2016; Escuder-Rodríguez et al., 2018), the only requirement to move with a rational approach to obtain the cellulase of interest.

9.3.1.6 Lipase and Esterase

Lipases are one of the most explored enzymes through metagenomic approaches (Voget et al., 2003; Lee et al., 2004; Rhee et al., 2005; Lee et al., 2006; Lopez-Lopez et al., 2014; Sahoo et al., 2020). These α/β hydrolases can be categorized into two groups that include true lipases (EC 3.1.1.3 triacylglycerol lipases) and esterases (EC 3.1.1.1 carboxylesterases). Metagenomic approaches have been employed to explore the lipases of unique features such as thermostable, halotolerant, unique enantioselectivity, and chemio-regioselectivity. Hot springs have been explored successfully to harness thermostable lipases/esterase (Sahoo et al., 2020; López-López et al., 2015). Sahoo et al. (2020) identified seven lipase-producing clones from the Taptapani hot spring, where one (Lip479) of them showed optimum activity at 65°C under alkaline pH of 8.0. Similarly, two thermostable esterases (EstE1 and Est 1) from thermal areas of Indonesia and Thailand showed optimum activity at 95°C and 70°C, respectively (Rhee et al., 2005; Tirawongsaroj et al., 2009). Compost soil-derived lipase/esterase has also been identified for thermophilic characteristics (Lu et al., 2019; Lu et al., 2021). Lu et al. (2019) retrieved two genes (Est1 and Est2) from compost metagenomes that on heterologous expression showed optimum activities at 80°C and 70°C, respectively. Soil (Almeida et al., 2019), and marine (Kennedy et al., 2011; Jeon et al., 2009; Jeon et al., 2012), and sponge environments (Su et al., 2015) are the other sources that have extensively been used for discovering lipases/esterases. Halophilic lipases find applications in the pharmaceutical industry, the functional metagenomic approaches have revealed few halophilic lipases that can withstand their stability at a higher salt concentration (Zhang et al., 2017). Similarly, high enantio- and regioselective lipases have been recovered from various metagenomes that exhibit applications in various pharmaceutical industries (Glogauer et al., 2011; López-López et al., 2015; Alnoch et al., 2015; Kumar et al., 2017). A metagenomic clone derived from an anaerobic lagoon DNA library showed very high activity towards triacylglycerol along with fair thermostable and halophilic properties (Glogauer et al., 2011). The presence of a specific motif 'GGG(A)X' among various lipases makes them suitable for their activity towards tertiary alcohols (Kourist et al., 2008).

9.3.1.7 Amylase

One of the highly relevant industrial enzymes that hydrolyze starch either by exo- or endo mode of cleavage. The α-amylases (EC 3.2.1.1) belong to the GH-13 family of glycosyl hydrolases that cleave the glycosidic bonds of α-1,4 linked backbone (Antranikian, 1992; Gupta et al., 2002). A plethora of amylases have been reported to date from a broad range of microorganisms, however, industries demand amylases of extreme properties. Metagenomic approaches revealed numerous amylases of varying properties from the amylases obtained from culturable sources. For example, a unique amylase gene (Amy13A) of 1461 bp was retrieved by screening a metagenomic library derived from the metagenome of biogas reactor (Jabbour et al., 2013). On the characterization of the expressed protein, it showed a temperature optimum at 80°C along with high salt tolerance of 25% NaCl (w/v). Similarly, an amylase from the Western ghat metagenomic library exhibited maximum activity at 60°C (Vidya et al., 2011). Thermo-acid stable amylases have immense applications in the pharmaceutical industries. Metagenomic approaches have uncovered thermophilic amylases that exhibit their optimum activity under acidic pH (Liu et al., 2011; Motahar et al., 2020). The discovery of several neutral (Sharma et al., 2010b; Nair et al., 2017) and alkaliphilic (Vidya et al., 2011; Wang et al., 2019) amylases signify the importance of metagenomics for discovering biocatalysts of varying properties. Recently, Nair and Bhat (2020) characterized a unique metagenomic amylase for its applications in the food and textile industries.

9.3.1.8 Protease

One of the industrially relevant enzymes that find applications in the leather and tanning industry along with detergent and food/feed industry (George et al., 2014). Proteases hydrolyze the protein by cleaving peptide bonds and release smaller polypeptides and monomeric amino acids (Singh et al., 2015). The functional as well sequence-based metagenomic approaches have successfully been employed for identifying several hidden genes of proteases from the diverse environments (Neveu et al., 2011; Wemheuer et al., 2013; Singh et al., 2015; DeCastro et al., 2016). Based on active amino acids participating in catalytic cleavage, proteases can be classified as serine, aspartic, cysteine, and metalloprotease (Singh et al., 2015). Of them, serine proteases have been abundantly recovered from culturable as well as metagenomic approaches (Pushpam et al., 2011; Neveu et al., 2011; Biver et al., 2013; Singh et al., 2015; Sun et al., 2020). A novel gene (pF1AL2) of serine protease has been recovered from marine sediment, that exhibited an optimum temperature of 80°C along with fair thermostability at 70°C for 6 hours by retaining 80% overall activity (Sun et al., 2020). Similarly, two highly thermo-alkali-stable proteases (DV1 and M30) have been uncovered from a dessert metagenome that showed maximum activity at 55°C and 40°C respectively (Neveu et al., 2011). Several proteases belong to aspartic proteases, metalloprotease, and cysteine proteases that were also explored from diverse environments such as soil (Biver et al., 2013), mangrove sediment (Pessoa et al., 2017), and hot springs (Singh et al., 2015; Wemheuer et al., 2013).

9.3.1.9 Miscellaneous Enzymes

Chitinases, pectinase, and asparaginase belong to other industrially relevant enzymes that have also been identified using metagenomic approaches. Chitinase (endo/exo- chitinases, β-N-acetyl glucosaminidase, and chitobiosidases) hydrolyzes the chitin substrates that are abundantly present in the cell wall of several fungi, insects, and nematodes, therefore, find application in pest management (Soares et al., 2015; Berini et al., 2017). Metagenomic approaches have uncovered several novel chitinases from diverse environments (Berini et al., 2019; Raimundo et al., 2021). Two metagenomic chitinase genes (Chi18H8 and 53D1) were successfully employed to disrupt the structural integrity of the domestic silkmoth *Bombyx mori* (Berini et al., 2019). Functional metagenomics of marine metagenome revealed potential chitin degradation pathways that illustrate the significance of microbial chitin utilizers in marine habitats (Raimundo et al., 2021). An antifungal chitobiosidase (Chi18H8) has been retrieved from a suppressive soil metagenome that showed an inhibitory effect in the growth *of Fusarium graminearum* and *Rhizoctonia solani* (Berini et al., 2017).

Pectinase enzymes find chief application in the juice extraction industry to dissolve the plant biomass in an ecofriendly manner without compromising the quality of the juice. Pectin represents one of the abundant polysaccharides in plant middle lamella and primary cell wall (Chung et al., 2014). To overcome the hurdles associated with available pectinases of cultivable microorganisms, metagenomic approaches have been employed where several pectinases have been uncovered (Sathya and Khan, 2014; Tian et al., 2015; Zhou et al., 2017). A microbiome-based study of compost metagenome signifies the abundance of pectinolytic genes in the sample that belong to several previously unrecognized bacteria (Zhou et al., 2017). Thermostable pectinases are preferred in industries over mesophilic homologs. Metagenomics has also revealed few thermostable pectinases (Singh et al., 2012; Sathya and Khan, 2014; Kaushal et al., 2018). A hyper-thermophilic pectinase was accessed from soil-based metagenomic library that sowed optimum activity at 70°C having a half-life of 5 h at 60°C (Singh et al., 2012). Similarly, Sathya and Khan (2014) identified a thermostable pectinase that showed activity in a broad range of pH (5–9) and temperature (20 to 60°C). Interestingly, both pectinases belong to GH-28 pectinases. More research is required to harness more pectinase using a metagenomic approach.

The asparagine hydrolyzing enzyme (L-asparaginase (EC 3.5.1.1)) finds application in the pharmaceutical industry for treating acute lymphoblastic leukemia (ALL). The majority of microbial L-asparaginases exhibit glutaminase activity, therefore, alternate metagenomic methods have also been made to achieve glutaminase free asparaginases. A handful of reports are available to retrieve the L-asparaginases using metagenomic approaches (Arjun et al., 2018; Sobat et al., 2021). A soil-based-metagenomic library revealed a potential L-asparaginase that was optimally active at 37°C under neutral pH conditions. Further characterization showed that this enzyme exhibits a strong affinity towards L-asparaginase and could be a better candidate over the commercial L-asparaginases from *E. coli* (Arjun et al., 2018). In another report, analysis of the Caspian-sea metagenome uncovered the presence of 87 putative candidate genes of L-asparaginase. Of them, three were fished out for their expression and functional activity, where two of them have been identified for having strong anti-proliferative activity against Jurkat cell lines (Sobat et al., 2021). Similarly, the gut microbiome of pediatric acute lymphoblastic leukemia (ALL) patients revealed a significant difference in microbial communities between decreased and increase asparaginase activity (Dunn et al., 2021). More efforts are required, as still we are very far from glutaminase-free halophilic and neutral asparaginases.

9.3.2 Antibiotics

Antibiotics are the secondary metabolites of microorganisms that are produced for their defense and survival in a competitive natural environment. However, humans have used these antibiotics like a drug to protect themselves against microbial-derived infections. Hence, antibiotics are considered as one of the most highly used microbial products in therapeutics. Due to the uncontrolled consumption of antibiotics, several microorganisms have developed resistance towards them. Therefore, limited stocks of antibiotics are challenging us to discover novel candidates. As the majority of the available antibiotics have been extracted from cultivable microorganisms that count less than 1% of total bacteria present in that environment. Therefore, metagenomic approaches have been employed to explore the remaining microflora of a habitat to uncover new antibiotics (Brady et al., 2001; DeCastro et al., 2016; Parsley et al., 2011). Due to the large insert size of gene clusters (30–100 Kb) that encode antibiotics (Donadio et al., 1991; Krause and Marahiel, 1988; Schwecke et al., 1995), the majority of the metagenomic libraries have been constructed either in fosmid/cosmid or BAC vectors (Lim et al., 2005; Voget et al., 2003; Rondon et al., 2000). Therefore, to achieve the large molecular weight DNA is the primary task to enhance the possibility of finding antibiotics. In a very early report, McNeil and colleagues (2001) have reported several molecules from soil-based metagenomic libraries that exhibited homology with indirubin. Similarly, another soil metagenome-based fosmid library also revealed several clones having an expression of indirubin and their homologs (Lim et al., 2005). Wang et al. (2000) constructed a unique metagenomic library by using Streptomycetes as a host. On screening, a unique novel compound Terragine has been identified that showed potential to have anti-mycobacterial activity. The discovery of Turbomycin A and B from soil metagenome further encourages recovering novel antibiotics (Gillespie et al., 2002). A fosmid-based soil metagenomic library revealed an array of ketosynthase domains that exhibit a role in the enzymatic activity of polykeitide synthases (Parsley et al., 2011). Diverse polyketide synthase genes have been recovered from metagenomic analysis of marine sponge metagenomes (Schirmer et al., 2005). Wrighton (2018) has reported malacidins from soil metagenome that showed calcium-dependent activity towards MRSA. The majority of the reports on discovering antibiotics using a metagenomic approach explored the soil environments (DeCastro et al., 2016). Therefore, exploitation of other habitats enhances the potential for discovering more novel candidates.

9.3.3 Vitamin Biosynthesis

Metagenomic approaches have also been successfully employed to harness vitamins from microbial sources (Entcheva et al., 2001; Das et al., 2019; Das et al., 2021). In a pioneer report, Entcheva et al. (2001) obtained seven different clones having sequences of biotin vitamins from a cosmid-based metagenomic library. Similarly, genes of 2,5-diketo-D-gluconic acid reductases (DKGRs) were identified using metagenomics that plays a significant role in the synthesis of vitamin C (Eschenfeldt et al., 2001). Recently Dash et al. (2021) revealed the abundance of vitamin K (menaquinones) pathways among the gut microbiome of diabetes mellitus patients. Similarly, gut microbiome has been affirmed as a rich source of B vitamins (Magnúsdóttir et al., 2015).

9.4 CONCLUSIONS

Metagenomics has emerged as an advanced revolutionary tool for exploring the hidden microorganisms from mesophilic as well as extremophilic environments. It is considered the second highly significant discovery to investigate the microorganisms after the invention of the microscope. Both functional, as well as for sequencing-based approaches, have their pros and cons. Functional-based screening is comparatively cost-effective over a sequence-based approach to get the novel homologs of proteins and other bioactive molecules. However, it demands a high throughput screening of the clones generated during the library construction. Substrate attached with dye facilitates for efficient screening of the clones. Robotic screening of clones using 96 well plate also boosts the screening multifold as compared to the manual screening. On the other hand, sequence-based screening includes the analysis of sequences of the entire metagenome, therefore, it provides the information of all the hidden sequences that can be explored for microbial diversity as well as for genes of interest. Extraction of high quality and high molecular weight metagenomic DNA from environmental samples is the requisite of the metagenome-based studies to achieve the metagenomic library of large insert size. It enhances the efficiency to obtain a good metagenomic library along with the possibility to get a greater number of positive clones of interest. Metagenomics has emerged as an advanced tool for analyzing the microbiome of a niche, where signature sequences of bacterial 16S rDNA are used to amplify the hypervariable regions of the inhabitant bacteria of an environmental sample. Microbiome-based analysis has revolutionized the understanding of microorganisms. The metagenomic approaches have opened Pandora's box to achieve the gene of interest and now microbiologists can explore the hidden chunk of microorganisms efficiently.

BIBLIOGRAPHY

Adesioye FA, Makhalanyane TP, Vikram S, Sewell BT, Schubert WD and Cowan DA (2018). Structural characterization and directed evolution of a novel acetyl xylan esterase reveals thermostability determinants of the carbohydrate esterase 7 family. Appl Environ Microbiol 84: e02695–17.

Almeida A, Mitchell AL, Boland M et al. (2019). A new genomic blueprint of the human gut microbiota. Nature 568:499–504.

Alnoch RC, Martini VP, Glogauer A, Costa ACdS, Piovan L, Muller-Santos M et al. (2015). Immobilization and characterization of a new regioselective and enantioselective lipase obtained from a metagenomic library. PLoS ONE 10(2):e0114945.

Amann RI, Ludwig W, Schleifer KH (1995). Phylogenetic identification and in situ detection of individual microbial cells without cultivation. Microbiol Rev 59(1):143–169.

Antranikian G (1992). Microbial degradation of starch. In Microbial Degradation of Natural Products. ed A Winkelmann (Germany: Weinheim VCH) 27–56.

Ariaeenejad S, Hosseini E, Maleki M, Kavousi K, Moosavi-Movahedi AA, Salekdeh GH (2019). Identification and characterization of a novel thermostable xylanase from camel rumen metagenome. Int J Biol Macromol 126:1295–1302.

Arjun JK, Aneesh BP, Kavitha T, Harikrishnan K (2018). Characterization of a novel asparaginase from soil metagenomic libraries generated from forest soil. Biotechnol Lett 40(2):343–348.

Bao L, Huang Q, Chang L, Sun Q, Zhou J and Lu H (2012). Cloning and characterization of two β-glucosidase/xylosidase enzymes from yak rumen metagenome. Appl Biochem Biotechnol 166:72–86.

Bashir Y, Singh SP, Konwar BK (2014). Metagenomics: An application-based perspective. Chin J Biol 2014:1–7.

Bastien G, Arnal G, Bozonnet S, Laguerre S, Ferreira F, Faure R et al. (2013). Mining for hemicellulases in the fungus-growing termite Pseudacanthotermes militaris using functional metagenomics. Biotechnol Biofuels 6:78.

Bazin A, Gautreau G, Médigue C, Vallenet D, Calteau A (2020). panRGP: A pangenome-based method to predict genomic islands and explore their diversity. Bioinformatics 36(2):i651–i658.

Berini F, Casartelli M, Montali A, Reguzzoni M, Tettamanti G and Marinelli F (2019). Metagenome-sourced microbial chitinases as potential insecticide proteins. Front Microbiol 10:1358.

Berini F, Casciello C, Marcone GL and Marinelli F (2017). Metagenomics: Novel enzymes from non-culturable microbes. FEMS Microbiol Lett 364:fnx211.

Biver S, Portetelle D, Vandenbol M (2013). Characterization of a new oxidant-stable serine protease isolated by functional metagenomics. Springerplus 2(1):410.

Brady SF, Chao CJ, Handelsman J, Clardy J (2001). Cloning and heterologous expression of a natural product biosynthetic gene cluster from eDNA. Org Lett 3(13):1981–1984.

Brennan Y, Allen W, Christoffersen L, Dupree P, Healay S, Hernandez M et al. (2004). Unusual microbial xylanases from insect guts. Appl Environ Microbiol 70:3609–3617.

Brito TL, Campos AB, Bastiaan von Meijenfeldt FA, Daniel JP, Ribeiro GB and Silva GGZ (2018). The gill-associated microbiome is the main source of wood plant polysaccharide hydrolases and secondary metabolite gene clusters in the mangrove shipworm Neoteredo reynei. PLoS One 13:e0200437.

Chai S, Zhang X, Jia Z et al. (2020). Identification and characterization of a novel bifunctional cellulase/hemicellulase from a soil metagenomic library. Appl Microbiol Biotechnol 104:7563–7572.

Chang L, Ding M, Bao L, Chen Y, Zhou J and Lu H (2011). Characterization of a bifunctional xylanase/endoglucanase from yak rumen microorganisms. Appl Microbiol Biotechnol 90:1933–1942.

Chung D, Pattathil SK, Biswal AK, Hahn MG, Mohnen D, Westpheling J (2014). Deletion of a gene cluster encoding pectin degrading enzymes in Caldicellulosiruptor bescii reveals an important role for pectin in plant biomass recalcitrance. Biotechnol Biofuels 7(1):147

Cowan DA (2000). Microbial genomes: The untapped resource. Trends Biotechnol 18:14–16.

Das P, Babaei P and Nielsen J (2019). Metagenomic analysis of microbe-mediated vitamin metabolism in the human gut microbiome. BMC Genomics 20:208.

Das T, Jayasudha R, Chakravarthy S et al. (2021). Alterations in the gut bacterial microbiome in people with type 2 diabetes mellitus and diabetic retinopathy. Sci Rep 11:2738.

DeCastro ME, Rodríguez-Belmonte E, González-Siso MI (2016). Metagenomics of thermophiles with a Focus on discovery of novel thermozymes. Front Microbiol 7:1521.

Delmont TO, Quince C, Shaiber A et al. (2018). Nitrogen-fixing populations of Planctomycetes and Proteobacteria are abundant in surface ocean metagenomes. Nat Microbiol 3:804–813.

Donadio S, Staver MJ, McAlpine JB, Swanson SJ and Katz L (1991). Modular organization of genes required for complex polyketide biosynthesis. Science 252:675–679.

Duan CJ, Xian L, Zhao GC, Feng Y, Pang H, Bai XL, Tang JL, Ma QS, Feng JX (2009). Isolation and partial characterization of novel genes encoding acidic cellulases from metagenomes of buffalo rumens. J Appl Microbiol 107:245–256.

Dunn KA, Connors J, Bielawski JP, Nearing JT, Langille MGI, Van Limbergen J, Fernandez CV, MacDonald T, Kulkarni K (2021). Investigating the gut microbial community and genes in children with differing levels of change in serum asparaginase activity during pegaspargase treatment for acute lymphoblastic leukemia. Leuk Lymphoma 62(4):927–936.

Ellila S, Bromann P, Nyyssonen M (2019). Cloning of novel bacterial xylanases from lignocellulose-enriched compost metagenomic libraries. AMB Express 9:124.

Entcheva P, Liebl W, Johann A, Hartsch T, Streit WR (2001). Direct cloning from enrichment cultures, a reliable strategy for isolation of complete operons and genes from microbial consortia. Appl Environ Microbiol 67:89–99.

Eschenfeldt WH, Stols L, Rosenbaum H, Khambatta ZS, Quaite-Randall E, Wu S, Kilgore DC, Trent JD, Donnelly MI (2001). DNA from uncultured organisms as a source of 2,5-diketo-D-gluconic acid reductases. Appl Environ Microbiol 67(9):4206–4214.

Escobar-zepeda A, De León AV, Sanchez-flores A (2015). The road to metagenomics: From microbiology to DNA sequencing technologies and bioinformatics. Front Genet 6:1–15

Escuder-Rodríguez JJ, DeCastro, ME, Cerdan ME, Rodríguez-Belmonte E, Becerra M, González-Siso MI (2018). Cellulases from thermophiles found by Metagenomics. Microorganisms 6(3):66.

Ferrer M, Golyshina OV, Chernikova TN, Khachane AN, Reyes-Duarte D, Santos VA, Strompl C, Elborough K, Jarvis G, Neef A, Yakimov MM, Timmis KN, Golyshin PN (2005). Novel hydrolase diversity retrieved from a metagenome library of bovine rumen microflora. Environ Microbiol 7:1996–2010.

Fichot EB, Norman RS (2013). Microbial phylogenetic profiling with the Pacific biosciences sequencing platform. Microbiome 1:10.

Garg R, Srivastava R, Brahma V et al. (2016). Biochemical and structural characterization of a novel halotolerant cellulase from soil metagenome. Sci Rep 6:39634.

George N, Chauhan PS, Kumar V, Puri N, Gupta N (2014). Approach to eco-friendly leather: Characterization and application of an alkaline protease for chemical free dehairing of skins and hides at pilot scale. J Cle Pro 1–9.

Ghadikolaei KK, Sangachini ED, Vahdatirad V, Noghabi KA, Zahiri HS (2019). An extreme halophilic xylanase from camel rumen metagenome with elevated catalytic activity in high salt concentrations. AMB Express 9:86.

Gillespie DE, Brady SF, Bettermann AD, Cianciotto NP, Liles MR, Rondon MR, Clardy J, Goodman RM, Handelsman J (2002). Isolation of antibiotics turbomycin A and B from a metagenomic library of soil microbial DNA. Appl Environ Microbiol 68:4301–4306.

Giovannoni SJ, Stingl U (2005). Molecular diversity and ecology of microbial plankton. Nature 437:343–348.

Glogauer A, Martini VP, Faoro H et al. (2011). Identification and characterization of a new true lipase isolated through metagenomic approach. Microb Cell Fact 10:54.

Gonçalves AC, Dos Santos AC, Dos Santos TF, Pessoa TB, Dias JC and Rezende RP (2015). High yield of functional metagenomic library from mangroves constructed in fosmid vector. Genet Mol Res 14:11841–11847.

Gupta R, Beg QK, Lorenz P (2002). Bacterial alkaline proteases: Molecular approaches and industrial applications. Appl Microbiol Biotechnol 59:15–32.

Healy FG, Ray RM, Aldrich HC, Wilkie AC, Ingram LO and Shanmugam KT (1995). Direct isolation of functional genes encoding cellulases from the microbial consortia in a thermophilic, anaerobic digester maintained on lignocellulose. Appl Microbiol Biotechnol 43(4):667–674.

Hettiarachchi SA, Kwon YK, Lee Y, Jo E, Eom TY, Kang YH, Kang DH, De Zoysa M, Marasinghe SD, Oh C (2019). Characterization of an acetyl xylan esterase from the marine bacterium Ochrovirga pacifica and its synergism with xylanase on beechwood xylan. Microb Cell Fact 18(1):122.

Huttenhower C, Gevers D, Knight R, Abubucker S, Badger JH, Chinwalla AT et al. (2012). Structure, function and diversity of the healthy human microbiome. Nature 486:207–214.

Jabbour D, Sorger A, Sahm K, Antranikian G (2013). A highly thermo-active and salt-tolerant α-amylase isolated from a pilot-plant biogas reactor. Appl Microbiol Biotechnol 97(7):2971–2978.

Jeon JH, Kim JT, Kang SG, Lee JH, Kim SJ (2009). Characterization and its potential application of two esterases derived from the arctic sediment metagenome. Mar Biotechnol 11(3):307–316.

Jeon JH, Lee HS, Kim JT, Kim SJ, Choi SH, Kang SG, Lee JH (2012). Identification of a new subfamily of salt-tolerant esterases from a metagenomic library of tidal flat sediment. Appl Microbiol Biotechnol 93(2):623–631.

Jeong YS, Na HB, Kim SK, Kim YH, Kwon EJ, Kim J et al. (2012). Characterization of xyn10J, a novel family 10 xylanase from a compost metagenomic library. Appl Biochem Biotechnol 166:1328–1339.

Jordan DB, Braker JD and Wagschal K (2016). Isolation and divalent metal activation of a β-xylosidase, RUM630-BX. Enzyme Microb Technol 82:158–163.

Kaushal G, Kumar J, Sangwan R S, Singh S P (2018). Metagenomic analysis of geothermal water reservoir sites exploring carbohydrate-related thermozymes. Int J Biol Macromol 119:882–895.

Kchouk M, Gibrat J, Elloumi M (2017). Generations of sequencing technologies: From first to next generation. Biol Med 9:1–8.

Kennedy J, O'Leary ND, Kiran GS, Morrissey JP, O'Gara F, Selvin J, Dobson AD (2011). Functional metagenomic strategies for the discovery of novel enzymes and biosurfactants with biotechnological applications from marine ecosystems. J Appl Microbiol 111 (4):787–799.

Kircher M, Kelso J (2010). High-throughput DNA sequencing – concepts and limitations. Bioessays 32(6):524–536.

Kourist R, Domínguez de María P, Bornscheuer UT (2008). Enzymatic synthesis of optically active tertiary alcohols: Expanding the biocatalysis toolbox. Chembiochem 9(4):491–498.

Krause M, Marahiel M A (1988). Organization of the biosynthesis genes for the peptide antibiotic gramicidin S. J Bacteriol 170:4669–4674.

Kuhad RC, Gupta R, Singh A (2011). Microbial cellulases and their industrial applications. Enz Res 2011:280696.

Kumar R, Banoth L, Banerjee UC, Kaur J (2017). Enantiomeric separation of pharmaceutically important drug intermediates using a Metagenomic lipase and optimization of its large-scale production. Int J Bio Macromol 95:995–1003.

Kumar V, Syal P, Satyanarayana T (2013a). Highly thermo-halo-alkali-stable β-1,4-endoxylanase from a novel polyextremophilic strain of Bacillus halodurans. Bioprocess Biosyst Eng 36:555–565.

Kumar V, Verma D, Satyanarayana T (2013b). Extremophilic bacterial xylanases: Production, characteristics and applications. Curr Biotechnol 2:380–399.

Kwon E J, Jeong Y S, Kim YH, Kim SK, Na HB, Kim J et al. (2010). Construction of a metagenomic library from compost and screening of cellulase- and xylanase-positive clones. J Appl Biol Chem 53:702–708.

Lee CC, Kibblewhite-Accinelli RE, Wagschal K, Robertson GH, Wong DWS (2006). Cloning and characterization of a cold-active xylanase enzyme from an environmental DNA library. Extremophiles 10:295–300.

Lee KT, Toushik SH, Baek JY, Kim JE, Lee JS, Kim KS (2018). Metagenomic mining and functional characterization of a novel KG51 bifunctional cellulase/hemicellulase from black goat rumen. J Agric Food Chem 66:9034–9041.

Lee SW, Won K, Lim HK, Kim JC, Choi GJ, Cho KY (2004). Screening for novel lipolytic enzymes from uncultured soil microorganisms. Appl Microbiol Biotechnol 65:720–726.

Leo VC, Morgan NV, Bern D et al. (2015). Use of next -generation sequencing and candidate gene analysis to identify underlaying defects in Patients With inherited platelet function disorders. J Thromb Haemost 13:643–650.

Lim HK, Chung EJ, Kim JC, Choi GJ, Jang KS, Chung YR, Cho KY, Lee SW (2005). Characterization of a forest soil metagenome clone that confers indirubin and indigo production on Escherichia coli. Appl Environ Microbiol 71:7768–7777.

Liu C, Zou G, Yan X, Zhou X (2018). Screening of multimeric β-xylosidases from the gut microbiome of a higher termite, Globitermes brachycerastes. Int J Biol Sci 14:608–615.

Liu J, Liu WD, Zhao XL et al.(2011). Cloning and functional characterization of a novel endo-β-1,4-glucanase gene from a soil-derived metagenomic library. Appl Microbiol Biotechnol 89:1083–1092.

Logares R, Haverkamp THA, Kumar S, Lanzén A, Nederbragt A J, Quince C et al. (2012). Environmental microbiology through the lens of high-throughput DNA sequencing: Synopsis of current platform and bioinformatics approaches. J Microbiol Methods 91:106–113.

López-López O, Cerdan ME, González Siso MI (2014). New extremophilic lipases and esterases from metagenomics. Current Prot Pept Sci 15(5):445–455.

López-López O, Knapik K, Cerdán ME, González-Siso MI (2015). Metagenomics of an alkaline hot spring in Galicia (Spain): Microbial diversity analysis and screening for novel lipolytic enzymes. Front Microbiol 6:1291.

Lu M, Schneider D, Daniel R (2021). Metagenomic screening for lipolytic genes reveals an ecology-clustered distribution pattern. bioRxiv 02.01.429111.

Lu M, Dukunde A, Daniel R (2019). Biochemical profiles of two thermostable and organic solvent – tolerant esterases derived from a compost metagenome. Appl Microbiol Biotechnol 103:3421–3437.

McNeil IA, Tiong CL, Minor C, August PR, Grossman TH, Loiacono KA et al. (2001). Expression and isolation of antimicrobial small molecules from soil DNA libraries. J Mol Microbiol Biotechnol 3:301–308.

Magnúsdóttir S, Ravcheev D, de Crécy-Lagard V, Thiele I (2015). Systematic genome assessment of B-vitamin biosynthesis suggests co-operation among gut microbes. Front Genet 6:148.

Maistrenko OM, Mende DR, Luetge M et al. (2020). Disentangling the impact of environmental and phylogenetic constraints on prokaryotic within-species diversity. ISME J 14:1247–1259

Matsuzawa T, Kaneko S, Yaoi K (2015). Screening, identification and characterization of a GH – 43 family β-xylosidase/α-arabinosidase useful for biomass saccharification from a compost microbial metagenome. Appl Microbiol Biotechnol 99:8943–8954.

Matsuzawa T, Kimura N, Suenaga H (2016). Screening, identification, and characterization of α-xylosidase from a soil metagenome. J Biosci Bioeng 122:393–399.

Mhuantong W, Charoensawan V, Kanokratana P, Tangphatsornruang S (2015). Comparative analysis of sugarcane bagasse metagenome reveals unique and conserved biomass-degrading enzymes among lignocellulolytic microbial communities. Biotechnol Biofuels 8:1–17.

Motahar SFS, Ali Khatibi, Maryam Salami, Shohreh Ariaeenejad, Zahra Emam-Djomeh, Hadi Nedaei et al. (2020). A novel metagenome-derived thermostable and poultry feed compatible α-amylase with enhanced biodegradation properties. Int J Biol Macromol 164:2124–2133.

Nair HP, Bhat SG (2020). Arabian Sea metagenome derived – α-amylase P109 and its potential applications. Ecol Genet and Genom 16:2405–9854.

Nair HP, Vincent H, Puthusseri RM, Bhat SG (2017). Molecular cloning and characterization of a halotolerant α-amylase from marine metagenomic library derived from Arabian Sea sediments. 3 Biotech, 7(1):65.

Neveu J, Regeard C, DuBow MS (2011). Isolation and characterization of two serine proteases from metagenomic libraries of the Gobi and Death Valley deserts. Appl Microbiol Biotechnol 91 (3):635–644.

Nguyen NH, Maruset L, Uengwetwanit T, Mhuantong W, Harnpicharnchai P, Champreda V et al. (2012). Identification and characterization of a cellulase-encoding gene from the buffalo rumen metagenomic library. Biosci Biotechnol Biochem 76 (6):1075–1084.

Nie Y, Zhao J, Tang Y, Guo P, Yang Y, Wu X, Hickey WJ (2016). Species divergence vs. functional convergence characterizes crude oil microbial community assembly. Front Microbiol 7:1–11.

Ohashi H, Hasegawa M, Wakimoto K, Miyamoto-Sato E (2015). Next-Generation Technologies for Multiomics Approaches Including Interactome Sequencing. BioMed Res Int:1–9.

Oliveira J, Martins M, Pinto Leite R, Sousa M, Santos R (2017). The new neuromuscular disease related with defects in the ASC-1 complex: Report of a second case confirms ASCC1 involvement. Clin Genet 92:434–439.

Oliver JD, Nilsson L, Kjelleberg S (1991). Formation of nonculturable Vibrio vulnificus cells and its relationship to the starvation state. Appl Environ Microbiol 57:2640–2644.

Pabbathi NPP, Velidandi A, Tavarna T et al. (2021). Role of metagenomics in prospecting novel endoglucanases, accentuating functional metagenomics approach in second-generation biofuel production: A review. Biomass Conv Bioref 7:1–28.

Pace NR (1997). A molecular view of microbial diversity and the biosphere. Science 276:734–740.

Pang H, Zhang P, Duan CJ, Mo XC, Tang JL, Feng JX (2009). Identification of cellulase genes from the metagenomes of compost soils and functional characterization of one novel endoglucanase. Curr Microbiol 58(4):404–408.

Parks D, Chuvochina M, Waite D et al. (2018). A standardized bacterial taxonomy based on genome phylogeny substantially revises the tree of life. Nat Biotechnol 36:996–1004.

Parks DH, Rinke C, Chuvochina M et al. (2017). Recovery of nearly 8,000 metagenome-assembled genomes substantially expands the tree of life. Nat Microbiol 2:1533–1542.

Parsley LC, Linneman J, Goode AM, Becklund K, George I, Goodman RM, Lopanik NB, Liles MR (2011). Polyketide synthase pathways identified from a metagenomic library are derived from soil acidobacteria. FEMS Microbiol Ecol 78(1):176–187.

Pasolli E, Asnicar F, Manara S, Zolfo M, Karcher N, Armanini F et al. (2019). Extensive unexplored human microbiome diversity revealed by over 150,000 genomes from metagenomes spanning age, geography, and lifestyle. Cell 176(3):649–662.e20.

Pelizzola M, Ecker JR (2011). The DNA methylome. FEBS Lett 585:1994–2000.

Pereira R, Barbosa T, Gales L, Oliveira E, Santos R, Oliveira J, Sousa M (2019). Clinical and genetic analysis of children with kartagener syndrome. Cells 8:900.

Pessoa TBA, Rezende RP, Marques ELS, Pirovani CP, Dos Santos TF et al. (2017). Metagenomic alkaline protease from mangrove sediment. J Basic Microbiol 57(11):962–973.

Pushpam PL, Rajesh T, Gunasekaran P (2011). Identification and characterization of alkaline serine protease from goat skin surface metagenome. AMB Express 1(1):3.

Rabbani B, Tekin M, Mahdieh N (2014). The promise of whole-exome sequencing in medical genetics. J Hum Genet 59:5–15.

Raimundo I, Silva R, Meunier L et al. (2021). Functional metagenomics reveals differential chitin degradation and utilization features across free-living and host-associated marine microbiomes. Microbiome 9:43.

Rhee JK, Ahn DG, Kim YG, Oh JW (2005). New thermophilic and thermostable esterase with sequence similarity to the hormone sensitive lipase family, cloned from a metagenomic library. Appl Environ Microbiol 71(2):817–825.

Rohman A, Dijkstra BW, Puspaningsih NNT (2019). β-xylosidases: Structural diversity, catalytic mechanism, and inhibition by monosaccharides. Int J Mol Sci 20:5524.

Rondon MR, August PR, Bettermann AD, Brady SF, Grossman TH et al. (2000). Cloning the soil metagenome: A strategy for accessing the genetic and functional diversity of uncultured microorganisms. Appl Environ Microbiol 66:2541–2547.

Roszak DB, Grimes DJ, Colwell RR (1984). Viable but nonrecoverable stage of Salmonella enteritidis in aquatic systems. Can J Microbiol 30:334–338.

Sae-Lee R, Boonmee A (2014). Newly derived GH43 gene from compost metagenome showing dual xylanase and cellulase activities. Folia Microbiol 59:409–417.

Sahoo RK, Das A, Sahoo K, Sahu A, Subudhi E. (2020). Characterization of novel metagenomic – derived lipase from Indian hot spring. Int Microbiol 23:233–240.

Sanchez-Flores A, Abreu-Goodger C (2014). A practical guide to sequencing genomes and transcriptomes. Curr Top Med Chem 14:398–406.

Sathya TA, Khan M (2014). Diversity of glycosyl hydrolase enzymes from metagenome and their application in food industry. J Food Sci 79:R2149–R2156.

Sato M, Suda M, Okuma J, Kato T, Hirose Y, Nishimura A et al. (2017). Isolation of highly thermostable β-xylosidases from a hot spring soil microbial community using a metagenomic approach. DNA Res 24:649–656.

Saxena R, Dhakan DB, Mittal P, Waiker P (2017). Metagenomic analysis of hot springs in Central India reveals hydrocarbon degrading thermophiles and pathways essential for survival in extreme environments. Front Microbiol 7:1–17.

Schirmer A, Gadkari R, Reeves CD, Ibrahim F, DeLong EF, Hutchinson CR (2005). Metagenomic analysis reveals diverse polyketide synthase gene clusters in microorganisms associated with the marine sponge Discodermia dissoluta. Appl Environ Microbiol 71(8):4840–4849.

Schmidt TM, DeLong EF, Pace NR (1991). Analysis of a marine picoplankton community by 16S rRNA gene cloning and sequencing. J Bacteriol 173:4371–4378.

Schwecke T, Aparicio JF, Molnár I, König A, Khaw LE et al. (1995). The biosynthetic gene cluster for the polyketide immunosuppressant rapamycin. Proc Natl Acad Sci USA 92:7839–7843.

Shah RK, Patel AK, Davla DM, Parikh IK, Subramanian RB, Patel KC, Jakhesara SJ, Joshi CG (2017). Molecular cloning, heterologous expression, and functional characterization of a cellulolytic enzyme (Cel PRII) from buffalo rumen metagenome. 3 Biotech 7(4):257.

Shao W, Wiegel J (1992). Purification and characterization of a thermostable b-xylosidase from Thermoanaerobacter ethanolicus. J Bacteriol 174:5848–5853.

Sharma S, Khan FG, Qazi GN (2010b). Molecular cloning and characterization of amylase from soil metagenomic library derived from Northwestern Himalayas. Appl Microbiol Biotechnol 86(6):1821–1828.

Shi H, Zhang Y, Li X, Huang Y, Wang L, Wang Y et al. (2013). A novel highly thermostable xylanase stimulated by Ca2+ from Thermotoga thermarum: Cloning, expression and characterization. Biotechnol Biofuels 6:26.

Shi W, Ding SY, Yuan JS (2011). Comparison of insect gut cellulase and xylanase activity across different insect species with distinct food sources. Bioenergy Res 4:1–10.

Singh KM, Reddy B, Patel D (2014). High potential source for biomass degradation enzyme discovery and environmental aspects revealed through metagenomics of Indian buffalo rumen. Biomed Res Int 26–7189.

Singh R, Chopra C, Gupta VK, Akhlaq B, Verma V, Rasool S (2015). Purification and characterization of CHpro1, a thermotolerant, alkali-stable and oxidation-resisting protease of Chumathang hot spring. Sci Bull (Beijing) 60(14):1252 1260.

Singh R, Dhawan S, Singh K, Kaur J (2012). Cloning, expression and characterization of a metagenome derived thermoactive/thermostable pectinase. Mol Biol Rep 39:8353–8361.

Soares FE de F, Queiroz JH de, Araújo JV de et al. (2015). Nematicidal action of chitinases produced by the fungus Monacrosporium thaumasium under laboratorial conditions. Biocontrol Sci Tech 25:337–344.

Sobat M, Asad S, Kabiri M, Mehrshad M (2021). Metagenomic discovery and functional validation of L-asparaginases with anti-leukemic effect from the Caspian Sea. iScience 24(1):101973.

Sousa FMO, Moura SR, Quinto CA, Dias JCT, Pirovani CP, Rezende RP (2016). Functional screening for cellulolytic activity in a metagenomic fosmid library of microorganisms associated with coral. Genet Molecul Res 15:gmr15048770.

Staley C, Sadowsky MJ (2018). Practical considerations for sampling and data analysis in contemporary metagenomics-based environmental studies. J Microbiol Methods 154: 14–18

Stein JL, Marsh TL, Wu KY, Shizuya H, DeLong EF (1996). Characterization of uncultivated prokaryotes: Isolation and analysis of a 40-kilobase-pair genome fragment from a planktonic marine archaeon. J Bacteriol 178(3):591–599.

Su, J, Zhang, F, Sun, W. et al. (2015). A new alkaline lipase obtained from the metagenome of marine sponge Ircinia sp. World J Microbiol Biotechnol 31:1093–1102.

Sun J, Li P, Liu Z et al. (2020). A novel thermostable serine protease from a metagenomic library derived from marine sediments in the East China Sea. Appl Microbiol Biotechnol 104:9229–9238.

Sun M, Zheng H, Meng L (2015). Direct cloning, expression of a thermostable xylanase gene from the metagenomic DNA of cow dung compost and enzymatic production of xylooligosaccharides from corncob. Biotechnol Lett 37:1877–1886.

Sunna A, Bergquist PL (2003). A gene encoding a novel extremely thermostable 1,4-β-xylanase isolated directly from an environmental DNA sample. Extremophiles 7:63–70.

Tebani A, Afonso C, Marret S, Bekri S (2016). Omics-based strategies in precision medicine: Toward a paradigm shift in inborn errors of metabolism investigations. Int J Mol Sci 17:1555.

Tian BY, Cao Y, Zhang KQ (2015). Metagenomic insights into communities, functions of endophytes and their associates with infection by root-knot nematode, Meloidogyne incognita, in tomato roots. Sci Rep 5:17087.

Tirawongsaroj P, Sriprang R, Harnpichamchai P, Thongaram T et al. (2009). Novel thermophilic and thermostable lipolytic enzymes from a Thailand hot spring metagenomic library. J Biotechnol 133:42–44.

Torsvik V, Goksøyr J, Daae FL (1990). High diversity in DNA of soil bacteria. Appl Environ Microbiol 56:782–787.

Van Rossum T, Ferretti P, Maistrenko OM et al. (2020). Diversity within species: Interpreting strains in microbiomes. Nat Rev Microbiol 18:491–506.

Venter JC, Remington K, Heidelberg JF, Halpern AL, Rusch D et al. (2004). Environmental genome shotgun sequencing of the Sargasso Sea. Science 304(5667):66–74.

Verma D, Satyanarayana T (2020). Xylanolytic extremozymes retrieved from environmental metagenomes: Characteristics, Genetic Engineering, and Applications. Front Microbiol 17(11):551109.

Verma D, Kawarabayasi Y, Miyazaki K and Satyanarayana T (2013). Cloning, expression and characteristics of a novel alkalistable and thermostable xylanase encoding gene (Mxyl) retrieved from compost-soil metagenome. PLoS One 8: e52459.

Verma D, Kumar R, Satyanarayana T (2019). Diversity in xylan-degrading prokaryotes and xylanolytic enzymes and their bio-prospects. In Microbial Diversity in Ecosystem Sustainability and Biotechnological Applications eds T. Satyanarayana, S. K. Das, and B. N. Johri (London: Springer Nature), 325–373.

Vidya J, Swaroop S, Singh SK et al. (2011). Isolation and characterization of a novel α-amylase from a metagenomic library of Western Ghats of Kerala, India. Biologia 66:939.

Voget S, Leggewie C, Uesbeck A, Raasch C, Jaeger KE, Streit WR (2003). Prospecting for novel biocatalysts in a soil metagenome. Appl Environ Microbiol 69:6235–6242.

Wagschal K, Heng C, Lee CC, Wong DW (2009). Biochemical characterization of a novel dual-function α-arabinofuranosidase/β-xylosidase isolated from a compost starter mixture. Appl Microbiol Biotechnol 81:855–863.

Wang G, Wang Y, Yang P, Luo H. Huang H, Shi P et al. (2010). Molecular detection and diversity of xylanase genes in alpine tundra soil. Appl Microbiol Biotechnol 87:1383–1393.

Wang GY et al. (2000). Novel natural products from soil DNA libraries in a Streptomycete host. Org Lett 2:2401–2404.

Wang H, Hart DJ, An Y (2019). Functional metagenomic technologies for the discovery of novel enzymes for biomass degradation and biofuel production. Bioenerg Res 12:457–470.

Wang Q, Luo Y, He B, Jiang LS, Liu JX, Wang JK (2015). Characterization of a novel xylanase gene from rumen content of Hu sheep. Appl Biochem Biotechnol 177:1424–1436.

Wang Z, Gerstein M, Snyder M (2009). RNA-Seq: A revolutionary tool for transcriptomics. Nat Rev Genet 10:57–63.

Wemheuer B, Taube R, Akyol P, Wemheuer F and Daniel R (2013). Microbial diversity and biochemical potential encoded by thermal spring metagenomes derived from the Kamchatka Peninsula. Archaea 2013:136714.

Whitman WB et al. (1998). Prokaryotes: The unseen majority. Proc Natl Acad Sci U. S. A 95:6578–6583.

Wrighton K (2018). Discovering antibiotics through soil metagenomics. Nat Rev Drug Discov 17:241.

Wu D, Wang S, Vinitchaikul P, Zhu Y, Zhou X, Gu Z et al. (2020). Directed modification of a ruminal cellulase gene (CMC-1) from a metagenomic library isolated from Yunnan gayal (Bos frontalis). Arch Microbiol 202(5):1117–1126.

Yasir M, Angelakis E, Bibi F, Azhar EI, Bachar D, Lagier JC et al. (2015). Comparison of the gut microbiota of people in France and Saudi Arabia. Nutr Diabetes 5: e153.

Zhang Y, Hao J, Zhang YQ, Chen XL, Xie BB, Shi M, Zhou BC, Zhang YZ, Li PY (2017). Identification and characterization of a novel salt-tolerant esterase from the deep-sea sediment of the south china sea. Front Microbiol 8:441.

Zhou J, Bao L, Chang L, Zhou Y and Lu H (2012). Biochemical and kinetic characterization of GH43 β-D-xylosidase/α-L-arabinofuranosidase and GH-30 α-L-arabinofuranosidase/β-D-xylosidase from rumen metagenome. J Ind Microbiol Biotechnol 39:143–152.

Zhou M, Guo P, Wang T et al. (2017). Metagenomic mining pectinolytic microbes and enzymes from an apple pomace-adapted compost microbial community. Biotechnol Biofuels 10:198.

10 Metagenomic Insights of Human Oral Microbiome

Sonu Bhatia, Vivek Tiwari, Neetu Sharma, Abhinashi Singh Sodhi and Navneet Batra

CONTENTS

10.1 INTRODUCTION

The advances and developments made in the field of molecular biology have become an essential part of biomedical research and have contributed a lot in the diagnosis and treatment strategies in the medical sector. During the last two decades several projects have been launched and

DOI: 10.1201/9781003220831-10

completed that have achieved new milestones in diagnosis of life-threatening diseases well in advance and their potential prevention and treatment schemes. In this context, human genome organization project had undertaken the sequence of around three billion chemical base pairs that make up the genetic material in humans. The term 'microbiome' is used for the millions of resident microorganisms colonized on human body. The various sites inhabited by microbes include skin, oral cavity, respiratory tract, gastrointestinal tract, sebaceous glands, etc. The different microflora presented at different sites develops different types of interactions among themselves and thus helps in maintaining their populations at constant levels and provides protection to humans from invading pathogens. The human microbiome contains a variety of bacteria, fungi, viruses, archaea, protists, and other microorganisms whose compositions are influenced by the environmental conditions at the time (Ursell et al., 2012). Different parts of the human body have distinct biomes in terms of environmental conditions and nutrient availability, which foster different populations. Even within a single body location, the microbiome composition can vary greatly between people in different states of health, with different lifestyles, or for a variety of other reasons. More than 700 species of bacteria have been identified in the oral cavity, the majority of which have a commensal relationship and are not associated with any disease, but the remainder can cause both. The Human Microbiome Project (HMP) helped us learn about the microbial phylotypes (both symbiotic and pathogenic microbes) that live in close proximity to humans by sequencing our extended genome (The Human Microbiome Project Consortium, 2012). HMP also helps us locate and classify our microbiome housing in various parts of the body. Each part of the body has its own microbial composition and distribution, which differs from person to person and may be related to genotype, hygiene, and diet, as well as health.

There are many complex mechanisms and reactions responsible for their role in providing health status to an individual. A part from that there are different factors which determine the population of resident microflora at a specific point in time such as age, diet, genetic makeup, climate, and personal hygiene. Any imbalance in the normal composition of microbiome at any site may lead to life-threatening diseases. In 2008, the Human Microbiome Project (HMP) was launched and presently contains sequenced data that is made available to the public (Turnbaugh et al., 2007). The project gave the first insight about microbiome of healthy human, which also includes about nine niches of the oral cavity (Methé et al., 2012).

This has further paved the way for the development of several microbial databases such as Human Oral Microbiome Database (HOMD) and Metagenomics of the Human Intestinal Tract (MetaHIT). The HOMD is aimed to provide information on taxonomic and genomic level. The genomic component provides updated observations and various bioinformatics tools and 16S rRNA gene reference sequences of oral community. The link for assessing HOMD website is www.homd.org (Chen et al., 2010). The information available at HOMD is collected on the basis of results obtained from sequencing of 16S rRNA genes of isolates or analysis of clone libraries. This information is used to make phylotypes which consists of members having approximately 98.5% similarity. Each species and phylotype cluster is assigned with a Human Oral Taxon (HOT) ID number.

Streptococcus mutans was reported to be the first organism to be isolated from oral cavity (Matsui and Cvitkovitch, 2010). The type of resident microbial species and their number vary among different populations but microbiome of each person is inimitable. The different sites of oral cavity invaded for colonization by bacterial species includes teeth surfaces, gingival surfaces, palates and tonsils. The various categories of oral disorders are known to affect people globally. As the oral route is believed to be the gateway to other organs hence lead to different other disorders. The supporting environment of the oral cavity harbours variety of organisms. One of the main diseases of oral cavity, gingivitis, affects gum tissue, while periodontitis is related to impact connective tissue and attributes to loss of teeth in affected individuals (Pihlstrom et al., 2005). The other class of disease is dental caries, which is caused by acid-producing species, including *Streptococcus mutans*, Streptococcus sobrinusand species of *Lactobacilli* (Takahashi and Nyvad,

2011). The development of biofilm harbors a variety of organisms responsible for dental diseases, and its formation can be avoided by following oral hygiene practices.

10.2 METAGENOMICS: CONCEPT AND METHODOLOGY

Antonie van Leeuwenhoek, who first observed microorganisms from dental plaques, first drew the knowledge about residential microorganisms in the 18th century. And it took about 400 years for us to discover all of our other cohabitants and not to fear that they could hurt us, before Robert Koch pioneered the practise of cultivating and isolating microorganisms using nutritional media, which has been widely used to isolate and classify microorganisms (Cho and Blaser, 2012; Grice and Segre, 2012). Jo Handelsman, Jon Clarly, and Robert M. Goodman coined the term 'metagenomics,' which was first published in 1998. Metagenomics is the study of metagenomes, which are genetic materials obtained directly from a sample. Meta means 'transcendent' (combination of different analyses) in Greek, and genomics is the study of the genome (the haploid set of chromosomes in a gamete or microorganism, or in each cell of a multicellular organism) (Handelsman, 2005). The genomic analysis of microbial DNA was performed directly from the microbial communities present in samples in metagenomics. This technique may be used to discover the large uncultured microbial diversity present in the environment for new molecules that can be used for therapeutic purposes and may include a variety of biotechnological applications. Additionally, studies have identified several novel microbial genes coding for metabolic pathways such as energy addition, carbon, and nitrogen metabolism in natural environments that were previously unknown.

10.2.1 Metagenomics Timeline and Milestones

- 1676: Leeuwenhoek reported his notings about oral microbiota.
- 1888: Robert Koch used solid cultures to isolate microorganisms.
- 1977: Fred Sanger developed DNA sequencing.
- 1977: Carl Woese used rRNA as taxonomic marker.
- 1980: Kary Mullis invented PCR.
- 1990: First microbial community study using 16S rRNA libraries.
- 1998: Handelsman et al. suggested the term metagenomics.
- 2005: Roche released the first NGS machine.
- 2006: Solexa released the first GA sequencer.
- 2008: Human microbiome project was published.

Culture methods have evolved from single-species culture-dependent experiments to diverse in vitro multispecies populations, and were used to identify microbes in the past. With technological advances, it is now possible to characterize the entire microbiotain vivo without using culture, as well as analyze the expression of individual genes for metagenomic research (Pozhitkov et al., 2011).

Metagenomics and microbiome activity in health and disease are two new fields of research that have arisen to detect and recognize the role of microbes in the body, as well as to comprehend the nature of microbiome development in health and disease (Mira, 2018). Metagenomics refers to a collection of techniques for detecting bacteria that are unable to be cultured. By applying metagenomic research to the entire population of microbes, it also recognizes the enormous genomic diversity of microbes. Through an analysis of metabolic pathway genes, metagenomics provides knowledge about the types of organisms present, and also explains about their functional potential. It also explains how to use databases of proteincoding sequences. It sequences the entirety of a sample's DNA (Baker et al., 2017). The oral microbiome is probably the most researched microbiome in humans to date, owing to the ease with which samples can be collected.

10.2.3 Sampling and Processing in Metagenomics (Thomas et al., 2012)

- The very first and most important step in metagenomics is sample processing.
- DNA extracted in order that it must contain all the cells presents in samples and its qualitymust be maintained for subsequent library production and sequencing.
- Then sample fractionation steps comes and it should be checked to make sure that sufficient amount of the required target is achieved and that minimum contamination of non-target material occurs.
- To maximize DNA yield and avoid co-extraction of enzymatic inhibitors that might interfere with subsequent processing physical seperation and isolation of samples is done.
- Direct lysis method over indirect lysis method of cells has a quantifiable bias in terms of microbial diversity, DNA yield, and resulting sequence fragment length.
- Some sample may often yield very small amounts of DNA but in library production formost sequencing technologies require high amounts of DNA (ng or µg), and hence amplification of starting material might be required to increase the yield.
- Multiple displacement amplification (MDA) using random hexamers and phage phi29 polymerase is done to increase DNA yields, and this method has been widely used in single-cell genomics/metagenomics (Figure 10.1).

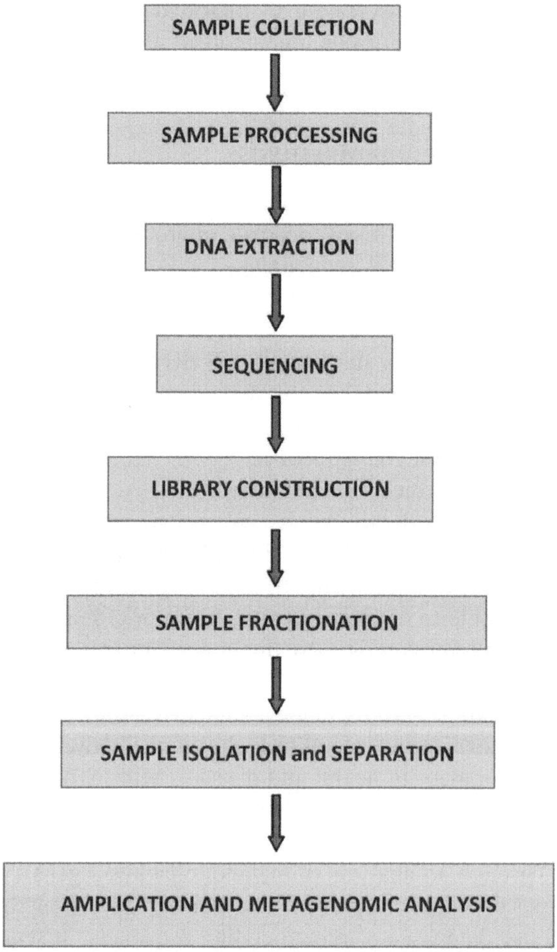

FIGURE 10.1 Sample processing in metagenomics.

10.3 SEQUENCING TECHNOLOGIES

Next-generation sequencing (NGS) has become increasingly important in taxonomic profiling, characterization, and study of microbial communities. Metagenomics samples are high-throughput, shortread sequences with a low cost of acquisition. These techniques are advantageous because they eliminate the need for cloning DNA fragments (Ansorge, 2009).

NGS technologies have progressed to accommodate a wide range of applications, costs, and capabilities (Scholz et al., 2012). The 454 Life Sciences (Roche) and Illumina systems (Solexa) are the most widely used techniques (Oulas et al., 2015). The 454 sequencing technology was the first commercially available next generation sequencing technology, and it is based on the pyrosequencing method, which allows for high throughput and low-cost research (Ronaghi, 2001). During the sequencing reaction, released pyrophosphate aids in the identification of the expanding chain by trapping the released pyrophosphate, which releases light as a result of an enzymatic reaction. Since different nucleotides are inserted in order, the light signal can be traced back to a particular nucleotide. Finally, the sequence information is gathered using the emitted light signals. The DNA fragments are amplified in the 454 pyrosequencer after being fixed on beads in a water-oil emulsion (Goodwin et al., 2016). Pyrosequencing has been used to analyse microbial diversity in a variety of settings, including marine environments and various soildideoxynucleotides environments.

The use of fluorescently labeled reversible terminator nucleotides is required by Illumina sequencing technology. Instead of being chemically modified to prevent further DNA synthesis as in Sanger sequencing, terminator nucleotides are attached to a blocking group that can be removed from the nitrogen base in one step. The synthesis of DNA is also dependenton a chip 9n to which primers are attached. The dyes attached to each nucleotide are excited by laser as the cycle progresses, accompanied by scanning of the inserted bases. A chemical reactionis used to eliminate the blocking group and dye before the next step can begin. It can be used to study microbial diversity in many environments (Fan et al., 2013; Lentini et al., 2014; Hong et al., 2015).

In addition to the techniques just described, metagenomic studies use a variety of other techniques. These include Applied Biosystems' SOLiD 5500 W Series, Pacific Biosciences' single-molecule realtime (SMRT) DNA sequencing, and Ion Torrent semiconductor sequencing (Oulas et al., 2015). Many more technologies, such as Oxford Nanopore Technologies' Strand sequencing technologies, allow the sequencing of intact DNA strands that pass through a protein nanopore. BioNano Genomics' IrysTechnology is one of the most exciting new developments in the genomics age (Figure 10.2).

The current trend in microbiome research is increasingly reliant on culture independent NGS technologies, regardless of the biome or ecosystem being studied, which arecosteffective in terms of both financial and computational costs, as well as the ongoing expansion of microbial genetic sequence databases. Only 57% of oral bacterial species have been officially named, and 13% have been cultured but are still unknown, and 30% are uncultured, according to the Human Oral Microbiome Database (HOMD). As a result, NGS techniques not only make analyses simple and convenient, but they also vastly expand our knowledge of microbiota that are unculturable or uncommon (Escapa et al., 2018).

However, it has been observed that the traditional microbial culturing approach introduces partisanship by selecting only microbes that thrive in isolation and under laboratory conditions. It wasn't long before it became clear that the culture-based approach is irrelevant for the majority of microbial communities. The use of the 16S rRNAgene technique in the 1990s improved microbial community recognition even further. With the advent of high-throughput sequencing technologies, metagenomics has become one of the most in-demand methods for microbial research, allowing whole genome sequencing to be performed ona variety of microbial species rather than just model organisms.

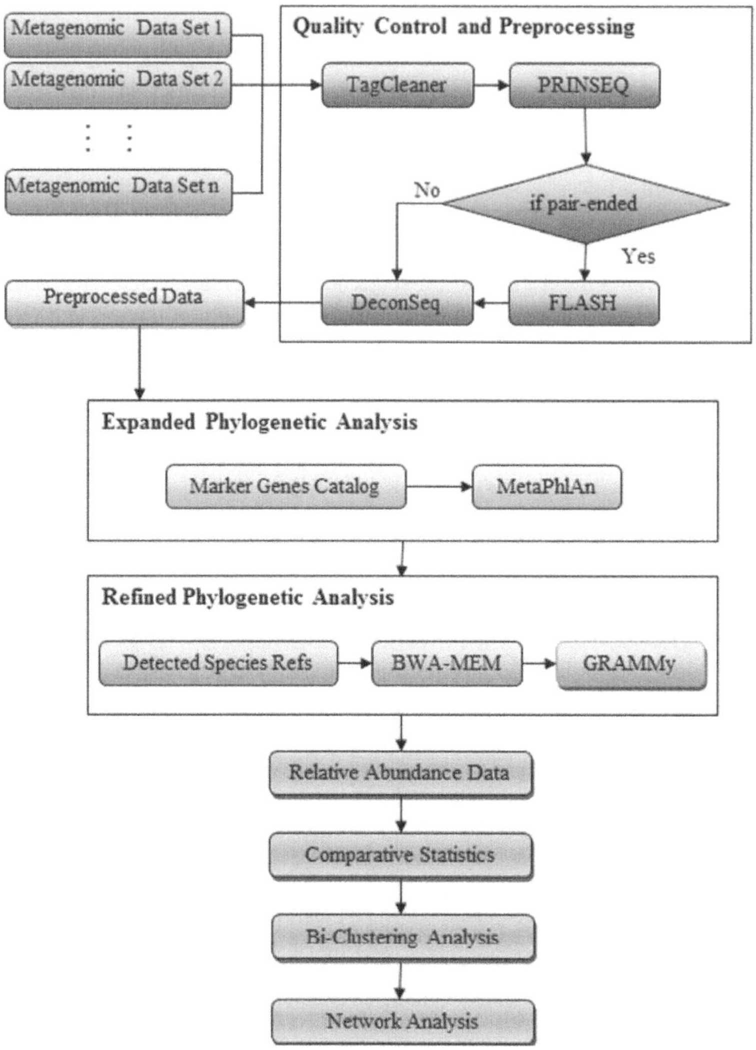

FIGURE 10.2 Integrated metagenomic analysis and data processing using bioinformatics.

Source: Reproduced from Ai et al. (2016)

10.3.1 16s rRNA Technology and NGS

For the past 15 years, Sanger sequencing of 16S rRNA clones has been commonly used to study the microbial population associated with dental caries, typically in combination with reverse-capture checkboard DNA hybridization. The findings of these studies showed a lot of diversity as well as a lot ofnew species and phylotypes.

They consistently found a connection between a variety of other microorganisms and dental caries, including *Propionibacterium* spp., *Bifidobacterium* spp., *Veillonella* spp., *Actinomyces* spp., and *Atopobium* spp. Candidate health associated bacterial organisms, such as Streptococcus mitis, have also been identified. Notably, the majority of these studies concentrated on primary tooth caries.

However, the high cost and time-consuming laboratory work associated with Sanger sequencing of 16S rRNA clones has limited the number of samples and clones that can be analyzed. Fortunately, the advent of next-generation sequencing has removed this restriction (NGS). NGS allows for unprecedented complexity and breadth of microbial community research at a lower rate. As a result, this provides an invaluable method for studying the oral microbiome. Because of their hypervariable region, which serves asgood markers of bacterial taxa in samples, NGS studies that use one or more regions of the 16S rRNA gene to identify microbial communities are common. This approach has recently been used to examine the microbiomes of dental caries in a series of tests, allowing researchers to gain a deeper understanding of the diversity of thecommunity linked to dental caries. On the place of methodological differences in sampling (saliva vs. supragingival plaque vs. carious dentine), hypervariable regions thatare chosen and sequenced, and the bioinformatic analysis uses a number of taxa, which includes *S. mutans*, *Lactobacillus* spp., *Propionibacterium* spp., *Veillonella* spp., and *Atopobium* spp., consistently showed an association with dental caries.

10.3.2 16S rRNA Technology and WMS (Whole Metagenome Sequencing)

The 16S rRNA gene sequencing of a targeted genome necessitates PCR amplification, which hasbeen shown to introduce errors such as nucleotide substitution, insertion and deletion, and chimaera formation, resulting in the identification of fraudulent species and bogus or false microbial diversity. Furthermore, PCR is prone to biases such as (1) limited primer coverage, which can lead to the failure toamplify certain taxa, especially novel ones, and (2) differential amplification of templates, which can alterthe relative abundance of species and thus distort the original microbial community structure. Gene amplification by PCR is not needed for shotgun whole metagenome sequencing (WMS) and thus allows for a higher resolution detection of microbial taxa in a culture than 16S rRNA-based sequencing; it also allows for exploration of microorganisms thatare not belonging to the category of bacteria, viruses, and fungi. Furthermore, WMS data also used to describe the microbial community's functional capacity (gene and pathway analysis).

WMS was used to characterize the functional capacity of the supragingival microbiome in the analysis by Belda-Ferre et al. (2012), and 16S rRNA sequences which are extracted from the data can be used to obtain taxonomic profiles. Because of limited sample size, there is no differential abundance analysis that can be conductedfurther (four subjects with caries and two without). Belstrom et al. (2017) with the help of metagenomics and metatranscriptomics investigated the oral microbiome of ten healthy subjects, ten with dental caries, and ten with periodontitis in a larger sample and found *Lactobacillus* spp., and *Veillonella* spp., as well as their transcripts, to be linked to dental caries, which is consistent with the literature. Unfortunately, saliva was used for the study in the place of supragingival plaque, which can explain the reason why functional analysis is not able to yield significant results.

There have been no attempts to use WMS data to obtain strain-level taxonomic assignments, examine interactions with microorganisms other than bacteria or perform in-depth functional analyses. The aim of this study was to use WMS to conduct multikingdom profiling at the strain stage and comprehensive functional characterization of the supragingival microbiome linked to dental caries in kids.

10.4 METHODS OF STUDYING THE ORAL MICROBIOME

10.4.1 Culture and Microscopy

Microscopy, biochemical and phenotypic measures, growth conditions, sugar utilization, and antibiotic sensitivity were among the ancient methods for identifying bacterial taxa (Krishnan et al., 2017). Culture-based approaches are unable to show the true diversity of the oral microbiome. Various researchers' efforts have now been cultured isolated, recognized, characterized, and categorized nearly half of the approximately 700 bacterial species that are found in the oral cavity (Aas et al., 2005).

Traditional culture and culture-based analytical technologies have a major flaw in that many bacterial species present in biological samples cannot be cultured, making them useless for research (Pozhitkov et al., 2011).

10.4.2 Gel-Based Techniques

Microbial communities' studies through high throughput screening method made it possible to many culture-independent techniques. Various techniques like denaturing gradient gel electrophoresis (DGGE), temperature gradient gel electrophoresis, and restriction fragment length polymorphism are among the techniques used (RELP) (Krishnan et al., 2017).

10.4.3 Polymerase Chain Reaction-Based Methods

The different PCR-based methods available for the identification of microbes include traditional polymerase chain reaction (PCR), real-time quantitative PCR, PCRDGGE, random amplified polymorphic DNA/arbitrarily primed PCR, multilocus sequence typing, PCR-RELP, and terminal-RELP (Aas et al., 2005).

10.4.4 DNA Microarrays

Phylogenetic DNA microarrays have been described as useful resources in the scientific community for highthroughput, bacterial communities systematic, and quantitative study in many microbial ecosystems, including the oral microbiota (Parolin et al., 2017).

10.5 16S RRNASEQUENCING

16S rRNA sequence analysis and metagenomics are two popular DNA sequencing approaches for studying uncultured oral microbial communities. Metagenomics follows whole genome shotgun sequencing, whereas 16S rRNA sequencing follows sequencing of the conserved 16S rRNA gene (WGS). A 'shotgun' procedure is used to shear all of the DNA samples at random. The sequencing is then carried out using either traditional Sanger sequencing or next generation sequencing (NGS). In most recent studies, 16S rDNA gene profiling is used to determine the species present in a sample, or shotgun metagenomics is used if complete gene content profiling in a given habitat is needed (Kuczynski et al., 2011). The 16S rRNA gene function has not changed over time, meaning that random sequence changes are a more reliable indicator of time (evolution); and the 16S rRNA gene (1500 bp) is large enough for informatics purposes (Tran et al., 2017). Since it is a highly conserved gene, it is superior to using the whole genome as a marker, since the reference gene in our database is less likely to differ from the gene in bacteria collected from environmental samples (Tran et al., 2017). 16S rRNA sequencing can only be used to determine the presence of bacterial organisms. As a result, scientists can only draw conclusions based on their findings, while shotgun metagenomics sequencing often shows the metabolic pathways at play (Zhang et al., 2015; Le Bars et al., 2017). Metagenomics WGS data can provide microbial communities with not only taxonomy but also biological functional profiles, whereas 16S rRNA profiling provides taxonomic composition (Moon et al., 2016).

10.6 NEXT-GENERATION SEQUENCING PLATFORMS

In the last decade, next generation sequencing techniques have revolutionized the study of microbial diversity. Large-scale sequencing projects can now be completed in a matter of days, if not hours (Figure 10.3).

The main NGS technologies are as follows:

- 454 pyrosequencing
- Applied Biosystems
- Illumina
- Pacific Biosciences
- Oxford Nanopore

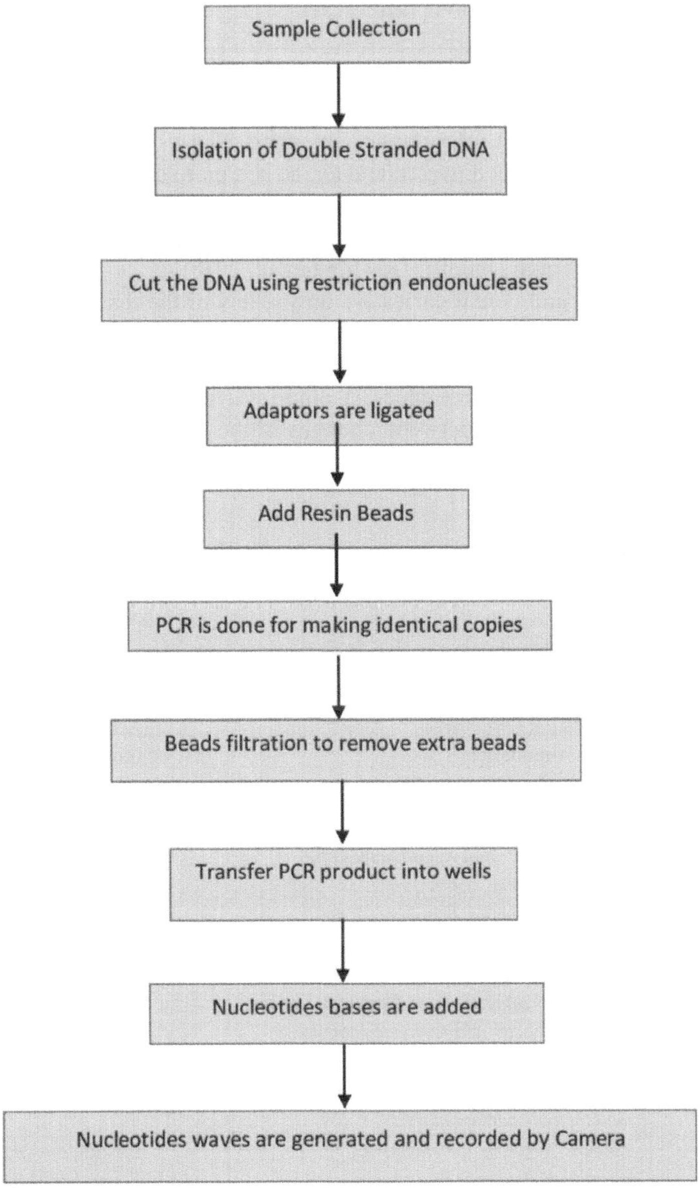

FIGURE 10.3 454 Pyrosequencing method.

Bioinformatics skills, such as data quality control, alignment, and mapping to good reference genomes, are needed for NGS research, filtering for high quality reads, eliminating chimacras, and normalizations across samples and populations for realistic interpretations (Krishnan et al., 2017). Researchers can now profile microbiomes and metagenomes at previously unimaginable depths thanks to these methods. The major advantages of NGS are its high throughput and the fact that complex taxa do not need to be targeted (Zaura et al., 2009). Regardless of the methods used to characterize them or the types of microorganisms targeted, the choices made at each point, from study design to analysis, can have an effect on the results (Bik, 2016).

10.7 ORAL MICROBIOME COMPOSITION

10.7.1 Bacteria

Bacteria, which include members of the phyla *Actinobacteria, Proteobacteria, Firmicutes, and Bacteroidetes*, are one of the most important components of the oral microbiome. Since the oral cavity is directly in contact with food, water, and air, so it is more difficult to distinguish the bacteria that live there. Since they lack specialized features like adhesions and receptors that help them attach to oral surfaces and form aggregates with other oral bacteria, many non-oral bacteria are unable to form colonies within the mouth (Dewhirst et al., 2010).

Periodontitis, gingivitis, and dental caries are only a few of the diseases that can damage the mouth, and they've all been linked to any kind of change in the oral microbiome (Table 10.1). As the mouth is an entry point for the respiratory and digestive systems and also highly vascularized, the oral microbiome may have consequences in other systemic diseases. Indeed, an increasing number of studies have found links between changes in the oral microbiome and other diseases (Table 10.2). This indicates the oral microbiota may be used as biomarkers to diagnose certain systemic diseases.

TABLE 10.1
Examples of Metagenomic Research on the Interaction between the Oral Microbiome and Oral Diseases

Disease	Associated Organisms (High Abundance)	Associated Organisms (Low Abundance)
Periodontitis	Phyla: *Spirochaetes, Synergistetes and Bacteroidetes* Classes: *Clostridia, Negativicutes and Erysipelotrichia* Genera: *Prevotella, Fusobacterium* Species: *Porphyromonas gingivalis, Treponema denticola, Tannerella forsythia, Filifactoralocis, Parvimonasmicra, Aggregatibacter actinomycetemcomitans* Archaea: *Methanobrevibacter oralis, Methanobacterium curvum/congolense, and Methanosarcina mazeii*	Phyla: *Proteobacteria* Classes: *Bacilli* Genera: *Streptococcus, Actinomyces, Granulicatella*
Dental caries	Genera: *Neisseria, Selenomonas, Propionibacterium* Species: *Streptococcus mutans, Lactobacillus spp.* Fungi: *Candida albicans*	Species: *non-mutans Streptococci, Corynebacterium matruchotii, Capnocytophaga gingivalis, Eubacterium* IR009, *Campylobacter rectus, Lachnospiraceae* sp. C1
Oral cancer	Species: *Capnocytophaga gingivalis, Prevotella melaninogenica and Streptococcus mitis, Peptostreptococcus stomatis, Streptococcus salivarius, Streptococcus gordonii, Gemella haemolysans, Gemella morbillorum, Johnsonellaignava and Streptococcus parasanguinis*	Species: *Granulicatella adiacens*
Esophageal cancer	Species: *Tannerella forsythia, Porphyromonas gingivalis*	Genera: *Neisseria* Sp., *Streptococcus pneumoniae*

TABLE 10.2
Metagenomic Studies and Example of the Oral Microbiome and Its Relationship to Systemic Diseases

Disease	Associated Organisms (High Abundance)	Associated Organisms (Low Abundance)
Colorectal cancer	Genera: *Lactobacillus, Rothia* Species: *Fusobacterium nucleatum*	NA
Pancreatic cancer	Genera: *Leptotrichia* (later in progression of disease) Species: *Porphyromonas gingivalis and Aggregatibacter actinomycetemcomitans* (at onset of disease)	Genera: *Leptotrichia* (at onset of disease) Species: *Porphyromonas gingivalis and Aggregatibacter actinomycetemcomitans* (later in progression of disease)
Cystic fibrosis	Species: *Streptococcus oralis* (depends on environmental conditions), *S. mitis, S. gordonii and S. sanguinis*	Species: *Streptococcus oralis* (depends on environmental conditions)
Cardiovascular disease	Species: *Campylobacter rectus Porphyromonas gingivalis, Porphyromonas endodontalis, Prevotella intermedia, Prevotella nigrescens* (oral commensals that were found on athersclerotic plaques – not necessarily at high abundance in oral cavity)	NA
Rheumatoid arthritis	Genera: *Veillonella, Atopobium, Prevotella, Leptotrichia* Species: *Rothia mucilaginosa, Rothia dentocariosa, Lactobacillus salivarius*	Genera: *Haemophilus, Neisseria* Species: *Porphyromonas gingivalis, Rothia aeria* sp
Alzheimers disease	Phyla: *Spirochaetes* Species: Porphyromonas gingivalis	NA
Diabetes	Genera: *Aggregatibacter, Neisseria, Gemella, Eikenella, Selenomonas, Actinomyces, Capnocytophaga, Fusobacterium, Veillonella, Streptococcus*	Genera: *Porphyromonas, Filifactor, Eubacterium, Synergistetes, Tannerella, Treponema*

10.7.2 Archaea

Archaeal domain members have been isolated from the oral cavity (Ferrari et al., 1994). The microflora, on the other hand, are significantly less common and diverse than oral bacteria. Despite the Archaea domain's widespread distribution and ability to thrive in a variety of ecological niches despite harsh environmental conditions, the oral cavity has been poorly maintained. They are underrepresented in a healthy mouth but over-represented in patients with periodontitis. One of the most common bacteria in the mouth is *Methanobrevibacter oralis* found in studies (Lepp et al., 2004; Vianna et al., 2006; Yamabe et al., 2008; Matarazzo et al., 2011). Other *Archaeaphylum* types, like *Methanobacterium curvum* and *Methanosarcina mazeii*, have also been found pathogenic and reported to be present in the oral microbiome (Wade, 2013).

Methanogenic Archaea has been isolated from the human mouth, gut, and vaginal area (Kulik et al., 2001). Despite their presence in the human microbiota, no connection has been established between disease severity and the relative abundance of members of the Archaea domain. Chronic periodontitis a polymicrobial infection that affects 35 percent of adults in the United States. Endocarditis, atherosclerosis, stroke, and preterm delivery of low-birthweight infants have all been linked to the disease, which can lead to tooth loss. Multiple members of the domain Bacteria have been linked to chronic periodontitis; however, nosingle member explains the majority of cases, nor the role of microbes in this disease. Simultaneously, little is known about the link between Archaea and chronic periodontitis. While methanogenic Archaea have been identified in the mouths of periodontitis patients, previous studies lacked fundamental controls and were not quantitative, making it impossible to create a substantial or clinically important connection between Archaea and disease. The current study establishes links between the prevalence of disease and archaeal DNA (Eckburg et al., 2003; Champagne et al., 2000; Paster et al., 2001; Brusa et al., 1987). The incidence of periodontal disease and the relative abundance of archaeal DNA in subgingival plaque, and the resolution of disease and decreased archaeal DNA abundance. Within the genus Methanobrevibacter, Archaea was discovered to be limited to a subset of human subjects and to be made up of two distinct rDNA phylotypes (Li et al., 2009). These findings have significant implications for the aetiology of a very common disease, as well as the function of methanogens in the human microbial ecosystem.

10.7.3 Protozoa

Many protozoan species, like *Trichomonas tenax* and *Entamoeba gingivalis*, are vastly present in people who have poor oral hygiene and gingival diseases (Wade, 2013). *Entamoeba gingivalis* is a parasitic worm that lives in the gingival tissues that surround the teeth and gums. It is rarely found in the tonsils. These are non pathogenic parasites that feed on mouth epithelial tissues, food waste, bacteria, and other things (Rousset and Lauvergeat, 1971; Feki and Molet, 1990). They are more common in people who have pyorrhoea, but they do not cause it. In patients with marginal chronic periodontitis, the commensal, anaerobic T. tenax is also common in the oral cavity (Hayawan and Bayoumy, 1992). Poor oral hygiene often results in a rise in bacteria and food waste, which serves as a source of nutrition for saprophytic protozoa. One explanation for the high prevalence of protozoa in people with oral disease and poor oral hygiene may be this (Wantland et al., 1958).

10.7.4 Fungi

Fungi are one of the most essential constituents of the oral microbiome, but are not associated with any symptoms in more than 50% of people. Candida species, on the other hand, have been related to a variety of chronic and acute infections. *Candida albicans* is the most dangerous, and it is mainly linked to oral mycoses. Other pathogenic species from HIV patients' oral cavities, such as *C. glabrata, C. tropicalis, C. krusei, and C. dubliniensis*, have recently been described (Weems et al., 1992; Sullivan et al., 1995). C. dubliniensishas also been found in the oral lesions of people suffering from other oral diseases such as periodontitis. Ghannoum et al. (2010) used a multitag pyrosequencing method to identify 85 fungal genera from the oral cavities of healthy people, including *Cladosporium, Sacharomycetales, Candida, Aureobasidium, Cryptococcus*, and *Fusarium*. They are considered as the most prevalent genera among the 85 fungal genera described.

10.7.5 Virus

The human mouth is a breeding ground for viruses linked to a variety of diseases. Viruses are most difficult to culture in media than other microbes which makes them less susceptible to conventional pathology techniques (Willner et al., 2009; Sedghizadeh et al., 2012). Before the discovery of PCR-based techniques, clinicians only considered viruses to be suspect when diagnosing aparticular disease disorder, and if a virus was found, it was considered to be the cause of the clinical condition (Pride et al., 2012; Robles-Sikisaka et al., 2013). However, experiments have shown that many viruses are more prevalent in the human body. Circoviruses, for example, are found in themajority of adults but are not linked to any health conditions. Similarly, polyomaviruses are found in the majority of the human population despite the fact that they are rarely responsible for causing symptoms. While a few viruses, like Herpesviruses, EpsteinBarr virus, and Cytomegaloviruses, are common in many people, they cause serious illness in a small percentage of them. Cold sores are caused by herpesviruses such as HSV1 and HSV2, while mononucleosis is caused by the Epstein Barr virus and cytomegaloviruses, which are transmitted by mouth-to-mouth communication. The rabies and mumps viruses, which affect the salivary glands, can also be found in diseased people's saliva. Human papillomaviruses contained in the mouth can causea variety of clinical conditions, including condylomas, papillomas, and focal epithelial hyperplasia (Moen et al., 2002; Jiang et al., 2009; Grinde and Olsen, 2010; Huang et al., 2011; Zhang et al., 2013; Naidu et al., 2014; Abeles et al., 2014; Edlund et al., 2015). The majority of oral viruses are bacteriophages, according to extensive studies on the oral viome using a metagenomics approach, which is not surprising given that the human mouth is denselycolonized by diverse bacteria. Despite this, periodontal disease patients have a higher prevalence of some bacteriophages (e.g., Myoviridae family viruses) (Santiago-Rodriguez et al., 2015).

10.8 THE HUMAN MICROBIOME PROJECT

The Human Microbiome Project (HMP) was launched by the National Institute of Health in 2008, explaining the importance of researching the human microbiome (Dewhirst et al., 2010). Breakthroughs in bioinformatics have made researching the human microbiome a lot simpler. These discoveries also sparked a flood of genomic and metagenomic research into microbe function in different environments (Gilbert et al., 2011). Rather than being a single initiative, the HMP is a series of many initiatives that are being launched simultaneously in several parts of the world, including the United States, the European Union, and Asia (Turnbaugh et al., 2007).

Microbial ecologists developed numerous in situ approaches for studying microorganisms, mainly by sequencing the 16S ribosomal RNA gene, after realizing that >99 percent of microbes fromthe atmosphere could not be easily cultured. 16S rRNA serves as a taxonomic and phylogenetic marker for identifying microbes from various microbial groups (Gevers et al., 2012). The advent of high-throughput DNA sequencing methods has resulted in a major increase in research into what constitutes a natural oral microbiome (Warinner, 2016). The tongue, rough palate, buccal mucosa, dorsum, keratinized gingiva or gums, palatine tonsils, neck, supra- and subgingival plaque, and saliva were chosen from healthy volunteers' oral cavity for HMPA of the HMP database, concluded that a relatively stable and small core oral microbiome was present in the many samples with low abundance (Ames et al., 2017).

10.9 HUMAN ORAL MICROBIOME DATABASE

The Human Oral Microbiome Database (HOMD) is a database that houses the sequences of oral bacterial genomes as well as a comprehensive resource that includes taxonomic details and a 16S rRNA identification tool (Wade, 2013). It's a one-of-akind database created by the National Institute of Dental and Craniofacial Research in 2010 to maintain track of oral-derived culturable and nonculturable microbe isolates (Chen et al., 2010). The expanded HOMD (eHOMD) was developed to provide the scientific community with a wide and carefully selected list of bacterial species found in the human aerodigestive tract (ADT), which consist the upper digestive and upper respiratory tracts, pharynx, sinuses, nasal passages, and oesophagus, also the oral cavity. For the ADT bacteria genomicsequences are being added to the eHOMDas they become usable, thanks to initiatives like the Human Microbiome Project, which is part of the HOMD project, and other sequencing projects (Chen et al., 2010).

The eHOMD database contains details on 772 prokaryotic organisms, 70% of which are cultivable and 30% of which are uncultivable, as well as wholegenome sequences for 482 taxa. The names of 57% of the 70% culturable species have already been assigned. *Firmicutes, Actinobacteria, Proteobacteria, Fusobacteria, and Spirochaetes* made up 96 percent of the total oral bacteria, according to safe oral cavity profilling of 16S rDNA (Verma et al., 2018).

10.10 APPLICATIONS OF METAGENOMICS IN ORAL MICROBIOME

10.10.1 Metagenomics in Oral Health

With the advancement in metagenomics in the last few years, anumber of techniques like Random Amplified Polymorphic DNA fingerprinting, Restriction Fragment Length Polymorphism, Quantitative Real-time Polymerase Chain Reaction, Denaturing Gradient Gel Electrophoresis, Next generation sequence technologies, etc., were employed for identification of unculturable species of oral microbiome. NGS aided in more accurate and fast identification of unexplored members of various environments. The information generated by NGS further aids in identification of novel metabolic pathways in various genes.

The earlier studies for identification of oral diversity make use of 16S rRNA sequence analysis due to simpler technique and low cost but was unable to provide complete information. The problem was overcome by employing principle of Metagenomics due to its ability to carry out rapid analysis, efficient data mining and fast reproducing results. With the development of new databases which are updated on regular basis and hence can provide a good platform for comparative analysis of new sequences (Xu and Gunsolley, 2014).

10.10.2 Microorganisms in Oral Infection

The composition of oral microorganisms is influenced by a variety of factors, including lifestyle (e.g., diet, oral hygiene), health (e.g., oral diseases, host immune responses, genetic susceptibility), and/or physical location in the oral cavity (tongue or tooth surfaces, as well as supragingival orsubgingival sites) (Palmer, 2014).

The nonspecific plaque hypothesis, proposed by Schultz-Haudt et al. in 1954 and Macdonald et al. in 1956, explains the role of microorganisms in oral diseases. According to this hypothesis, aggregation of any microorganisms (or nonspecific species) in the biofilm played a role in causing inflammation. In the absence of tooth brushing, Theilade (1986) discovered a complex biofilm on the gingival tissues of patients, which included over 200 different organisms. As destructive periodontitis progressed, the number of bacterial cells increased and the distribution of bacterial species changed, but no single species appeared in active or inactive sites. The specific plaque theory, suggested by Loesche in 1979, describes how periodontal disease was caused by specific species. Multiple studies had identified specific bacterial species that were found in large numbers in samples from diseased tissues of patients, according to the hypothesis. Immune responses specific to these bacteria were also discovered in the patients.

The ecological plaque hypothecus, suggested by Marsh in 1994, discusses bacterial functions incaries and periodontal diseases, and this theory proposed that periodontal disease was caused by both microorganisms and environmental factors. As a result of environmental changes, a relatively healthy composition of bacterial species in biofilm (microbial homeostasis) may be substantially changed, contributing to disease. Changes in sugar intake, diet, or pH, for example, can alter the microbiota balance, causing the biofilm composition to shift toward cariogenic organisms. The ratio of potentially cariogenic species, such as *Streptococci mutans* and *Lactobacilli*, will increase in a lowpH setting. Oral diseases can be prevented not only by targeting pathogenic organisms, but also by interfering with the mechanisms that cause homeostasis to break down (Marsh, 1994).

PSD, or polymicrobial synergy and dysbiosis, is a fourth theory for understanding bacterial functions in periodontal diseases that was recently discovered (Hajishengallis, 2012). Rather than specific oral pathogens, this hypothesis proposes that oral diseases are initiated by a synergistic and dysbiotic microbial group. Different members of polymicrobial groups, each of which represents a unique genetic mix, collaborate to form and stabilize the microbiota that causes disease. Some infectious bacterialorganisms, known as "keystone pathogens," are thought to play important roles in modulating the host response and disrupting homeostasis, resulting in dysbiosis and periodontal disease, via interactions with other pathogens.

10.10.3 Metagenomics in Studying Oral Microorganisms

According to current evidence, humans are 'super-organisms.' In the human body, more than a thousand individual microorganisms coexist with the same number of body cells (Backhed et al., 2005). The entire human microbiome is made up of this collection of microorganisms. Oral microbiota are a major component of the human microbiome, and they are essential for human health since they influence microbiota in other areas of the body, such as the respiratory and gastrointestinal tracts. Many molecular biological methods have been used to recognize and classify uncultivable oral microbial organisms, for example, Restriction Fragment Length Polymorphism (RFLP) (Lepine et al., 1995), Random Amplified Polymorphic DNA fingerprinting (RPAD) (Menard et al., 1995), Denaturing Gradient Gel Electrophoresis (DGGE) (Zijnge et al., 2003), Quantitative Real-time Polymerase Chain Reaction (qPCR) (Blome et al., 2008), Microarray Chip (Colombo et al., 2009), Checkerboard Hybridization (Sakamoto et al., 2005), and others (Nyvad et al., 2013).

Without the use of bacterial culture, next generation sequence technologies (NGS) (Metzker, 2010) enable thestudy of a large number of microorganisms in various niches. Studies using NGS technologies todissect the microbiota in oral diseases have yielded a number of findings (Jenkinson, 2011; Zaura, 2012; Burne et al., 2012; Wade, 2013; Han and Wang, 2013; Nyvad et al., 2013). Using DNA sequencing to investigate various environmental niches, several new species of microbes have been discovered. New organisms have been discovered in samples obtained from various locations, such as the ocean, a low complexity microbial biofilm culture of acidic water (Tyson et al., 2004), and soil (Trings et al., 2005). New genes and metabolic pathways have been observed through analysis of these nucleotide sequences.

10.10.4 Metagenomics Provides Insight of Bacterial Profile

Metagenomics has been used to examine bacterial profiles in the oral microbiome. Metagenomicanalysis yields longer, larger 16S rRNA genes, which must be assembled from sequences used inbacterial species classification. In comparison to the shorter amplicons of particular regions (400bp) of 16S rRNA genes obtained in 16S rRNA sequence analysis, the assembled sequences provide longer 16S rRNA genes by metagenomic analysis, allowing for full taxonomic analysis of bacterial organisms. Furthermore, the ability to use all loci in metagenomics to identify bacterial species provides more detailed taxonomy. Due to the limited number of bacterial and patient samples, metagenomic analysis may also have less power to associate bacterial species with diseases.

Several metagenomics studies have looked at bacterial profiles. Xie and colleagues (2010) used metagenomics to look at biofilm samples from a caries-free, periodontally healthy subject and found 12 well-characterized phyla, including members of the TM7 and BRC1 clades, from a total of 860 megabases (Mb) of sequence. The presence of pathogens as well as opportunistic pathogens in the samples supports the ecological plaque theory of oral diseases.

10.10.5 Metagenomics Identifies Specific Metabolic Pathways and Genes Associated with Oral Diseases

Liu and colleagues used metagenomics to investigate metabolic pathways in bacterial populations from stable and periodontal disease samples (Liu et al., 2012). The Kyoto Encyclopedia of Genes and Genomes (KEGG) database was used to organize sequence reads into contigs, which were then translated and assigned to metabolic functions. The metabolic pathway review revealed that stableand diseased samples had different metabolic profiles.

The genes associated with periodontal disease samples display metabolic pathway genes associated with 'a parasitic lifestyle,' for example, they use degraded nutrients from lysed host cells andother bacterial cells, and genes associated with an intracellular lifestyle, the healthy genes as well as intracellular periodontal genes undergo fatty acid metabolism, and genes that are associated with "a parasitic lifestyle," for example, they use degraded nutri ents from lysed host cells and other bacterial cells, and genes that are associated with 'a parasitic lifestyle' (Liu et al., 2012). Several pathways, including fatty acid biosynthesis, aspartate metabolism, and glycerol-3-phosphate metabolism, were also found to be enriched in the healthy samples.

Certain fatty acids are also beneficial to periodontal health (Campan et al., 1997; Kesavalu et al., 2007). By looking at enrichment forgenes involved in homoserine metabolism in healthy samples, Liu et al. (2012) had important findings. Homoserine acts as a signaling molecule in quorum sensing (a bacterial mechanism of stimulus and response associated with environmental population density), indicating that various metabolic activity for quorum sensing occurs within the stable microbiome. Wang and colleagues (2013) compared the genes of metabolic pathways in stable and periodontal disease samples. They found under representation of genes for carbohydrate metabolism, amino acid metabolism, energy metabolism, lipid metabolism, membrane transport, and signal transduction in the subgingival plaques of chronic periodontal patients, but over representation of genes for glycan biosynthesis and cell motility.

In samples from active caries, Belda-Ferre et al. (2012) discovered more genes for mixedacid fermentation and competence, as well as more genes for antibacterial peptides, periplasmic stress response genes, capsular and extracellular polysaccharides, and bacitracin stress response gene (Belda-Ferre et al., 2012). As can be seen from the aforementioned metagenomic research, various metabolic pathways and genes are linked to oral diseases.

Xie and colleagues (2010) discovered antibiotic and toxic compound resistance genes in oral biofilm samples, accounting for 2.8 percent of the overall predicted genes that code for proteins. Antibiotic resistance genes were discovered for a variety of antibiotic groups, including lactams, aminoglycosides, fluoroquinolones, and the peptide antibiotic bacitracin, which all have different resistant mechanisms. Resistance genes for general multi-drug or heavy metal roles, such as efflux pumps, are also included. A variety of genes linked to antibiotic and metal resistance were also discovered by Liu and colleagues (mercury and cobalt-zinc-cadmium) (Liu et al., 2012). The presence of these numerous resistance genes suggests that the oral cavity, which houses the oral microbiota, is a key gene pool for antibiotic resistance in the human microbiome.

10.10.6 Metagenomics Connects Bacterial Species with Metabolic Pathways

Belda-Ferre et al. (2012) and colleagues investigated and compared the bacterial and genomic profiles of caries and healthy samples to determine the basic roles of individual caries bacterial organisms. They investigated gene categories using the COG method, and discovered that genes belonging to different taxonomic groups were not fairly distributed, and that particular taxonomic groups were correlated with specific functions. *Bacilli*, for example, have more genes for defence mechanisms, signalling, and carbohydrate metabolism, while *Clostridia* have more genes for cell motility. The association of various bacterial species with healthy tissues (*Bacilli*) or with caries (Bacilli) is shown by the different gene distributions indicating different functions (Clostridia). They also proposed that people who have never had caries might have a genetic reservoir of new anticaries compounds and probiotics in their dental biofilm. The competition between *S. Sanguinis and S. mutans* in the oral cavity is one known example in which *S. sanguinis* produces H_2O_2, which inhibits *S. mutans*, while *S. mutans* produces mutacins, which inhibits *S. Sanguinis* (Kreth et al., 2005; Chen et al., 2011).

10.11 CONCLUSIONS

The major challenge in studying the diversity of any site pertains to the inability of most of the microbes to grow on petriplate. But with the developments made in molecular biology and availability of culture free methods like metagenomics and next-generation sequencing (NGS) one can have any easy access to these unexplored population of microflora. The introduction of high-throughput sequencing techniques helped to explore the oral microbiome.

BIBLIOGRAPHY

Aas, J.A., Paster, B.J., Stokes, L.N., Olsen, I. and Dewhirst, F.E., 2005. Defining the normal bacterial flora of the oral cavity. *Journal of Clinical Microbiology*, 43(11): 5721–532.

Abeles, S.R., Robles-Sikisaka, R., Ly, M., Lum, A.G., Salzman, J., Boehm, T.K. and Pride, D.T., 2014. Human oral viruses are personal, persistent and gender-consistent. *The ISME Journal*, 8(9): 1753–1767.

Ames, N.J., Ranucci, A., Moriyama, B. and Wallen, G.R., 2017. The human microbiome and understanding the 16S rRNA gene in translational nursing science. *Nursing Research*, 66(2): 184.

Ansorge, W.J., 2009. Next-generation DNA sequencing techniques. *New Biotechnology*, 25(4): 195–203.

Backhed, F., Ley, R.E., Sonnenburg, J.L., Peterson, D.A. and Gordon, J.I., 2005. Host-bacterial mutualism in the human intestine. *Science*, 307(5717): 1915–1920.

Baker, J.L., Bor, B., Agnello, M., Shi, W. and He, X., 2017. Ecology of the oral microbiome: beyond bacteria. *Trends in Microbiology*, 25(5): 362–374.

Belda-Ferre, P., Alcaraz, L.D., Cabrera-Rubio, R., Romero, H., Simon-Soro, A., Pignatelli, M. and Mira, A., 2012. The oral metagenome in health and disease. *The ISME Journal*, 6(1): 46–56.

Belstrom, D., Constancias, F., Liu, Y., Yang, L., Drautz-Moses, D.I., Schuster, S.C., Kohli, G.S., Jakobsen, T.H., Holmstrup, P. and Givskov, M., 2017. Metagenomic and metatranscriptomic analysis of saliva reveals disease-associated microbiota in patients with periodontitis and dental caries. *NPJ Biofilms and Microbiomes*, 3(1): 1–8.

Bik, E.M., 2016. Focus: microbiome: the hoops, hopes, and hypes of human microbiome research. *The Yale Journal of Biology and Medicine*, 89(3): 363.

Blome, B., Braun, A., Sobarzo, V. and Jepsen, S., 2008. Molecular identification and quantification of bacteria from endodontic infections using real-time polymerase chain reaction. *Oral Microbiology and Immunology*, 23(5): 384–390.

Brusa, T., Conca, R., Ferrara, A., Ferrari, A. and Pecchioni, A., 1987. The presence of methanobacteria in human subgingival plaque. *Journal of Clinical Periodontology*, 14(8): 470–471.

Burne, R.A., Zeng, L., Ahn, S.J., Palmer, S.R., Liu, Y., Lefebure, T., Stanhope, M.J. and Nascimento, M.M., 2012. Progress dissecting the oral microbiome in caries and health. *Advances in Dental Research*, 24(2): 77–80.

Campan, P., Planchand, P.O. and Duran, D., 1997. Pilot study on n-3 polyunsaturated fatty acids in the treatment of human experimental gingivitis. *Journal of Clinical Periodontology*, 24(12): 907–913.

Champagne, C.M., Madianos, P.N., Lieff, S., Murtha, A.P., Beck, J.D. and Offenbacher, S., 2000. Periodontal medicine: emerging concepts in pregnancy outcomes. *Journal of the International Academy of Periodontology*, 2(1): 9–13.

Chen, L., Ge, X., Dou, Y., Wang, X., Patel, J.R. and Xu, P., 2011. Identification of hydrogen peroxide production-related genes in Streptococcus sanguinis and their functional relationship with pyruvate oxidase. *Microbiology*, 157: 13.

Chen, T., Yu, W.H., Izard, J., Baranova, O.V., Lakshmanan, A. and Dewhirst, F.E., 2010. The human oral microbiome database: a web accessible resource for investigating oral microbe taxonomic and genomic information. *Database*, 2010: 1–10. https://doi.org/10.1093/database/baq013

Cho, I. and Blaser, M.J., 2012. The human microbiome: at the interface of health and disease. *Nature Reviews Genetics*, 13(4): 260–270.

Colombo, A.P.V., Boches, S.K., Cotton, S.L., Goodson, J.M., Kent, R., Haffajee, A.D., Socransky, S.S., Hasturk, H., Van Dyke, T.E., Dewhirst, F. and Paster, B.J., 2009. Comparisons of subgingival microbial profiles of refractory periodontitis, severe periodontitis, and periodontal health using the human oral microbe identification microarray. *Journal of Periodontology*, 80(9): 1421–1432.

Dewhirst, F.E., Chen, T., Izard, J., Paster, B.J., Tanner, A.C., Yu, W.H., Lakshmanan, A. and Wade, W.G., 2010. The human oral microbiome. *Journal of Bacteriology*, 192(19): 5002–5017.

Eckburg, P.B., Lepp, P.W. and Relman, D.A., 2003. Archaea and their potential role in human disease. *Infection and Immunity*, 71(2): 591–596.

Edlund, A., Santiago-Rodriguez, T.M., Boehm, T.K. and Pride, D.T., 2015. Bacteriophage and their potential roles in the human oral cavity. *Journal of Oral Microbiology*, 7(1): 27423.

El Hayawan, I.A. and Bayoumy, M.M., 1992. The prevalence of entamoebagingivalisand trichomonastenax in periodontal disease. *Journal of the Egyptian Society of Parasitology*, 22(1): 101–105.

Escapa, I.F., Chen, T., Huang, Y., Gajare, P., Dewhirst, F.E. and Lemon, K.P., 2018. New insights into human nostril microbiome from the expanded human oral microbiome database (eHOMD): a resource for the microbiome of the human aerodigestive tract. *Msystems*, 3(6).

Fan, W., Huo, G., Li, X., Yang, L., Duan, C., Wang, T. and Chen, J., 2013. Diversity of the intestinal microbiota in different patterns of feeding infants by Illumina high-throughput sequencing. *World Journal of Microbiology and Biotechnology*, 29(12): 2365–2372.

Feki, A. and Molet, B., 1990. Importance of trichomonastenaxand entamoebagingivalis protozoa in the human oral cavity. *Revue D'odonto-stomatologie*, 19(1): 37–45.

Ferrari, A., Brusa, T., Rutili, A., Canzi, E. and Biavati, B., 1994. Isolation and characterization of Methanobrevibacteroralis sp. nov. *Current Microbiology*, 29(1): 7–12.

Gevers, D., Knight, R., Petrosino, J.F., Huang, K., McGuire, A.L., Birren, B.W., Nelson, K.E., White, O., Methé, B.A. and Huttenhower, C., 2012. The human microbiome project: a community resource for the healthy human microbiome. *PLoSBiol*, 10(8): e1001377.

Ghannoum, M.A., Jurevic, R.J., Mukherjee, P.K., Cui, F., Sikaroodi, M., Naqvi, A. and Gillevet, P.M., 2010. Characterization of the oral fungal microbiome (mycobiome) in healthy individuals. *PLoSpathog*, 6(1): e1000713.

Gilbert, J.A. and Dupont, C.L., 2011. Microbial metagenomics: beyond the genome. *Annual Review of Marine Science*, 3: 347–371.

Goodwin, S., McPherson, J.D. and McCombie, W.R., 2016. Coming of age: ten years of next-generation sequencing technologies. *Nature Reviews Genetics*, 17(6): 333.

Grice, E.A. and Segre, J.A., 2012. The human microbiome: our second genome. *Annual Review of Genomics and Human Genetics*, 13: 151–170.

Grinde, B. and Olsen, I., 2010. The role of viruses in oral disease. *Journal of Oral Microbiology*, 2(1): 2127.

Hajishengallis, G., Darveau, R.P. and Curtis, M.A., 2012. The keystone-pathogen hypothesis. *Nature Reviews Microbiology*, 10(10): 717–725.

Han, Y.W. and Wang, X., 2013. Mobile microbiome: oral bacteria in extra-oral infections and inflammation. *Journal of Dental Research*, 92(6): 485–491.

Handelsman, J., 2005. Metagenomics: application of genomics to uncultured microorganisms. *Microbiology and Molecular Biology Reviews*, 69(1): 195–195.

Hong, C., Si, Y., Xing, Y. and Li, Y., 2015. IlluminaMiSeq sequencing investigation on the contrasting soil bacterial community structures in different iron mining areas. *Environmental Science and Pollution Research*, 22(14): 10788–10799.

Huang, R., Li, M. and Gregory, R.L., 2011. Bacterial interactions in dental biofilm. *Virulence* 2: 435–444.

The Human Microbiome Project Consortium, 2012. Structure, function and diversity of the healthy human microbiome. *Nature*, 486: 207–214.

Jenkinson, H.F., 2011. Beyond the oral microbiome. *Environ Microbiol*, 13: 3077–3087. PMID:21906224. http://dx.doi.org/10.1111/j.1462-2920.2011.02573.x

Jiang, M., Abend, J.R., Johnson, S.F. and Imperiale, M.J., 2009. The role of polyomaviruses in human disease. *Virology*, 384(2): 266–273.

Kesavalu, L., Bakthavatchalu, V., Rahman, M.M., Su, J., Raghu, B., Dawson, D., Fernandes, G. and Ebersole, J.L., 2007. Omega-3 fatty acid regulates inflammatory cytokine/mediator messenger RNA expression in Porphyromonasgingivalis-induced experimental periodontal disease. *Oral Microbiology and Immunology*, 22(4): 232–239.

Kreth, J., Merritt, J., Shi, W. and Qi, F., 2005. Competition and coexistence between Streptococcus mutans and Streptococcus sanguinis in the dental biofilm. *Journal of Bacteriology*, 187(21): 7193–7203.

Krishnan, K., Chen, T. and Paster, B.J., 2017. A practical guide to the oral microbiome and its relation to health and disease. *Oral Diseases*, 23(3): 276–286.

Kuczynski, J., Stombaugh, J., Walters, W.A., González, A., Caporaso, J.G. and Knight, R., 2011. Using QIIME to analyze 16S rRNA gene sequences from microbial communities. *Current Protocols in Bioinformatics*, 36(1): 10–17.

Kulik, E.M., Sandmeier, H., Hinni, K. and Meyer, J., 2001. Identification of archaealrDNA from subgingival dental plaque by PCR amplification and sequence analysis. *FEMS Microbiology Letters*, 196(2): 129–133.

Le Bars, P., Matamoros, S., Montassier, E., Le Vacon, F., Potel, G., Soueidan, A., Jordana, F. and de La Cochetière, M.F., 2017. The oral cavity microbiota: Between health, oral disease, and cancers of the aerodigestive tract. *Canadian Journal of Microbiology*, 63(6): 475–492.

Lentini, V., Gugliandolo, C., Bunk, B., Overmann, J. and Maugeri, T.L., 2014. Diversity of prokaryotic community at a shallow marine hydrothermal site elucidated by Illumina sequencing technology. *Current Microbiology*, 69(4): 457–466.

Lepine, G., Savett, D.A. and Progulske-Fox, A., 1995. Restriction fragment length polymorphism analysis of two hemagglutinin loci, serotyping and agglutinating activity of *Porphyromonas gingivalis* isolates. *Oral Microbiology and Immunology*, 10(1): 1–7.

Lepp, P.W., Brinig, M.M., Ouverney, C.C., Palm, K., Armitage, G.C. and Relman, D.A., 2004. MethanogenicArchaea and human periodontal disease. *PNAS*, 101(16): 6176–6181.

Li, C.L., Liu, D.L., Jiang, Y.T., Zhou, Y.B., Zhang, M.Z., Jiang, W., Liu, B. and Liang, J.P., 2009. Prevalence and molecular diversity of Archaea in subgingival pockets of periodontitis patients. *Oral Microbiology and Immunology*, 24(4): 343–346.

Liu, B., Faller, L.L., Klitgord, N., Mazumdar, V., Ghodsi, M., Sommer, D.D., Gibbons, T.R., Treangen, T.J., Chang, Y.C., Li, S. and Stine, O.C., 2012. Deep sequencing of the oral microbiome reveals signatures of periodontal disease. *PLoS One*, 7(6): e37919.

Loesche, W.J., 1979. Clinical and microbiological aspects of chemotherapeutic agents used according to the specific plaque hypothesis. *Journal of Dental Research*, 58(12): 2404–2412.

Macdonald, J.B., Sutton, R.M., Knoll, M.L., Madlener, E.M. and Grainger, R.M., 1956. The pathogenic components of an experimental fusospirochetal infection. *The Journal of Infectious Diseases*: 15–20.

Marsh, P.D., 1994. Microbial ecology of dental plaque and its significance in health and disease. *Advances in Dental Research*, 8(2): 263–271.

Matarazzo, F., Ribeiro, A.C., Feres, M., Faveri, M. and Mayer, M.P.A., 2011. Diversity and quantitative analysis of Archaea in aggressive periodontitis and periodontally healthy subjects. *Journal of Clinical Periodontology*, 38(7): 621–627.

Matsui, R. and Cvitkovitch, D., 2010. Acid tolerance mechanisms utilized by streptococcus mutans. *Future Microbiology*, 5(3): 403–417.

Menard, C. and Mouton, C., 1995. Clonal diversity of the taxon porphyromonasgingivalis assessed by random amplified polymorphic DNA fingerprinting. *Infection and Immunity*, 63(7): 2522–2531.

Methé, B.A., Nelson, K.E., Pop, M., Creasy, H.H., Giglio, M.G., Huttenhower, C., Gevers, D., Petrosino, J.F., Abubucker, S., Badger, J.H. and Chinwalla, A.T., 2012. A framework for human microbiome research. *Nature*, 486(7402): 215.

Metzker, M.L., 2010. Sequencing technologies – the next generation. *Nature Reviews Genetics*, 11(1): 31–46.

Mira, A., 2018. Oral microbiome studies: potential diagnostic and therapeutic implications. *Advances in Dental Research*, 29(1): 71–77.

Moen, E.M., Huang, L. and Grinde, B., 2002. Molecular epidemiology of TTV-like mini virus in Norway. *Archives of Virology*, 147(1): 181–185.

Moon, J.H. and Lee, J.H., 2016. Probing the diversity of healthy oral microbiome with bioinformatics approaches. *BMB Reports*, 49(12): 662.

Naidu, M., Robles-Sikisaka, R., Abeles, S.R., Boehm, T.K. and Pride, D.T., 2014. Characterization of bacteriophage communities and CRISPR profiles from dental plaque. *BMC Microbiology*, 14(1): 1–13

Nyvad, B., Crielaard, W., Mira, A., Takahashi, N. and Beighton, D., 2013. Dental caries from a molecular microbiological perspective. *Caries Research*, 47(2): 89–102.

Oulas, A., Pavloudi, C., Polymenakou, P., Pavlopoulos, G.A., Papanikolaou, N., Kotoulas, G., Arvanitidis, C. and Iliopoulos, L., 2015. Metagenomics: tools and insights for analyzing next-generation sequencing data derived from biodiversity studies. *Bioinformatics and Biology Insights*, 9:BBI-S12462.

Palmer Jr, R.J., 2014. Composition and development of oral bacterial communities. *Periodontology 2000*, 64(1): 20–39.

Parolin, C., Giordani, B., Palomino, R.A.N., Biagi, E., Severgnini, M., Consolandi, C., Caredda, G., Storelli, S., Strohmenger, L. and Vitali, B., 2017. Design and validation of a DNA-microarray for phylogenetic analysis of bacterial communities in different oral samples and dental implants. *Scientific Reports*, 7(1): 1–12.

Paster, B.J., Boches, S.K., Galvin, J.L., Ericson, R.E., Lau, C.N., Levanos, V.A., Sahasrabudhe, A. and Dewhirst, F.E., 2001. Bacterial diversity in human subgingival plaque. *Journal of Bacteriology*, 183(12): 3770–3783.

Pihlstrom, B.L., Michalowicz, B.S. and Johnson, N.W., 2005. Periodontal diseases. *The Lancet*, 366(9499): 1809–1820.

Pozhitkov, A.E., Beikler, T., Flemmig, T. and Noble, P.A., 2011. High-throughput methods for analysis of the human oral microbiome. *Periodontology* 2000, 55(1): 70–86.

Pride, D.T., Salzman, J., Haynes, M., Rohwer, F., Davis-Long, C., White, R.A., Loomer, P., Armitage, G.C. and Relman, D.A., 2012. Evidence of a robust resident bacteriophage population revealed through analysis of the human salivary virome. *The ISME Journal*, 6(5): 915–926.

Robles-Sikisaka, R., Ly, M., Boehm, T., Naidu, M., Salzman, J. and Pride, D.T., 2013. Association between living environment and human oral viral ecology. *The ISME Journal*, 7(9): 1710–1724.
Ronaghi, M., 2001. Pyrosequencing sheds light on DNA sequencing. *Genome Research*, 11(1): 3–11.
Rousset, J.J. and Lauvergeat, J.A., 1971. Oral protozoa. Epidemiologic inquiry. *La Pressemedicale*, 79(33): 1495–1497.
Sakamoto, M., Umeda, M. and Benno, Y., 2005. Molecular analysis of human oral microbiota. *Journal of Periodontal Research*, 40(3): 277–285.
Santiago-Rodriguez, T.M., Ly, M., Bonilla, N. and Pride, D.T., 2015. The human urine virome in association with urinary tract infections. *Frontiers in Microbiology*, 6: 14.
Scholz, M.B., Lo, C.C. and Chain, P.S., 2012. Next generation sequencing and bioinformatic bottlenecks: the current state of metagenomic data analysis. *Current Opinion in Biotechnology*, 23(1): 9–15.
Schultz-Haudt, S., Bruce, M.A. and Bibby, B.G., 1954. Bacterial factors in nonspecific gingivitis. *Journal of Dental Research*, 33(4): 454–458.
Sedghizadeh, P.P., Yooseph, S., Fadrosh, D.W., Zeigler-Allen, L., Thiagarajan, M., Salek, H., Farahnik, F. and Williamson, S.J., 2012. Metagenomic investigation of microbes and viruses in patients with jaw osteonecrosis associated with bisphosphonate therapy. *Oral Surgery, Oral Medicine, Oral Pathology and Oral Radiology*, 114(6): 764–770.
Sullivan, D.J., Westerneng, T.J., Haynes, K.A., Bennett, D.E. and Coleman, D.C., 1995. Candida dubliniensis sp. nov.: phenotypic and molecular characterization of a novel species associated with oral candidosis in HIV-infected individuals. *Microbiology*, 141(7): 1507–1521.
Takahashi, N. and Nyvad, B., 2011. The role of bacteria in the caries process: ecological perspectives. *Journal of Dental Research*, 90(3): 294–303.
Theilade, E., 1986. The non-specific theory in microbial etiology of inflammatory periodontal diseases. *Journal of Clinical Periodontology*, 13(10): 905–911.
Thomas, T., Gilbert, J. and Meyer, F., 2012. Metagenomics-a guide from sampling to data analysis. *Microbial Informatics and Experimentation*, 2(1): 1–12.
Tran, Q., Pham, D.T. and Phan, V., 2017. Using 16S rRNA gene as marker to detect unknown bacteria in microbial communities. *BMC Bioinformatics*, 18(14): 155–161.
Tringe, S.G., Von Mering, C., Kobayashi, A., Salamov, A.A., Chen, K., Chang, H.W., Podar, M., Short, J.M., Mathur, E.J., Detter, J.C. and Bork, P., 2005. Comparative metagenomics of microbial communities. *Science*, 308(5721): 554–557.
Turnbaugh, P.J., Ley, R.E., Hamady, M., Fraser-Liggett, C.M., Knight, R. and Gordon, J.I., 2007. The human microbiome project. *Nature*, 449(7164): 804–810.
Tyson, G.W., Chapman, J., Hugenholtz, P., Allen, E.E., Ram, R.J., Richardson, P.M., Solovyev, V.V., Rubin, E.M., Rokhsar, D.S. and Banfield, J.F., 2004. Community structure and metabolism through reconstruction of microbial genomes from the environment. *Nature*, 428(6978): 37–43.
Ursell, L.K., Metcalf, J.L., Parfrey, L.W. and Knight, R., 2012. Defining the human microbiome. *Nutrition Reviews*, 70(suppl_1): S38–S44.
Verma, D., Garg, P.K. and Dubey, A.K., 2018. Insights into the human oral microbiome. *Archives of Microbiology*, 200(4): 525–540.
Vianna, M.E., Conrads, G., Gomes, B.P.F.A. and Horz, H.P., 2006. Identification and quantification of archaea involved in primary endodontic infections. *Journal of Clinical Microbiology*, 44(4): 1274–1282.
Wade, W.G., 2013. The oral microbiome in health and disease. *Pharmacological Research*, 69(1): 137–143.
Wang, J., Qi, J., Zhao, H., He, S., Zhang, Y., Wei, S. and Zhao, F., 2013. Metagenomic sequencing reveals microbiota and its functional potential associated with periodontal disease. *Scientific Reports*, 3(1): 1–10.
Wantland, W.W., Wantland, E.M., Remo, J.W. and Winquist, D.L., 1958. Studies on human mouth protozoa. *Journal of Dental Research*, 37(5): 949–950.
Warinner, C., 2016. Dental calculus and the evolution of the human oral microbiome. *Journal of the California Dental Association*, 44(7).
Weems Jr, J.J., 1992. Candida parapsilosis: epidemiology, pathogenicity, clinical manifestations, and antimicrobial susceptibility. *Clinical Infectious Diseases*, 14(3): 756–766.
Willner, D., Furlan, M., Haynes, M., Schmieder, R., Angly, F.E., Silva, J., Tammadoni, S., Nosrat, B., Conrad, D. and Rohwer, F., 2009. Metagenomic analysis of respiratory tract DNA viral communities in cystic fibrosis and non-cystic fibrosis individuals. *PLoS One*, 4(10): e7370.

Xie, G., Chain, P.S.G., Lo, C.C., Liu, K.L., Gans, J., Merritt, J. and Qi, F., 2010. Community and gene composition of a human dental plaque microbiota obtained by metagenomic sequencing. *Molecular Oral Microbiology*, 25(6): 391–405.

Xu, P. and Gunsolley, J., 2014. Application of metagenomics in understanding oral health and disease. *Virulence*, 5(3): 424–432.

Yamabe, K., Maeda, H., Kokeguchi, S., Tanimoto, I., Sonoi, N., Asakawa, S. and Takashiba, S., 2008. Distribution of Archaea in Japanese patients with periodontitis and humoral immune response to the components. *FEMS Microbiology Letters*, 287(1): 69–75.

Zaura, E., 2012. Next-generation sequencing approaches to understanding the oral microbiome. *Advances in Dental Research*, 24(2): 81–85.

Zaura, E., Keijser, B.J., Huse, S.M. and Crielaard, W., 2009. Defining the healthy "core microbiome" of oral microbial communities. *BMC Microbiology*, 9(1): 1–12.

Zhang, F., He, S., Jin, J., Dong, G. and Wu, H., 2015. Exploring salivary microbiota in AIDS patients with different periodontal statuses using 454 GS-FLX Titanium pyrosequencing. *Frontiers in Cellular and Infection Microbiology*, 5: 55. https://doi.org/10.3389/fcimb.2015.00055

Zhang, Q., Rho, M., Tang, H., Doak, T.G. and Ye, Y., 2013. CRISPR-Cas systems target a diverse collection of invasive mobile genetic elements in human microbiomes. *Genome Biology*, 14(4): 1–15.

Zijnge, V., Harmsen, H.J.M., Kleinfelder, J.W., Van Der Rest, M.E., Degener, J.E. and Welling, G.W., 2003. Denaturing gradient gel electrophoresis analysis to study bacterial community structure in pockets of periodontitis patients. *Oral Microbiology and Immunology*, 18(1): 59–65.

11 Role and Application of CRISPR-Cas System in Genetic Disorders

Ravi Kr. Gupta, Shilpi Yadav, Ruchi Sankhwar and Abhishek Kumar

CONTENTS

11.1 INTRODUCTION

Clustered regularly interspaced short palindromic repeats, termed as CRISPRs, were discovered in prokaryotic genome as a repetitive sequence and functionally responsible for the conversion of alkaline phosphatase (Ishino et al., 1987). Later on, discovery of CRISPR associated sequences, termed as Cas protein, which work as nucleases and help to cleave complementary sequences of bacteriophages formed the CRISPR-Cas system. Till now, CRISPR-Cas system is the most advanced tool for gene editing and is used for therapeutic interventions to treat human genetic disorders. Increasing human population and anthropogenic activities in the destruction of the environment will create selective pressure on the environment and human health that may alter the human gene pool. This may eventually lead to the generation of various human genetic disorders. Hence, conventional and traditional approaches for the treatment of these human diseases will be no longer sufficient and we need to develop new approaches for the betterment of human health (Chen et al., 2019).

Theoretically, gene editing is not a new concept, because our cells do this job almost thousands of times in a day to repair our DNA through DNA-repair mechanisms during replication, making our genome integrated. Many endonucleases are programmed to alter the DNA sequences and led to the discovery of meganucleases, zinc finger nucleases (ZFN), Transcription activator-like effector nucleases (TALENs), and CRISPR-associated (Cas) nucleases (Urnov, 2018; Manghwar

DOI: 10.1201/9781003220831-11

et al., 2019). CRISPR-Cas system is gaining popularity day by day because Cas enzymes are RNA guided, work on Watson-Crick base pairing, and can be programmed to alter the DNA sequences according to our needs (Brouns et al., 2008), whereas other endonucleases work on protein-DNA interactions. CRISPR-based therapeutic approaches are widely accepted because of their broad applications as a gene editing tool and clinical corrections of human genetic disorders (Doudna, 2020). Although CRISPR-Cas system has great potential as a gene therapeutic, it might not work in some genetic diseases, for example inserting and replacing a mutated gene copy with the corrected exogenous gene copy might not produce the regulatory elements required for the proper expression and function of a gene. More research needs to be done for the improvement of current technologies. Since CRISPR-Cas system is not present in eukaryotes, scientists use this genetic tool to edit, delete, add, and modify diseased genes in humans. In this chapter we have systematically discussed the origin, mode of action, and application of CRISPR-Cas system in human genetic diseases.

11.2 ORIGIN OF CRISPR-CAS SYSTEM

The discovery of CRISPR happened accidentally in 1987 by Japanese scientist Yoshizumi Ishino during the study of phosphatase enzyme in *E. coli* (Ishino et al., 1987). These spacer sequences were further used for strain identification in Mycobacterium tuberculosis by van Embden and his co-workers (1993), who named it spoligotyping (Groenen et al., 1993). Eventually many repetitive sequences were identified in various archaea and bacteria and were thought to play an important role in DNA repair mechanisms (Makarova et al., 2015). It was found that bacterial strains which consists of these homologous spacer sequences, similar to the sequences present in bacteriophages (viruses which infect the bacteria), are immune to the infection by bacteriophages (Ishino et al., 2018). This study suggested the role of CRISPR-Cas system as a defense mechanism that gives adaptive immunity to bacteria. Advancement of genetic tools led the researchers to use the CRISPR-Cas system for gene editing (Mali et al., 2013). There were several genes identified which encode DNA repair proteins associated with CRISPR, termed cas genes, which encode Cas proteins. There are four cas genes discovered in prokaryotes, named Cas1 to Cas4. Cas1 and Cas2 did not present any similarity to known proteins but Cas3 resembles to helicases and Cas4 belongs to RecB exonucleases (Jansen et al., 2002). University of Paris has developed a web server to identify CRISPR and associated cas genes in any published microbial genome or user-submitted sequence (Grissa et al., 2007; https://crispr.i2bc.paris-saclay.fr). This has opened a new way to analyze known genome for the identification and mining of new CRISPR loci. Till today, as of May 15, 2021, a total 7014 genomes are analyzed and 8939 CRISPRs are found, in which 870 CRISPRs are present in archaea and 8069 are present in bacterial genomes.

11.3 CLASSIFICATION OF CRISPR-CAS SYSTEM

In the early genomics era many CRISPR locus were identified in archaeal genomes specifically in thermophilic and hyperthermophilic organisms that suggested its role in adaptation of these organisms at high temperature but later on many CRISPR locus were identified in mesophilic bacteria that reveal some different function of this locus (Jansen et al., 2002). Mojica et al. identified the spacer sequences in CRISPR locus were adapted from bacteriophages and prophages (Mojica et al., 2005). This was the first evidence for CRISPR involvement in a bacterial immune defense mechanism. The universal immune defense mechanism was present in most of the bacteria and archaea. Comparative genomics reveals substantial differences in cas genes and effector molecules that create diversity of CRISPR system. In any given genome the number of CRISPR arrays may vary from 1 to 18, and repeat sequences present in CRISPR may be from 2 to 374 in number (Marraffini and Sontheimer, 2010). It is to be noted that some closely related bacterial strains have substantial differences in their CRISPR content and a few distinct bacterial strains may have almost identical CRISPR content. For example, Mycobacterium tuberculosis has distinct CRISPR sequences but Mycobacterium leprae did not reported to have any CRISPR sequences. Furthermore, E. coli and Mycobacterium avium have almost identical CRISPR content (Jansen et al., 2002).

The classification of CRISPR-Cas system is based on the encoded effector proteins, i.e. Cas proteins in different stoichiometry. Typically, CRISPR system has been divided into two major families, namely class 1 and class 2 CRISPR system. Class 1 systems have multi-effector complexes with 4 to 7 Cas proteins, whereas, class 2 systems have a single multidomain effector protein (Makarova, 2011) (Figure 11.1). The majority of archaeal organisms consist in class 1 CRISPR-Cas system; however, most bacterial strains contain class 2 CRISPR-Cas system in nature (Burstein et al., 2017). Based on the organization of the cas genes and signature sequences present in it, the two classes of CRISPR-Cas system consist of three types, namely type-I, type-III, and type-IV in class 1 and type II, V, and VI in class 2 CRISPR system. Subtle differences in the locus organization and signature sequences present in cas genes further classified each type into subtypes: type-I consists of Cas3 signature protein, type-II consists of Cas9, and type-III consists of Cas10 protein. Type I and type II play major roles in bacterial immune defense mechanism and type II, having the only large domain protein, has been used for gene editing technologies (Koonin et al., 2017; Makarova et al., 2015) (Figure 11.1). The processing of precursor CRISPR RNA (pre-crRNA) in class 1 and class 2 is different. Class 1 system has multiple Cas proteins that belong to subtype I to E, whereas, in class 2, system processing is catalyzed by RNAseIII along with tracrRNA (trans-acting CRISPR RNA); tracrRNAs belong to subtype V-B. The pre-crRNA processing in types V and VI have been performed by uncharacterized effector protein (East-Seletsky et al., 2016).

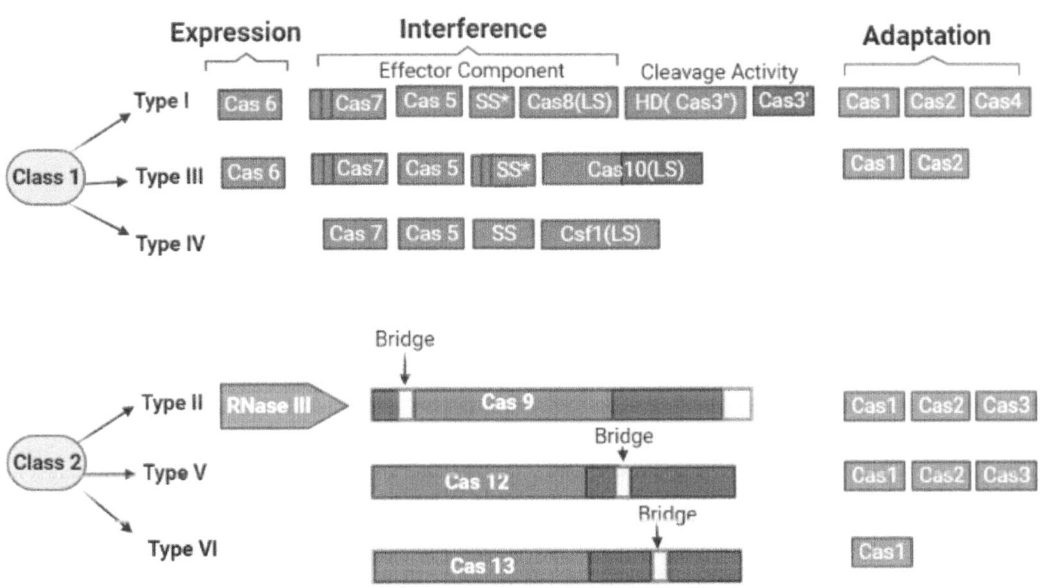

FIGURE 11.1 Classification and diversity of CRISPR-Cas system. Two classes of CRISPR-Cas system have been defined on the basis of sequence analysis and cas gene organization. Both the classes consist of three different types and subtypes based on the organization of cas genes and their effector molecules.

11.4 MODE OF ACTION OF CRISPR-CAS SYSTEM

Cas proteins are conserved and associated with CRISPR system, Casl and Cas2 help the bacteria to identify new spacer sequence from viral genome and integrate them into CRISPR array. Cas3 protein resembles helicases with seven motifs and Cas4 is similar to RecB exonucleases, hence Cas3 and Cas4 are involved in DNA repair and recombination. The role of Cas proteins in CRISPR system was analytically defined by Makarova et al. (2006). They have defined the role of Cas proteins similar to RNAi of eukaryotic system where protein complex binds to the gene and silences it. Similarly, Cas9 cleaves DNA and Casl3 cleaves RNA. There are two to six base pair short sequences in the viral genome termed as **PAM**s (**p**rotospacer **a**djacent **m**otifs) found near the Cas nuclease target sites. Therefore, PAMs are necessary for the functioning of Cas proteins. This protects the integrity of CRISPR array so that only the foreign viral nucleic acids are cleaved, not its own DNA (Shah et al., 2013).

Class 2 CRISPR system has been exploited for genome editing technology because of the simple stoichiometry of effector complexes. There are numerous effector proteins which are discovered in class 2 CRISPR system, i.e. Cas9 in type II, Casl3a and 13b in type VI, and Casl2a and 12b in type V. Cas9 is a crRNA-dependent endonuclease. The mode of action of CRISPR-Cas system can be divided into three stages: (A) adaptation: where Casl and Cas2 proteins help to insert the new spacer sequences homologous to foreign DNA/RNA into the CRISPR repeat arrays; (B) expression: where expression and processing of pre-crRNA into crRNA takes place; and (C) interference: by homology identification of spacer sequence present in crRNA and foreign nucleic acid, Cas protein having nuclease activity forms a complex and degrades the foreign DNA/RNA (Figure 11.2) (Koonin and Makarova, 2013). Class 2 CRISPR system, having the long Cas9 protein, consist of nuclease domains RuvC and HNH, which cleave the nontarget and target DNA respectively. In a nut shell, CRISPR-Cas mechanism consists of three important components, i.e.

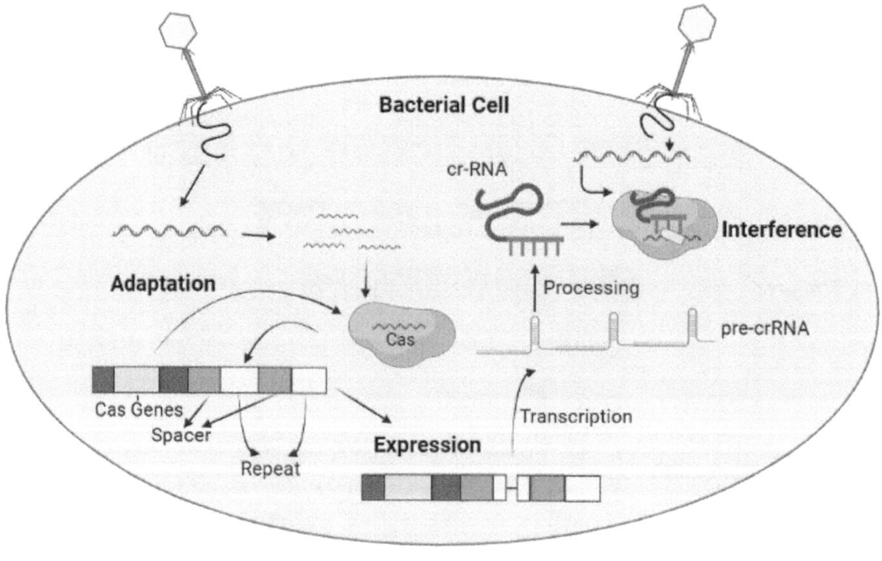

FIGURE 11.2 Adaptive immune system developed in bacteria by CRISPR-Cas system. (A) adaptation: invading bacteriophage short DNA sequences are recognized by Cas proteins and incorporated into the spacer region of CRISPR array; (B) expression: pre-crRNA is expressed and processed to crRNA having homology sequences; (C) interference: if the same bacteriophage infects the cell due to homology sequences present in the crRNA, foreign DNA is captured and forms CRISPR complex with Cas protein which cleaves the DNA.

natural guide RNA, Cas 9 endonucleases, crRNA (CRISPR-RNA) and tracrRNA (transactivation CRISPR-RNA) (Nishimasu et al., 2014). Type II Cas proteins have the ability to cleave foreign genetic material that is attached to crRNA (Datsenko et al., 2012).

11.5 APPLICATION OF CRISPR-CAS IN GENETIC DISORDERS

CRISPR-Cas system was initially used for spoligotyping in Mycobacteria and genetic markers in various bacterial strains (Groenen et al., 1993). Thereafter, its potential application comes out when it was used for cleavage of nucleic acid. It was used as a valuable solution for antibiotic-resistant pathogenic bacterial strains. CRISPR was successfully used to kill antibiotic resistant Staphylococcus aureus, which is a major pathogen of human skin and soft tissue infection (Bikard et al., 2014). The major hindrance of CRISPR-Cas application is the delivery of the system in in vivo settings. Although various delivery methods using nanotechnology have shown positive results, more research is needed in this direction. The most advanced application of CRISPR-Cas system is in the treatment of human genetic disorders such as Hemoglobinopathies, inherited eye disease, muscular genetic disease, genetic liver disease, skin disease, respiratory disorders, cystic fibrosis, Huntington's disease, and cancer.

Till now, 14 clinical trials are completed or are undergoing on various genetic disorders with the help of CRISPR-Cas technology (Table 11.1). The studies include genetic intervention at primary cultured cells, stem cells, or biologicals. The major genetic disorders included in the clinical trials are Kabuki syndrome 1, gastrointestinal neoplasms, β-Thalassemia, Sickle Cell Disease, Neurofibromatosis Type 1, Hereditary Transthyretin Amyloidosis, Leber Congenital Amaurosis 10, Familial Mediterranean Fever, and Rubinstein-Taybi Syndrome (Table 11.1).

Theoretically, CRISPR-Cas system led us to edit, delete, mutate, or add any gene in the human genome and correct any diseased gene or hereditary disorder. This has started a new era of therapeutic role of CRISPR not only in genetic diseases but also in various environmental, public health, and pathogenic diseases. In our opinion, this is just the beginning of new therapeutic approaches that will advance day-by-day with new scientific research. Next we discuss some of the important genetic diseases which are currently under way in CRISPR-Cas research.

11.5.1 Cancer

The therapeutic goal of CRISPR-Cas system in cancer research is to remove oncogenes or replace the mutated gene with the correct gene sequence. Several animal and cell models are developed to show the application of CRISPR system as a gene therapeutic. For example, primary hematopoietic stem cells (HSPCs) were transfected with lentivirus containing Cas9-sgRNA system to activate several oncogenes and develop myeloid neoplasia for the establishment of a new model for the study of cancer (Heckl et al., 2014). The generation of disease models is required for the comprehensive study of disease mechanisms and developing new therapeutic plans of action. CRISPR/Cas has been largely used for creating disease-related cellular models, such as Duchenne muscular dystrophy (DMD) (Roux et al., 2018), Aniridia-related keratopathy (ARK) (Hosseini et al., 2019), Brittle bone, X-linked Adrenoleukodystrophy (X-ALD) (Paquet et al., 2016), Alzheimer's disease (Heywood et al., 1964, Egorova et al., 2019), Atherosclerosis (Roh et al., 2018), Obesity, and Diabetes (Markossian et al., 2018).

Scientists are also using the CRISPR-Cas system for the treatment of cancer in humans. CRISPR was used to modify the immune T-cells; the major breakthrough was achieved by the development of chimeric antigen receptor-T cell therapy (CAR-T), cell therapy for the treatment of lymphoma and leukemia. In these patients, tumor cells bind to the programmed cell death (PD-1) protein expressed on T-cells and inhibit the host immune response against the cancer. Researchers have used CRISPR-Cas9 tool to delete the gene which is responsible for the production of protein PD-1 so that T-cells could not be recognized by tumor cells. Some clinical trials have been done in lung cancer patients and results were positive with some limitations (Lu et al., 2018). Scientists had also

TABLE 11.1
Current Clinical Trials of Genetic Disorders as per the Search Results from US National Library of Medicine (https://clinicaltrials.gov/ct2/home)

S.No	Targeted Genetic Disorder	Identifier Number	Title of the Clinical Trial	Interventions
1	Kabuki Syndrome 1	NCT03855631	Exploiting epigenome editing in Kabuki Syndrome: a new route towards gene therapy for rare genetic disorders	Genetic: Intervention on primary cultured cells
2	Gastrointestinal Epithelial Cancer	NCT04426669	A Study of Metastatic Gastrointestinal Cancers Treated with Tumor Infiltrating Lymphocytes in Which the Gene Encoding the Intracellular Immune Checkpoint CISH Is Inhibited Using CRISPR Genetic Engineering	Drug: Cyclophosphamide|Drug: Fludarabine|Biological: Tumor-Infiltrating Lymphocytes (TIL)|Drug: Aldesleukin
3	Beta-Thalassemia	NCT03655678	A Safety and Efficacy Study Evaluating CTX001 [CTX001 is an investigational, autologous, gene edited hematopoietic stem cell therapy for patients suffering from severe hemoglobinopathies] in Subjects with Transfusion-Dependent β-Thalassemia	Biological: CTX001
4	Beta-Thalassemia	NCT04208529	A Long-term Follow-up Study in Subjects Who Received CTX001 [CTX001 is an investigational, autologous, gene edited hematopoietic stem cell therapy for patients suffering from severe hemoglobinopathies]	Biological: CTX001
5	Sickle Cell Disease	NCT03167450	Examining the Knowledge, Attitudes, and Beliefs of Sickle Cell Disease Patients, Parents of Patients with Sickle Cell Disease, and Providers Towards the Integration of CRISPR in Clinical Care	Not defined
6	Sickle Cell Disease	NCT03745287	A Safety and Efficacy Study Evaluating CTX001 [CTX001 is an investigational, autologous, gene edited hematopoietic stem cell therapy for patients suffering from severe hemoglobinopathies] in Subjects with Severe Sickle Cell Disease	Biological: CTX001
7	Sickle Cell Disease	NCT04774536	Transplantation of Clustered Regularly Interspaced Short Palindromic Repeats Modified Hematopoietic Progenitor Stem Cells (CRISPR_SCD001) in Patients with Severe Sickle Cell Disease	Drug: CRISPR_SCD001
8	Thalassemia	NCT03728322	iHSCs (hematopoietic stem cells) with the Gene Correction of HBB Intervent Subjests with β-thalassemia Mutations	Biological: iHSCs treatment group
9	Neurofibromatosis Type 1 (Tumors of the Central Nervous System)	NCT03332030	Stem Cells in NF1 Patients with Tumors of the Central Nervous System	Diagnostic Test: Collection of Stem Cells
10	Sickle Cell Disease	NCT04819841	Gene Correction in Autologous CD34+ Hematopoietic Stem Cells (HbS to HbA) to Treat Severe Sickle Cell Disease	Genetic: GPH101 Drug Product

S.No	Targeted Genetic Disorder	Identifier Number	Title of the Clinical Trial	Interventions
11	Hereditary Transthyretin Amyloidosis	NCT04601051	Study to Evaluate Safety, Tolerability, Pharmacokinetics, and Pharmacodynamics of NTLA-2001 in Patients with Hereditary Transthyretin Amyloidosis with Polyneuropathy (ATTRv-PN)	Biological: NTLA-2001
12	Blindness (Leber Congenital Amaurosis)	NCT03872479	Single Ascending Dose Study in Participants With LCA10 (Leber Congenital Amaurosis)	Drug: EDIT-101
13	Familial Mediterranean Fever (MEFV Gene Mutation)	NCT04478409	Characterization of a Functional Test for Mediterranean Family Fever Screening – 2	Biological: one additional blood sample during a planned blood test
14	Rubinstein-Taybi Syndrome	NCT04122742	Diagnosis of RSTS: Identification of the Acetylation Profiles as Epigenetic Markers for Assessing Causality of CREB Binding Protein and EP300 Protein Coding Gene) Variants	Procedure: skin biopsy for the primary fibroblast culture and a 15 ml blood sample (three unnamed samples of 5 ml) in each of the 4 SRT patients included.|Other: Generation of Induced Pluripotent Stem Cells (iPSC) from fibroblasts obtained by skin biopsy|Other: Histone acetylation profiles of cells of SRT patients with CREBBP mutations|Other: Functional involvement of identified epigenetic alterations|Biological: Culture of lymphoblastoid line from blood sample

used the CRISPR-Cas9 system to delete three genes (TRAC, TRBC, and PDCD) responsible for the evasion of cancer cells from host immune system and add another gene (NY-ESO-1) to help the immune cells to recognize tumor cells (Stadtmauer et al., 2020). These clinical experiments strongly suggest the application of CRISPR-Cas system in cancer therapeutics. Moreover, CRISPR therapeutics company is presently working on clinical trial on blood cancer patients with CRISPR engineered CAR-T cell therapy.

11.5.2 Hemoglobinopathies (Genetic Blood Disorders)

Hemoglobinopathies are a group of genetic blood disorders such as sickle cell disease (SCD), β-thalassemia, and Hemophilia. In β-thalassemia several mutations including small insertions, single point mutation, or deletions in β-globin gene are reported, resulting in loss or reduced β-globin synthesis (Telen et al., 2019). Sickle cell disease is caused by amino acid Glu to Val mutation in β-globin subunit of hemoglobin (Canver et al., 2015), leading to abnormal hemoglobin expression. In both of these blood disorders, a key transcription factor BCL11A is targeted by CRISPR-Cas9 system to edit the erythroid enhancer region in hematopoietic stem and progenitor cells (HSPCs), producing CTX001. The clinical data suggested that use of CTX001 resulted in less or no blood transfusions required in β-thalassemia patients and no bleeding episodes reported in SCD (Frangoul et al., 2020) (Table 11.1). Another application of CRISPR technology was applied in hemophilia blood disorder in which the gene responsible for blood clotting factor IX (FIX) is missing or has very low expression. In preclinical studies, CRISPR technology has been used to insert gene encoding FIX clotting factor into nonhuman primates that resulted in the expression of a normal level of blood clotting factor FIX.

11.5.3 Inherited Eye Disease

Most inherited eye blindness is due to a specific gene mutation which can be easily cured by CRISPR-Cas technique. It is to be noted that the immune system has limited activity in human eyes, hence, the problem of rejection of CRISPR treatment is reduced. Therefore, it is easy to manipulate a mutation related to congenital eye disorder. CRISPR-Cas9 technology has been successfully used for the treatment of Leber Congenital Amaurosis (LCA), which is the major cause of childhood blindness, and there was no treatment available till now. Currently, phase II clinical trial is underway by Editas Medicine, USA, directly on the human body rather than ex vivo. The aim of the study is to edit the gene using CRISPR-Cas system and restore the function of light sensitive cells before the patients completely lose their sight. Raun and his coworkers have demonstrated a cellular model of LCA10 by introducing the CEP290 splice mutation into human embryonic kidney cells using CRISPR/Cas9 technology. LCA10 is a severe retinal dystrophy caused by a mutation in the CEP290 gene. There is an intronic mutation that generated cryptic splice donor site, due its large size targeted genomic deletion is the best therapeutic approach for LCA10 patients. Scientists restored the splice mutation of CEP290 gene in LCA10 patients by using SpCas9 technique.

11.5.4 Muscular Dystrophy

Duchenne muscular dystrophy (DMD) is a muscular genetic disease caused by multiple mutations in dystrophin gene that interfere with the production of muscle protein and needed for the production of healthy muscle fibers. It is the most common form of progressive muscular dystrophy, and is identified by muscle weakness, loss of ambulation, and premature death. CRISPR/cas9 technology is the best tool for the treatment of Duchenne muscular dystrophy, which includes correction of point mutations, knock-in of the full missing exon, and restoring DMD's reading frame by deletion of the mutated exons. CRISPR/Cas9-mediated HDR (homology directed repair) approach has been used for the correction of point mutation (Zhu et al., 2017) and the addition of missing exon (Li et al., 2015). Deletion of one or more exons from the mutated transcript can restore the reading frame and generate a truncated but yet functional protein. This approach has been used to remove exon 23, exons 20–23, exons 50–54, and exons 45–55 by using multiple gRNAs that target regions surrounding desired exon(s) to generate more than one double strand break (DSB) simultaneously and removing large segments (EL Refaey et al., 2017).

Limb-girdle muscular dystrophies types 2B (LGMD2B), types 2D (LGMD2D) and Myotonic dystrophy type 1 (DM1) are other muscular dystrophy that are targeted by CRISPR/Cas9 technology (Turan et al., 2016; Van Agtmaal et al., 2017). Recently, a new approach has been developed for DMD which potentially corrects 3000 mutations that causes DMD by a single cut along the DMD gene mutation hotspots (Long et al., 2018). Scientists have developed 12 guide RNAs coupled with Cas9 enzyme to find the mutation hotspots in DMD gene which can restore the normal function of cardiac muscles. This strategy of single cut editing may be more useful than deleting multiple exons which may interrupt the proper expression of gene and forms nonfunctional protein.

11.5.5 Genetic Liver Diseases

The hereditary tyrosinemia type I (HTI) is a liver genetic disease in which mutation in gene-Fah responsible for the synthesis of FAH (fumarylacetoacetate hydrolase) occurs. This is the last enzyme of tyrosine catabolic pathway in liver cells. Mutation in this gene resulted in the accumulation of toxic substances in the liver cells that leads to the liver damage (Yinet al.2014; Yin et al.,

2016). Researchers have shown the hydrodynamic injection of CRISPR-Cas9 DNA and short single stranded DNA HDR template in mouse model can correct the Fah mutation in mouse liver however, the correction rate is very low (Yin et al., 2014). Yin and his coworkers used nanolipid delivery of Cas9 mRNA and Adeno-associated viral delivery of sgRNA/HDR template into Fah mutant mice model. By this method the correction rate of Fah mutations increased and resulted in reduced disease symptoms in mouse model (Yin et al., 2016).

11.5.6 Cystic Fibrosis

Cystic fibrosis **t**ransmembrane **c**onductance **r**egulator (CFTR) is a life-threatening disease caused by mutation in CFTR gene. This gene encodes a protein responsible for chloride transport in the cells and help in the water retention. Due to mutation in CFTR gene water content decreases and mucus that surrounds the cells becomes thick and sticky. This eventually effect the functioning of lungs by clogging the airways and make them prone to infection and breathing problem. Geurts and coworkers have developed a new way to correct nonsense mutation in Cystic fibrosis patients. They used fusion complex of CRISPR-SpCas9-ABE (PAM recognition sequence: NGG) which allows the base editing from A-T to G-C (A-G base editing or ABE), however it is a very tricky technique because of small window of single stranded DNA during the Cas9 binding to the target sequences (Gaudelli et al., 2017; Koblan et al., 2018).

Inherited surfactant protein (SP) deficiency is another important lung disease caused by genetic mutations in SFTPB, SFTPC, and ATP-binding cassette protein member 3 (ABCA3) genes. These proteins are involved in the regulation of hemostasis of air-liquid interference at alveoli in the lungs (Nogee LM, 2004). Mutation in SFTPB or SFTPC in alveolar type 2 cells (AT2) causes distress on oxygen exchange in the lungs due to lower surface tension (Alapati et al., 2017). Scientists have successfully engineered CRISPR/Cas9 to edit SP associated genes by single base substitution. They deliver the gene editing reagents using adeno-associated viral vector (AAV) intranasally or intratracheally in mouse model (Mahiny et al., 2015).

11.5.7 Epidermal/Skin Disorders

Skin genetic disorders are easier to treat compared to other genetic disorders because ofdirect exposure to the reagents, easy invitro culture of epithelial cells and manipulation through iPSC technology. There are several skin genetic disorders which are targeted by gene therapy with the help of CRISPR/Cas9 system such as monogenic inherited epidermolysis bullosa (EB) disorders, Epidermolytic palmoplantar keratoderma (EPPK) and Melanoma.

Epidermolysis bullosa is a blistering disease that is caused by mutations in twenty different genes of cutaneous membrane region (Uitto et al., 2018). EB Simplex (EBS) subtype is caused due to dominant missense mutation in the keratin genes. Scientist used CRISPR-Cas9-induced HDR to edit the missense mutation of keratin gene (KRT14) allele in EBS patients' keratinocytes (Kocher et al., 2017).

EPPK is an autosomal keratin disease which causes abnormal thickening of palm and sole skin (Has et al., 2016). This is caused by missense mutation in KRT9 gene which forms mutated protein. Since, it is localized in specific places CRISPR/Cas9 can easily use for its treatment. Researchers have used sgRNA and Cas9 directed to mutant KRT9 allele and deliver it locally through lentiviral vectors. Cas9 disrupt the formation of mutated K9 protein in mouse model and suggested its role as therapeutics (Luan, 2018).

11.5.8 Neurological Genetic Disorders

Several neurological disorders have been targeted by CRISPR-Cas technology by making new animal models or editing the genes responsible for disorders. The first research using CRISPR technology was done in prion diseases (Mehrabian et al., 2014). Onwards several other disorders such as Alzheimer's disease (AD), Parkinson's disease (PD), Huntington's disease (HD) and Spinocerebellar ataxia type 2 (SCA2) have been targeted by CRISPR/Cas technology.

Alzheimer's disease (AD) is most common neurodegenerative disease in elder persons and leading cause of death. Several genetic mutations have been linked with AD such as mutation in amyloid precursor protein (APP), apolipoprotein-E (ApoE), gamma-secretase gene presenillins (PSEN1 and PSEN2) (Hutton and Hardy 1997; Chow et al., 2010). Scientists used CRISPR-Cas9-sgRNA complex to target BACE1 gene responsible for amyloid plaques formation (Chow et al., 2010).

Parkinson's disease is another neurodegenerative disorder characterized by death of neurons by the formation of synuclein protein accumulation in neurons. Mutations in genes PINK1 (PTEN-induced kinase 1), LRRK2 (Leucine-rich repeat kinase 2) and SNCA (alpha-synuclein) have been reported in PD. Using zebra fish model Lee et al., induced frameshift mutation in MCU gene (Mitochondrial uniporter) with CRISPR/Cas9 technology which resulted in the restoration of mitochondrial activity (Lee et al., 2018).

Huntington's disease is characterized by overexpression of cytosine-adenine-guanine (CAG) repeats in huntingtin gene (HTT). The numbers of repeats reciprocate the onset of disease symptoms. Since, it is not easy to manipulate the genes in the brain with CRISPR technology scientists develop an enzyme Cas9 Nickases to precisely cut the effector gene (Dabrowska et al., 2018).

Amyotrophic lateral sclerosis is also a neurodegenerative disease which has shown to linked with the mutation in genes such as SOD1 (super oxide dismutase1), PfN1 (Profilin1) and FUS (RNA/DNA binding protein). Researchers used the CRISPR-Cas9 system to edit the mutation in these genes in induced pluripotent stem cells (iPSCs) from fibroblast of familial ALS patients (Wang et al., 2017).

11.6 CONCLUSION

CRISPR-Cas technology has been discovered 30 years ago and till now many research articles have been published to show its application in various fields such as agriculture, environment, microorganisms and human health. The most promising outcome of CRISPR technology was in DNA/RNA editing, manipulation, deletion and addition. As discussed above in this chapter we have explained the CRISPR technology in details emphasizing its role in human genetic disorders. Recently, CRISPR-Cas/13a gene editing system has been successfully used to inhibit Coronavirus replication in COVID positive patients at Stanford University, USA. The results were promising and showed 90% decrease in the total viral load in the lungs of COVID-positive patients (Abbott et al., 2020). The latest discovery of modified Cas9 enzyme known as dCas9 can be used to target epigenome with guideRNA. Scientists have linked the dCas9 enzyme with many epigenetic affecting enzymes such as demethylase, methylase or acetyltransferase which can activate or repress the transcription of genes without affecting the genome. More research is needed for application of CRISPR technology to regulate and control gene expression in human cells.

Biggest challenge for this useful technology is to translate these research outcomes into successful therapeutics to cure human diseases. Although there are many challenges in implementation of this technology in human body; such as how the human body will react with foreign DNA, delivery of CRISPR-Cas system into human, host immune response is also a major hindrance, risk of mutation in off target genes and many unknown effects. However, scientists all over the world are working on improvement of this technology day by day and may come out with new solutions. Despite having many challenges, CRISPR-Cas system is a state-of-the-art technique and an empowering scientific tool which is giving a new hope in the world of science.

BIBLIOGRAPHY

Alapati D, Morrisey EE. 2017. Gene editing and genetic lung disease. Basic research meets therapeutic application. *Am J Respir Cell Mol Biol.* 56(3):283–290.

Bikard D, Euler CW, Jiang W, Nussenzweig PM, Goldberg GW, Duportet X, Fischetti VA, Marraffini LA. 2014. Exploiting CRISPR-Cas nucleases to produce sequence-specific antimicrobials. *Nat Biotechnol.* 32:1146–1150.

Brouns SJ, Jore MM, Lundgren M, Westra ER, Slijkhuis RJ, Snijders AP, Dickman MJ, Makarova KS, Koonin EV, van der Oost J. 2008. Small CRISPR RNAs guide antiviral defense in prokaryotes. *Science.* 321:960–964.

Burstein D, Harrington LB, Strutt SC, Probst AJ, Anantharaman K, Thomas BC, Doudna JA, Banfield JF. 2017. New CRISPR-Cas systems from unculti- vated microbes. *Nature.* 542:237–241.

Canver MC, Smith EC, Sher F, Pinello L, Sanjana NE, Shalem O, et al. 2015. BCL11A enhancer dissection by Cas9-mediated in situ saturating mutagenesis. *Nature.* 527:192–7

Chen K, Wang Y, Zhang R, Zhang H, Gao C. 2019. CRISPR/Cas genome editing and precision plant breeding in agriculture. *Annu. Rev. Plant Biol.* 70:667–697.

Chow VW, Mattson MP, Wong PC, Gleichmann M. 2010. An overview of APP processing enzymes and products. *Neuromolecular Med.* 12:1–12.

Dabrowska M, Juzwa W, Krzyzosiak WJ, Olejniczak M. 2018. Precise excision of the CAG tract from the huntingtin gene by cas9 nickases. *Front. Neurosci.* 12:75.

Datsenko KA, Pougach K, Tikhonov A, Wanner BL, Severinov K, Semenova E. 2012. Molecular memory of prior infections activates the CRISPR/Cas adaptive bacterial immunity system. *Nat. Commun.* 3:945.

Doudna JA. 2020. The promise and challenge of therapeutic genome editing. *Nature.* 578:229–236.

East-Seletsky A, O'Connell MR, Knight SC, Burstein D, Cate JH, Tjian R, Doudna JA. 2016 Two distinct RNase activities of CRISPR-C2c2 enable guide-RNA processing and RNA detection. *Nature.* 538:270–273.

Egorova TV, Zotova ED, Reshetov DA, Polikarpova AV, Vassilieva SG, Vlodavets DV, et al. 2019. CRISPR/Cas9-generated mouse model of Duchenne muscular dystrophy recapitulating a newly identified large 430 kb deletion in the human DMD gene. *Dis Model Mech.* 12.

El Refaey M, Xu L, Gao Y, Canan BD, Adesanya TMA, Warner SC, Akagi K, Symer DE, Mohler PJ, Ma J, Janssen PML, Han R. 2017. In vivo genome editing restores dystrophin expression and cardiac function in Dystrophic mice. *Circ Res.* 121:923–929.

Haydar Frangoul, Yael Bobruff, Maria Domenica Cappellini, Selim Corbacioglu, Christine Marie Fernandez, Josu de la Fuente, Stephan A Grupp, Rupert Handgretinger, Tony W. Ho, Suzan Imren, Antonis Kattamis, Julie Lekstrom-Himes, Franco Locatelli, Yimeng Lu, Markus Y Mapara, Mariane de Montalembert, Sarah D. Mulcahey, Damiano Rondelli, Niraj Shanbhag, Sujit Sheth, Sandeep Soni, Martin H. Steinberg, Michael J. Weinstein, John K. Wu, Donna Wall; Safety and Efficacy of CTX001 in Patients with Transfusion-Dependent β-Thalassemia and Sickle Cell Disease: Early Results from the Climb THAL-111 and Climb SCD-121 Studies of Autologous CRISPR-CAS9-Modified CD34+ Hematopoietic Stem and Progenitor Cells. Blood 2020; 136 (Supplement 1): 3–4. doi: https://doi.org/10.1182/blood-2020-139575

Gaudelli NM, Komor AC, Rees HA, Packer MS, Badran AH, Bryson DI, Liu DR. 2017. Programmable base editing of A-T to G-C in genomic DNA without DNA cleavage. *Nature.* 551:464–471

Grissa I, Vergnaud G, Pourcel C. 2007 Jul. CRISPRFinder: a web tool to identify clustered regularly interspaced short palindromic repeats. *Nucleic Acids Res.* 35.

Groenen PM, Bunschoten AE, van Soolingen D, van Embden JD. 1993. Nature of DNA polymorphism in the direct repeat cluster of Mycobac terium tuberculosis; application for strain differentiation by a novel typ- ing method. *MolMicrobiol.* 10:1057–1065.

Has C, Technau-Hafsi K. 2016 Feb. Palmoplantar keratodermas: clinical and genetic aspects. *J Dtsch Dermatol Ges.* 14(2):123–139.

Heckl D et al. 2014. Generation of mouse models of myeloid malignancy with combinatorial genetic lesions using CRISPR-Cas9 genome editing. *Nat. Bio-Technol.* 32:941–946.

Heywood JD, Karon M, Weissman S. 1964. Amino Acids: Incorporation into α- and β-Chains of Hemoglobin by Normal and Thalassemic Reticulocytes. *Science.* 146:530.

Hosseini Far H, Patria YN, Motazedian A, Elefanty AG, Stanley EG, Lamande SR, et al. 2019. Generation of a heterozygous COL1A1 (c.3969_3970insT) osteogenesis imperfecta mutation human iPSC line, MCRIi001-A-1, using CRISPR/Cas9 editing. *Stem Cell Res.* 37:101449.

Hutton M, Hardy J. 1997. The presenilins and Alzheimer's disease. *Hum Mol Genet.* 6:1639–1646.

Ishino Y, Krupovic M, Forterre P. 2018. History of CRISPR-Cas from encounter with a mysterious repeated sequence to genome editing technology. *J Bacteriol.* 200:e00580–17.

Ishino Y, Shinagawa H, Makino K, Amemura M, Nakata A. 1987. Nucle- otide sequence of the iap gene, responsible for alkaline phosphatase isozyme conversion in Escherichia coli, and identification of the gene product. *J Bacteriol.* 169:5429–5433.

Jansen R, Embden JD, Gaastra W, Schouls LW. 2002. Identification of genes that are associated with DNA repeats in prokaryotes. *Mol Micro- biol.* 43:1565–1575.

Koblan L, Doman J, Wilson C et al. 2018. Improving cytidine and adenine base editors by expression optimization and ancestral reconstruction. *Nat Biotechnol.* 36:843–846.

Kocher T, Peking P, Klausegger A, Murauer EM, Hofbauer JP, Wally V, Lettner T, Hainzl S, Ablinger M, Bauer JW, Reichelt J, Koller U. 2017 Nov 1. Cut and paste: Efficient homology-directed repair of a dominant negative KRT14 mutation via CRISPR/Cas9 nickases. *Mol Ther.* 25(11):2585–2598.

Koonin EV, Makarova KS. 2013. CRISPR-Cas. *RNA Biology.* 10(5):679–686.

Koonin EV, Makarova KS, Zhang F. 2017. Diversity, classification and evolution of CRISPR-Cas systems. *Curr. Opin. Microbiol.* 37:67–78.

Lee KS, Huh S, Lee S, Wu Z, Kim AK, Kang HY, et al. 2018. Altered ER-mitochondria contact impacts mitochondria calcium homeostasis and contributes to neurodegeneration in vivo in disease models. *Proc Natl Acad Sci USA.* 115:E8844–E8853

Li HL, Fujimoto N, Sasakawa N, Shirai S, Ohkame T, Sakuma T, Tanaka M, Amano N, Watanabe A, Sakurai H, Yama- moto T. 2015. Precise correction of the dystrophin gene in duchenne muscular dystrophy patient induced pluripotent stem cells by TALEN and CRISPR-Cas9. *Stem Cell Rep.* 4:143–154.

Long C, Li H, M Tiburcy, C Rodriguez-Caycedo, V Kyrychenko, H Zhou, Y Zhang, Yi-Li Min, J Shelton, P P. A. Mammen, N Y. Liaw, W Zimmermann, RB-Duby, J W. Schneider, E N. Olson. 2018. Correction of diverse muscular dystrophy mutations in human engineered heart muscle by single-site genome editing. *Sci Adv.* 4(1):eaap9004

Lu Y, Xue J, Deng T, Zhou X, Li W. 2018. A phase I trial of PD-1 deficient engineered T cells with CRISPR/Cas9 in patients with advanced non-small cell lung cancer. *J Clin Oncol.* 36:3050–3050.

Luan XR, Chen XL, Tang YX, Zhang JY, Gao X, Ke HP, Lin ZY, Zhang XN. 2018 Sep 7. CRISPR/Cas9-mediated treatment ameliorates the phenotype of the epidermolytic palmoplantar keratoderma-like mouse. *Mol Ther Nucleic Acids.* 12:220–228.

Mahiny AJ, Dewerth A, Mays LE, Alkhaled M, Mothes B, Malaeksefat E, Loretz B, Rottenberger J, Brosch DM, Reautschnig P, et al. 2015. In vivo genome editing using nuclease-encoding mRNA corrects SP-B deficiency. *Nat Biotechnol.* 33:584–586

Makarova KS, Grishin NV, Shabalina SA, Wolf YI, Koonin EV. 2006. A putative RNA-interference-based immune system in prokaryotes: com- putational analysis of the predicted enzymatic machinery, functional analogies with eukaryotic RNAi, and hypothetical mechanisms of action. *Biol Direct.* 1:7

Makarova KS, Haft DH, Barrangou R, Brouns SJ, Charpentier E, Horvath P, Moineau S, Mojica FJ, Terns RM, Terns MP, White MF, Yakunin AF, Garrett RA, van der Oost J, Backofen R, Koonin EV. 2011. Evolution and classifi- cation of the CRISPR-Cas systems. *Nat Rev Microbiol.* 9:467–477.

Makarova KS, Wolf YI, Alkhnbashi OS, Costa F, Shah SA, Saunders SJ. 2015. An updated evolutionary classification of CRISPRCas systems. *Nature Rev Microbiol.* 13(11):722–736.

Mali P, Yang L, Esvelt K, Aach J, Guell M, DiCarlo J, et al. 2013. RNA-guided human genome engineering via Cas9. *Science.* 339:823–826.

Manghwar H, Lindsey K, Zhang X, Jin S. 2019. CRISPR/Cas system: Recent advances and future prospects for genome editing. *Trends Plant Sci.* 24(12):1102–1125.

Markossian S, Guyot R, Richard S, Teixeira M, Aguilera N, Bouchet M, et al. 2018. CRISPR/Cas9 editing of the mouse thra gene produces models with variable resistance to thyroid hormone. *Thyroid.* 28:139–150.

Marraffini LA, Sontheimer EJ. 2010. CRISPR interference: RNA-directed adaptive immunity in bacteria and archaea. *Nat Rev Genet.* 11:181–190.

Mehrabian M, Brethour D, MacIsaac S, Kim JK, Gunawardana CG, Wang H, et al. 2014. CRISPR-Cas9-based knockout of the prion protein and its effect on the proteome. *PLoS One.* 9:e114594.

Mojica FJM, Díez-Villaseñor C, García-Martínez J, Soria E. 2005. Interven- ing sequences of regularly spaced prokaryotic repeats derive from for- eign genetic elements. *J Mol Evol.* 60:174–182.

Nishimasu H, Ran FA, Hsu Patrick D, Konermann S, Shehata Soraya I, Dohmae N, Ishitani R, Zhang F, Nureki O. 2014. Crystal structure of cas9 in complex with guide RNA and target DNA. *Cell.* 156:935–949.

Nogee LM. 2004 Jun 8. Genetic mechanisms of surfactant deficiency. *Biol Neonate*. 85(4):314–318. doi: 10.1159/000078171

O'Geen H, Ren C, Nicolet CM, Perez AA, Halmai J, Le VM, Mackay JP, Farnham PJ, Segal DJ. dCas9-based epigenome editing suggests acquisition of histone methylation is not sufficient for target gene repression. Nucleic Acids Res. 2017 Sep 29;45(17):9901–9916. doi: 10.1093/nar/gkx578. PMID: 28973434; PMCID: PMC5622328.

Paquet D, Kwart D, Chen A, Sproul A, Jacob S, Teo S, et al. 2016. Efficient introduction of specific homozygous and heterozygous mutations using CRISPR/Cas9. *Nature*. 533:125–129.

Roh JI, Lee J, Park SU, Kang YS, Lee J, Oh AR, et al. 2018. CRISPR-Cas9-mediated generation of obese and diabetic mouse models. *Exp Anim*. 67:229–237.

Roux LN, Petit I, Domart R, Concordet JP, Qu J, Zhou H, et al. 2018. Modeling of aniridia-related keratopathy by CRISPR/Cas9 genome editing of human limbal epithelial cells and rescue by recombinant PAX6 protein. *Stem Cells*. 36:1421–1429.

Shah SA, et al. 2013. Protospacer recognition motifs: mixed identities and functional diversity. *RNA Bio*. 10(5):891–899.

Stadtmauer EA, Fraietta JA, Davis M M, Cohen A D, Weber K L, Lancaster E, Mangan P, Kulikovskaya I, Gupta M, Chen F, Tian L, Gonzalez V, Xu J, Jung In-y, Melenhorst J. J, Plesa G, Shea J, Matlawski T, Cervini A, Gaymon AL, Desjardins S, Lamontagne A, Salas-Mckee J, Fesnak A, Siegel D, Levine BL, Jadlowsky KJ, R M. Young, Chew A, Hwang W, Hexner EO, Carreno BM, Nobles CM, Bushman FD, Parker KM, Qi Y, Satpathy AT, Chang HY, Zhao Y, Lacey S, June CH. 2020 Feb. CRISPR-engineered T cells in patients with refractory cancer. *Science*. 28.

Telen MJ, Malik P, Vercellotti GM. 2019. Therapeutic strategies for sickle cell disease: towards a multi-agent approach. *Nat Rev Drug Discov*. 18:139–158.

Turan S, Farruggio AP, Srifa W, Day JW, Calos MP. 2016. Precise correction of disease mutations in induced pluripotent stem cells derived from patients with limb girdle muscular Dystrophy. *Mol Ther*. 24:685–696.

Uitto J, Bruckner-Tuderman L, McGrath JA, Riedl R, Robinson C. 2018 May. EB2017-progress in epidermolysis bullosa research toward treatment and cure. *J Invest Dermatol*. 138(5):1010–1016.

Urnov FD. 2018. Genome Editing B.C. (Before CRISPR): Lasting lessons from the "old testament". *CRISPR J*. 1:34–46.

Van Agtmaal EL, Andre LM, Willemse M, Cumming SA, van Kessel IDG, Wjaa van den Broek G, Gourdon D, Furling V, Mouly DG, Monckton DG, Wansink DG. 2017. CRISPR/Cas9-Induced (CTGCAG)n repeat instability in the myo-tonic Dystrophy type 1 locus: implications for therapeutic genome editing. *Mol Ther*. 25:24–43.

van Embden JD, Cave MD, Crawford JT, Dale JW, Eisenach KD, Gicquel B, Hermans P, Martin C, McAdam R, Shinnick TM, et al. Strain identification of Mycobacterium tuberculosis by DNA fingerprinting: recommendations for a standardized methodology. *J Clin Microbiol*. 1993 Feb;31(2):406–9. doi: 10.1128/jcm.31.2.406-409.1993.

Wang L, Yi F, Fu L, Yang J, Wang S, Wang Z, Suzuki K, Sun L, Xu X, Yu Y, Qiao J, Belmonte JCI, Yang Z, Yuan Y, Qu J, Liu GH. 2017 May. CRISPR/Cas9-mediated targeted gene correction in amyotrophic lateral sclerosis patient iPSCs. *Protein Cell*. 8(5):365–378. doi: 10.1007/s13238-017-0397-3

Yin H, Song CQ, Dorkin JR, Zhu LJ, Li Y, Wu Q, et al. 2016. Therapeutic genome editing by combined viral and non-viral delivery of CRISPR system components in vivo. *Nat Biotechnol*. 34:328–333.

Yin H, Xue W, Chen S, Bogorad RL, Benedetti E, Grompe M, et al. 2014. Genome editing with Cas9 in adult mice corrects a disease mutation and phenotype. *Nat Biotechnol*. 32:551–553.

Zhu P, Furen W, Mosenson J, Zhang H, He TC, Wen-Shu W. 2017. CRISPR/Cas9-mediated genome editing corrects dystrophin mutation in skeletal muscle stem cells in a mouse model of muscle dystrophy. *Mol Ther Nucl Acids*. 7:31–41.

12 Prospects of Genomics Tools in Conservation of Farm Animal Breeds

SK Niranjan and Manishi Mukesh

CONTENTS

12.1 INTRODUCTION

Animal breeds have evolved as the culminated effects of artificial selection for specific need and utilization along with the contribution of natural selection for specific habitat and climatic conditions. Such breeds and their subtypes, including strain, varieties, or ecotypes represent a specific gene pool wherein the alleles contributing towards traits selected are either very high in frequency or get fixed at some of the loci. Local adaptation along with artificial selection has molded the gene pool as a whole of the domesticated breeds more drastically. Continued selection of animals in specific habitats, as well as demographic changes, molded the gene pools distinctly, which have further culminated in developing unique morphological, physiological features in populations; and overall, the distinct breed emerged.

FAO has recognized about 8800 distinct breeds and other populations of various livestock and poultry species in the world (FAO, 2021a). Most of the breeds are local and native to the region, whereas some are regional and international breeds. FAO's status of trends of animal genetic resources 2020 (FAO, 2021a) reports that about 32 percent of the livestock breeds are at risk, globally. Similarly, the unknown status of more than 50 percent of breeds is also a great concern towards impending threat to a large portion of the global animal genetic resources diversity. In the wider perspective, a large proportion of flora and fauna in the globe is seriously under threat; and widely assumed that it would be the sixth-largest event of extinction of species, which would be mainly due to human-based activities complexed causing climatic changes and increased fragility of ecosystems. The Global Assessment Report on Biodiversity and Ecosystem Services 2019 of the United Nations has estimated that about one million species are at risk of extinction.

To preserve all kinds of biodiversity in general and the farm animal biodiversity specifically, conservation efforts have been initiated internationally for a long time. In situ conservation, i.e.

DOI: 10.1201/9781003220831-12

conservation of the population, its habitat is known as the best method of conservation. Although, it is the most traditional method of conservation of animal breeds across the globe, it is quite participatory in nature and could be carried out by the local society itself. Other methods include ex situ conservation of live animals and cryopreservation of germplasm for posterity. It is recommended that the country should adopt in situ conservation complemented by cryopreservation activities.

12.2 CONSERVATION GENOMICS: A NEW ROLE

Genetics helps in conservation in characterizing as well as assessing the genetic diversity of the animal breeds. Conservation genetics is a decades-old field which encompasses genetic analyses for the conservation of flora and fauna. Molecular genetic tools have been largely used for characterizing and assessing the diversity in farm and wild animals for the last few decades. The use of molecular data for describing livestock, including genetic characterization started in the 1990s. Initial molecular approaches were largely based on the use of microsatellite markers for diversity analysis of the populations. For uniform using of the markers, FAO has also recommended the battery of markers for various livestock and poultry species. The microsatellite markers are mainly located in the neutral region, which is differently evolved than the functional region of the genome. In farm animals, because of the presence of intense selection pressure, the functional regions of the genome are much evolved and modulated. In wild animals, the genetic parameters of a population remain uncorrelated when analyzed based on neutral microsatellite markers and functional immunological markers (MHC).

Microsatellite-based analyses largely indicate about genetic drift and related demographic changes, whereas functional markers define the adaptive processes due to selection pressure; which shows different interpretations in both of the approaches. Generally, a breed is considered distinct if it evolved largely due to adaptive processes instead of genetic drift-associated processes (Grueber, 2015). After the advent of whole-genome sequencing, the availability of genome-wide SNP markers, both functionally active and neutral parts, has largely defended these biases in population genetic analysis.

Now, as whole genome sequencing has become highly economical, the new era of genomics has revolutionized the characterization and diversity analysis of the livestock breeds across the globe. Whole genome sequences are available for a number of species, including domestic animals. Whole genome-based tools can be used for a number of purposes, important for conservation, like species identity, degree of hybridization, genetic diversity, demographic history, population decline or expansion, and effective population size. All this was possible earlier also with microsatellite markers; however, it is now more accurate after inclusion of genomics data. Genome-wide SNP markers, either identified through WGS and subsequent biostatistical tools or typed through SNP chip now useful for solving many molecular mysteries about the livestock breeds and populations. Most important aspects included genetic constitution, including diversity, admixture, inbreeding of the breed, assessing the homogeneity of a breed, differentiation and phylogenetic relationship of breeds or populations, gene flow and introgression of new population, along with assessing the divergence of domesticated animals from their wild ancestors. Studies on adaptative changes, production, disease resistance tolerance, and susceptibility are the new dimensions of the whole genome-based association studies in livestock. Conservation genomics is now has become much evolved science, applicable to the genomic analysis to the preservation of the biodiversity. Genomic Comparison has the most usefulness for the conservation of the fauna, wherein the related populations/breeds/species are studied for their evolution. In domestic animals, comparative genomics add another facet for studying the relatedness of the breeds along with breed formation events. Reduced representation libraries, RNAseq transcriptomic data may be useful genomic tools for comparative genomics in farm animals. In farm animals, the use of SNP chip is much pronounced to assess the diversity and emergence of various breeds.

12.3 CONVENTIONAL MOLECULAR TOOLS

In early days of conservation genetics, some molecular techniques have been used to type polymorphic loci across the genome. Prominent techniques included restriction fragment length polymorphism (RFLP), amplified fragment length polymorphism (AFLP), and short tandem repeats (STR), i.e. microsatellite. The AFLP technology has the capability to detect various polymorphisms in different genomic regions, simultaneously. It is also highly sensitive and reproducible. RFLP based variation detection in genomes is a vital tool in genome mapping and genetic disease analysis also the basis for early methods of characterization of genetic diversity in animal populations. Among these, the microsatellite was found to be a robust method of detecting genetic differences between breeds. On the other hand, only SNPs with a frequency of one SNP in 1 kb genomic sequence are the most abundant markers in genome. The SNPs are now considered important for genotyping and are useful in epidemiological studies, gene discovery, gene mapping, candidate gene polymorphism testing, diagnostics and risk profiling, homogeneity testing, and parentage verification. These markers were utilized in various forms, from small to large scale genotyping; including Denaturing Gradient Gel Electrophoresis (DGGE), Amplification Refractory Mutation System (ARMS), KASP assays, Taqman assay, Sequenom and high-resolution melting.

12.4 HIGH THROUGHPUT GENOMIC TOOLS

SNPs markers localized in a smaller genomic region are utilized for various structural and functional analysis of a genomic region. However, unlike microsatellites, these techniques could not be utilized for diversity analysis due to their local presence. After NGS, the availability of genome-wide SNPs data has gained momentum in livestock to be utilized for various purposes in recent time. These markers have been proved to be better than and even replacing microsatellite markers. The use of genome-wide SNP markers is much more versatile for both of the major fields of population genetics and genetic improvement. Polymorphic SNP markers are identified either through direct whole genome sequencing or the use of genotyping SNP arrays. The arrays-based genotyping is much more economical than whole genome sequencing and can be implemented in large samples with comparatively less time. Whole genome SNP markers are more suitable than microsatellites for high throughput genetic analysis, using DNA microarray technology.

SNP Genotyping Arrays: Genotyping arrays or SNP Chip have been designed for almost all livestock species, including cattle, buffalo, sheep, goat, horse, pig, dromedary camel, Bactrian camel, and chicken. These arrays have been developed mainly on two platforms – Affymetrix Axiom and Illumina Infinium. These chips, with low, medium, and high density, are utilized for various purposes like breed assignment, diversity analysis, biodiversity mapping, high resolution genetic mapping, parentage verification, genomic selection, trait selection, GWAS, QTL analysis, identification of deleterious mutations, traceability, etc.

Whole genome sequencing: Whole genome sequencing derives the large SNP data set for any population, wherein SNPs across the genome are assessed. Not only identifying all kinds of variants like SNPs, indels, CNVs, large genomic insertions and deletions, are captured through the WGS, it also provides nearly true picture of variability of the genome as well population. High cost and more complexed bioinformatics are among the few constraints which hinder the use of the technology on a large scale. In fact, WGS covers the highest number of SNPs than any of the available technology used for SNP typing. WGS with long read sequencing with higher depth carried out on Pacific Bioscience platform may lead to constructing the De novo assembly, specifically of new species. However, for population analysis, whole genome resequencing with short and long read sequencing at lower depth and being economical and with lesser bioinformatic analysis, is more preferred for studying large numbers of animals in a population.

Reduced representation whole genome techniques: Genotyping by sequencing (GBS) is another approach of reduced representation of genome. This also enables the identification of a large fraction of SNPs of the populations. A number of GBS-based methods have evolved, among which ddRAD sequencing is more popular. This decreases the cost of resequencing significantly and is most useful to the unique species or breeds. It is also useful for the species for which SNP arrays are not available commercially. Almost all kinds of population genetic analyses, estimating relatedness, linkage disequilibrium, parentage assignment can be carried out by using GBS data.

12.5 CONSERVATION GENOMICS FOR DIVERSITY ASSESSMENT

Genetic diversity is always essential for natural as well as artificial selection. Assessing the genome wide genetic diversity within the population is the fundamental step of conservation programe. This diversity can be measured by using the genome wide markers; the most common being microsatellite and the SNPs. For most of the species the assessment of endangerment is now based on assessment of genomic diversity, either at targeted regions or across the genome. Many species and breeds have been evaluated for the traits related with the fitness and resilience. Now it has been observed that the fitness of any population is a complex phenomenon of factors like genetic diversity of the population, population dynamics or demographic history, as well as ecological relationship. Low genetic diversity is often linked with increased inbreeding as well as genetic drift, which largely culminates into decreased survival and health risk. Such effects are more evident in a small population. To tackle the consequences of small population, many programs are oriented towards increasing the genetic diversity by introducing the animals from other populations. Genetic diversity can be measured by using neutral markers like microsatellite and SNPs analysis. Since there is little effect of selection on these markers, the variation in these markers can be attributed to the genetic drift, gene flow, and mutation. However, neutral marker cannot provide the information on the fitness, adaptive genetic variation as well as evolutionary potential of the population (Teixeira and Huber, 2021).

12.6 ASSESSING GENOMIC DIVERSITY OF POPULATION

Diversity within the population is the foremost requirement in conservation programs. Phenotypic diversity in a population largely indicates its genetic diversity, however identifying the traits which should be considered for assessing phenotypic diversity is not an easy task. Moreover, when animal breeds tend to show very high phenotypic homogeneity among the individuals, estimating the genetic diversity becomes most essential for assessing the overall health of the population. Optimum genetic diversity at genome level is always desirable for a long-running conservation programe. If it is low or very high, it also entails about the breeding strategies to be adopted to make it optimum, specifically in conserved populations.

Assessing genetic diversity within the population provides an opportunity to assess the genetic fitness of the population as well as formulating the conservation programs for making them more successful. Most of the time, the genetic diversity of a population decreases with the population size, however, it may be more pronounced due to apparently low effective population size, breeding preferences and methods of the population as well as the random drift within the genome. Similarly, genomic regions having the functional importance specially related with the immunity are much important. Any breed which becomes adaptive to a specific environment may also need the genetic variability in the regions regulating to these, for providing better opportunity for selection of variants in a changed environment. Therefore, for an effective conservation program, true assessment of genetic diversity is essentially required.

Genomic molecular tools help in explaining the genetic structure of a population, including observed and expected homozygosity, F statistics, as well as bottleneck events, origin, migration and dispersal of a breed, inbreeding level, kinship analysis (FAO, 2021b). This becomes more important for the species and breed which could not be assessed due to many reasons. When we compare the diversity of genomic markers of various breeds or populations it also assesses the comparative analysis of these populations. Importantly, it also provides an opportunity for identifying breeds which are genetically close. Molecular phylogeny of various populations also provides an opportunity for identifying the most diverged and unique population, which prioritizes our conservation program. The most diverged breeds may be conserved first. Observed heterozygosity (H_o) of the population is the simplest indicator of the diversity of a breed. It is compared with the expected heterozygosity (H_e) to identify the departure of the population from HW equilibrium and the factor causing the departure. H_o lower than H_e indicates inbreeding or subdivision of the populations, whereas a larger value hints of the admixing of the population. Inbreeding coefficient (F) can also be estimated by Fisher statistics (F_{IS}) using observed heterozygosity and expected heterozygosity. Estimation of inbreeding coefficient is important in assessing the apparent threat to the population in lieu of inbreeding depression and may be important for deciding the conservation strategies. It provides much of the information like relatedness of individuals or parents, mating type, population structure including recent changes; all are important for deciding the conservation strategies. Inbreeding causes the increased homozygosity, especially at recessive deleterious alleles, as well as decreased heterozygosity to the allele which may have heterozygote advantage. Conventionally, pedigree based analysis through assessing the probability of identity either of pedigreed animals or gametic combinations. Now many marker-based estimation methods have been designed to assess the F value or say genomic inbreeding. Genomic based best method depends on a number of markers/alleles/individuals, relatedness among individuals and their mating structure. Whole genome sequence data seems to evaluate different inbreeding estimators based on accurately estimated genome wide heterozygosity. Assessing Linkage Disequilibrium (LD) may be important for identifying the recent as well as old demographic processes occurring in a population (Alemu et al., 2021). LD is generated due to non-random association between two alleles or SNPs, due to their location in close proximity on same chromosome. High value of LD for quite separated SNPs can be observed in genetic drift of small population, or due to hitchhiking effect of selection process. To make the markers independent and in linkage equilibrium, pruning of markers in LD is required for estimating genetic distances, PCA, and ancestral analysis. The LD also becomes essential for the calculation of haplotype-based runs of homozygosity for assessing the selection sweeps or signatures in the population.

Further, comparative genomics in farm animals is also useful for assessing the genetic influences due to species hybridization and crossbreeding. In case of hybridization, whole genome studies can determine the causes of recently man-made or ancient genetic changes due to natural introgression. This also quantifies the degree of introgression or the degree of ancient or recent hybridization between threatened or non-threatened species. Identifying the degree of hybridization can be helpful to recover the pure lineage. Conservation planning can be made to recover the pure breed animals from hybridized through adopting proper breeding strategies, including selective breeding (Simčič et al., 2015). Similarly, comparing various populations may also be utilized for prioritizing the breeds for conservation, particularly during resource constraints. The breed with ancient lineage with evolutionary significance generally requires more attention for the conservation during comparative analysis or phylogeny of related breeds. Here, genome wide data may provide more accurate and reliable phylogenetic tree for breed differentiation as well as identifying the most distinct or unique breed or populations. The analysis also explains that the differentiation in populations is due to the evolutionary processes for demographic processes.

It is difficult to maintain the genetic variability in a small population. A small population continuously faces inbreeding and genetic drift, which remains as a threat to that population. In a small population genetic drift predominates over the selection and mutation, making the genome of individuals more unpredictable sometimes. Increased inbreeding level with lower genetic diversity and lower adaptability increases the level of risk of extinction by reducing the fitness traits like survivability, reproduction, and growth at individual level. Sometime, environmental modifications of the isolated small populations also enhance the susceptibility for these effects. All these factors virtually diminish the genetic variability over the generations and severely affect the fitness as well as adaptability traits in individuals (Frankham, 1998). High inbred animals tend to have low fitness, longevity, production, and reproduction due to homozygosity of undesirable and deleterious alleles. All this makes it desirable to assess the genome at individual levels also. At an individual level, the observed heterozygosity H_o, as a proportion of heterozygous genotype in an individual values indicated about the hybridization or crossbreeding, if it is very high (more than +3SD than average population), similarly if the value is very low it indicates the inbreeding of that individual. Both of the cases are not desirable for the conservation point of view. An animal with very high H_o has diluted its genome due to mixing of alleles of two highly diverged populations, at large proportion of the loci. H_o intends to be high when there are mixing of higher diverged populations than that of similar populations.

12.7 IMMUNE-RELATED GENOMIC REGIONS: ASSESSING ADAPTABILITY

Genetic variability in immune response genes is always desirable for increased adaptability and fitness of any population. Allelic diversity, especially in the form of heterozygosity at many immune related genic regions like MHC, TLRs, NLRs, immunoglobin families are desirable for the fitness of any population in vertebrates. In fact, variabilities in these regions are also used for assessing the fitness of the populations, particularly in wild animal species. In some instances, such a population, showing extreme monomorphism at microsatellite markers, may possess optimum fitness by carrying sufficient genetic variation at the MHC region. Much of the conservation programs in mammalian populations are, now, aimed towards increasing the diversity at MHC class II loci with an objective to increase population fitness. On the other hand, MHC diversity directly influences the survival of individuals against infectious disease and is thought to have a significant impact on population fitness. Decreased MHC variability in animals may predispose the species to be extinct in the time of certain disease occurrence. There are certain examples when the low MHC diversity caused the extinction of populations after occurrence of diseases. Therefore, variation at genetic regions of immune-related genes is a new focus to measure the genetic variability as well as fitness of the population and is regarded as a potentially important index in conservation genetics (Yasukochi et al., 2012).

12.8 GENOMICS FOR GENETIC IMPROVEMENT

Genetic improvement for utility traits may be another approach for conserving the breed or populations. In the livestock field, genomics is now intensively used to increase productivity, which ultimately helps in animal conservation. Such genome-based technologies can be applied to enhance the genetic progress through increasing the selection intensity, reducing generation interval, and sometimes increasing genetic variation. Some DNA-based technologies are being used for selecting the animals with desirable traits. Marker assisted selection (MAS) desires to increase the proportion of animals exhibiting marker allele(s) that is/are linked to a favorable trait. A number of molecular markers have been identified in livestock, which are related to growth and milk production. Candidate gene and quantitative trait loci (QTL) are the approaches for MAS. Candidate gene based markers can be used more successfully and reliably compared to QTLs. Genomic selection (GS) involves selection of animals for breeding on the basis of their genotype for tens of thousands of random SNPs, covering an entire genome. With the availability of SNP chips for livestock species, the technology is being utilized for whole genome selection, as thousands of SNPs can be genotyped simultaneously. Transgenics opens new avenues for creation of transgenic livestock by genetic engineering and hence provides a viable platform for enhancing production and disease resistance. Gene introgression/transgression in a specific population has also been practiced in some of the cases and has given desirable results, and may be utilized for conservation purposes also.

12.9 BIOTECHNOLOGY FOR CONSERVATION

Several new biotechnologies have emerged during the last few years that may be equally useful and applicable for the conservation efforts. However, the popularity of these technologies for conservation is still not great in most part of the world. The biotechnological tools like cryo-preservation and micro-manipulation of gametes and embryo, embryo cloning, and gene transfer could be supporting technologies to increase the efficiency of conventional conservation techniques. Reproductive techniques such as Multiple Ovulation Embryo Transfer (MOET) technology, embryo manipulation, cloning and semen sex sorting offer possibilities for faster multiplication of superior germplasm. Cloning techniques have been successfully used in at least 22 different species and have contributed to conserve a number of wild species. Stem cells undergo self-renewal throughout life and are capable of transferring genetic information to the offspring, therefore has potential in conserving the germplasm. Super-ovulation, estrus synchronization, and embryo transfer (ET) have provided the platform to propagate the superior female for faster genetic improvement. Embryo cloning and splitting can give the advantage to produce a greater number of animals in case of impending threat. All these biotechnologies can be well utilized for the conservation of the breeds or species, the population of which has been reached almost up to extinction. Cryopreservation of germplasm in the form of semen, ova, DNA, somatic cells, gonads, embryo, blood, serum in the cryobank purpose is also another way for conservation.

12.10 CONCLUSION

Genomic diversity analyses may be important as prior information for conservation of the breed. The attributes like level of diversity within the population, genetic similarities among the breeds, level of crossbreeding and hybridization, consanguinity, inbreeding depression, distinctive genetic traits, and genetic defects are important for conservation of any breed, which can be accurately assessed based on genomics. Although genomic tools alone would not be sufficient, at present, for conservation; it would be needed to be combined with other traditional methods of conservation. The genomic approach would only be helping in prioritizing and properly managing and achieving the goals of conservation in the earliest possible way. Therefore, integration of molecular genetic markers with conventional conservation strategies would be the approach for conserving the rich biodiversity in years to come.

BIBLIOGRAPHY

Alemu SW, Kadri NK, Harland C, Faux P, Charlier C, Caballero A and Druet T. 2021. An evaluation of inbreeding measures using a whole-genome sequenced cattle pedigree. *Heredity*. 126:410–423.

FAO. 2021a. *Status and trends of animal genetic resources – 2020*. FAO. Rome.

FAO. 2021b. *Genomic characterization of animal genetic resources-Draft updated guidelines*. 11th Session of Intergovernmental Technical Working Group on AnGRFA, FAO, Rome.

Frankham R. 1998. Inbreeding and extinction: Island populations. *Conservation Biology*. 12:665–675. https://doi.org/10.1111/j.1523-1739.1998.96456.x

Grueber CE. 2015. Comparative genomics for biodiversity conservation. *Computational and Structural Biotechnology Journal*. 13:370–375.

Simčič M, Smetko A, Sölkner J, Seichter D, Gorjanc G, Kompan D, et al. 2015. Recovery of native genetic background in admixed populations using haplotypes, phenotypes, and pedigree information – using Cika cattle as a case breed. *PLoS One*. 10:e0123253.

Teixeira JC and Huber CD. 2021. The inflated significance of neutral genetic diversity in conservation genetics. *PNAS*. 118(10):e2015096118. https://doi.org/10.1073/pnas.2015096118

Yasukochi Y, Kurosaki T, Yoneda M, Koike H, and Satta Y. 2012. MHC class II DQB diversity in the Japanese black bear, Ursus thibetanus japonicus. *BMC Evol Biol*. 12:230. https://doi.org/10.1186/1471-2148-12-230

13 Genomic Evaluation of Dairy Sires

The Process, Challenges, and Prospects

Supriya Chhotaray and Vikas Vohra

CONTENTS

13.1 INTRODUCTION

Sustainable dairy production depends on the methods used to select the Sire (male) and Dam (female) to be parents of next-generation in an organized dairy herd, as superior genetics leads to superior production. This mainly depends on methods used to select the Sires, as genetically, a Sire is considered as more than half of the herd. In Animal Genetics and Breeding, there are four paths which lead to genetic improvement in a dairy herd: Sire to Son, Sire to Daughter, Dam to Son and Dam to Daughter. The Sire to Son path contributes towards the maximum genetic gain. Thus, it is necessary to understand the importance of superior dairy Sire as well as the various methods used to evaluate the genetics of dairy Sire. Till date, various methods of dairy animal selection have been used, which were largely the indices of economic traits-based selection of female animals. Selection of dairy Sires are progeny based or based on the performance of the bulls' daughter, where males are selected for dairy traits based on the performance of their daughters. Improved animal genetics and breeding methods have been largely focused on selection of superior Sire. The methods available now for selection of a Sire is largely based on both conventional progeny testing programs and much recently the genome-based markers have also been used across the World to judge the genetic potential of a probable Sire. In agrarian countries like India, and unlike the developed countries, small size of animal holding in a dairy herd renders any selection strategy inefficient and difficult to implement, but still considerable genetic improvement has been achieved using conventional methods in dairy Sire selection. In absence of a defined breeding plan and well-defined national genetic evaluation system, the breeding decision of dairy Sire depends solely on the will of an individual livestock breeder/farmer. Thus, for a sustainable dairy development program, it is imperative to understand the developments that took place in the methods of Sire evaluation. The present chapter is an attempt to comprehend and synthesize the available information on

DOI: 10.1201/9781003220831-13

dairy Sire selection across the World, from the beginning of the 20th century till today, covering various aspects and accuracy of different processes, methods, challenges and prospects of dairy Sire evaluation.

13.2 HISTORY OF SIRE EVALUATION

Dairy production is the key element of livestock production that focuses on maximum production gain in next-generation through the selection of bulls with high genetic potential. From the early 1900s to the 1960s was the Era of Sires' Daughter-Averages through which genetic evaluation of dairy bulls was taken up. In 1905, record-keeping systems were developed and introduced in North America with limited use of computing systems such as adding machines. Quantities such as performance averages, variances, and co-variances were mostly calculated. Later on, the development of indices to evaluate bulls started when Rice (1933) proposed useful criteria for an index. At earliest dairy bulls were compared based on Daughter-Dam Comparisons. Eventually, between 1925 and 1944, many different equations for combining daughter and Dam averages were proposed. Influenced by the work of Sewall Wright and Sir Ronald A. Fisher, Dr. Jay L. Lush constructed the selection index in the course of the 1920s and 30s and laid the foundation stone for the development of scientific animal breeding methods, henceforth, Prof. Lush is regarded as the father of animal breeding. Wright (1934) and Hazel (1943) proposed a method of selection index-based on a single trait, which was widely used and modified several times, and was known by various names, viz., 'index of combined selection' or 'family selection' (Lush, 1947; Osborne, 1957; Gibson, 1995). Harvey and Lush proposed a selection index for dairy cattle for the very first time in 1952, for simultaneous improvement of production and body conformation traits. Developments were made for within farm selection of cows for improved milk production using records from the Dam and its families, using the index proposed by Legates and Lush (1954). When Lush (1944) developed the selection index method, he proposed the weighting of mate's information in the daughter-Dam comparisons. In the early to mid-1950s the contemporary comparison method and the herd-mate comparisons method of Sire evaluation were initiated, which provided opportunities to accelerate genetic improvement in production traits in several countries. Animal scientists Robertson and Rendel are credited as the developer of the Contemporary Comparison based methods, while Herd-mate Comparison was supposedly proposed by Henderson, Carter, and Godfrey, both published in 1954. Robertson and Rendel also proposed the Least Squares methods of calculating 'Breeding Value' of bulls for the first time in 1954. Here it is important to mention that 'Breeding Value' is a core concept of Animal Breeding methods, and can be defined as an average value of an individual judged by the mean value of its progeny for a quantitative trait. Breeding value estimate also determines the Progeny Transmitting Ability (PTA) and 'Average effect of Gene Substitution' which are also an important concept in animal evaluation. Searle (1964), Cunningham (1965), and Henderson (1952, 1963) extended the same idea to develop several procedures for the estimation of fixed effects (i.e., model parameters are non-random) from weighted-Least Squares.

Charles R. Henderson graduated under the guidance of L. N. Hazel and Jay L. Lush from Iowa State University, and made immense contributions in evaluation of dairy animals. He modified Least-Squares equation for the condition where means are unknown, which otherwise is similar to Lush's selection index. However, only after meeting with Searle, Henderson became interested in matrix algebra and developed Best Linear Unbiased Prediction (BLUP) estimates from mixed model solutions, which stems the present-day basis of Sire evaluation in animal breeding.

Henderson had taught the animal model to his students as 'individual cow model' during the 1960s and for 'Breeding Value' estimation using BLUP Henderson wrote Fortran programs which were used at Cornell University for genetic evaluation using a Sire Model in 1972–76. The programs were run on an IBM 360 machine with 128K of memory. The cow model was later converted as the 'Animal Model' by Quaas and Pollak (1980). Today, the Animal Model is used in almost every country in the World that applies mixed model equations for genetic evaluation of livestock species.

The applicability of the mixed model equations (MME) has now been extended to Genomic data, which indicates the versatility of MME in Animal Breeding. Hence, C. R. Henderson, who made the huge contribution to animal genetic evaluations in all species of livestock, should be rightfully acknowledged, though our words fall short for it.

Charles R. Henderson (1953) also developed three methods (known as Henderson's Methods 1, 2, and 3) for estimating variance components. These methods were easy to calculate and yielded unbiased estimates, but estimates were often negative. However, these methods were extensively used until the 1970s and then were replaced by likelihood methods. Along with BLUP and MME, one of the major discoveries by Henderson was the method to invert a matrix of additive genetic relationships (Henderson, 1976). Several models such as the reduced Animal Model, repeated record model, test day models, and multi-trait models have been developed since then, which are being used for the improvement of various livestock species.

In dairy cattle, most of the important economic traits are expressed only in females, hence, historically evaluation and characterization of Sire was being done based on the daughters' performance records. This is called progeny testing; and it has been the sole way to accurately determine estimated breeding values (EBV) based on phenotypic information (traits measured on an individual, like milk yield, body height) and to select Sires based on their progeny performance. Sattler (2013) described progeny testing as a three-fold program that provides information on Sire selection for artificial insemination (AI), data needed for effective genetic evaluations, and generates genetically proven young bulls for wide-scale use in AI programs. However, the technique has its own disadvantages that the selection decisions in a breeding program are postponed in the progeny-testing (PT) program and generation interval is lengthened. Evaluation through PT programs is relatively slow since considerable time is required to gather adequate daughter records for genetic evaluations of bulls with high accuracy. There comes the role of genomic selection in genetic improvement of dairy Sires. In conventional PT programs, the generation interval of Sires and Dams of bulls are 7 and 4 years, respectively, while it is significantly reduced to approximately 2.5 years for both the Sires and Dams of bulls with the use of genomic markers in genetic evaluations (García-Ruiz et al., 2016). The generation interval refers to the average age of Sire at which it becomes a parent. In a breeding program, no Sire can be used for breeding the cows unless its genetic merit is proven, thus, in order to prove a Sire, i.e., by the time the daughter of a Sire performs, a large time lapses thereby increasing the generation interval, but under genomic selection strategy the time required to select the Sire is considerably low, as low as age of sexual maturity in buls. Since genomic breeding value (GBV) can be estimated in a dairy herd as early as the day an animal is born.

13.3 DEVELOPMENT OF GENOMIC SELECTION

Several conventional dairy Sire selection methods discussed in the previous section have played a major role in improvement of dairy production. Best linear unbiased prediction (BLUP) method of genetic evaluation eventually became the method of choice worldwide for evaluation and selection of Sires and was rapidly incorporated in methodology of progeny testing (PT) programs.

All of these genetic evaluation methods have assumed the Infinitesimal Model described by Fisher (1918) that assumes that production performance is controlled by an infinite number of loci (genes) and each of them adds an infinitesimally small quantity to the total genetic worth of an individual. The genetic merit of an animal is estimated as the sum total of the effects exerted by the infinite number of loci as the correct number of loci that were involved and their effect size haven't been fully determined yet. Scientists are continually working on the discovery of loci with large effects and the actual number of loci involved in the expression of a phenotype immediately after the molecular structure of DNA, the genetic material, was determined by Crick and Watson (1953).

Improvement of economically important quantitative traits in Livestock have been relied on BLUP estimation of 'Breeding Values' based on phenotypes and pedigree records. Through this strategy, average milk yield has increased by 5000 kg in the past 40 years in the American dairy cattle population of which 60% is due to genetic improvement (Zhang et al., 2011). However, traditional selection methods faced limitation in efficiency, particularly for the traits which are lowly heritable, longevity, sex-limited, and carcass traits. To overcome these limitations, it was assumed that the discovery of useful genetic markers might tackle the situation.

During the post-genomic era, the advancements in the field of molecular genetics have led to extensive use of genetic markers in the past 20 years for genetic improvement through marker-assisted selection (MAS). The accuracy of selection is likely to increase by MAS programs. However, the application of MAS in animal breeding practices has been very limited due to several reasons. Molecular geneticists used markers that were large DNA segments (100–1000 base pairs) but were not very close to the actual genes. As the success of MAS depends on the closeness of marker and genes, the term linkage disequilibrium (LD) was used to denote the closeness. High LD indicates more closeness of the marker to the gene and less chances of recombination between the marker allele and gene allele across generations. MAS requires LD of 30% or more between markers and genes, to act markers to the genes. In practice, very few markers have some significant effects on the economic traits and a marginal proportion of genetic variation is explained by each of these markers. Determination of marker genotypes often lacks consistency; consequently, selection on marker genotypes was not as successful as expected. Determination of DNA markers mostly relied on restriction fragment length polymorphisms (RFLP), however, the discovery of microsatellites that were much smaller segments of DNA proved to be a better marker, but only a few microsatellites could be found for quantitative traits in Livestock.

In 1984, the Human Genome Project was conceived with the aim to sequence all of DNA and was finally published in April 2003. Scientists were able to identify gene sequences and finally concluded that approximately 25,000 genes exist in the human genome and discovered single nucleotide polymorphisms (SNPs) by comparing genome sequences of different individuals. The genome was found to be full of millions of SNPs. By 2004, an SNP panel containing 5000 SNPs at one time was developed to genotype dairy bulls or cows. Genotyping of dairy bulls in Canada was initially started with a 5000 SNP panel, known as the 5K chip. Meuwissen et al. (2001), through a simulation of thousands of evenly spaced SNPs across the genome, could estimate each SNP allele effect on the target trait. Furthermore, those estimates could be applied to a different herd of bulls that have not been evaluated yet have genotype information, but not any daughter performance record. Hence, they showed that in the simulations, the breeding value could be estimated with an accuracy of 0.81, which is much larger than the accuracy of EBV obtained through a simple parent average that is generally less than 0.40. Schaeffer (2006) outlined the advantages of genomic estimation of breeding value and how this strategy could double the genetic progress in dairy cattle by saving 5 years of progeny testing a bull.

The first of genomic studies in Canada involving less than 500 proven (PT) bulls and 10K SNP chip, revealed that in the validation group accuracy was 0.50 to 0.60 instead of 0.81 as in Meuwissen et al. (2001). The genetic gain was due to the selection decision made on young bulls 5 years earlier than the ongoing progeny test schemes and not from the increased accuracy of the Genomic Estimated Breeding Value (GEBV). Around the year 2006, several thousand dairy bulls from various geographical regions were genotyped through joint ventures of different countries and USDA. The project also involved the development of a new 50K chip. More than 3000 bulls were genotyped during this project and calculated GEBV were made official in the United States and Canada.

13.3.1 The Process of Genomic Prediction

The Human Genome Project when published the first draft of genome sequence, the use of biallelic markers called SNPs (pronounced as 'Snips') came into limelight and its use in conventional BLUP models led to development of genomic selection. Till date, several SNP chip panels have been developed using various methods of next generation sequencing to identify new SNPs, most recent one being the deep sequencing methods. SNPs are the most common genetic variants used today in genomic based prediction of performance in Livestock, and dairy Sire evaluation in a selection program (k/a genomic selection).

The major success of genomic selection lies in the estimation of individual SNP allele effects on the trait of economic interest in Sire selection. To estimate SNP effects, the requirements is a reference population, also called training data, is used (Meuwissen, 2007). The reference population is comprised of at least 1000 individuals with genotype as well as phenotype information from own performance records (refer to Figure 13.1), deregressed proofs (VanRaden et al., 2009), and daughter-yield deviations (González-Recio et al., 2008). Genotypic and phenotypic information are combined to estimate the effects of each SNPs. Genome estimated breeding values for young selection candidates are estimated using the estimates from the reference population. Several approaches have been discovered to decide which animals to include in the reference population. According to De Roos et al. (2007), VanRaden et al. (2009) proven (PT) bulls with national breeding values can be used as a reference group. Furthermore, the close relationships between young selection candidates and animals in the reference population render the Breeding Values more reliable (Habier et al., 2007). Theoretically, to allow reliable prediction, an optimal reference population should comprise the whole range of genotypes and phenotypes. However, practically this is not possible in a real biological population, so the reference population is designed such as it reflects the whole range of phenotypes and genotypes as close as possible. A reference population composed of animals sharing recent pedigree with the young selection candidates yields better accuracy in prediction (Habier et al., 2007; De Roos et al., 2008) as LD persists across families or breeding lines used for developing prediction equation. This requires a high density of markers across the genome with high LD between markers to accurately predict the GEBVs, but the constraint lies in the cost of genotyping to generate millions of SNP genotypes in dairy Sire evaluation.

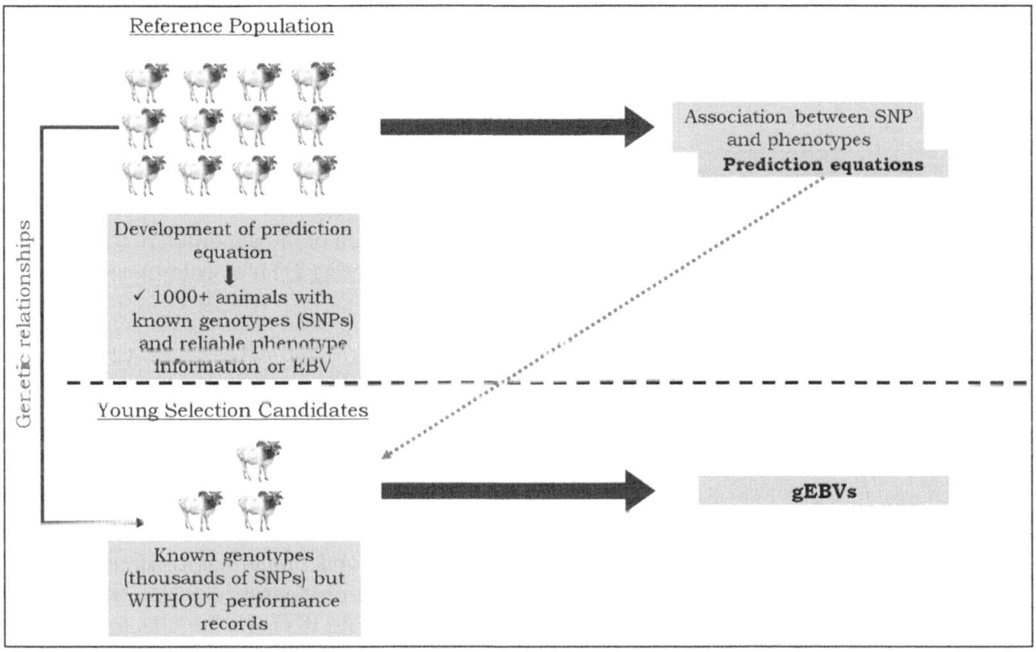

FIGURE 13.1 Schematic diagram representing the overall process of prediction of GEBVs for genomic selection programs

13.3.2 Estimation of GEBVs

Current progress in next-generation sequencing technologies has successfully created a vast pool of SNP genotype information, which along with phenotypic data and pedigree records form the simplest model to predict GEBVs. The incorporation of pedigree records helps in deriving an additive relationship matrix (Calus and Veerkamp, 2007). However, in the absence of reliable pedigree information, the additive relationship matrix is constructed directly from the genotype data (Fernando, 1998). Pedigree information combined with genotypes of individuals helps in outlining possible errors in pedigree and genotype information generated. Marker haplotypes can also be derived by combining these two sources of information (Meuwissen and Goddard, 2001)..

Prediction of SNP effects from reference population can be done by the simplest model (Calus, 2010):

$$Y_i = \mu + \text{animal}_i + \Sigma\,(\text{SNP}_{ijk}) + e_i$$

Where

Y_i = trait of interest or national EBV or daughter-yield deviation or deregressed proofs or average daughter performance of animal_i

μ = fixed effects or simply the mean

animal_i = polygenic random effect

$\Sigma\,(\text{SNP}_{ijk})$ = effects SNP, summed across allele_k ($k = 1, 2$) at loci_j for animal_i

Then, the GEBVs of animal_i can be computed in the test population: $\text{GEBV}_i = \text{animal}_i + \Sigma\,(\text{SNP}_{ijk})$

Estimation of the effect of each SNP allele and fitting the allele effects ignores the recombination that often breaks the linkage between QTL and marker alleles. Incorporation of marker haplotypes that are assumed to carry the same QTL alleles can be used in the model instead may solve the problem to a certain extent (Meuwissen et al., 2001; Villumsen and Janss, 2008). The marker haplotypes are considered identical-by-descent (IBD) that have the same marker alleles inherited from a common ancestor. Haplotypes become equivalent when IBD between them approaches unity, thus it reduces the number of effects and increases the estimation power (Yu et al., 2005; Calus, 2010). Quantitative trait loci (QTL) distributions suggest that there are many loci with minor effects while very few loci exert major effects (Hayes and Goddard, 2001). Haplotype or SNP effects estimated in reference population should ideally resemble the QTL effects' underlying distribution, however, often the estimated marker effects don't capture all of the QTL effects. Additionally, smaller QTLs are not captured in QTL mapping studies and several QTLs and many SNPs linked to QTLs constitute an overall underlying distribution of the quantitative trait. Hence, care should be taken while using the prior information of underlying QTL distribution to estimate the marker effects.

13.3.3 Various Methods of Solving the Model

The simple assumption behind the incorporation of SNPs or SNP haplotypes in genetic evaluation model describing underlying QTL distribution has been explained in the previous section.

Several new methods have been developed in the past two decades for the prediction of GEBVs using genomic information. The very first method developed was a multistep model in which direct genomic values (DGV) were computed using distinct reference and validation populations. These direct genomic values could then be used in combination with conventional EBVs or simply as correlated traits (Kachman et al., 2013). This approach was used for the genomic selection of dairy cattle in the USA. It has also been applied in genomic evaluation of the beef cattle population (Saatchi et al., 2011; Snelling et al., 2011). However, this method includes only genotyped animals in the model, and it is unmanageable to genotype all the animals.

Misztal et al. (2009) proposed the single-step method of GBLUP (ssGBLUP) that combines information on phenotypes, pedigree, and genotypes in a single evaluation. The ssGBLUP can be implemented in maternal effect and multi-trait models. It ensures that genotypic, phenotypic, and pedigree information are not double-counted, and each information source is properly weighted in the model. The ssGBLUP can also be used with a fewer number (sample) of genotyped animals. Whereas multi-step method lacks all this flexibility and has lesser accuracy and more genomic inflation. Aguilar et al. (2010), Christensen and Lund (2010) used ssGBLUP for the first time in dairy population having more than 6000 genotyped animals.

Major challenges faced in solving the model is to accurately estimate the SNP effects from phenotypes of a few thousand animals (n) and over ten thousand to millions of SNPs (p). This results in the famous 'small (n) big (p) problem' or over-parameterized model; this problem is where the numbers of parameters are more than that can be estimated from the data, it can be overcome by implementing ridge regression BLUP (RR-BLUP) or SNPBLUP. This model assumes a normal distribution of SNP effects and constant variance that implies the same proportion of variance is expressed by many if not all the SNPs, which is practically unreal. Another challenge is to eliminate the loci having very small or nearly zero effect on the distribution of QTL effects and to determine the actual number of loci whose effects need to be estimated. Weller et al. (2012) proposed the use of a subset of SNPs with large estimated effects that are incorporated in an animal model and the animal effects represent the remaining polygenic effects after the SNPs. Selecting tag-SNPs or a subset of SNPs, which explain much of the genomic variation in the complete set of SNPs genotyped can solve the 'small n big p' problem. Selection of the subset of SNPs can be achieved through various methods such as the association of SNPs with a trait of interest, feature selection using machine learning procedures (Long et al., 2007), forward stepwise regression (Habier et al., 2007) and principal component analysis (Solberg, 2008).

Another important approach is to determine whether SNPs are truly associated with a QTL exerting some effect on phenotype, and implement the same for the prediction of GEBV at once. Meuwissen et al. (2001) successfully implemented this approach in a Bayesian models (BayesA, B, C and π).

BayesA model assumes that all SNPs have a certain effect on the trait of interest and very few SNPs have large effects while many have small effects. Therefore, variances assumed for each SNPs differ from each other. However, Bayesian methods are non-linear and are often affected by shrinkage that makes small effects more smaller and large effects even larger.

BayesB model uses a Metropolis-Hastings algorithm for every iteration to determine whether a locus has any effect on the trait or not. This method assumes that a proportion (π) of the SNPs have zero effect while 1-π SNPs have a nonzero effect. Many loci are rendered with zero variance because of this assumption. BayesB becomes BayesA when π is zero. However, this model relies mostly on the considered prior information to estimate the SNP effects that creates a major concern (Gianola et al., 2006).

BayesC model: As an alternative to BayesB, a mixture of SNP-BLUP and BayesB methods can be done. It assumes a constant variance for a fraction of SNPs having an effect on the trait as in SNP-BLUP and a fraction (π) of SNPs have zero effect as in BayesB (Meuwissen and Goddard, 2004; Calus et al., 2008; Janss et al., 2008). This method is called BayesC where Gibbs sampling is applied and the Metropolis-Hastings step is avoided. QTLs with the distribution of large effects pick up the same SNPs as the BayesB model, while the SNPs with smaller effects still explain small and equal of the variance as in the SNP-BLUP model (Calus et al., 2008).

BayesCπ is a modified BayesC with an additional step to estimate the proportion of SNPs not having any effect on the trait.

Other than SNP effects, the genomic relationship matrix (GRM) instead can be used in the model that assumes each of the SNP contributes equally to the additive genetic variance. Such model is known as GBLUP (Meuwissen et al., 2001). A GRM constructed over all the markers can be used in a regular mixed model instead of an additive relationship matrix (Goddard, 2009).

VanRaden (2008) proposed that the contribution of each SNP to the GRM could be weighted to account for differences in variances. This method is like SNP-BLUP, but Genomic Breeding Values are estimated instead of SNP effects. The effectiveness of GBLUP depends on the closeness of GRM to the realized genetic relationships. Appropriate quality control of genomic data before constructing the GRM can avoid much of biases in any estimate. Variance not represented by the markers can be explained by including an extra polygenic effect. All these models decrease the dimensionality of the large SNP data set.

In a span of a decade, several alternative methods such as non-parametric kernel (Gianola et al., 2006; Gianola and Van Kaam, 2008), principal component analysis, PCA (Solberg, 2008), partial least squares (PLS) regression (Raadsma et al., 2008; Solberg, 2008), genetic algorithms, and Bayesian-LASSO (de los Campos et al., 2009) have been proposed. Non-parametric kernel methods are as effective as Bayesian methods when only additive model is assumed (Bennewitz and Meuwissen, 2008), but is more effective than the Bayesian models when non-additive (dominance and epistatic effects) effects are also considered in the model (Gonza´lez-Reci et al., 2008). PCA and PLS both helps to reduce the dimensionality of the genotype matrix by identifying a set of markers explaining maximum proportion of variance from the available genotypes. However, PLS depends on the trait for the same while PCA acts independently. Apart from additive effects, dominance and epistatic (non-additive) effects can also be included in the model as suggested by Gianola et al. (2006) and Gianola and van Kaam (2008).

There exists a difference between Genome-based predicted estimated breeding values (GEBVs) and phenotype or conventional EBV predicted; the former (GEBV) uses pedigree and marker information while EBVs are predicted from progeny information. In genomic evaluations, biases may exist for the estimated effects, especially, (i) when QTLs are estimated as fixed effects, and (ii) when there is insufficient LD between marker and QTL as the latter may lead to underestimation of the effects. Practically, this arises as all the SNPs present on a SNP chip are not utilized for the prediction purpose. VanRaden et al. (2013) suggested that about 40% of the total SNPs present on a chip are useful for prediction and explain the reasons of a smaller number of usable SNPs as their low call rates, lower minor allele frequency (MAF), absence of sufficient polymorphism, and redundancy of the markers.

An important concern in genomic evaluation is to decide the nature of the marker effect-whether it should be included in model as fixed or random effects. SNPs are often considered as a fixed effect in various studies (Baruch and Weller, 2008), especially when the sample genotyped is very small. However, recent studies show that when marker effects are treated as random, it eliminates the problem of over-parameterization (when number of genotypes outnumber the records/sample) in the prediction. The random markers effect also takes into account the co-variances among the marker effects and minimizes the 'noise' or null effect of markers on the trait of interest (Weller, 2016).

13.3.4 Accuracies of Genomic Predictions

The accuracy of the GEBV prediction depends upon several factors and these should be taken into consideration while developing the prediction model and designing the genomic evaluation programs. Of course, there are some ways to improve the accuracy of a genomic prediction.

Independent of the prediction approach, one such major factor is the linkage disequilibrium, LD, between the marker and the QTL and its consistency in the reference as well as test population (Goddard, 2009). The LD between marker and QTL mostly depends on the marker density. If the LD is incomplete or broken in further generations, re-estimation of marker effects is required; but some of the QTL variance still can't be captured and is left out even after re-estimation attempt is made, thus incomplete LD leads to lower accuracy. In addition to the preceding, the accuracy of the marker effects estimated depends on the number of phenotypic records included in the reference population (Hayes et al., 2009; VanRaden et al., 2009), selection of animals for reference population (Habier et al., 2007) and the heritability of the target trait (Calus et al., 2008; Hayes et al., 2009).

It is imperative that the accuracy can be improved by including more animals in the reference population (Meuwissen et al., 2001). The nature of sampling, by selecting the individuals within family, for both reference and test population in such a way that the same genetic relationship exists between the individuals of two populations can potentially improve the accuracy of genomic prediction (Legarra et al., 2008). The accuracy of GEBVs can also be increased by increasing the marker density (Meuwissen et al., 2001; Calus et al., 2008; Solberg et al., 2008) and by increasing the heritability of the trait (Calus et al., 2008; Solberg, 2008).

In dairy-Sire evaluation with the real data set, when unequal variance for each SNP is assumed the GEBV predicted is slightly more reliable than equal variance assumption (Hayes et al., 2009; VanRaden et al., 2009). BayesB gives more reliable estimates (VanRaden et al., 2009). Single step GBLUP (ssGBLUP) yields least biased GEBVs, more consistent solutions, and better accuracy as it integrates phenotype, genotype, and pedigree data. However, its limitation lies in its assumption of equal variances for each of the SNPs.

An additional gain in accuracy can be obtained by optimizing weights for SNPs in SNP-weighted ssGBLUP (WssGBLUP). National EBVs or Degressed proofs can be used instead of raw phenotypes to increase the accuracy as the heritability is effectively increased per animal due to the weighting of the offspring contribution to bull's EBV (Fikse and Banos, 2001). Accuracies obtained in several studies based on the real data set for genomic evaluation of dairy cattle across the globe are listed in Table 13.1.

TABLE 13.1
Accuracy of GEBVs Predicted through Various Methods for Different Traits in Dairy Cattle

Country	Cattle Breed	Trait(s)/Phenotype	Method	Accuracy	Reference
Australia	Australian Holstein Friesian	Australian Selection index	BayesR	0.41	Moser et al. (2009)
	Australian Holstein Friesian	Protein percentage	BayesR	0.6	
US	Holstein Friesian	Final score	ssGBLUP	0.37–0.41	Aguilar et al. (2010)
Denmark	Nordic Red	Milk	One step blending approach	0.479	Su et al. (2012)
		Fat		0.486	
		Protein		0.488	
		Fertility		0.447	
		Birth index		0.398	
		Calving Index		0.392	
		Udder health		0.461	
		Other diseases		0.441	
		Body conformation		0.428	
		Feet and legs		0.384	
		Udder conformation		0.45	
		Milking ability		0.439	
		Temperament		0.404	
		Longevity		0.46	
		Yield		0.49	
		Average		0.443	
Israel	Israeli Holstein	Milk yield	ssGBLUP	0.18	Lourenco et al. (2014)

(*Continued*)

TABLE 13.1 (Continued)

Country	Cattle Breed	Trait(s)/Phenotype	Method	Accuracy	Reference
			WssGBLUP	0.18	
			Bayes C	0.2	
		Fat percentage	ssGBLUP	0.39	
			WssGBLUP	0.42	
			Bayes C	0.41	
Germany	Brown Swiss	General temperament	Two-step approach	0.37	Kramer (2014)
		Milking temperament		0.2	
		Aggressiveness		0.19	
		Rank order in herd		0.27	
		Milking speed		0.48	
		Udder depth		0.45	
		Position of labia		0.36	
		Days to first heat		0.12	
China	Chinese Holstein	Milk yield	One step blending approach	0.54	Li et al. (2014)
		Fat yield		0.73	
		Fat percentage		0.7	
		Protein yield		0.61	
		Protein percentage		0.74	
		Average		0.66	
Denmark	Danish Holstein	CFHA (Interval from calving to first high activity)	ssGBLUP	0.09	Ismael et al. (2017)
		CFI (Interval from calving to first insemination)	ssGBLUP	0.08	
Netherlands	Dutch Holstein	Number of oocytes	ssGBLUP	0.54	Cornelissen et al. (2017)
		Number of MOET embryos	ssGBLUP	0.52	

Note: The table has been adapted from INVITED REVIEW: Genomic selection and its significance in Indian dairying by Vohra (2018).

13.4 CHALLENGES AND FUTURE PROSPECTS

Accuracy of GEBVs predicted is high, to obtain maximum genetic gain these GEBVs should be incorporated into National Evaluation Programs without further delay (Harris et al., 2008). This depends on the adaptability of National Evaluation Services and dairy industries to include genomic information for prediction of Breeding Value and selection of Sires with high genetic merit.

When we consider the dairy industry then the industry-wide scale of combining pedigree, phenotype, and genomic information to predict GEBVs poses a considerable challenge. One of the major challenges lies in the limited number of animals genotyped than present in the population. In dairy advanced countries, genomic selection (GS) has resulted in an increased rate of genetic gains in dairy cattle and the production of genomically proven bulls. The credit for their huge success (implementation of GS) can be attributed to the fact that a well-established conventional genetic evaluation system was already in place long before the introduction of GS. On the other hand, developing countries are lagging in implementation of GS, may be due to lack of proper breeding infrastructure, pedigree and systematic phenotypic recording and absence of computational and analytical tools, which are fundamental for the success of any conventional genetic evaluation programs. Ribaut et al. (2010) indicated the upsurge in information and communication technology has created opportunities to counter some of the challenges specific to developing countries by establishing global virtual platforms. The rapid developments in high-throughput next-generation sequencing technologies for SNP genotyping have reduced the genotyping costs. There are few consortiums working solely for genomic evaluation in dairy cattle in developing countries. Cost-effective genotyping services provided by several companies may boost the implementation of genomic evaluations, yet it is necessary to establish consortiums for genomic evaluation programs on a large scale in developing countries of South-Asian and African subcontinents (Mrode et al., 2019).

Genomic selection programs should also include tools for parentage verification, determination of breed composition, breed characterization and computation of genomic inbreeding coefficients for sustainable utilization of genomic information. Such an approach enhances the cost-effectiveness of genotyping and provides maximum benefit. Higher returns from the genomic selection are expected by implementing artificial reproduction technologies, such as AI and IVF in dairy cattle that would aid in the dissemination of superior germplasm, and gender-biased semen from genomically proven bulls. This shall augment the productivity and profitability of farmers having small animal holdings. Framers, breeder associations should come forward and should be encouraged to participate in a nationwide Sire evaluation program. Hand holding of developing countries by dairy developed nations for the implementation of genomics-based Sire selection should be promoted, use of a Sire produced and evaluated in one country, can be easily used in female population across the nations as nowadays transportation of bull's germplasm is few hours distance. The role of governments to formulate the breeding policy favoring the GS, regulatory frameworks for initiating long-term genomic breeding programs, all these shall play a decisive role in the overall success of the GS programs. Further, it is proposed that the limited genotype information available with each country could be pooled together based on the geographical distribution of the breed/transboundary breeds, which would possibly increase the accuracy at the initial stages (Mrode et al., 2018). Simultaneously, the issues related with data ownership and accessibility from pooling the genomic data should be addressed. Thus, governing bodies formed by joint efforts of various countries should manage data sharing and float the confidentiality agreement.

13.5 SUMMARY

Genomic selection is the ultimate approach that exploits molecular markers, SNPs, to evaluate dairy animals and design novel breeding programs. Genomic breeding values for young selection candidates can be predicted from genotype information and prediction model can be developed and tested in the reference populations. The gain in accuracies due to genomic evaluations are up to 0.31 higher than that of pedigree-based indices. Several methods have been reported to estimate marker effects and predict the GEBV. However, the accuracy of the prediction depends upon LD between marker and QTL, size of the reference population, the genetic relationship between reference and test population, heritability of the trait, prediction model used, and number of animals genotyped. There has been an increase in reliabilities of GEBV over conventional EBV (2–20%) for young bulls with no daughter records. It has been seen that the higher genetic gain in genomic selection programs is consequent to reduced generation interval when compared to conventional progeny testing programs. Although genomic selection has yielded fruitful results in the past decade, but challenges and opportunities both remains in implementation of GS programs. In times to come, it is envisaged that with the advent of biotechniques for direct gene manipulation through CRISPER/Cas9 technology and others, the possibility also exists to change the genotypes of embryos in farm animals. GS and other technologies have a bearing on future Sire evaluation and selection in a breeding program. The story of 'omics' has just began.

BIBLIOGRAPHY

Aguilar I, Misztal I, Johnson DL, Legarra A, Tsuruta S and Lawlor TJ (2010). Hot topic: A unified approach to utilize phenotypic full pedigree and genomic information for genetic evaluation of Holstein final score. *Journal of Dairy Science* 93(2): 743–752.

Baruch E and Weller JI (2008). Incorporation of discrete genotype effects for multiple genes into animal model evaluations when only a small fraction of the population has been genotyped. *Journal of Dairy Science* 91(11): 4365–4371.

Bennewitz J and Meuwissen TH (2008). Genomic breeding value estimation using kernel regression and additive models. In *12th Quantitative Trait Locus and Marker Assisted Selection Workshop*, Uppsala, Sweden 34.

Calus MP (2010). Genomic breeding value prediction: methods and procedures. *Animal* 4(2): 157–164.

Calus MPL and Veerkamp RF (2007). Accuracy of breeding values when using and ignoring the polygenic effect in genomic breeding value estimation with a marker density of one SNP per cM. *Journal of Animal Breeding and Genetics* 124(6): 362–368.

Calus MPL, Meuwissen THE, De Roos APW and Veerkamp RF (2008). Accuracy of genomic selection using different methods to define haplotypes. *Genetics* 178(1): 553–561.

Christensen OF and Lund MS (2010). Genomic prediction when some animals are not genotyped. *Genetics Selection Evolution* 42(1): 1–8.

Cornelissen MAMC, Mullaart E, Van der Linde C and Mulder HA (2017). Estimating variance components and breeding values for number of oocytes and number of embryos in dairy cattle using a single-step genomic evaluation. *Journal of Dairy Science* 100(6): 4698–4705.

Crick F and Watson J (1953). A structure for deoxyribose nucleic acid. Nature 171(737–738): 3.

Cunningham EP (1965). The evaluation of Sires from progeny test data. *Animal Science* 7(2): 221–231.

de los Campos G, Naya H, Gianola D, Crossa J, Legarra A, Manfredi E et al. and Cotes JM (2009). Predicting quantitative traits with regression models for dense molecular markers and pedigree. *Genetics* 182(1): 375–385.

De Roos APW, Hayes BJ, Spelman RJ and Goddard ME (2008). Linkage disequilibrium and persistence of phase in Holstein – Friesian Jersey and Angus cattle. *Genetics* 179(3): 1503–1512.

De Roos APW, Schrooten C, Mullaart E, Calus MPL and Veerkamp RF (2007). Breeding value estimation for fat percentage using dense markers on Bos taurus autosome 14. *Journal of Dairy Science* 90(10): 4821–4829.

Fernando RL (1998 January). Genetic evaluation and selection using genotypic phenotypic and pedigree information. In *Proceedings of the 6th World Congress on Genetics Applied to Livestock Production* 26: 329–336. Armidale.

Fikse WF and Banos G (2001). Weighting factors of Sire daughter information in international genetic evaluations. *Journal of Dairy Science* 84(7): 1759–1767.

Fisher RA (1918). The correlation between relatives on the supposition of Mendelian Inheritance. *Trans Royal Soc Edinburgh* 52: 399–433.

García-Ruiz A, Cole JB, VanRaden PM, Wiggans GR, Ruiz-López FJ and Van Tassell CP (2016). Changes in genetic selection differentials and generation intervals in US Holstein dairy cattle as a result of genomic selection. *Proceedings of the National Academy of Sciences* 113(28): E3995–E4004.

Gianola D and van Kaam JB (2008). Reproducing kernel Hilbert spaces regression methods for genomic assisted prediction of quantitative traits. *Genetics* 178(4): 2289–2303.

Gianola D, Fernando RL and Stella A (2006). Genomic-assisted prediction of genetic value with semiparametric procedures. *Genetics* 173(3): 1761–1776.

Gibson JP (1995). *An Introduction to the Design and Economics of Animal Breeding Strategies.* Guelph, Canada.

Goddard M (2009). Genomic selection: prediction of accuracy and maximisation of long term response. *Genetica* 136(2): 245–257.

González-Recio O, Gianola D, Long N, Weigel KA, Rosa GJ and Avendano S (2008). Nonparametric methods for incorporating genomic information into genetic evaluations: an application to mortality in broilers. *Genetics* 178(4): 2305–2313.

Habier D, Fernando RL and Dekkers JC (2007). The impact of genetic relationship information on genome-assisted breeding values. *Genetics* 177(4): 2389–2397.

Harris BL, Johnson DL, Spelman RJ and Sattler JD (2008 June). Genomic selection in New Zealand and the implications for national genetic evaluation. In *Proc Interbull Meeting*, Niagara Falls, Canada.

Harvey WR and Lush JL (1952). Genetic correlation between type and production in Jersey cattle. *Journal of Dairy Science* 35(3): 199–213.

Hayes BEN and Goddard ME (2001). The distribution of the effects of genes affecting quantitative traits in livestock. *Genetics Selection Evolution* 33(3): 1–21.

Hayes BJ, Bowman PJ, Chamberlain AJ and Goddard ME (2009). Invited review: Genomic selection in dairy cattle: Progress and challenges. *Journal of Dairy Science* 92(2): 433–443.

Hazel LN (1943). The genetic basis for constructing selection indexes. *Genetics* 28(6): 476–490.

Henderson CR (1952). Specific and general combining ability. *Heterosis* 352–370.

Henderson CR (1953). Estimation of variance and covariance components. *Biometrics* 9(2): 226–252.

Henderson CR (1963). Selection index and expected genetic advance. In *Statistical Genetics and Plant Breeding*, National Academy of Sciences No. 982, Washington, DC: National Research Council Publication, 141–163.

Henderson CR (1976). A simple method for computing the inverse of a numerator relationship matrix used in prediction of breeding values. *Biometrics*: 69–83.

Ismael A, Løvendahl P, Fogh A, Lund MS and Su G (2017). Improving genetic evaluation using a multitrait single-step genomic model for ability to resume cycling after calving measured by activity tags in Holstein cows. *Journal of Dairy Science* 100(10): 8188–8196.

Janss L, GregersenV, Bendixen C and Lund M (2008). Validation of genomic predictions in pigs using medium-dense marker coverage. In *Book of Abstracts of the 59th Annual Meeting of the EAAP*, Vilnius Lithuania.

Kachman SD, Spangler ML, Bennett GL, Hanford KJ, Kuehn LA, Snelling WM et al.and Pollak EJ (2013). Comparison of molecular breeding values based on within-and across-breed training in beef cattle. *Genetics Selection Evolution* 45(1): 1–9.

Kramer M (2014). *Genomic breeding value estimation for novel functional traits in Brown Swiss Cattle*. PhD, Georg-August-University, Gottingen, Germany.

Legarra A, Robert-Granié C, Manfredi E and Elsen JM (2008). Performance of genomic selection in mice. *Genetics* 180(1): 611–618.

Legates JE and Lush JL (1954). A selection index for fat production in dairy cattle utilizing the fat yields of the cow and her close relatives. *Journal of Dairy Science* 37(6): 744–753.

Li X, Wang S, Huang J, Li L, Zhang Q and Ding X (2014). Improving the accuracy of genomic prediction in Chinese Holstein cattle by using one-step blending. *Genetics Selection Evolution* 46(1): 1–5.

Long N, Gianola D, Rosa GJ, Weigel KA and Avendaño S (2007). Machine learning classification procedure for selecting SNPs in genomic selection: application to early mortality in broilers. *Journal of Animal Breeding and Genetics* 124(6): 377–389.

Lourenco DAL, Misztal I, Tsuruta S, Aguilar I, Ezra E, Ron M et al.and Weller JI (2014). Methods for genomic evaluation of a relatively small genotyped dairy population and effect of genotyped cow information in multiparity analyses. *Journal of Dairy Science* 97(3): 1742–1752.

Lush JL (1944). The optimum emphasis on Dams' records when proving dairy Sires. *Journal of Dairy Science* 27(11): 937–951.

Lush JL (1947). Family merit and individual merit as bases for selection Part I. *The American Naturalist* 81(799): 241–261.

Meuwissen T (2007). Genomic selection: marker assisted selection on a genome wide scale. *Journal of Animal Breedingand Genetics* 124: 321–322.

Meuwissen TH and Goddard ME (2001). Prediction of identity by descent probabilities from marker-haplotypes. *Genetics Selection Evolution* 33(6): 1–30.

Meuwissen TH and Goddard ME (2004). Mapping multiple QTL using linkage disequilibrium and linkage analysis information and multitrait data. *Genetics Selection Evolution* 36(3): 261 279.

Meuwissen TH, Hayes BJ and Goddard ME (2001). Prediction of total genetic value using genome-wide dense marker maps. *Genetics* 157(4): 1819–1829.

Misztal I, Legarra A and Aguilar I (2009). Computing procedures for genetic evaluation including phenotypic full pedigree and genomic information. *Journal of Dairy Science* 92(9): 4648–4655.

Moser G, Tier B, Crump RE, Khatkar MS and Raadsma HW (2009). A comparison of five methods to predict genomic breeding values of dairy bulls from genome-wide SNP markers. *Genetics Selection Evolution* 41(1): 1–16.

Mrode R, Coffey M, Ojango J, Mujibi D, Okeyo M, Strucken EM et al.and Aliloo H (2018). The impact of modelling and pooled data on the accuracy of genomic prediction in small holder dairy data. In *Proceedings of the World Congress on Genetics Applied to Livestock Production.* Electronic Poster Session – Methods and Tools – Prediction, 615, Auckland.

Mrode R, Ojango JM, Okeyo AM and Mwacharo JM (2019). Genomic selection and use of molecular tools in breeding programs for indigenous and crossbred cattle in developing countries: current status and future prospects. *Frontiers in Genetics* 9: 694.

Osborne R (1957). The use of Sire and Dam family averages in increasing the efficiency of selective breeding under a hierarchical mating system. *Heredity* 11(1): 93–116.

Quaas RL and Pollak EJ (1980). Mixed model methodology for farm and ranch beef cattle testing programs. *Journal of Animal Science* 51(6): 1277–1287.

Raadsma HW, Moser G, Crump RE, Khatkar MS, Zenger KR, Cavanagh JAL and Tier B (2008). Predicting genetic merit for mastitis and fertility in dairy cattle using genome wide selection and high density SNP screens. *Animal Genomics for Animal Health* 132: 219–223.

Ribaut JM, De Vicente MC and Delannay X (2010). Molecular breeding in developing countries: challenges and perspectives. *Current Opinion in Plant Biology* 13(2): 213–218.

Rice VA (1933). Which is the best index. *Guernsey Breeders J* 43: 238–239.

Robertson A and Rendel J (1954). The performance of heifers got by artificial insemination. *The Journal of Agricultural Science* 44(2): 184–192.

Saatchi M, McClure MC, McKay SD, Rolf MM, Kim J, Decker JE et al.and Taylor JF (2011). Accuracies of genomic breeding values in American Angus beef cattle using K-means clustering for cross-validation. *Genetics Selection Evolution* 43(1): 1–16.

Sattler CG (2013). Progeny testing and genomics: Where are we and where are we going? In *Proc Dairy Cattle Reprod*, Coun Ann Meeting Indianapolis 98–103.

Schaeffer LR (2006). Strategy for applying genome-wide selection in dairy cattle. *Journal of Animal Breeding and Genetics* 123(4): 218–223.

Searle SR (1964). Review of sire-proving methods in New Zealand Great Britain and New York State. *Journal of Dairy Science* 47(4): 402–413.

Snelling WM, Allan MF, Keele JW, Kuehn LA, Thallman RM, Bennett GL et al.and Rolfe KM (2011). Partial-genome evaluation of postweaning feed intake and efficiency of crossbred beef cattle. *Journal of Animal Science* 89(6): 1731–1741.

Solberg TR (2008) *Methods for prediction of genome-wide breeding values using dense marker genotyping.* PhD, Norwegian University of Life Sciences.

Solberg TR, Sonesson AK, Woolliams JA and Meuwissen THE (2008). Genomic selection using different marker types and densities. *Journal of Animal Science* 86(10): 2447–2454.

Su G, Madsen P, Nielsen US, Mäntysaari EA, Aamand GP, Christensen OF and Lund MS (2012). Genomic prediction for Nordic Red Cattle using one-step and selection index blending. *Journal of Dairy Science* 95(2): 909–917.

VanRaden PM (2008). Efficient methods to compute genomic predictions. *Journal of Dairy Science* 91(11): 4414–4423.

VanRaden PM Null DJ Sargolzaei M Wiggans GR Tooker ME Cole JB and Doak GA (2013). Genomic imputation and evaluation using high-density Holstein genotypes. *Journal of Dairy Science* 96(1): 668–678.

VanRaden PM Van Tassell CP Wiggans GR Sonstegard TS Schnabel RD Taylor JF and Schenkel FS (2009). Invited review: Reliability of genomic predictions for North American Holstein bulls. *Journal of Dairy Science* 92(1): 16–24.

Vohra V (2018). Genomic selection and its significance in Indian dairying. *Indian Journal of Dairy Science* 71(6): 539–545.

Villumsen TM and Janss L (2008). Genomic selection focusing on haplotype length and prior settings. In *12th Quantitative Trait Locus and Marker Assisted Selection Workshop*, Uppsala, Sweden 41.

Weller JI (2016). *Genomic Selection in Animals.* Hoboken, NJ: John Wiley and Sons Incorporated.

Weller JI, Glick G, Shirak A, Ezra E, Zeron Y and Ron M (2012). Predictive ability of selected subsets of single nucleotide polymorphisms (SNPs): For moderately sized dairy cattle populations. *Proceedings of the 2012 Interbull Bulletin* No. 46. Cork, Ireland.

Wright S (1934). *Systems of Mating and other Papers*. Ames, IA: Iowa State College Press.
Yu K, Xu J, Rao DC and Province M (2005). Using tree-based recursive partitioning methods to group haplotypes for increased power in association studies. *Animals of Human Genetics* 69(5): 577–589.
Zhang Z, Zhang Q and Ding X (2011). Advances in genomic selection in domestic animals. *Chinese Science Bulletin* 56(25): 2655–2663.

14 Advancement in Reproductive Biotechnologies in Livestock

Dharmendra Kumar, Meeti Punetha, Seema Dua, Pradeep Kumar and PS Yadav

CONTENTS

14.1 INTRODUCTION

Reproduction is a prime biological process of producing offspring, identical to their parents. Animal agriculture relies heavily on the ability of animals to reproduce effectively. Reproductive failure, on the other hand, is amongst the most significant factors which limit productivity of animal production systems, culminating in huge economic losses yearly. Scientific knowledge advances our understanding of the fundamental biology of reproduction, facilitating the development of cutting-edge management strategies that maximize reproductive efficiency while minimizing economic losses. Recently, new findings and techniques related to reproductive biotechnology have manifold applications to boost the animal productivity economically. Using reproductive biotechnology, genetic resources of animals can be preserved via sperm and embryo banking. It also contributes to improve agriculturally important traits in livestock, like propagation of superior genetics, production of offspring with predetermined sex and increasing herd quality, and employing shipment of frozen embryos in inconsiderably less time than traditional breeding methods. Faster multiplication of superior germplasm has also been achieved by dissemination of elite stock through artificial insemination (AI). Cloning, transgenic animal production, embryo transfer, embryo sexing, and in vitro fertilization (IVF) are examples of specialized reproductive technologies which continue to offer new possibilities for consistent enhancement and safety of products derived from commercially valuable food animals, as well as the conservation of rare species. These powerful new approaches have also emerged for conservation of endangered species, through in situ and ex situ conservation programs. More recently, assisted reproductive technology (ART) encompasses all current and anticipated uses of technology in animal reproduction which entails an array of appliances and procedures that

DOI: 10.1201/9781003220831-14

enable the realization of safe, improved, and healthier reproduction. ART has had a tremendous impact on animal reproduction since its introduction.

Recent advances in reproductive biotechnologies have shown important developments in the livestock industry as well as in biomedical sciences such as gene transfer, which can be used to create new products like novel proteins in milk of livestock animals (Brink et al., 2000). Animal agriculture would need to take advantage of these technological advancements in order to produce increasingly efficient and profitable livestock in order to meet the demands of a constantly rising global human population. The potential for significant breakthroughs in animal agriculture via reproductive biotechnology has been recognized in several aspects in animal science. This chapter provides an overview of recent advances in reproductive biotechnology applications in animal agriculture, as well as opportunities to enhanced productivity traits in livestock species.

14.2 AVAILABLE REPRODUCTIVE BIOTECHNOLOGIES

14.2.1 Artificial Insemination

AI is the practice of manually transferring semen from a male animal into the reproductive tract of a female animal. The first AI was attempted by Ilya Ivanovich Ivanoff in 1899 in domestic farm animals, dogs, rabbits, and poultry. Since then this technology was attempted worldwide in many other species, including human beings. In India, the very first buffalo calf using this technology was born in 1943 (Ranjhan and Pathak, 1993). Afterward, plenty have been born, and the practice has gained popularity amongst farmers in rural areas. To date, AI is the most effective reproductive biotechnology that has revolutionized animal breeding. This technique is most commonly used for the distribution of superior genetic material of superior males, addition of new genetic materials by importing frozen semen at a lower cost than live animals, progeny testing of males to increase the rate and quality of genetic selection, the use of frozen semen even after the death of a bull, and prevention of venereal diseases (Foote et al., 2002; Singh et al., 2013a). The AI technique allows efficient use of semen (through dilution and appropriate doses) over large numbers of females as well as using semen from males with superior genotype and also accelerates rates of conception via proper heat detection, reproductive organ management, and medication. One of the most important applications of AI in conservation is to prevent genetic depression in free-living organisms due to group heterogeneity. But apart from the immense potential of AI in conservation projects, AI technology in wildlife has considerable limitations. Although most non-domestic mammals' sperm can be collected by artificial vagina, vaginal condoms, electroejaculation (under anaesthesia), and digital masturbation of the penile bulb, some animals, such as rhinos, great apes, non-domestic equids, canids, and marsupials, find these processes difficult (Pukazhenthi and Wildt, 2004), as it is crucial to understand the appropriate sperm deposition site (vagina, cervix, or uterus) and the proper time during estrus for AI. AI aids in the prevention of infectious or viral diseases that can be transmitted when animals are in close proximity or share the same habitat, increases the rate of genetic growth and production value by using semen of superior males of high genetic merit, and also allows breeding between animals in different geographic locations at different times, even after the death of the superior male. When combined with other reproductive biotechnologies, including sperm cryopreservation, conservation of rare breeds, or endangered species and sperm sexing, AI becomes a potent tools capable of delivering substantial productive gains.

14.2.2 Sperm Sexing

Determination of prenatal sex is a long-standing goal of the reproductive biologist. Sex sorting is the technology in which predetermined sexed sperm are being used to produce offspring of the desired sex, either male or female. Due to this, the demand for sexed semen has been increased dramatically from the last decade. Of all the several attempts and approaches to spermatozoa sexing, the only method that has shown any promise is the method of sorting using DNA content by means of flow cytometry. For animal breeders, utilizing sex-sorted sperm to select the sex of the progeny has proven to be beneficial (Johnson, 2000; Seidel, 2015). The flow cytometry principle distinguishes spermatozoa with fluorescent labeled X-chromosomes from sperms with fluorescently labeled Y-chromosomes. Sperm cells of interest are sorted with the speed of 15 million spermatozoa per hour (Garner, 2006) and accuracy of predicting the sex of calves is between 85% and 95% accurate (Garner, 2001; Espinosa-Cervantes and Córdova-Izquierdo, 2013). Flow cytometry sperm sexing has not lived up to its initial success in the bovine breeding industry. Sexed semen doses are too costly for its broader usage, and their fertility is lower than those of unsexed semen doses. Almost all sexed bovine sperm are preserved and then used in AI or IVF. Sexed semen can be used to improve progeny testing, reducing the risk of sex-related diseases, and, as well as to conserve superior and rare animals. This technology has been quickly embraced by the dairy and beef industries. Other animal industries are gradually putting it to the test for its use in their own niches and habitats (Diers et al., 2020). One of the major drawbacks of this technology is the low number of sexed sperm produced per unit of time, and sexed sperm exhibit a variety of damages, including destabilization of the sperm membrane and capacitation-like changes, thereby decreasing the longevity of sorted spermatozoa in the female genital tract (Gosálvez et al., 2011; Diers et al., 2020). More recently, nanotechnology and microfluidics-based technologies emerged as game-changing approaches for sperm sex sorting, which may enhance the fertility potential of sex-sorted semen samples (Neculai-Valeanu and Ariton, 2021).

14.2.3 In Vitro Embryo Production

Another reproductive technology that endeavors female gamete pool to improve animal germplasm is in vitro embryo production (IVEP). IVEP by means of IVF has drawn the interest of innumerable researchers as it can salvage the genetic potential from infertile animals and can yield large numbers of embryos from the ovaries of slaughtered animals. In vitro maturation (IVM), fertilization (IVF), and culture (IVC) are sequentially performed to produce embryos in vitro (Kumar and Anand, 2012). The first IVF born offspring in mammalian species was produced in a rabbit (Chang, 1968). The birth of a calf following transfer of IVF produced a 4-cell stage embryo in a recipient cow in 1982 (Brackett et al., 1982). The first calves produced entirely from IVM, IVF, and IVC were born in 1987 (Fukuda et al., 1990). Madan et al. reported the birth of an in vitro-derived buffalo calf in 1991 (Madan et al., 1991). The bodies of literature involving researches on this aspect have dramatically become heavier in the subsequent years. Oocytes are collected from the live animals using transvaginal ultrasound-guided retrieval, popularly known as Ovum Pick-Up or OPU (Pieterse et al., 1988; Marin et al., 2019). This technique has widened the application of IVEP by merging assisted reproduction with genetic improvement in livestock. Alternatively, oocytes can also be collected from slaughter-house ovaries by aspiration in suitable medium.

Ultrasound-guided aspiration of immature oocytes from cattle and buffalo is widely used and also can be performed efficiently in horses. Laparoscopy or laparotomy is carried out in pigs, sheep, and goats for oocyte retrieval. Regarding IVM and IVF procedures, the success rate is high for cattle, buffalo, sheep, and goats, but less in pigs due to polyspermy (Casillas et al., 2018). About the offspring produced by the IVEP method, a few points are noteworthy:

- Overall male:female sex ratio remains almost 1:1,
- Birth weights of IVF calves are heavier than those of AI calves, and
- Birth weights of IVEP-derived lambs are affected by in vitro culture condition and environmental milieu of the recipient.

The application of IVEP is not only confined within the genesis of calves, lambs, or kids but the emerging technologies of modern days like embryonic stem cell production, transgenesis, and gene targeting are also dependent on this technique. Numerous factors such as donors, sperm, and breed can affect the efficiency of IVEP, thus affecting the yield and gestational viability of the embryos. IVF is a most encouraging technology as it provides the best opportunity to overcome certain types of infertility, simultaneous production of more numbers of calves, accurate assessment of the functional performance of female and male gametes, and provision of pronuclear stage ova for nuclear transfer and manipulation (Ferré et al., 2020). There are still unresolved bottlenecks of IVEP that limit the wider application of the technology, including reduced oocyte quality after IVM, lower embryo cryo-tolerance and reduced pregnancy rates. Further, communication between embryo and maternal environment is probably an important research target to understand and use maternal factors that can positively regulate embryo development. So research on this basic area to optimize the yield of good quality embryos is needed for the sake of progress of advanced research.

14.2.4 Genome Editing

Genome editing of an animal refers to altering its genetic material artificially through targeted modifications (addition, removal of certain DNA or RNA sequence or replacement) at a specific site in the genome using nucleases (also known as molecular scissors). Desired genome modifications of an animal are required for medical research in the study of complex physiological and pathophysiological processes, creating models for human disease, and improving breeding value and disease resistance in animals (Kumar and Kues, 2020). Genome editing is currently expanding rapidly, representing increasing numbers of genetically engineered animals with an agriculture application, including sheep (Zhang et al., 2019), goats (He et al., 2018), pigs (Whyte and Prather, 2011), and cows (van Berkel et al., 2002; Wang et al., 2017). Genome editor includes protein guided Zinc Finger Nucleases (ZFNs) and Transcriptional Activator-Like Effector Nucleases (TALENs) and RNA guided clustered regularly interspaced short palindromic repeats/CRISPR-associated protein 9 (CRISPR/Cas9) systems. These genome editors mediate targeted genetic alterations by enhancing the DNA mutation rate via induction of double-strand breaks at a predetermined genomic site. Among the important three genome editors, the most significantly evolving of the current generation of genome editing technologies is the class of RNA-guided endonucleases known as Cas9 from the microbial adaptive immune system. CRISPR has quickly become one of the most common methods for genome editing due to its simplicity, adaptability, and its capability of targeting multiple genomic sites in one shot (Bharati et al., 2020).

Genome editing system gives a profound opportunity to make specific and precise changes to the genome of an animal to improve animal welfare, performance, productivity, and disease resistance. The myostatin gene (MSTN) is a common target for research which leads to increase growth and muscle development in farm animal species and has a significant effect on a trait of economic importance. To date, the farm animals in which the myostatin gene has been modified include cattle, pigs, sheep, and goats (Ni et al., 2014; Proudfoot et al., 2015; Crispo et al., 2015; Wang et al., 2015). By mutating or silencing genes, genome editing technology has greatly improved the efficiency of producing transgenic organisms, disease models for drug discovery, discovering novel drug targets, and functional validation of a gene (Paul et al., 2019; Punetha et al., 2021). The diagnostic possibilities inherent in CRISPR have also been identified, and, now, a new diagnostic tool called SHERLOCK and DETECTR is all set to change clinical laboratory diagnostics in a foundational way (Mustafa and Makhawi, 2021).

The insertion of mouse gene SP110 to a targeted site via transcription activator-like effector (TALE) nickase in the bovine genome helped in production of anti-tuberculosis transgenic cattle (Wu et al., 2015). The life-changing technology of genome editing by CRISPR/Cas9 have been widely adopted to control economically important animal diseases such as porcine respiratory and reproductive system syndrome (PRRS) and African swine fever (ASF), which causes great economic losses in the pig industry. CRISPR/Cas9 technology was used to prevent the expression of the gene responsible for the production of CD163 protein after piglets were born in order to keep the

PRRS virus from spreading within the piglets bodies (Whitworth et al., 2016). Transgenic sheep with anti-brucellosis abilities have also now been developed (Li et al., 2016).

On a modest scale, genome-modified cattle are being produced; however, there has yet to be a broad effect on cow genetic improvement. To date, cattle have been developed to provide milk with pharmaceutically useful human proteins (Bauman et al., 2006; Robl et al., 2007), mastitis resistance (Liu et al.,2014), and polled features (Carlson et al., 2016). With genome-editing technological advancements, more disease-resistant livestock will arise in the future. The gene-edited animals have shown major improvements in lean meat yield, disease tolerance, and other desirable characteristics. Aside from financial incentives for farmers, genome-edited livestock with precise sequence changes holds great promise for genetic improvement of livestock using the new understanding of the phenotypic influence of genetic sequence variants (Bharati et al., 2020; Dua et al., 2021a). While genome-editing technology has the potential to address many challenges in traditional livestock breeding, it should be used with caution, as with any new technology.

14.2.5 Animal Cloning

Somatic cell nuclear transfer (SCNT), commonly known as animal cloning, is an advanced ART technique used to produce genetically identical copies of an animal without natural or artificial mating. Cloning is a more advanced form of selective assisted breeding in animal husbandry that entails isolating diploid donor somatic cells and their fusion to enucleated oocytes to produce blastocyst stage embryos (Shah et al., 2008; Keefer, 2015). The produced blastocyst is reprogrammed for development of new offspring. SCNT can provide remarkable and fascinating experimental systems for conservation of endangered species, dissemination of desirable genotypes, and investigation of the biological processes involved during early embryonic development (Singla et al., 2015; Dua et al., 2021b).

Since the production of the first cloned animal, Dolly the sheep, more than 20 species have been cloned utilizing SCNT procedures using micromanipulator or handmade nuclear transfer (Keefer, 2015). The idea of handmade cloning, also known as micromanipulator-free SCNT, was introduced in 1998 in which recipient oocytes were enucleated using microblade followed by fusion of two enucleated oocytes with one donor cell (Vajta et al., 2001). Researchers have cloned many livestock species like cows, pigs, buffalo, goats, and sheep using HMC technique (Vajta et al.2008; Selokar et al., 2018). Animal cloning offers a promising alternative to the conventional breeding to speed up the dissemination of elite germplasm, however, the results are not always encouraging, due to poor conception rate, placental abnormalities, early embryonic loss and development abnormality of cloned animals, which are the major challenges (Arnold et al., 2008). Extensive research on farm cloning indicates that a major factor which affects the efficiency of SCNT is incomplete reprogramming of the donor nucleus leading to dysregulation of gene expression with long-lasting effects into adulthood in some surviving somatic cell clones (Yamanaka et al., 2018). To improve the success of animal cloning by minimizing the epigenetic error, the choice of appropriate donor cell type appears to be a major factor (Saini and Selokar, 2018; Sharma et al., 2018). Different types of donor cells, such as cumulus cells, semen-derived epithelial cells, neuron, trophoblastic cells, primordial germ cells, embryonic stem cells, fetal fibroblast and adult fibroblast granulosa cells can be used to produce cloned animals as shown in Figure 14.1. Among the various cells types, skin fibroblast are the most commonly used donor cells due to their easy isolation and robust in nature.

Animal cloning is now being adopted by various private companies in the United States, China, Korea, Argentina, and Australia for commercial production of cloned animal. On January 15, 2008, the USDA approved the utilization of meat and milk procured from cloned animals, and there is no restriction on their movement into the food supply for human and animal consumption. India's first cloned animal – buffalo – the black gold was brought into the world in 2009 (Sammrupa), and thereafter successful pregnancy and production of cloned calves, viz. 'Garima 1', Garima-11 Mahima, Shresth Swaran, Hisar-Gaurav, and Sach-Gaurav have been accomplished using the

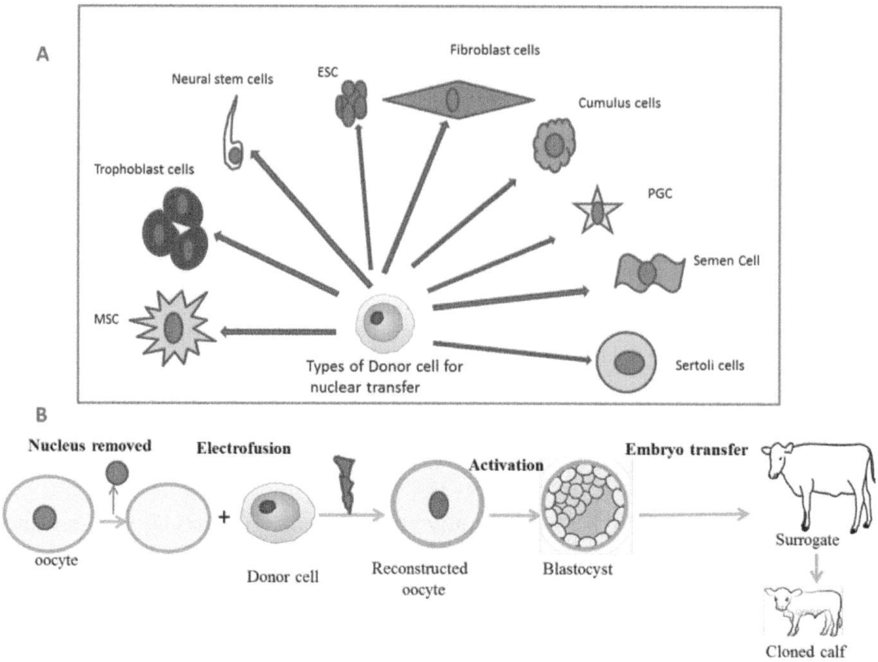

FIGURE 14.1 Different types of donor cells for somatic cell nuclear transfer. (A) Different types of diploid cells such as Sertoli cells, primordial germ cells (PGC), mesenchymal stem cells (MSC), fibroblast cells, embryonic stem cells (ESC), neural stem cells, kidney cells, and trophoblast cells can be utilized in animal cloning and therapeutic research. (B) Somatic cell nuclear transfer process of farm animals.

handmade cloning technique (Singla et al., 2015; Selokar et al., 2019a, 2019b; Yadav et al., 2020). Recently, we have produced seven clones from a single superior buffalo bull and also have recloned an earlier cloned buffalo bull named 'Hisar-Gaurav' (Yadav et al., 2020). In a recent study, the fertility potential of cloned bulls was compared with breeding bulls and found normalcy in their in vitro and in vivo fertilization ability (Saini et al., 2020).

Animal cloning offers potential applications to consumers, farmers, and endangered species, as it allows farmers and stakeholders for rapid multiplication of their desired livestock that produce healthier and safer food product. The advanced application of cloning has opened novel opportunities for genetic engineering, xenotransplantation, animal genetic diversity conservation, and tissue regeneration. Creating clone animals carrying genetic defects can mimic human genetic disorders and are also good models for evaluating long-term responses of the drug treatments. In future, the scope of cloning can be in therapeutics, generating genetically modified animal organs which are suitable models for the study and treatment of human illness.

14.2.6 Stem Cell Technology

Stem cells are unspecialized and undifferentiated cells with the potential to develop into several other types of cells of the body and have the ability of self-renewal. Based on their potency, stem cells are categorized into five (1) totipotent stem cells which can differentiate into all cells of the organism and can form both embryo and extra-embryonic structures, e.g. zygote; (2) pluripotent stem cells which can differentiate into cells of three different germ layers but not extra-embryonic structures, e.g. embryonic stem (ES) cell; (3) multipotent stem cells which can produce only closely related cell of specific lineage, e.g. haematopoietic stem cell: (4) oligopotent stem cells can differentiate into a few different cell types, e.g. myloid stem cell; (5) unipotent stem cells give rise to only one cell type, e.g. dermatocytes (Choudhary et al., 2016). Likewise based on their source, they have been categorized as ES cells, fetal stem cells, and adult stem cells (Kumar et al., 2009; Kumar et al., 2015a). Among these, ES cells are the most potent one, which can give rise to all three embryonic germ layers and possesses the pluripotency markers, i.e. Oct4, Nanog, and sox2. Because of its pluripotent nature, ES cells are regarded as one of the most efficient donor cells in cloning (Shah et al., 2014). Mesenchymal stem cells (MSCs) are adult multipotent stem cells with well-defined characteristics for self-regeneration and differentiation into functional cell type. In recent years, cloning with MSCs has been investigated in different farm animal species (Table 14.1), such as porcine, bovine, caprine, ovine, and equine (Song et al.2015; Gugjoo et al.2018; Gugjoo et al.2019; Zolfaghar et al., 2020). The developmental ability of embryos reconstructed using MSCs has been associated with their high telomeric activity, improved epigenetic reprogramming and pluripotency gene expression profile, which were similar to IVF embryos. Hematopoietic stem cells are best understood adult multipotent stem cells situated in the bone marrow. They can give rise to all other blood cell types. Furthermore, there are fetal derived stem cells from fetal origin, i.e. amniotic fluid, umbilical cord, and blood (Singh et al., 2013; Mann et al., 2013; Ghosh et al., 2015; Ghosh et al., 2016). These cells showed limited potency to maintain in culture and are not able to differentiate to form chimera and did not show germ-line transmission, which categorizes them as stem cell-like cells instead of true stem cells.

More recently, another class of stem cells derived from reprogramming of somatic cells called induced pluripotent stem (iPS) cells using ectopic expression of a set of key reprogramming factors such as Oct4, Nanog, c-Myc, KLF4, etc. (Takahashi and Yamanaka, 2006). These cells are similar to ES cells, showing unlimited self-renewal ability in vitro, morphology, stemness, epigenetic expression, and their capability to differentiate into all somatic cell types including the germ-line. Till date, iPS cells have been effectively produced from different types of cells including fibroblasts, keratinocytes, leukocytes, hepatocytes, neural stem cells, pancreas cells, cord blood cells, and other cell types. This strategy has been quickly and generally received in livestock species and successfully generated iPS cells from pig, sheep, goat, dog, horse, cattle, and buffalo (for review, Kumar et al., 2015b; Navarro et al., 2019; Kumar et al., 2020; Kumar et al., 2021). The iPS cell lines produced from farm animals are practically indistinguishable from their ES cell counterparts,

TABLE 14.1
Different Types of Stem Cells Used as Donor Cell in Animal Cloning

Species	Tissue Origin	Cell Type	Blastocyst Rate	Embryo Transferred Recipient	Pregnancy	Reference
Pig	Bone marrow	HSC	-	302	2	Inoue et al. (2006)
	Adipose tissue	MSC	29.6	14	2	Yang et al. (2016)
	Embryo	EGC	34.5	200	12	Sechar et al. (2017)
	WL-PNF	iPSC	36.7	126	4	Sechar et al. (2017)
Cattle	Amniotic fluid	MSC	17.27	8	1	da Silva et al. (2016)
	Adipose tissue	MSC	25	-	-	Hill et al. (2019)
Buffalo	IVF embryo	ESC	27.3	14	1	George et al. (2011)
Horse	Bone marrow	MSC	45	297	37	Olivera et al. (2018)
Dog	Adipose tissue	MSC	-	82	2	Oh et al. (2011)

WL-PNF = White/Landrace perinatal fibroblast line, IVF = in vitro fertilization, HSC = haematopoietic stem cell, MSC = mesenchymal stem cell, EGC = embryonic germ cell line, iPSC = induced pluripotent stem cell, ESC = embryonic stem cell.

sharing regular signs of pluripotency like colony morphology, unlimited self-renewal ability, and in vitro and in vivo differentiation potentials; however some preliminary reports propose that chimeras and germ-line transmission can likewise be accomplished in sheep and pig (West et al., 2011; Sartori et al., 2012). The successful utilization of stem cells in farm species could be an ideal biological material for progress of veterinary and regenerative medicine. Due to the self-renewal and differentiation abilities of stem cells make it magnificent game changer for innovative medicine, disease modeling, cell-based therapies, drug testing, breed conservation, organ generation, and in vitro meat production (Kumar et al., 2021). These stem cells show great promise in regenerative medicine both in human and livestock species and may consider the best model for study human disease. The number of studies has expanded essentially, and this has been accompanied by breakthroughs in various fields like personalization of therapy, tissue engineering, animal cloning, to preserve endangered species, and chronic and regenerative disease mitigation (Gugjoo et al., 2018). Stem cell is a potential candidate to be used as donor nuclei that minimize the obstacles in SCNT experiments. More extensive research must be performed using stem cells from animals which are presumed to be evolutionarily closer to humans to provide a therapy for a pernicious disease (Kumar et al., 2017; Zakrzewski et al., 2019).

14.2.7 Next-Generation Technique of Cryopreservation

14.2.7.1 Sperm Vitrification

Cryopreservation is a physical mechanism which uses ultra-low temperatures to preserve cell function and its structure via reducing their metabolic activities (Kumar et al., 2016). Cryopreservation is a multi-step and slow process which induces many unfavorable changes in spermatozoa, such as physical damage caused by the intra- and extracellular ice formation, and chemical damage due to osmotic stress. These damages are associated with the use of a variety of cell-permeating cryoprotectants such as glycerol, glycol, dimethylacetaldehyde, and dimethyl sulfoxide which leads to cell injury and also directly affects the spermatozoa's DNA integrity (Ozkavukcu et al., 2008; Kadirvel et al. 2012). Cooling and thawing is also an integral part of the cryopreservation process which have a significant impact on the cell's sustainability through a variety of mechanisms, such as damage to the cell's cytoskeleton, plasma membranes or organelles, or, by directly damaging DNA (John Morris et al., 2012). The overall main limitations of the currently popular slow freezing method of sperm cryopreservation are low conception rate and cryoinjury to damage sperm functionality (Kumar et al., 2017; Jang et al., 2017). Commonly, 20 million sperm are cryopreserved in the traditionally slow freezing method. But, due to huge demand of frozen semen for increasing the coverage of AI of breedable population of the animals in the country, it is the hour of need to reduce the number of spermatozoa per insemination, which ultimately increases the total production of semen utilizing existing facilities. It was observed that when semen has diluted to less than 20 million sperm per dose a dilution effect impacted on the functionality of sperm (Patil et al., 2020; Arjun et al., 2021). The cold shock during long equilibration period (4 h) and ice crystal formation during slow freezing method resulted in poor conception rates of frozen-thawed sperm. A number of studies have been attempted on one or the other semen extenders as well as in cryopreservation protocols of slow freezing method but could not overcome the constant problem of losing about half of the initially motile population after thawing displayed compromised sperm functional activity (Hezavehei et al., 2018; Dalal et al., 2019; Dalal et al., 2020). This ultimately limits the optimum use of a high genetic merit bull for a widespread genetic improvement plan, much required for boosting the dairy economy of country like India.

Recently, an alternative method of sperm cryopreservation was developed which maintained a balance between an increasing concentration of dissolved substances in the freezing solution and intracellular ice crystal formation called 'vitrification'. The first vitrification was performed by Luyet and Hodapp over eight decades ago, in 1938, in frog sperm in liquid air (Luyet and Hodapp, 1938). Following investigation of Luyet and Hodapp, the promising findings of effective vitrification in frog (Hoagland and Pincus, 1942; Smith, 1961), and in fowl (Schaffner, 1942) spermatozoa were not confirmed, and study in this direction was halted for many years. Again in 1985, vitrification was successfully applied in mouse embryos (Rall and Fahy, 1985) using very high concentrations of permeable cryoprotectants and now it is well established and routinely applied for vitrification of embryo and oocytes. In particular, this method entails putting the cells in a medium containing cryoprotectants before exposing them to liquid nitrogen at -196°C to maintain cell functionality and metabolism (Shah et al., 2019). For decades, this method has been used to cryopreserve mammalian gametes, mostly female gametes and embryos. The principal of vitrification is solidification of liquid directly into a glassy state without use of permeable cryoprotectants, which are the main obstacle posed by traditional cryopreservation. In this procedure, there is neither need to equilibrate the sperm for a long period before vitrification nor to remove cryoprotectants after warming/thawing of extended semen. Therefore, researchers are now moving from slow freezing method to much more rapid vitrification to overcome the limitation of slow freezing methods. This was called 'first wind' of vitrification. Unfortunately, this technology failed when applied to vitrification of sperm due to sperm cell inability to tolerate such high levels of cryoprotectants (Gilmore et al., 1997; Liebermann et al., 2002). Therefore, it is not possible to extrapolate the conventional vitrification technique for sperm vitrification.

The 'second wind' of the vitrification was brought with successful vitrification of human spermatozoa without use of any permeable cryoprotectants (like Glycerol, DMSO etc.) by the Isachenkos and their colleagues in Cologne, Germany (Isachenko et al., 2004). This technique has now been named 'kinetic sperm vitrification'. The 'kinetic sperm vitrification' is different with conventional vitrification associated with oocytes and embryos (Katkov et al., 2006). Isachenko group standardized the kinematic sperm vitrification through using cryoloop (Isachenko et al., 2004), and droplet methods (Isachenko et al., 2008). Further, progress has been made to maintain aseptic condition of vitrified sperm in liquid nitrogen; they had developed novel capillary method of sperm vitrification (Isachenko et al., 2012). Using aseptic capillary method of sperm vitrification of human spermatozoa documented birth of two healthy babies following intracytoplasmic sperm injection (ICSI) and a healthy baby born after intrauterine insemination with vitrified spermatozoa (Sánchez et al., 2012). The capillary method was only suitable for IVF and ICSI but not suitable for AI. Hence, they again modified the protocol of kinematic sperm vitrification in 0.25 (Sánchez et al., 2012) and 0.5 ml (Isachenko et al., 2011) straws. After that, many groups of scientists confirmed the kinetic sperm vitrification for human spermatozoa. But the technology was not standardized for domesticated animals like dogs, boar, stallions, rams, goats, cow bulls, and buffalo bulls. However, only a few groups of researchers tried to vitrify dog (Caturla-Sánchez et al., 2018; Cerdeira et al., 2020), boar (Arraztoa et al., 2017 a, b), and ram spermatozoa (Jiménez-Rabadán et al., 2015; Arando et al., 2019; Bóveda et al., 2020) but they failed to vitrify sperm except in a stallion (Hidalgo et al., 2018).

14.2.7.2 Directional Freezing of Spermatozoa

Semen freezing is a process that involves loading semen samples into plastic straws, cooling them at ultra-low temperature, and then storing them in liquid nitrogen tanks. When large numbers of straws are filled, cooled, frozen, fail in evaluation standard, and then ultimately are discarded, it is indeed a costly endeavor. The large number of straws often takes up a lot of space in the liquid nitrogen tanks, which must be maintained on a regular basis. Huge volumes of sperm may be frozen in more condensed suspensions in reusable 'Hollow Tubes' as an alternative (Arav and Saragusty, 2013). The tube can then be washed and sterilized for potential use after the samples are thawed and vacated. The frozen samples can be thawed, repackaged in tubes, and frozen once again. Later on, this technology was called directional freezing. Directional freezing of spermatozoa relies on a basic thermodynamic principle in which ice crystals are precisely regulated via the sample by controlling the velocity of the sample movement through the predetermined temperature gradient using Multi-Thermal Gradient (MTG) device. The MTG device allows for precise control of ice crystal propagation during the freezing process resulting in a significant reduction in cell damage caused during freezing. The directional freezing technique can be used to freeze larger volumes. The cooling rate (G V = 330° C/min) and ice front propagation are precisely regulated by moving the semen in the test tube at a constant velocity (V) through a linear temperature gradient (G = 5.5° C/mm, from 5 to -50°C). At −50°C, seeding is performed automatically for 100 seconds. The semen sample is advanced at a constant velocity via a linear temperature gradient after the initial seeding point and the tubes are plunged into liquid nitrogen (Arav and Saragusty, 2016).

Directional freezing technique has been successfully demonstrated in cattle and rabbit sperm cryopreservation and termed 'double freezing'. Appropriate conception rates in cattle have been demonstrated on a small scale (Arav et al., 2002) and on a large scale (Saragusty et al., 2009). After the second round of freezing, rabbit fertility and kindling rates decreased, but they remained within reasonable limits. Directional freezing can become advantageous under certain conditions. Cogent, the first commercial bovine sperm sex-sorting center, preferred to freeze their sorted spermatozoa using this technology, which is also known as Harmony Freeze. They discovered that when they used this technology instead of traditional freezing in straws, they were able to obtain 16% more live cells.

A number of studies have shown that directional freezing is superior to traditional freezing in straws due to its ability to regulate ice crystal morphology and efficiently dissipate released heat. Even the sperm of stallions which is regarded as "bad freezers" are successfully freezed using the directional freezing method, with 88% of the ejaculates (N = 97) having acceptable post-thaw motility (>35%) compared to just 59% of the ejaculates frozen in straws (Saragusty et al., 2007). To put it another way, the freezing method influences the concept of bad, average, and good freezers in stallions to a large extent. Another research (Zirkler et al., 2005) found no difference between directional and traditional freezing in straws. When the directional freezing technology was used in rabbits, better results were obtained, with fertility rates comparable to those obtained with fresh semen (Si et al., 2006). Using this technology, excellent results of up to 40% post-thaw motility were obtained in a variety of other domestic animals, including domestic fowl, pig, donkey, goat, and ram (Arav et al., 2002; Gacitua and Arav, 2005). Because of its ability to monitor the ice crystal morphology, the homogeneous cooling rate over a large volume sample and continuous seeding during the freezing process, directional freezing is very beneficial for conservation of endangered species germplasm including cryopreservation of ovary.

14.2.7.3 Embryo Cryopreservation

Embryo cryopreservation is the process of preserving an embryo at an ultra-low temperature for maintaining its cellular structure and viability. It is an important step for the preservation and widespread of valuable high-merit animals. The ability to store embryos for essentially indefinite periods of time is a valuable tool for overcoming the fundamental barrier of ET programs, which is the scarcity of suitable recipients. Embryo cryopreservation came into eminence when Whittingham succeeded in cryopreservation of eight-cell mouse embryos in 1972 (Whittingham et al., 1972). Following that, a large number of embryos from a variety of mammalian species were frozen, thawed, and successfully transferred, demonstrating the usefulness of this ART technique. The main advantage of freezing embryos instead of sperm or oocytes is that the embryo comprises the entire genome, i.e., the total number of chromosomes for the individual, and it can be transferred to a foster mother with a known or unknown genetic history without risk of genetic alteration.

Cryopreservation of embryos at low temperature can lead to formation of intracellular and extracellular ice crystal formation and can cause osmotic stress which can damage cellular structure and thus reducing viability (Vajta, 2000). To cryopreserve embryos, a variety of techniques have been devised, including slow or rapid warming, slow freezing, rapid cooling, and vitrification. Traditional equilibrium methods involving freezing machines with controlled cooling rates and cryoprotectants such as glycerol and DMSO are commonly used for embryo cryopreservation. The slow-cooling freezing method cool the embryo from around −5 to −7°C to around −30 to −35°C at a rate of 0.5°C/min permits ice to develop outside the cells as cryoprotectant solutions replace intracellular water before quick submission into liquid nitrogen, reducing the quantity of intracellular ice development (Moore and Hasler, 2017). Currently, vitrification is the most preferred cryopreservation technique, as it is known to provide better embryo and oocyte viability after thawing than any other cryopreservation methods. Vitrification permits cells to dehydrate at ambient temperature and cool rapidly, thus reducing their exposure to ice-nucleation temperatures and preventing the formation of ice crystals (Stoecklein et al., 2021). During vitrification, cells are processed at room temperature with a series of solutes having increasing concentrations of cryoprotectants, and thus allowing the cell to be adequately dehydrated before freezing; after which the cell is immediately plunged into liquid nitrogen, allowing the glassy environment to develop both intracellularly and extracellularly (Argyle et al., 2016).

Embryo cryopreservation is becoming an important part of many embryo transfer initiatives. In fact, 70% of cattle embryos are frozen rather than transported immediately after harvesting. Embryo cryopreservation is a critical tool for the prolonged storage and exchange of elite livestock and endangered species. The cryopreservation of embryos and oocytes has been established to allow global transportation and re-introduction of endangered species. Embryocryo-banks allow for secure and affordable long-term storage, as well as eliminate the requirement for the national and international transportation animals. The switch from fresh to frozen-thawed embryo transfer resulted in a 10–13 percentage point drop in conception rates (Hasler, 2001). Following the transfer of frozen-thawed embryos, the stage of embryo development was also demonstrated to affect conception rates (Hasler, 2011). Embryos selected for cryopreservation should be of the highest possible quality and at the correct stage of cleavage. Furthermore, pregnancy rates associated with frozen/thawed in vitro produced embryos remain lower than those acquired using in vivo counterparts, necessitating the development of a standardized cryopreservation procedure.

14.3 FUTURE PERSPECTIVE

Biotechnology is a major technology for ensuring proper food security for the world's ever-increasing population. For decades, biotechnology research in animal science has focused on genetics and reproduction as basic prerequisite for effective livestock production. Superovulation, AI, cloning, IVEP, cryopreservation, sperm sexing, and embryo transfer are examples of ARTs that have aided animal breeding operations by permitting faster transmission of valuable characteristics in livestock populations in a shorter timeframe than traditional methods. A valuable tool for enhancing the animal breeding environment is the combination of genome editing and cutting-edge reproductive technologies. However, for a variety of reasons, the overall efficiency of modern reproductive technologies remains very low. One of the greatest hurdles for strengthening ART is a poor understanding of germ cell and embryo biology. The study of secretome in combination with metabolomics, proteomics, and genomics will provide a clear understanding of embryonic and cellular biology which can further assist in improving the IVEP. Although several other 'omics' technologies have initially been used, information regarding germ cell biology and embryo protein production is still lacking. Understanding donor nucleus reprogramming, effective cryopreservation procedures for oocytes and embryos, and advancements in the technology for separating X and Y spermatozoa can all aid in the successful implementation of reproductive biotechnological developments in livestock for the benefit of environment, humans, and animals.

BIBLIOGRAPHY

Arando A, Delgado JV, Arrebola FA, León JM, Alcalá CJ, Pérez-Marín CC. Vitrification induces critical subcellular damages in ram spermatozoa. *Cryobiology*. 2019;87:52–59.

Arav A, Saragusty J. Directional freezing of sperm and associated derived technologies. *Anim Reprod Sci*. 2016;169:6–13.

Arav A, Saragusty J. Directional freezing of spermatozoa and embryos. *Reprod Fertil Dev*. 2013;26(1):83–90.

Arav A, Yavin S, Zeron Y, Natan D, Dekel I, Gacitua H. New trends in gamete's cryopreservation. *Mol Cell Endocrinol.* 2002;187(1–2):77–81.

Argyle CE, Harper JC, Davies MC. Oocyte cryopreservation: where are we now? *Hum Reprod Update.* 2016;22(4):440–449.

Arjun V, Kumar P, Dutt R, et al. Is addition or removal of seminal plasma able to compensate for the dilution effect of buffalo semen? *Andrologia.* 2021:e14123. doi: 10.1111/and.14123.

Arnold DR, Fortier AL, Lefebvre R, Miglino MA, Pfarrer C, Smith LC. Placental insufficiencies in cloned animals – a workshop report. *Placenta.* 2008;29Suppl A:S108–S110.

Arraztoa CC, Miragaya MH, Chaves MG, et al. Porcine sperm vitrification I: cryoloops method. *Andrologia.* 2017a;49(7). doi: 10.1111/and.12706

Arraztoa CC, Miragaya MH, Chaves MG, Trasorras VL, Gambarotta MC, Neild DM. Porcine sperm vitrification II: spheres method. *Andrologia.* 2017b;49(8). doi: 10.1111/and.12738

Bauman DE, Mather IH, Wall RJ, Lock AL. Major advances associated with the biosynthesis of milk. *J Dairy Sci.* 2006 Apr;89(4):1235–1243.

Bharati J, Punetha M, Kumar BS, Vidyalakshmi GM, Sarkar M, Michael JD, Singh RK. Genome editing in animals: an overview. In: Genomics and biotechnological advances in veterinary, poultry, and fisheries. *Academic Press* 2020;75–104.

Bóveda P, Toledano-Díaz A, Castaño C, et al. Ultra-rapid cooling of ibex sperm by spheres method does not induce a vitreous extracellular state and increases the membrane damages. *PLoS One.* 2020;15(1):e0227946.

Brackett BG, Bousquet D, Boice ML, Donawick WJ, Evans JF, Dressel MA. Normal development following in vitro fertilization in the cow. *Biol Reprod.* 1982;27(1):147–158.

Brink MF, Bishop MD, Pieper FR. Developing efficient strategies for the generation of transgenic cattle which produce biopharmaceuticals in milk. *Theriogenology.* 2000;53(1):139–148.

Carlson DF, Lancto CA, Zang B, et al. Production of hornless dairy cattle from genome-edited cell lines. *Nat Biotechnol.* 2016;34(5):479–481.

Casillas F, Betancourt M, Cuello C, et al. An efficiency comparison of different in vitro fertilization methods: IVF, ICSI, and PICSI for embryo development to the blastocyst stage from vitrified porcine immature oocytes. *Porcine Health Manag.* 2018;4:16.

Caturla-Sánchez E, Sánchez-Calabuig MJ, Pérez-Gutiérrez JF, Cerdeira J, Castaño C, Santiago-Moreno J. Vitrification of dog spermatozoa: effects of two cryoprotectants (sucrose or trehalose) and two warming procedures. *Cryobiology.* 2018;80:126–129.

Cerdeira J, Sánchez-Calabuig MJ, Pérez-Gutiérrez JF, Hijon M, Castaño C, Santiago-Moreno J. Cryopreservation effects on canine sperm morphometric variables and ultrastructure: comparison between vitrification and conventional freezing. *Cryobiology.* 2020;95:164–170.

Chang MC. In vitro fertilization of mammalian eggs. *J Anim Sci.* 1968;27Suppl 1:15–26.

Choudhary KK, Kavya KM, Jerome A, Sharma RK. Advances in reproductive biotechnologies. *Vet World.* 2016;9(4):388–395.

Crispo M, Mulet AP, Tesson L, et al. Efficient generation of myostatin knock-out sheep using CRISPR/Cas9 technology and microinjection into zygotes. *PLoS One.* 2015;10(8):e0136690.

da Silva CG, Martins CF, Cardoso TC, et al. Production of Bovine Embryos and Calves Cloned by Nuclear Transfer Using Mesenchymal Stem Cells from Amniotic Fluid and Adipose Tissue. *Cell Reprogram.* 2016;18(2):127–136.

Dalal J, Chandolia RK, Pawaria S, et al. Low-density lipoproteins protect sperm during cryopreservation in buffalo: unraveling mechanism of action. *Mol Reprod Dev.* 2020;87(12):1231–1244.

Dalal J, Kumar P, Chandolia RK, et al. A new role for RU486 (mifepristone): it protects sperm from premature capacitation during cryopreservation in buffalo. *Sci Rep.* 2019;9(1):6712.

Diers S, Heise J, Krebs T, Groenewold J, Tetens J. Effect of sexed semen on different production and functional traits in German Holsteins. *Vet Anim Sci.* 2020;9:100101.

Dua S, Bajwa KK, Prashar A, Bansal S, Beniwal M, Kumar P, Pinetha M, Selokar NL, Yadav PS, Kumar D. Empowering of reproductive health of farm animals through genome editing technology. *J Reprod Healthc Med.* 2021a;2(4):1–8.

Dua S, Sharma P, Saini M, et al. Cryobanking of primary somatic cells of elite farm animals – A pilot study in domesticated water buffalo (Bubalusbubalis). *Cryobiology.* 2021b;98:139–145.

Espinosa-Cervantes R, Córdova-Izquierdo A. Sexing sperm of domestic animals. *Trop Anim Health Prod.* 2013;45(1):1–8.

Ferré LB, Kjelland ME, Strøbech LB, Hyttel P, Mermillod P, Ross PJ. Review: recent advances in bovine in vitro embryo production: reproductive biotechnology history and methods. *Animal.* 2020;14(5):991–1004.

Foote RH, Brockett CC, Kaproth MT. Motility and fertility of bull sperm in whole milk extender containing antioxidants. *AnimReprod Sci.* 2002;71(1–2):13–23.

Fukuda Y, Ichikawa M, Naito K, Toyoda Y. Birth of normal calves resulting from bovine oocytes matured, fertilized, and cultured with cumulus cells in vitro up to the blastocyst stage. *BiolReprod.* 1990;42(1):114–119.

Gacitua H, Arav A. Successful pregnancies with directional freezing of large volume buck semen. *Theriogenology.* 2005;63(3):931–938.

Garner DL. Flow cytometric sexing of mammalian sperm. *Theriogenology.* 2006;65(5):943–957.

Garner DL. Sex-Sorting mammalian sperm: concept to application in animals. *J Androl.* 2001;22(4):519–526.

George A, Sharma R, Singh KP, et al. Production of cloned and transgenic embryos using buffalo (Bubalusbubalis) embryonic stem cell-like cells isolated from in vitro fertilized and cloned blastocysts. *Cell Reprogram.* 2011;13(3):263–272.

Ghosh K, Kumar R, Singh J, et al. Buffalo (*Bubalus bubalis*) term amniotic-membrane-derived cells exhibited mesenchymal stem cells characteristics in vitro. *Vitro Cell DevBiol Anim.* 2015;51(9):915–921.

Ghosh K, Selokar NL, Gahlawat SK, Kumar D, Kumar P, Yadav PS. Amnion epithelial cells of buffalo (*Bubalus bubalis*) term placenta expressed embryonic stem cells markers and differentiated into cells of neurogenic lineage in vitro. *AnimBiotechnol.* 2016;27(1):38–43.

Gilmore JA, Liu J, Gao DY, Critser JK. Determination of optimal cryoprotectants and procedures for their addition and removal from human spermatozoa. *Hum Reprod.* 1997;12(1):112–118.

Gosálvez J, Ramirez MA, López-Fernández C, et al. Sex-sorted bovine spermatozoa and DNA damage: I. Static features. *Theriogenology.* 2011;75(2):197–205.

Gugjoo MB, Amarpal A, Sharma GT. Mesenchymal stem cell basic research and applications in dog medicine. *J Cell Physiol.* 2019;234(10):16779–16811.

Gugjoo MB, Amarpal, Chandra V, Wani MY, Dhama K, Sharma GT. Mesenchymal stem cell research in veterinary medicine. *Curr Stem Cell Res Ther.* 2018;13(8):645–657.

Hasler JF. Effect of embryo stage on pregnancy rate following direct transfer of bovine embryos frozen in ethylene glycol. *ReprodFertil Dev.* 2011;24(1):131.

Hasler JF. Factors affecting frozen and fresh embryo transfer pregnancy rates in cattle. *Theriogenology.* 2001;56(9):1401–1415.

He Z, Lu R, Zhang T, et al. A novel recombinant human plasminogen activator: efficient expression and hereditary stability in transgenic goats and in vitro thrombolytic bioactivity in the milk of transgenic goats. *PLoS One.* 2018;13(8):e0201788.

Hezavehei M, Sharafi M, Kouchesfahani HM, et al. Sperm cryopreservation: a review on current molecular cryobiology and advanced approaches. *Reprod Biomed Online.* 2018;37(3):327–339.

Hidalgo M, Consuegra C, Dorado J, et al. Concentrations of non-permeable cryoprotectants and equilibration temperatures are key factors for stallion sperm vitrification success. *AnimReprod Sci.* 2018;196:91–98.

Hill ABT, Bressan FF, Murphy BD, Garcia JM. Applications of mesenchymal stem cell technology in bovine species. *Stem Cell Res Ther.* 2019;10(1):44.

Hoagland H, Pincus G. Revival of mammalian sperm after immersion in liquid nitrogen. *J Gen Physiol.* 1942:25(3):337–344.

Inoue K, Ogonuki N, Miki H, et al. Inefficient reprogramming of the hematopoietic stem cell genome following nuclear transfer. *J Cell Sci.* 2006;119(Pt 10):1985–1991.

Isachenko E, Isachenko V, Weiss JM, et al. Acrosomal status and mitochondrial activity of human spermatozoa vitrified with sucrose. *Reproduction.* 2008;136(2):167–173.

Isachenko V, Isachenko E, Katkov II, et al. Cryoprotectant-free cryopreservation of human spermatozoa by vitrification and freezing in vapor: effect on motility, DNA integrity, and fertilization ability. *BiolReprod.* 2004;71(4):1167 1173.

Isachenko V, Maettner R, Petrunkina AM, et al. Cryoprotectant-free vitrification of human spermatozoa in large (up to 0.5 mL) volume: a novel technology. *Clin Lab.* 2011;57(9–10):643–650.

Isachenko V, Maettner R, Petrunkina AM, et al. Vitrification of human ICSI/IVF spermatozoa without cryoprotectants: new capillary technology. *J Androl.* 2012;33(3):462–468.

Jang TH, Park SC, Yang JH, et al. Cryopreservation and its clinical applications. *Integr Med Res.* 2017;6(1):12–18.

Jiménez-Rabadán P, García-Álvarez O, Vidal A, et al. Effects of vitrification on ram spermatozoa using free-egg yolk extenders. *Cryobiology.* 2015;71(1):85–90.

John Morris G, Acton E, Murray BJ, Fonseca F. Freezing injury: the special case of the sperm cell. *Cryobiology.* 2012;64(2):71–80.

Johnson, L.A., 2000. Sexing mammalian sperm for production of offspring: the state-of-the-art. *Anim Reproduct Sci.* 60:93–107.

Kadirvel G, Periasamy S, Kumar S. Effect of cryopreservation on apoptotic-like events and its relationship with cryocapacitation of buffalo (*Bubalus bubalis*) sperm. *ReprodDomest Anim.* 2012;47(1):143–150.

Katkov II, Isachenko V, Isachenko E, Kim MS, Lulat AG-MI, Mackay AM, Levine F. Low- and high- temperature vitrification as a new approach to bio-stabilization of reproductive and progenitor cells. *Int J Refrigeration.* 2006;29(3):346–357.

Keefer CL. Artificial cloning of domestic animals. *ProcNatlAcadSci U S A.* 2015;112(29):8874–8878.

Kumar D, Anand T, Kues WA. Clinical potential of human-induced pluripotent stem cells: perspectives of induced pluripotent stem cells. *Cell Biol Toxicol.* 2017;33(2):99–112.

Kumar D, Anand T, Lalaji SN, et al. Buffalo embryonic, fetal and adult stem cells: progress and challenges. *Agric Res.* 2015a;4:7–20.

Kumar D, Anand T, Singh MK, Chauhan MS, Manik RS. Characterization of embryonic stem cells: a special focus on farm animals. *Indian J Biotechnol.* 2009;8(1):23–32.

Kumar D, Anand T, Talluri TR, Kues WA. Potential of transposon-mediated cellular reprogramming towards cell-based therapies. *World J Stem Cells.* 2020;12(7):527–544.

Kumar D, Anand T. In vitro embryo production in buffalo: basic concepts. *J Buffalo Sci.* 2012;1(1):50–54.

Kumar D, Kumar P, Singh P, Yadav SP, Yadav PS. Assessment of sperm damages during different stages of cryopreservation in water buffalo by fluorescent probes. *Cytotechnology.* 2016;68(3):451–458.

Kumar D, Talluri TR, Anand T, Kues WA. Induced pluripotent stem cells: mechanisms, achievements and perspectives in farm animals. *World J Stem Cells.* 2015b;7(2):315–328.

Kumar D, Talluri TR, Selokar NL, Hyder I, Kues WA. Perspectives of pluripotent stem cells in livestock. *World J Stem Cells.* 2021;13(1):1–29.

Kumar D, Kues WA. Application of genome editing in farm animals. In: Genomics and biotechnological advances in veterinary, poultry, and fisheries. *Academic Press* 2020:131–149.

Li Y, Lian D, Deng S, et al. Efficient production of pronuclear embryos in breeding and nonbreeding season for generating transgenic sheep overexpressing TLR4. *J AnimSciBiotechnol.* 2016;7:38.

Liebermann J, Nawroth F, Isachenko V, Isachenko E, Rahimi G, Tucker MJ. Potential importance of vitrification in reproductive medicine. *BiolReprod.* 2002;67(6):1671–1680.

Liu X, Wang Y, Tian Y, et al. Generation of mastitis resistance in cows by targeting human lysozyme gene to β-casein locus using zinc-finger nucleases. *ProcBiol Sci.* 2014;281(1780):20133368.

Luyet BJ, Hodapp EL. Revival of frog's spermatozoa vitrified in liquid air. *Exp Biol Med.* 1938;39:433–434.

Madan ML, Singla SK, Jailkhani S, Ambrose JD. In vitro fertilization in buffalo and birth of first ever IVF buffalo calf. *Proc III World Buffalo Cong.* 1991;7:11–17.

Mann A, Yadav RP, Singh J, Kumar D, Singh B, Yadav PS. Culture, characterization and differentiation of cells from buffalo (*Bubalus bubalis*) amnion. *Cytotechnology.* 2013;65(1):23–30.

Marin DFD, de Souza EB, de Brito VC, et al. In vitro embryo production in buffaloes: from the laboratory to the farm. *Anim Reprod.* 2019;16(2):260–266.

Moore SG, Hasler JF. A 100-year review: reproductive technologies in dairy science. *J Dairy Sci.* 2017;100(12):10314–10331. doi: 10.3168/jds.2017-13138

Mustafa MI, Makhawi AM. SHERLOCK and DETECTR: CRISPR-Cas systems as potential rapid diagnostic tools for emerging infectious diseases. *J Clin Microbiol.* 2021;59(3):e00745–20.

Navarro M, Soto DA, Pinzon CA, Wu J, Ross PJ. Livestock pluripotency is finally captured in vitro. *Reprod Fertil Dev.* 2019;32(2):11–39.

Neculai-Valeanu AS, AritonAM. Game-changing approaches in sperm sex-sorting: microfluidics and nanotechnology. *Animals (Basel).* 2021;11(4):1182.

Ni W, Qiao J, Hu S, et al. Efficient gene knockout in goats using CRISPR/Cas9 system. *PLoS One.* 2014;9(9):e106718.

Oh HJ, Park JE, Kim MJ, et al. Recloned dogs derived from adipose stem cells of a transgenic cloned beagle. *Theriogenology.* 2011;75(7):1221–1231.

Olivera R, Moro LN, Jordan R, et al. Bone marrow mesenchymal stem cells as nuclear donors improve viability and health of cloned horses. *Stem Cells Cloning.* 2018;11:13–22.

Ozkavukcu S, Erdemli E, Isik A, Oztuna D, Karahuseyinoglu S. Effects of cryopreservation on sperm parameters and ultrastructural morphology of human spermatozoa. *J Assist Reprod Genet.* 2008;25(8):403–411.

Patil S, Kumar P, Singh G, et al. 'Semen dilution effect' on sperm variables and conception rate in buffalo. *Anim Reprod Sci*. 2020;214:106304.

Paul A, Bharati J, Punetha M, et al. Transcriptional regulation of thrombospondins and its functional validation through CRISPR/Cas9 mediated gene editing in corpus luteum of water buffalo (Bubalus Bubalis). *Cell Physiol Biochem*. 2019;52(3):532–552.

Pieterse MC, Kappen KA, Kruip TA, Taverne MA. Aspiration of bovine oocytes during transvaginal ultrasound scanning of the ovaries. *Theriogenology*. 1988;30(4):751–762.

Proudfoot C, Carlson DF, Huddart R, et al. Genome edited sheep and cattle. *Transgenic Res*. 2015;24(1):147–153.

Pukazhenthi BS, Wildt DE. Which reproductive technologies are most relevant to studying, managing and conserving wildlife? *Reprod Fertil Dev*. 2004;16(1–2):33–46.

Punetha M, Kumar S, Paul A, et al. Deciphering the functional role of EGR1 in Prostaglandin F2 alpha induced luteal regression applying CRISPR in corpus luteum of buffalo. *Biol Res*. 2021;54(1):9.

Rall WF, Fahy GM. Ice-free cryopreservation of mouse embryos at −196°C by vitrification. *Nat Commun*. 1985;313:573–575.

Ranjhan SK, Pathak NN. *Textbook on buffalo production* (No. Ed. 3). New Delhi: Vikas Publishing House Pvt. Ltd, 1993.

Robl JM, Wang Z, Kasinathan P, Kuroiwa Y. Transgenic animal production and animal biotechnology. *Theriogenology*. 2007;67(1):127–133.

Saini M, Selokar NL. Approaches used to improve epigenetic reprogramming in buffalo cloned embryos. *Indian J Med Res*. 2018;148 (Suppl):S115–S119.

Saini M, Sheoran S, Vijayalakshmy K, et al. Semen parameters and fertility potency of a cloned water buffalo (*Bubalus bubalis*) bull produced from a semen-derived epithelial cell. *PLoS One*. 2020;15(8):e0237766.

Sánchez R, Risopatrón J, Schulz M, Villegas JV, Isachenko V, Isachenko E. Vitrified sperm banks: the new aseptic technique for human spermatozoa allows cryopreservation at -86 °C. *Andrologia*. 2012;44(6):433–435.

Saragusty J, Gacitua H, Pettit MT, Arav A. Directional freezing of equine semen in large volumes. *Reprod Domest Anim*. 2007;42(6):610–615.

Saragusty J, Gacitua H, Zeron Y, Rozenboim I, Arav A. Double freezing of bovine semen. *Anim Reprod Sci*. 2009;115(1–4):10–17.

Sartori C, DiDomenico AI, Thomson AJ, et al. Ovine-induced pluripotent stem cells can contribute to chimeric lambs. *Cell Reprogram*. 2012;14(1):8–19.

Secher JO, Liu Y, Petkov S, et al. Evaluation of porcine stem cell competence for somatic cell nuclear transfer and production of cloned animals. *Anim Reprod Sci*. 2017;178:40–49.

Seidel GE Jr. Lessons from reproductive technology research. *Annu Rev Anim Biosci*. 2015;3:467–487.

Selokar NL, Saini M, Kumar D, et al. 10 years after the birth of India's first cloned farm animal, where is buffalo cloning heading. *Indian J Biotechnol*. 2019a;18:191–192.

Selokar NL, Sharma P, Krishna A, et al. Establishment of a somatic cell bank for Indian buffalo breeds and assessing the suitability of the cryopreserved cells for somatic cell nuclear transfer. *Cell Reprog*. 2018;20(3):157–163.

Selokar NL, Sharma P, Saini M, et al. Successful cloning of a superior buffalo bull. *Sci Rep*. 2019b;9(1):11366.

Shaffner CS. Longevity of fowl spermatozoa in frozen condition. *Science*. 1942;96(2493):337.

Shah D, Rasappan, Shila, Gunasekaran K. A simple method of human sperm vitrification. *MethodsX*. 2019;6:2198–2204.

Shah RA, George A, Singh MK, et al. Hand-made cloned buffalo (*Bubalus bubalis*) embryos: comparison of different media and culture systems. *Cloning Stem Cells*. 2008;10(4):435–442.

Shah SM, Saini N, Ashraf S, Kaur R, Chauhan MS. Stem cell and cloning research in farm animals: a special focus on buffalo. *J Med Sci*. 2014;1(4):163–173.

Sharma P, Yadav AS, Selokar NL, Kumar D, Dhaka SS, Yadav PS. Epigenetic status of buffalo fibroblasts treated with sodium butyrate a chromatin remodeling agent. *Tissue Cell*. 2018;50:51–58.

Si W, Hildebrandt TB, Reid C, et al. The successful double cryopreservation of rabbit (Oryctolagus cuniculus) semen in large volume using the directional freezing technique with reduced concentration of cryoprotectant. *Theriogenology*. 2006;65(4):788–798.

Singh J, Mann A, Kumar D, Duhan JS, Yadav PS. Cultured buffalo umbilical cord matrix cells exhibit characteristics of multipotentmesenchymal stem cells. *In Vitro Cell Dev Biol Anim*. 2013b;49(6):408–416.

Singh P, Kumar D, Kumar P. et al. Cryopreservation and quality assessment of buffalo bull semen collected from farmer's doorstep. *Agric Res*. 2013a;2:148–152.

Singla SK, Selokar NL, Saini M, Palta P, Chauhan MS, Manik RS. Buffalo cloning: what we have achieved so far. *Current Sci.* 2015;109(4):670–671.

Smith AU. *Biological effects of freezing and supercooling.* London: Edward Arnold Ltd, 1961.

Song Z, Cong P, Ji Q, et al. Establishment, Differentiation, Electroporation and Nuclear Transfer of Porcine Mesenchymal Stem Cells. *Reprod Domest Anim.* 2015;50(5):840–848.

Stoecklein KS, Ortega MS, Spate LD, Murphy CN, Prather RS. Improved cryopreservation of in vitro produced bovine embryos using FGF2, LIF, and IGF1. *PLoS One.* 2021;16(2):e0243727.

Takahashi K, Yamanaka S. Induction of pluripotent stem cells from mouse embryonic and adult fibroblast cultures by defined factors. *Cell.* 2006;126(4):663–676.

Vajta G, Korösi T, Du Y, et al. The Well-of-the-Well system: an efficient approach to improve embryo development. *Reprod Biomed Online.* 2008;17(1):73–81.

Vajta G, Lewis IM, Hyttel P, Thouas GA, Trounson AO. Somatic cell cloning without micromanipulators. *Cloning.* 2001;3(2):89–95.

Vajta G. Vitrification of the oocytes and embryos of domestic animals. *Anim Reprod Sci.* 2000;60–61:357–364.

van Berkel PH, Welling MM, Geerts M, et al. Large scale production of recombinant human lactoferrin in the milk of transgenic cows. *Nat Biotechnol.* 2002;20(5):484–487.

Wang K, Ouyang H, Xie Z, et al. Efficient Generation of Myostatin Mutations in Pigs Using the CRISPR/Cas9 System. *Sci Rep.* 2015;5:16623.

Wang M, Sun Z, Yu T, et al. Large-scale production of recombinant human lactoferrin from high-expression, marker-free transgenic cloned cows. *Sci Rep.* 2017;7(1):10733.

West FD, Uhl EW, Liu Y, et al. Brief report: chimeric pigs produced from induced pluripotent stem cells demonstrate germline transmission and no evidence of tumor formation in young pigs. *Stem Cells.* 2011;29(10):1640–1643.

Whittingham D, Leibo S, Mazur P. Survival of mouse embryos frozen to −196 and −269 C. *Science.* 1972;178(4059):411–414.

Whitworth KM, Rowland RR, Ewen CL, et al. Gene-edited pigs are protected from porcine reproductive and respiratory syndrome virus. *Nat Biotechnol.* 2016;34(1):20–22.

Whyte JJ, Prather RS. Genetic modifications of pigs for medicine and agriculture. *Mol Reprod Dev.* 2011;78(10–11):879–891.

Wu H, Wang Y, Zhang Y, et al. TALE nickase-mediated SP110 knockin endows cattle with increased resistance to tuberculosis. *Proc Natl Acad Sci U S A.* 2015;112(13):E1530-E1539.

Yadav PS, Selokar NL, Kumar D, et al. ICAR-CIRB produces seven clones of a superior buffalo breeding bull. *Current Sci.* 2020;119(7):1077.

Yamanaka KI, Yamashita K, Khatun H, et al. Normal DNA methylation status in sperm from a somatic cell cloned bull and their fertilized embryos. *Anim Sci J.* 2018;89(10):1406–1414.

Yang Z, Vajta G, Xu Y, et al. Production of pigs by hand-made cloning using mesenchymal stem cells and fibroblasts. *Cell Reprogram.* 2016;18(4):256–263.

Zakrzewski W, Dobrzyński M, Szymonowicz M, Rybak Z. Stem cells: past, present, and future. *Stem Cell Res Ther.* 2019;10(1):68.

Zhang R, Liu F, Hunt P, et al. Transcriptome analysis unraveled potential mechanisms of resistance to Haemonchuscontortus infection in Merino sheep populations bred for parasite resistance. *Vet Res.* 2019;50(1):7.

Zirkler H, Gerbes K, Klug E, Sieme H. Cryopreservation of stallion semen collected from good and poor freezers using a directional freezing device (Harmony CryoCare – Multi thermal Gradient 516). *Anim Reprod Sci.* 2005;89:291–294.

Zolfaghar M, Mirzaeian L, Beiki B, et al. Wharton's jelly derived mesenchymal stem cells differentiate into oocyte like cells in vitro by follicular fluid and cumulus cells conditioned medium. *Heliyon.* 2020;6(10):e04992.

15 Proteomics-Based Advancements in Research toward Sustainable Production of Dairy Livestock

Sudarshan Kumar, K Mohanty, JK Kauhsik, M Mukesh and Nikunj Tyagi

CONTENTS

15.1 INTRODUCTION

The livestock sector is growing leaps and bound globally. The contribution from livestock is multifarious and it is difficult to quantitate the benefits from different ramifications of the livestock sector. Dairy industries, piggeries, poultry farming, beef industries, goat and sheep rearing, etc., have been progressing at speeds which have never been seen before. The dairy sector is the major contributor in GDP from livestock.

Global milk production in 2019 remained a record 395,000 million tons per annum. The largest producer of milk, India contributes more than 175 million metric tons every year (2019). In India, this wouldn't have been possible without the progressively large number of milk-producing animals across different states. According to the livestock census of 2019, the number of domestic bovine animals, including cattle, buffalo, mithun (gayal) and yak, has reached around 303 million. The demand for milk is increasing with an ever-increasing global population and with ever-increasing improvement in social and health standards. The per capita availability of milk in India is 337 gram/day as compared to average world per capita availability of 229 gram/day. The recommended milk intake ranges from 300 to 750 ml per day in different countries (WHO, 2003; USDA, 2010). The Indian Council of Medical Research (ICMR) recommends an average daily intake of 300 ml per day of milk while the national average per capita consumption is hardly about 185 gm and 146 gm a day respectively in urban and rural India. Based on estimates by the National Dairy Development Board (NDDB), the demand for milk is likely to reach 180 million tons by 2022.

DOI: 10.1201/9781003220831-15

The data on milk production, its future production potential, demand, availability, and real consumption indicates that there is a serious need to understand the whole system engaged in milk production and consumption. Though we may produce enough milk to feed the global population, unless the purchasing power of the individual is strong, the deficit between per capita availability and per capita consumption will remain higher.

15.2 CHALLENGES IN ENHANCING FUTURE PRODUCTION

Through the decades of the 70s to 90s the country rode on the initiatives of operation flood, which thoroughly overhauled the country's production potential, which was based on adoption of artificial insemination, cross breeding, establishing cooperative structures of channels between rural milk producers and urban consumers, and modernization of feed processing art. Those decades witnessed the growth rate in milk sector to the tune of 4–5% per annum with an exceptional growth of 6.28% during 2014–16, which was three times more than the world average growth of 2.2%. However, the growth has slumped down to less than 3% in recent years. There are several issues which may be obvious or non-obvious. For example, the dairy industry has seen a serious fall in investment in last few years. The shift to urban life has resulted in a dwindling number of dairy holders in rural areas. There has been scarcity of quality feed and fodder because of shrinkage in arable area available for grazing and growing fodder.

To meet the future demand it is expected that there should be an average incremental increase of 5 million tons per annum over the next 15 years. There has been a tremendous shift in our approaches to accelerate the production in our country, which are multidimensional in nature. The liberalization policy on the economic front from the year 2002 onward has encouraged private players to invest in the dairy sector. The involvement of Reliance, Pepsi and Coca-Cola are noteworthy.

Among the various factors regulating the milk production in India, the increasing demand is at top position which dictates other subordinate factors (Figure 15.1). The increasing population and the inclination for a healthy and diet conscious lifestyles are the spontaneous trigger for the increased demand of milk. Government policies are critical; they are based on nationwide sensors of demand and challenges. It addresses the problem at all fronts, whether it is at the production level or consumer level or at the market level. It perceives the problem and enables focused and intensive measures to be taken at a higher level. The white revolution as outcome of the operation flood is a glaring example of government policy-based adoption of cross breeding in the country.

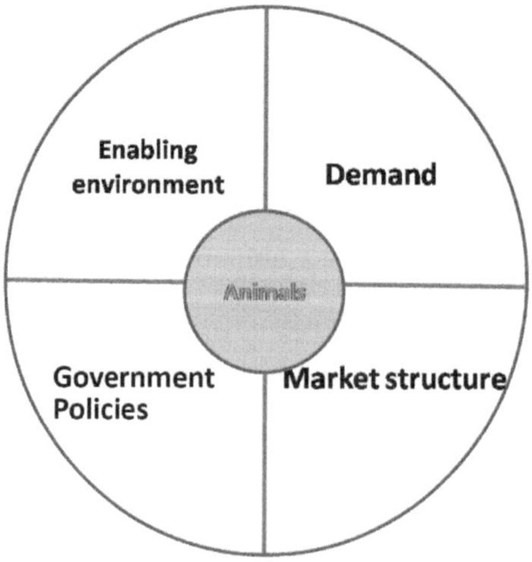

FIGURE 15.1 Factors regulating the milk production in India.

15.3 IMPORTANCE OF INFORMATION-BASED POLICY

Research at various fronts is the most important dimension among the regulatory factors. For any progressive country, sustained success is dependent on its ability to adapt to change. And the adaption to change is never possible without research. Research is a knowledge-based exercise which requires financial investment and time. The world is engaged in various researches at every front to improve its skill to fully tap the potential of the existing and dwindling resources. In the dairy sector, although researches are happening in all the front areas of milk production, marketing and consumption, it is beyond the scope of this chapter to discuss all of them. Here we are discussing the salient research in the area of proteomics-based approaches to understand the problem of milk production and answers to the significant problems faced by the dairy sector (Figure 15.2).

Proteomics is a branch of emerging science that allows the study of a spectrum of proteins present in a given tissue or fluid (the proteome). It has contributed significantly in our understanding of the dairy animals and its physiology, milk production and processing in finer ways than it was possible through gross observation. It is encouraging that researchers in animal sciences have started employing proteomics for generating novel information about the animal production potential, which includes health, diagnostics, milk processing, etc. In Europe there has been creation of a European Cooperation in Science and Technology (COST) Action on Farm Animal Proteomics (FAP) (www.cost-faproteomics.org), which has formed a network among the researchers demonstrating that proteomics can and will have a highly significant role to play in farm animal management for the production of quality milk to meet the nutritional needs of the world. In India, there are no such initiative; however, researchers are actively using proteomics to address the issues related to milk production in indigenous animals for pregnancy, estrous, and infectious diseases.

15.4 PROTEOMICS IN FEED, FODDER, AND RUMINANT DIGESTION

Feed and fodder plays a critical role in dairy husbandry. The productivity of an animal is largely dependent on the type of nutrition it receives. In contrast to the initial periods, where there was no dearth of pasture land and plenty of agricultural waste was available, animals used to feed on quality grass, feed and fodder, resulting in the production of very high quality milk. However, with shrinkage in pasture land, dairy husbandry became more intensive, where most of the feeds were in processed form, including cakes, silage, and hay and industry-made mineral mixtures, etc. The

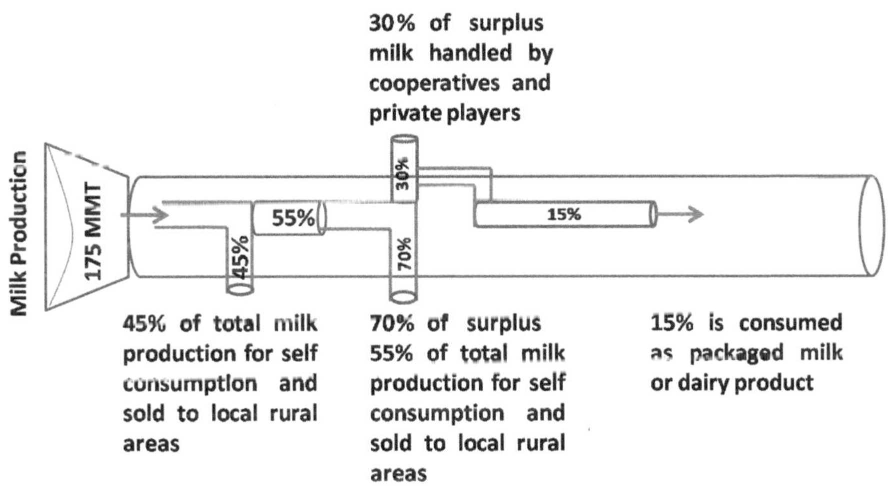

FIGURE 15.2 The schematic representation of milk production and utilization in India in 2019.

nutritional balance and the dietary metabolism in ruminants required a thorough re-evaluation of feeding practices. The recent 15–20 years have produced extensive studies on the nutrition, metabolism, and digestive physiology of dairy animals. Proteomics-based studies have really contributed a lot in the formulation of healthy and nutritious feed for the animals in a more closed husbandry practice. Dai et al. (2017) published a report in the journal *Scientific Reports* and concluded that rice straw diets are of poor quality and severely affect at least four crucial biological processes responsible for reduced milk synthesis in cows, one among them being decreased abundance of transcripts/proteins related to cell growth/development, which impairs growth of the bovine mammary gland. So this study highlights the significance of adaptability in our approach to obtain maximum productivity from our animals by choosing the right kind of balance where mammary gland adapts to low nutrient availability and feeding strategies adapt to more efficiently utilize low-quality forage. Protein is an important component in bovine milk. The milk protein is not only vital for calf growth and development but also offers bioactive proteins and peptides which, when consumed, enhance the health of animals and humans. Research efforts are focusing on factors, such as nutrition, that can influence the quantity and profile of proteins within the bovine milk proteome. A study conducted on 16 jersey cows revealed that cows grazed on alternate forage crops (AFC): a base pasture with strip-tilled with oat (Avena sativa), buckwheat (Fagopyrum esculentum), and chickling vetch (Lathyrus sativus) so that the AFC species comprised 10% of the AFC group's pasture DM intake (DMI), cows grazed on pastures containing AFC had higher αs1-CAS content ($P = 0.005$), and higher relative-abundances of seven low-abundance proteins within the skim milk and MFGM fractions ($P \leq 0.05$) (Scuderi et al., 2018). Negative energy balance (NEB) in cattle directly influences milk composition, with respect to both proteins and lipids. Lu et al. (2013) applied a proteomic and metabolomic approach to study NEB in milk. These authors showed that milk from these cows has an increased quantity of acute-phase proteins, galactose-1-phosphate and unsaturated fatty acid, in comparison with milk from cattle with a good energy balance. These observations have provided some insight into the potential mechanisms involving stomatin and galactose-1-phosphate in NEB dairy cow's milk and demonstrate a relationship between metabolism and the quality of milk.

15.5 PROTEOMICS FOR LOW-YIELDING INDIGENOUS AND EXOTIC COWS

Indigenous animals are comparatively low yielding. The genotypic background is responsible for it. But this factor cannot be regarded as demerit because the same genotype also is responsible for conferring disease resistance and better adaptability to harsh environments and poor nutrition. However, the proteomics-based studies are trying to understand how nutrition, environment, and management can be best utilized to improve the production potential of indigenous dairy animals.

15.6 PROTEOMICS FOR DISEASES IN DAIRY ANIMALS

Mastitis is a major problem in dairy animals. There have been lots of studies on minimizing the occurrence of mastitis and developing new strategies, including both antibiotics and antibiotic-free systems. In India, the studies on indigenous animals are much less. Smolenski et al., 2007, used mass spectrometry-based proteomics to understand the changes in expression of proteins in milk in clinical mastitis. They found at least 15 proteins which were involved in host defense. Dairy industries, especially the milk production system, are seriously affected with certain diseases in cows like mastitis, endometritis (retained placenta, prolapsed uterus), Johne's disease, salmonellosis, and bovine respiratory complex. A large number of production diseases such as, ketosis and milk fever, are also being investigated through proteomics. In addition, the stress-related changes and adaptation to changing climate are at the front of proteomics-based research. Mass spectrometry-aided proteomics that includes 2DE followed by matrix-assisted laser desorption/ionization (MALDI) time-of-flight (TOF)/MS and liquid chromatography coupled with tandem mass spectrometry

(LC-MS/MS), has revolutionized the research approaches leaving past the candidate-specific antibody-based approach. Thus, the understanding of the bovine system as a whole has resulted in the application of systems biology strategies in animal science which was limited to highly sophisticated advanced research in humans.

Our knowledge about the mastitis disease, the host pathogen relationship, and the milk quality deterioration during mastitis has increased significantly due to application of proteomics. Both naturally occurring infections (Hogarth et al., 2004; Smolenski et al., 2007) and experimentally induced infections (Boehmer et al., 2008, 2010a, 2010b; Danielsen et al., 2010) have been studied in depth. In India, National Dairy Research Institute, Karnal, have investigated to find the biomarkers for subclinical mastitis diagnosis (Bathla et al., 2020). These data indicate the involvement of the acute-phase response, oxidative stress, complement activation, protease inhibition and lipid metabolism by the innate immune system to combat infection by pathogens. Milk exosomes (Reinhardt et al., 2012), milk fat globular membrane proteins (Reinhardt and Lippolis, 2008) have been studied intensively using proteomics.

15.7 PROTEOMICS FOR DIAGNOSTICS AND THERAPEUTICS IN COWS

Currently, blood is considered as easy and suitably ample for a diagnosis. However, if milk constituents could be profiled appropriately it has the potential of replacing blood for many diagnostic purposes. Biomarkers for physiology-related changes like pregnancy or estrous, and in various diseases, are now being searched and developed in milk. Pregnancy diagnosis in bovine is still an unmet goal to the dairy farmers. The discovery of protein-based biomarkers in blood, milk, and saliva of dairy animals has been greatly benefited by the application of proteomics. They first time discerned the urine of karan fries cows and found more than 1550 proteins being excreted (Bathla et al., 2015). They also reported a list of proteins in urine that can serve as potential targets for the development of pregnancy diagnostic kit (Rawat et al., 2016).

Understanding the nature and pathogenesis of pathogens involved in infectious diseases is of paramount importance in developing therapeutic strategies. Searching for solutions to the emergence of antibiotic resistance is becoming a challenge to the practioners, farmers, and researchers. Again, proteomics has enlightened us in a better way than the conventional ways and means. Readers are advised to read the reports of Taverna et al. (2007) and Tedeschi et al. (2009), which discuss pathogen response during intramammary infections in cows. A panel of peptides has been reported by Mansor et al. (2013) and Albalat et al. (2014) that can be used for diagnosis of mastitis and identification of causative pathogens.

15.8 PROTEOMICS TO TAP THE FULL POTENTIAL OF MILK

There has been a tremendous shift in our approach to consume milk from its raw state to processed and packaged state, which is still likely to increase in the time to come. Industries engaged in milk processing are challenged with the variations and fluctuation in the milk composition and constituents. The functional attributes of milk are strongly correlated with the type of constituents and their composition. The composition is subject to change due to factors like breeding, feed, age, season, and stage of lactation. Proteomics has enabled looking at the milk constituents in various dairy species not only qualitatively but quantitatively as well.

In a study (D'Ambrosio et al., 2008) on buffalo milk, whey, and milk fat globule membrane fractions, 72 proteins were identified using 2-DE/MALDI-TOF PMF and 1-DE/muLC-ESI-IT-MS-MS. Many of these proteins were identified as being critical in several process industry parameters like coagulation/cheese-making processes. Detection of minor proteins in milk was enabled through highly advanced mass spectrometry-based proteomics that emphasized the importance of low abundant proteins in offering nutrition to the newborn through its major components, and promotion of development of digestive tract and in conferring protection to the neonate.

Traditionally, we are aware of the unique properties of milk from goat, equine, and camel. The science of proteomics has enabled an understanding of the active components and their mechanism of action towards such beneficial attributes. Yang et al. (2013) compared the proteome of cow, yak, buffalo, goat, and camel milk whey. The results revealed significant differences among these animals. Of particular importance were the signature proteins in these species as whey acidic protein (WAP) and quinine oxidoreductase for camel milk, Biglycan for goat milk and F1MK50 for yak milk and clusterin for buffalo milk. Similarly, the effect of heating has been studied on camel milk (Benabdelkamel et al., 2017; Felfoul et al., 2017), which bears important repercussions on milk processing industries. Calvano et al. (2013) described a rapid and sensitive method to detect adulteration in milk, in particular to detect mixtures of powdered milk in liquid milk, both in raw and processed products. Caira et al. (2013) reviewed and described different typical peptides useful to detect different types of milk and adulterations through the detection of α_{S1}-CN variants. Several recent reviews have presented the application of proteomics in milk science, from description of a bovine Peptide Atlas (Bislev et al., 2012) to milk production, as well as on the quality and safety of milk (Roncada et al., 2012).

15.9 MILK SAFETY, ADULTERANTS, AND MILK-BORNE PATHOGENS

Proteomics can help improve the detection of pathogens in food samples – for example, MALDI-TOF MS for diagnostic microbiology is a successful example of how proteomics can answer this global challenge. Geng et al. (2006) used proteomics to detect one of the very dangerous food-borne pathogens, Listeria monocytogenes, directly on selective-enrichment broth. Cheese quality heavily depends on the type and quality of protein in the milk. Hinz et al. (2012) used a 2DE-based proteomic approach to correlate the differences in the proteolysis of milk proteins during lactation stages. While comparing the proteome of cheddar cheese made from the milk collected at different times of lactation, the authors identified some proteins that could be useful to assess influence of season on the quality. Wedholm et al. (2008) described protein markers of the cheese yield through proteomic analysis. In particular, they highlighted several proteins related to the production as a specific β-CN fragment, an isoform B of β lactoglobulin, and other whey proteins as lactoferrin and vitamin D-binding proteins. Regarding the role of casein fragments in poorly coagulating milk for cheese production, another interesting proteomic work has been published (Jensen et al., 2012). Reduced levels of phosphorylation of α_{S1}-CN form (α_{S1}-CN 8P) are related to poorly coagulating and non-coagulating milk, together with decreased levels of glycosylated κ-CN forms.

15.10 CONCLUSION

In a rapidly growing world, ensuring food security and quality nutrition is challenging. With constantly dwindling resources, it is even more challenging to sustain the acceleration in the dairy sector. Working in micro environments has generated new insight into our understanding of lactation physiology. Milk yield from indigenous animals is relatively less in comparison to the exotic breeds, but, with the use of advanced technology like proteomics, Mass spectrometry, next generation sequencing, and several omics platforms, we are optimistic that the production potential can be increased to a desirable level. The major policy-level decision similar to the one as initiated in Europe will definitely exert a powerful push to the production and processing industry.

BIBLIOGRAPHY

Albalat A, Husi H, Siwi J, Nally J, Eckersall PD and Mullen W. 2014. Capillary electrophoresis interfaced with a mass spectrometer (CE-MS): technical considerations and applicability for biomarker studies in animals. *Curr Prot Pept Sci* 15, 23–35.

Bathla S, Rawat P, Baithalu R, Yadav ML, Naru J, Tiwari A, Kumar S, Balhara AK, Singh S, Chaudhary S, Kumar R, Lotfan M, Behare P, Phulia SK, Mohanty TK, Kaushik JK, Nallapeta S, Singh I, Ambatipudi

SK, Mohanty AK. 2015. Profiling of urinary proteins in Karan Fries cows reveals more than 1550 proteins. *J Proteomics* 127, 193–201.

Bathla S, Sindhu A, Kumar S et al. 2020. Tandem Mass Tag (TMT)-based quantitative proteomics reveals potential targets associated with onset of Sub-clinical Mastitis in cows. *Sci Rep* 10, 9321. https://doi.org/10.1038/s41598-020-66211-6

Benabdelkamel H, Masood A, Alanazi IO, Alzahrani DA, Alrabiah DK, AlYahya SA and Alfadda AA. 2017 Apr. Proteomic profiling comparing the effects of different heat treatments on camel (camelus dromedarius) milk whey proteins. *Int J Mol Sci* 18(4), 721. https://doi.org/10.3390/ijms18040721

Bislev SL, Deutsch EW, Sun Z, Farrah T, Aebersold R, Moritz RL, Bendixen E and Codrea MC. 2012. A bovine peptideatlas of milk and mammary gland proteomes. *Proteomics* 12, 2895–2899.

Boehmer J, Bannerman D, Shefcheck K and Ward J. 2008. Proteomic analysis of differentially expressed proteins in bovine milk during experimentally induced Eschcrichia colimastitis. *J Dairy Sci* 91, 4206–4218.

Boehmer J, Ward J, Peters R, Shefcheck K, McFarland M and Bannerman D. 2010a. Proteomic analysis of the temporal expression of bovine milk proteins during coliform mastitis and label-free relative quantification. *J Dairy Sci* 93, 593–603.

Boehmer JL, DeGrasse JA, McFarland MA, Tall EA, Shefcheck KJ, Ward JL and Bannerman DD. 2010b. The proteomic advantage: label-free quantification of proteins expressed in bovine milk during experimentally induced coliform mastitis. *Vet Immunol Immunopathol* 138, 252–266.

Caira S, Pinto G, Balteanu VA, Chianese L and Addeo F. 2013. A signature protein-based method to distinguish Mediterranean water buffalo and foreign breed milk. *Food Chem* 141, 597–603.

Calvano CD, Monopoli A, Loizzo P, Faccia M and Zambonin C. 2013. Proteomic approach based on MALDI-TOF MS to detect powdered milk in fresh cow's milk. *J Agricult Food Chem* 61, 1609–1617

Dai W, Chen Q, Wang Q, White RR, Liu J and Liu H. 2017. Complementary transcriptomic and proteomic analyses reveal regulatory mechanisms of milk protein production in dairy cows consuming different forages. *Sci Rep* 7, 44234. https://doi.org/10.1038/srep44234

D'Ambrosio C, Arena S, Salzano AM, Renzone G, Ledda L and Scaloni A. 2008 Sep. A proteomic characterization of water buffalo milk fractions describing PTM of major species and the identification of minor components involved in nutrient delivery and defense against pathogens. *Proteomics* 8(17), 3657–366. https://doi.org/10.1002/pmic.200701148

Danielsen M, Codrea MC, Ingvartsen KL, Friggens NC, Bendixen E and Rontved CM. 2010. Quantitative milk proteomics – host responses to lipopolysaccharide-mediated inflammation of bovine mammary gland. *Proteomics* 10, 2240–2249.

Felfoul I, Jardin J, Gaucheron F, Attia H, Ayadi MA. 2017. Proteomic profiling of camel and cow milk proteins under heat treatment. *Food Chem* 216, 161–169. https://doi.org/10.1016/j.foodchem.2016.08.007

Geng T, Hahm BK and Bhunta AK. 2006. Selective enrichment media affect the antibody-based detection of stress-exposed Listeria monocytogenes due to differential expression of antibody-reactive antigens identified by protein sequencing. *J Food Prot* 69, 1879–1886.

Hinz K, O'Connor PM, O'Brien B, Huppertz T, Ross R and Kelly AL. 2012. Proteomic study of proteolysis during ripening of Cheddar cheese made from milk over a lactation cycle. *J Dairy Res* 79, 176–184.

Hogarth CJ, Fitzpatrick JL, Nolan A, Young FJ, Pitt A and Eckersall PD. 2004. Differential protein composition of bovine whey: a comparison of whey from healthy animals and from those with clinical mastitis. *Proteomics* 4, 2094–2100.

Jensen H, Poulsen N, Andersen K, Hammershoj M, Poulsen H and Larsen L. 2012. Distinct composition of bovine milk from Jersey and Holstein-Friesian cows with good, poor, or noncoagulation properties as reflected in protein genetic variants and isoforms. *J Dairy Sci* 95, 6905–6917.

Lu J, Antunes Fernandes E, Páez Cano AE, Vinitwatanakhun J, Boeren S, van Hooijdonk T, van Knegsel A, Vervoort J and Hettinga KA. 2013. Changes in milk proteome and metabolome associated with dry period length, energy balance, and lactation stage in postparturient dairy cows. *J Proteome Res* 12(7), 3288–3296.

Mansor R, Mullen W, Albalat A, Zerefos P, Mischak H, Barrett DC, Biggs A and Eckersall PD. 2013. A peptidomic approach to biomarker discovery for bovine mastitis. *J Proteomics* 85, 89–98.

Rawat P, Bathla S, Baithalu R, Yadav ML, Kumar S, Ali SA, Tiwari A, Lotfan M, Naru J, Jena M, Behere P, Balhara AK, Vashishth R, Singh I, Dang A, Kaushik JK, Mohnaty TK and Mohanty AK. 2016. Identification of poteintial protein biomarkers for early detection of pregnancy in cow urine using 2D DiGE and Label free quantitation. *Clin Proteom* 13(1), 1–14.

Reinhardt TA and Lippolis JD. 2008. Developmental changes in the milk fat globule membrane proteome during the transition from colostrum to milk. *J Dairy Science* 91(6), 2307–2318.

Reinhardt TA, Lippolis JD, Nonnecke BJ and Sacco RE. 2012. Bovine milk exosome proteome. *J Proteom* 75(5), 1486–1492.

Roncada P, Piras C, Soggiu A, Turk R, Urbani A and Bonizzi L. 2012. Farm animal milk proteomics. *J Proteom* 75, 4259–4274.

Scuderi RA. 2018. *The effects of diet on the bovine milk proteome*. Graduate College Dissertations and Theses, 846. https://scholarworks.uvm.edu/graddis/846

Smolenski G, Haines S, Kwan FY-S, Bond J, Farr V, Davis SR, Stelwagen K and Wheeler TT. 2007. Characterisation of host defence proteins in milk using a proteomic approach. *J Proteome Res* 6, 207–215.

Taverna F, Negri A, Piccinini R, Zecconi A, Nonnis S, Ronchi S and Tedeschi G. 2007. Characterization of cell wall associated proteins of aStaphylococcus aureusisolated from bovine mastitis case by a proteomic approach. *Vet Microbiol* 119, 240–247.

Tedeschi G, Taverna F, Negri A, Piccinini R, Nonnis S, Ronchi S and Zecconi A. 2009. Serological proteome analysis of Staphylococcus aureusisolated from sub-clinical mastitis. *Vet Microbiol* 134, 388–391.

U.S. Department of Agriculture and U.S. Department of Health and Human Services. 2010. *Dietary guidelines for Americans*. 7th Edition, Washington, DC: U.S. Government Printing Office, December 2010. Available at www.dietaryguidelines.gov. Accessed on 8 November 2011.

Wedholm A, Moller H, Stensballe A, Lindmark-Mansson H, Karlsson A, Andersson R, Andren A and Larsen L. 2008. Effect of minor milk proteins in chymosin separated whey and casein fractions on cheese yield as determined by proteomics and multivariate data analysis. *J Dairy Sci* 91, 3787–3797.

WHO. 2003. *Global strategy for infant and young child feeding*. Available at www.who.int/child_adolescent_health/topics/prevention_care/child/nutrition/global/en/. Accessed on 8 November 2011.

Yang Y, Bu D, Zhao X, Sun P, Wangand J and Zhou L. 2013. Proteomic analysis of cow, yak, buffalo, goat and camel milk whey proteins: Quantitative differential expression patterns. *J Proteome Res* 12(4), 1660–1667. https://doi.org/10.1021/pr301001m

16 Buffalo Cloning
Multiplying India's Black Gold

Naresh L Selokar, Manoj Kumar Singh, Dharmendra Kumar, Rakesh Sharma, Manmohan Singh Chauhan and Prem Singh Yadav

CONTENTS

16.1 INTRODUCTION

Domesticated buffaloes (*Bubalus bubalis*) contribute significantly to global milk and meat production. According to the latest statistics, the world's buffalo population is 208 million heads, of which 96.79% are present in Asian countries such as India, China, and Pakistan (Minervino et al., 2020). Globally, the buffalo has contributed 15.14% of the fresh whole milk; whereas, cattle has contributed 80.05% (FAOSTAT, 2019). In Asia, the buffalo has contributed 35.59% of the fresh whole milk; whereas, cattle have contributed 59% (FAOSTAT, 2019). With huge contributions, the buffalo is considered a key animal for livestock farming in Asian countries.

In India, buffalo plays a significant role in agricultural economies and provides employment to millions of small and marginal farmers. Most Indian farmers rear one or two buffalo under the mixed crop-livestock farming system. In 2018–19, India produced 187.7 million tons of milk, of which 49% (91.7 million tons) was contributed by buffalo (Annual report 2019–20, Department of Animal Husbandry Dairying & Fisheries (DADF) of India, http://dahd.nic.in). In addition to a significant share of India's milk production, buffalo also contributed significantly to the meat industry. Annually, India has been earning more than INR 20,000 Crores (3,000,000,000 USD Millions) through the export of buffalo meat and its products. In 2021–22, India exported the buffalo meat and its products, worth INR 24,613.24 Crores (3,303.31 USD Millions) (Data obtained from Agricultural and Processed Food Products Export Development Authority (APEDA) of India, http://apeda.gov.in.). A huge contribution of buffalo makes this bovine species a preferred farm animal to boost India's white (milk) and pink (meat) revolution.

To augment the productivity of farm animals, in 2013, the Indian government formulated the National Livestock Policy. This policy aims to increase the productivity, mainly milk, of the livestock in a sustainable manner, while protecting biodiversity and ensuring bio-security and farmers' livelihoods (National Livestock Policy draft was accessed from a website of Department of Animal Husbandry Dairying& Fisheries (DADF) of India, http://dahd.nic.in.). For the buffalo, it has been suggested to upgrade the low producers through breeding with high producers. Also, the nondescript buffalo population is to be upgraded through breeding with pure indigenous breeds. To implement the objectives of the livestock policy, there is need of large-scale coverage of artificial insemination (AI).

DOI: 10.1201/9781003220831-16

For achieving high AI coverage, the high genetic merit disease-free bulls need to be produced at a large number, either through conventional breeding or through the use of assisted reproductive technologies (ARTs). Among the ARTs, AI is the only technology that has a wider reach at the village level and has been used to produce high productive next-generation animals. In recent years, other ARTs such as embryo transfer (ET), Ovum Pick-Up and in vitro fertilization (OPU-IVF) have been exploited for the faster multiplication of elite animals, particularly breeding bulls and bull mothers. Animal cloning, one of the ARTs, also offers faster multiplication of known genotype animals. In India, over the last few years, the Indian Council of Agricultural Research (ICAR) has been working on the development of animal cloning technology in buffalo. India has made significant progress in the field of buffalo cloning technology (BCT) and produced 25 cloned buffaloes (Selokar et al., 2018; Yadav et al., 2020). The current aim of BCT is to produce multiple genetic copies of elite buffalo bulls. In this chapter, we discuss the status of BCT in India and its applications in buffalo breeding.

16.2 JOURNEY OF BUFFALO CLONING TECHNOLOGY

In the late 1990s, BCT research in India was initiated by researchers of ICAR-National Dairy Research Institute (NDRI), Karnal (Singla et al., 1997). Through dedicated efforts for over a decade, in 2009, NDRI's researchers produced the world's first cloned riverine buffalo, and afterward, several cloned buffaloes were produced by using different types of somatic cells (Selokar et al., 2018). This technology involves electrofusion of a somatic cell with an enucleated oocyte (nucleus free oocyte), followed by chemical activation. The electrofused embryos are cultured in vitro under a control gas incubator for seven to eight days. The cultured embryos then developed to blastocysts, which are transfered into recipient buffaloes (each buffalo receives one or two blastocysts) to produce cloned calves (Selokar et al., 2018).

A produced cloned calf has similar genetic makeup as the animal, from which somatic cells are used as nuclear donors. To produce cloned buffaloes, we adopted a simple and efficient technology called 'hand-made cloning' (HMC), which does not require costlier and skilled manpower-based micromanipulators (steps of buffalo HMC are presented in Figure 16.1). Over the decades, we

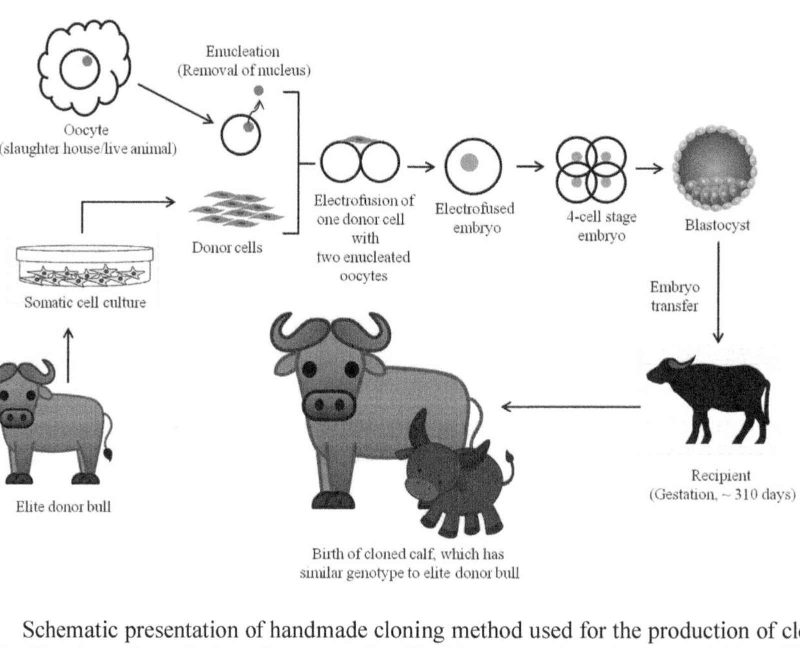

FIGURE 16.1 Schematic presentation of handmade cloning method used for the production of cloned buffaloes. This method involves electrofusion of one somatic cell with two enucleated oocytes, followed by chemical activation and in vitro culture of activated embryos for seven to eight days. Cultured embryos then developed to blastocysts, which transfer into recipient buffaloes (each buffalo receives one or two blastocysts) to produce cloned calves. This technology is well-established and several cloned buffaloes have been produced in India.

made several modifications, including optimization of in vitro culture condition (Shah et al., 2008) and electrofusion conditions (Selokar et al., 2012b), cell cycle synchronization (G1/G0 stage) of donor cells before nuclear transfer (Selokar et al., 2012a), use of low oxygen tension for embryo culture (Saini et al., 2015), brilliant cresyl blue staining to select competent oocytes (Mohapatra et al., 2015a), and selection of early cleaved embryos for embryo transfer (Kaith et al., 2015). These modifications have improved the invitro blastocyst production rates, which are now 30–35% of electrofused embryos. In a search of best donor cell types for large-scale application, we used several donor cell types such as skin-derived fibroblast cells (Shah et al., 2009; Selokar at al., 2019), urine-derived epithelial cells (Madheshiya et al., 2015), milk-derived epithelial cells (Golla et al., 2012), blood-derived lymphocytes (Jyotsana et al., 2016). semen-derived epithelial cells (Selokar et al., 2014), and embryo-derived stem cells (George et al., 2011; Mohapatra et al., 2015b). In our studies, we found that skin-derived fibroblast cells are the most suitable donor cell type for BCT applications. At present, we use fibroblast cells derived from a skin biopsy from an underneath part of the tail just above the anus; this part is the least exposed to sunlight. By incorporation of several modifications that are mentioned earlier, recently, we could achieve live birth rates of up to 15 to 20% (Yadav et al., 2020; Shyam et al., 2020). In Table 16.1, we summarize cloned buffaloes available in India.

Currently, two research organizations, namely NDRI and ICAR-Central Institute for Research on Buffaloes, Hisar (CIRB), are working on BCT in India. Very recently, CIRB's researchers

TABLE 16.1
Buffalo Clones Available in India and Their Status

Sex of Donor	Age of Donor	Cell Type	Month of Birth	Current Status	Reference
Female	Embryonic stem cell	IVF embryonic stem cell	August 2010	Delivered calves	George et al. (2011)
	Adult	Ear skin fibroblast	May 2014		Jyotsana et al. (2015)
	Adult	Urine-derived epithelial cell	February 2015		Madheshiya et al. (2015)
Male	Fetal	Ear skin fibroblast	August 2010	Producing semen and progenies	Saha et al. (2013)
	Adult	Fresh semen-derived epithelial cell	March 2013		Selokar et al. (2014)
	Adult	Frozen thawed semen-derived epithelial cell	July 2014	Producing semen	Raja et al. (2019)
	Adult	Tail skin fibroblast	December 2015	Producing semen and progenies	Selokar et al. (2019)
	Adult*	Assamese buffalo tail skin fibroblast	December 2017	Producing semen	Selokar et al. (2018)
	Adult	Tail skin fibroblasts	Seven cloned copies of a superior bull. These were produced between October 2019 and March 2020	Producing semen	Yadav et al. (2020)
	Adult		Re-cloned calf of a cloned bull (born in January 2020)	Producing semen	Yadav et al. (2020)
	Adult		Three cloned calves of superior bulls (These clones were born from June to December 2020)	Producing semen	Unpublished data (NDRI)

At present, India has 16 cloned bulls which can produce more than two million semen doses for breeding.
*represent a cloned bull of Assamese buffalo, rest clones belong to Murrah breed which is a famous for high milk productivity.

FIGURE 16.2 Seven clones (left to right) of a single breeding bull and a recloned calf (rightmost) of cloned bull. These clones were born between October 2019 and January 2020.

reported two commendable successes in buffalo cloning: (1) production of seven cloned copies of a single buffalo bull, and (2) successful birth of a recloned calf of a cloned bull. These achievements are the first published reports in the field of BCT research worldwide (Figure 16.2). At present, six cloned bulls are available at the NDRI and ten cloned bulls are available at the CIRB (Table 16.1). Recent long-term scientific studies demonstrated that the semen of cloned bulls has qualified semen fertility tests and healthy calves have been produced using AI (Selokar et al., 2019; Saini et al., 2020). Also, cloned females have normal reproduction cycles and delivered calves through conventional breeding. Results of these studies indicate that cloned buffaloes can be used as founder animals to produce next-generation offspring using the application of ARTs (AI, ET, and OPU-IVF). According to authors' personal experiences, clones that attended maturity have normal growth, health, and reproduction; however, more scientific data on clones' longevity and performance, and progeny performance, are needed before recommending their uses in classical breeding. Our efforts are continuing to generate more data on clones and their progeny.

16.3 APPLICATION OF BUFFALO CLONING TECHNOLOGY

India bestows the world best buffalo breeds (e.g. Murrah, one of the most prestigious buffalo breeds). According to the livestock population census report of 2019, India has109 million buffaloes, that accounts for over 56% of the world's buffalo population. The average milk yield of Indian buffalo is around 5 to 6 kgs per day; however, some buffaloes have high milk productivity, 20 to 25 kgs per day. Farmers who are engaged in buffalo husbandry prefer the breeding of their female buffaloes with the semen of superior bulls that have good pedigree records. In India, the National Dairy Plan (NDP) has been initiated to increase the productivity of dairy animals (buffaloes and cows). As per NDP, higher productivity can be achieved by expanding coverage of AI, from an existing 30% coverage to 80–100% coverage. It is anticipated that approximately 100 million frozen doses will be required annually (900 bulls) during the first phase, and 140 million frozen doses annually (1200 bulls) by the end of the second phase (Selokar 2018). Therefore, there is a huge requirement for quality bulls for semen production. In addition, the livestock population census of 2019 (population data was accessed from http://dahd.nic.in/about-us/divisions/statistics) was alarming in that the male buffalo population had significantly decreased by 42.35% compared to the previous population census (in 2012, 16 million males, which are now shrunken to 9 million in 2019). However, the population of milk producing buffaloes has increased by 4.34% (38.16 million in 2019). If the decreasing trend in the male population continues, then, there will be an acute shortage of males for natural mating and artificial insemination. Therefore, urgent scientific and technological interventions are needed to save the male population that could later be used as bulls.

Since, India has well-established BCT that can be used to make multiple copies of breeding bulls in the shortest possible time to mitigate semen demand. In Figure 16.3, we represent a schematic

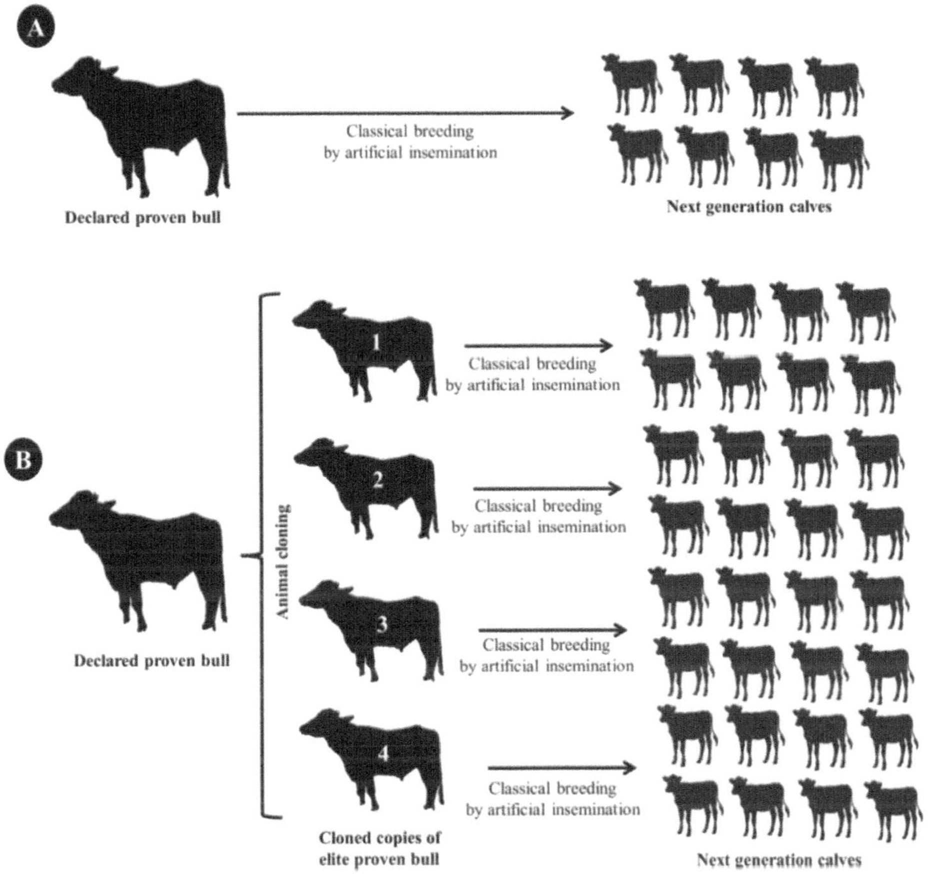

FIGURE 16.3 Use of buffalo cloning technology (BCT) in buffalo breeding. (A) In classical breeding, a bull is selected based on pedigree records of its parents, and its semen is used to produce next-generation calves. Based on the performance of daughters, a bull is ranked and recommended for nominated mating to produce next-generation elite animals. This process, called progeny testing, requires eight to ten years to declare a bull to be proven. Also, a limited calf crop is achieved through this breeding strategy. (B) BCT-based breeding, in which a proven bull can be selected to produce its multiple copies. Since the cloned copies have identical genotype to a proven bull, they can directly be used to produce next-generation calf crop through classical breeding. This way, a large number of superior genotype calves can be produced in the shortest possible time. To address the issue of inbreeding, the cloned copies should not be used in the locations where a proven donor bull is already used, and the locations that have a large population of nondescript buffalo can be targeted to upgrade them to pure population. Importantly, BCT is the only available method to restore the full genotype of any declared proven bull, even after death.

Source: Image from Selokar (2018)

presentation of the use of the BCT for buffalo breeding. There are some specific advantages of bull cloning over the females, such as (1) thousands of progeny can be produced from clones of elite bulls in a year to have a significant impact on faster dissemination of superior genetics, (2) cloned copies of elite bulls can compensate for the demand of their semen, (3) restoration of the genetics of elite bulls that had culled due to aging or lameness, and (4) restoration of progeny-tested bulls that died years ago, but, their semen is available in the semen bank.

To understand the impact of BCT on genetic improvement, for an example, 50 cloned bulls (five copies each of ten top-ranked breeding bulls) can produce more than five million semen doses that are sufficient to produce two to three million next-generation buffaloes. It has been speculated that the use of cloned animals can increase the inbreeding rates in the population. To address the inbreeding issues, we suggest that pure breed (for example, Murrah) cloned bulls can be appropriately used to upgrade the nondescript buffalo population (which is humungous, 43% of total buffalo population, approximately 46 million) sharing 14–15% of total milk contributed by buffalo. This way BCT can play a key role in the upgrade of nondescript buffalo to pure population. There is need to encourage efforts towards the use of BCT to upgrade the nondescript buffalo for enhancing the economics of Indian buffalo farmers.

16.4 CHALLENGES AND OPPORTUNITIES

Despite the potential application of BCT, this technology has suffered from poor success in terms of live, healthy births. The major challenges are low conception rates, high pregnancy losses, and mortality of born clones that need to be overcome through scientific interventions (Saini et al., 2018). The major factor that affects the success of cloning is abnormal genomic reprogramming (Saini et al., 2016). The genomic defects such as abnormal DNA methylation and histone modifications have been observed in cloned embryos of buffalo (Saini et al., 2016). Also, cloned embryos have higher apoptosis rates as compared to invitro produced IVF embryos (Saini et al., 2017; Selokar et al., 2017; Agrawal et al., 2018). Therefore, attempts have been made to correct the genomic defects in the cloned embryos by culturing them under appropriate culture conditions (for example, low O2 tension), treatment of donor cells/embryos with epigenetic modifiers, and pathway modulators (Saini et al., 2017; Selokar et al., 2017; Agrawal et al., 2018; Rashmi et al., 2019; Sah S et al., 2020).

A very recent study, Shyam et al. (2020) reported that Dickkopf-1, an inhibitor of canonical WNT signaling pathway, treatment of cloned embryos increased the blastocyst, conception, and live birth rates. In this study, a 25% pregnancy rate has been reported; however, the exact cause of abnormal genomic reprogramming is yet to be identified. Therefore, further attempts are required to improve the quality of produced embryos for higher overall success rates. We observed that some born clones had died within the first three months of their age and they displayed various organ abnormalities such as enlarged liver, heart, and kidney, and were easily prone to infection (Saini et al., 2018). To have a healthy post-natal life, there is a need to develop a package of practices for care and management of born clones so that they can survive and not easily succumb to any infections.

In addition to the production of quality embryos, the appropriate selection of the recipient animals is essential to achieve more success. In our personal experience, due to the paucity of quality recipients, many embryos are forced to transfer into reproductively unsound animals to perceive clone pregnancies. Previous studies suggested that the poor health condition of recipients significantly compromised the conception rates of transferred embryos (Jacob et al., 2012; Hansen et al., 2020). Therefore, the selection of recipients is the most crucial to achieve higher success rates. Buffalo cloning researchers have been relying on scientific information that is available in cattle and pigs, despite the fact that buffalo is a different species in terms of evolution, genetics, physiology, and reproduction. There should be consistent efforts are required to generate buffalo-specific data that will help to unravel the cellular and molecular mechanisms behind the abnormal genomic reprogramming in cloned embryos.

In addition to the application of BCT in classical breeding, to date, animal cloning is the sole technology used to produce genetically modified designer animals. Recent genome editing (GE) tools such as zinc-finger nucleases, transcription activator-like effector nucleases, and clustered regularly interspaced short palindromic repeats (CRISPR)/Cas9 have opened new avenues to produce designer animals with improved production and health (Lee et al., 2020). Some of the examples are the production of polled (hornless) cattle (Young et al., 2020), improvement of udder health through the expression of human lysozyme gene in the mammary gland (Liu et al., 2014), and improvement of wool fiber quality in goats and sheep by modifying the FGF5 gene (Li et al., 2017, 2019). Very recently, in buffalo, GE tools have also been used to modify the GDF8 gene, which is a negative regulator of muscle growth, in cultured fibroblast cells (Su et al., 2018). In cattle, sheep, goats, and pigs, it has been proved that knockout of the GDF8 gene led to double muscle characteristics (Proudfoot et al., 2015; He et al., 2018; Wu et al., 2018; Ding et al., 2020). Keeping the potentials of GE tools with BCT, more attempts are needed to explore GE tools to produce buffalo model animals for profitable farming in future.

16.5 CONCLUSION

Buffalo contributed significantly to make India rank first in global milk production. However, this valued species has been neglected for a long period due to poor breeding management and lack of awareness. The genetic improvement of buffalo using the application of ARTs is less implemented as compared to cattle. However, in India, during the last two decades, buffalo species have gained significant importance in the field of ARTs due to the development of BCT. BCT is a well-established technology and several cloned buffaloes have been produced. BCT has a great potential to propagate selected genetics for improving milk and beef production. Research studies reported that buffalo clones have normal growth, health, fertility, and production. However, there is a lack of data regarding the application of BCT worldwide; primarily due to the fact that very few laboratories have been working on BCT. Considering the requirement of quality semen in India, we believe that BCT can be used for large-scale production of breeding bulls for breeding, especially to upgrade the nondescript buffalo population. Indian government must look carefully toward this promising technology and can start developing strategies for how to implement BCT for the improvement of the buffalo population. Also, other countries, where the buffalo population dominates, can also explore BCT for the production of productive animals.

16.6 ACKNOWLEDGEMENTS

Buffalo cloning work in India is supported by Indian Council of Agricultural Research, Delhi, Department of Biotechnology, Delhi, and Department of Science and Technology, Delhi. Authors highly acknowledged contributions of several students and research staff that had been working in two Indian institutions, ICAR-National Dairy Research Institute, Karnal, and ICAR-Central Institute for Research on Buffaloes, Hisar.

BIBLIOGRAPHY

Agrawal H, Selokar NL, Saini M, Singh MK, Chauhan MS, Palta P, Singla SK, Manik RS. (2018). m-carboxycinnamic acid bishydroxamide improves developmental competence, reduces apoptosis and alters epigenetic status and gene expression pattern in cloned buffalo (Bubalusbubalis) embryos. *Reprod. Domest. Anim.* 53(4):986–996.

Ding Y, Zhou SW, Ding Q, Cai B, Zhao Xe., Zhong S, Miao Jin M, Wang XL, Bao-hua MA, Chen YL. (2020). The CRISPR/Cas9 induces large genomic fragment deletions of MSTN and phenotypic changes in sheep. *J. Integr. Agric.* 19:1065–1073.

FAO. FAOSTAT. (2019). Available online at: www.fao.org/faostat/en/# data/QA (accessed May 15, 2020).

George A, Sharma R, Singh KP, Panda SK, Singla SK, Palta P, Manik R, Chauhan MS. (2011). Production of cloned and transgenic embryos using buffalo (Bubalus bubalis) embryonic stem cell-like cells isolated from in vitro fertilized and cloned blastocysts. *Cell Reprogram.* 13:263–272.

Golla K, Selokar NL, Saini M, Chauhan MS, Manik RS, Palta P, Singla SK. (2012). Production of nuclear transfer embryos by using somatic cells isolated from milk in buffalo (Bubalusbubalis). *Reprod. Domest. Anim.* 47(5):842–848.

Hansen PJ. (2020). The incompletely fulfilled promise of embryo transfer in cattle-why aren't pregnancy rates greater and what can we do about it? *J Anim Sci.* 1;98(11):skaa288.

He Z, Zhang T, Jiang L, Zhou M, Wu D, Mei J, Cheng Y. (2018). Use of CRISPR/Cas9 technology efficiently targetted goat myostatin through zygotes microinjection resulting in double-muscled phenotype in goats. *Biosci. Rep.* 38:1–8.

Jacob JC, Haag KT, Santos GO, Oliveira JP, Gastal MO, Gastal EL. (2012). Effect of embryo age and recipient asynchrony on pregnancy rates in a commercial equine embryo transfer program. *Theriogenology.* 1;77(6):1159–1166.

Jyotsana B, Sahare AA, Raja AK, Singh KP, Nala N, Singla SK, Chauhan MS, Manik RS, Palta P. (2016). Use of peripheral blood for production of buffalo (Bubalus bubalis) embryos by handmade cloning. *Theriogenology.* 86(5):1318–1324.

Jyotsana B, Sahare AA, Raja AK, Singh KP, Singla SK, Chauhan MS, Manik RS, Palta P (2015). Handmade cloned buffalo (Bubalus bubalis) embryos produced from somatic cells isolated from milk and ear skin differ in their developmental competence, epigenetic status, and gene expression. *Cell Reprogram.* 17(5):393–403.

Kaith S, Saini M, Raja AK, Sahare AA, Jyotsana B, Madheshiya P, Palta P, Chauhan MS, Manik RS, Singla SK. (2015). Early cleavage of hand-made cloned buffalo (Bubalusbubalis) embryos is an indicator of their developmental competence and quality. *Reprod. Domest. Anim.* 50(2):214–220.

Lee K, Uh K, Farrell K. (2020). Current progress of genome editing in livestock. *Theriogenology.* 1;150:229–235.

Li G, Zhou S, Li C, Cai B, Yu H, Ma B, Huang Y, Ding Y, Liu Y, Ding Q, He C, Zhou J, Wang Y, Zhou G, Li Y, Yan Y, Hua J, Petersen B, Jiang Y, Sonstegard T, Huang X, Chen Y, Wang X. (2019). Base pair editing in goat: nonsense codon introgression into FGF5 results in longer hair. *FEBS J.* 286:4675–4692.

Li WR, Liu CX, Zhang XM, Chen L, Peng XR, He SG, Lin JP, Han B, Wang LQ, Huang JC, Liu MJ. (2017). CRISPR/Cas9-mediated loss of FGF5 function increases wool staple length in sheep. *FEBS J.* 284(17):2764–2773.

Liu X, Wang Y, Tian Y, Yu Y, Gao M, Hu G, Su F, Pan S, Luo Y, Guo Z, Quan F, Zhang Y. (2014). Generation of mastitis resistance in cows by targeting human lysozyme gene to β-casein locus using zinc-finger nucleases. *Proc. Biol. Sci.* 19;281(1780):20133368.

Madheshiya PK, Sahare AA, Jyotsana B, Singh KP, Saini M, Raja AK, Kaith S, Singla SK, Chauhan MS, Manik RS, Palta P. (2015). Production of a cloned buffalo (Bubalus bubalis) calf from somatic cells isolated from urine. *Cell. Reprogram.* 17:160–169.

Minervino AHH, Zava M, Vecchio D, Borghese A (2020). Bubalus bubalis: a short story. *Front. Vet. Sci.* 7:570413.

Mohapatra SK, Sandhu A, Neerukattu VS, Singh KP, Selokar NL, Singla SK, Chauhan MS, Manik RS, Palta P. (2015a). Buffalo embryos produced by Hand-made cloning from oocytes selected using BCB staining have better developmental competence and quality, and are closer to embryos produced by in vitro fertilization in terms of their epigenetic status and gene expression pattern. *Cell. Reprogram.* 17:141–150.

Mohapatra SK, Sandhu A, Singh KP, Singla SK, Chauhan MS, Manik R, Palta P. (2015b). Establishment of trophectoderm cell lines from buffalo (Bubalus bubalis) embryos of different sources and examination of in vitro developmental competence, quality, epigenetic status and gene expression in cloned embryos derived from them. *PLoS One.* 10(6):e0129235.

Proudfoot C, Carlson DF, Huddart R, Long CR, Pryor JH, King TJ, Lillico SG, Mileham AJ, McLaren DG, Whitelaw CB, Fahrenkrug SC. (2015). Genome edited sheep and cattle. *Transgenic. Res.* 24:147–153.

Raja AK, Sahare AA, Jyotsana B, Priya D, Palta P, Chauhan MS, Manik RS, Singla SK. (2019). Reducing the cytoplasmic volume during hand-made cloning adversely affects the developmental competence and quality, and alters relative abundance of mRNA transcripts and epigenetic status of buffalo (Bubalus bubalis) embryos. *Anim. Reprod. Sci.* 208:106136.

Rashmi Sah S, Shyam S, Singh MK, Palta P. (2019). Treatment of buffalo (Bubalus bubalis) SCNT embryos with microRNA-21 mimic improves their quality and alters gene expression but does not affect their developmental competence. *Theriogenology.* 126:8–16.

Sah S, Sharma AK, Singla SK, Singh MK, Chauhan MS, Manik RS, Palta P. (2020). Effects of treatment with a microRNA mimic or inhibitor on the developmental competence, quality, epigenetic status and gene expression of buffalo (Bubalus bubalis) somatic cell nuclear transfer embryos. *Reprod. Fertil. Dev.* 32:508–521.

Saha A, Panda SK, Chauhan MS, Manik RS, Palta P, Singla SK. (2013). Birth of cloned calves from vitrified-warmed zona-free buffalo (Bubalus bubalis) embryos produced by hand-made cloning. *Reprod. Fertil. Dev.* 25:860–865.

Saini M, Selokar NL, Agrawal H, Singla SK, Chauhan MS, Manik RS, Palta P. (2015). Low oxygen tension improves developmental competence and reduces apoptosis in Hand-made cloned buffalo (Bubalusbubalis) embryos. *Livest. Sci.* 172:106–109.

Saini M, Selokar NL, Agrawal H, Singla SK, Chauhan MS, Manik RS, Palta P. (2016). Treatment of buffalo (Bubalus bubalis) donor cells with trichostatin and 5-aza-2'-deoxycytidine alters their growth characteristic, gene expression and epigenetic status and improves the in-vitro developmental competence, quality and epigenetic status of cloned embryos. *Reprod. Fertil. Dev.* 28:824–837.

Saini M, Selokar NL, Agrawal H, Singla SK, Chauhan MS, Manik RS, Palta P. (2017). Treatment of donor cells and reconstructed embryos with a combination of trichostatin-A and 5-aza-2'-deoxycytidine improves the developmental competence and quality of buffalo embryos produced by handmade cloning and alters their epigenetic status and gene expression. *Cell. Reprogram.* 19:208–215.

Saini M, Selokar NL, Palta P, Chauhan MS, Manik RS, Singla SK. (2018). An update: reproductive handmade cloning of water buffalo (Bubalus bubalis). *Anim. Reprod. Sci.* 197:1–9.

Saini M, Sheoran S, Vijayalakshmy K, Rajendran R, Kumar D, Kumar P, Sharma RK, Kumar R, Mohnaty TK, Selokar NL, Yadav PS. (2020). Semen parameters and fertility potency of a cloned water buffalo (Bubalus bubalis) bull produced from a semen-derived epithelial cell. *PLoS One.* 21;15(8):e0237766.

Selokar NL (2018). Cloning of breeding buffalo bulls in India: initiatives & challenges. *Indian J. Med. Res.* 148(Suppl S1):120–124.

Selokar NL, Saini M, Agrawal H, Palta P, Chauhan MS, Manik R, Singla SK. (2017). Valproic acid increases histone acetylation and alters gene expression in the donor cells but does not improve the in vitro developmental competence of buffalo (Bubalus bubalis) embryos produced by hand-made cloning. *Cell. Reprogram.* 19:10–18.

Selokar NL, Saini M, Muzaffer M, Krishnakanth G, Saha AP, Chauhan MS, Manik R, Palta P, Madan P, Singla SK. (2012a). Roscovitine treatment improves synchronization of donor cell cycle in G0/G1 stage and in vitro development of handmade cloned buffalo (Bubalus bubalis) embryos. *Cell. Reprogram.* 14:146–154.

Selokar NL, Saini M, Palta P, Chauhan MS, Manik R, Singla SK. (2014). Hope for restoration of dead valuable bulls through cloning using donor somatic cells isolated from cryopreserved semen. *PLoS One.* 9:e90755.

Selokar NL, Saini M, Palta P, Chauhan MS, Manik RS, Singla SK. (2018). Cloning of buffalo, a highly valued livestock species of south and Southeast Asia: any achievements? *Cell. Reprogram.* 20:89–98.

Selokar NL, Shah RA, Saha AP, Muzaffar M, Saini M, Chauhan MS, Manik RS, Palta P, Singla SK. (2012b). Effect of post-fusion holding time, orientation and position of somatic cell-cytoplasts during electrofusion on the development of handmade cloned embryos in buffalo (Bubalus bubalis). *Theriogenology.* 78:930–936.

Selokar NL, Sharma P, Saini M, Sheoran S, Rajendran R, Kumar D, Sharma RK, Motiani RK, Kumar P, Jerome A, Khanna S, Yadav PS. (2019). Successful cloning of a superior buffalo bull. *Sci. Rep.* 6;9(1):11366.

Shah RA, George A, Singh MK, Kumar D, Anand T, Chauhan MS, Manik RS, Palta P, Singla SK. (2009). Pregnancies established from handmade cloned blastocysts reconstructed using skin fibroblasts in buffalo (Bubalus bubalis). *Theriogenology.* 71:1215–1219.

Shah RA, George A, Singh MK, Kumar D, Chauhan MS, Manik R, Palta P, Singla SK. (2008). Hand-made cloned buffalo (Bubalus bubalis) embryos: comparison of different media and culture systems. *Cloning. Stem. Cell.* 10:435–442.

Shyam S, Goel P, Kumar D, Malpotra S, Singh MK, Lathwal SS, Chand S, Palta P. (2020). Effect of Dickkopf-1 and colony stimulating factor-2 on the developmental competence, quality, gene expression and live birth rate of buffalo (Bubalus bubalis) embryos produced by hand-made cloning. *Theriogenology.* 157:254–262.

Singla SK, Manik RS, Madan ML. (1997). Micromanipulation and cloning studies on buffalo oocytes and embryos using nucleus transfer. *Indian J. Exp. Biol.* 1997;35:1273–1283.

Su X, Cui K, Du S, Li H, Lu F, Shi D, Liu Q. (2018). Efficient genome editing in cultured cells and embryos of Debao pig and swamp buffalo using the CRISPR/Cas9 system. *Vitro Cell. Dev. Biol. Anim.* 54:375–383.

Wu M, Du L, Liu R, Wei C, Wang Y, Yang L, Yang L, Liu J, Wang Y, Wang C. (2018). Double-muscled phenotype in mutant sheep directed by the CRISPRCas9 system. *Cloning Transgenes.* 7:3–7.

Yadav PS, Selokar, NL, Kumar D, Sharma RK, Kumar P, Kumar R, Saini M, Dua S. (2020). ICAR-CIRB produces seven clones of a superior buffalo breeding bull. *Curr. Sci.* 119(7):1077.

Young AE, Mansour TA, McNabb BR, Joseph T, Josephine B, Titus C, Alison VE. (2020). Genoic and phenotypic analyses of six offspring of a genome-edited hornless bull. *Nat. Biotechnol.* 38:225–232.

17 Applications of Biotechnological Techniques in Mitigating Environmental Concerns

RC Sobti, Apurav Sharma and Sanjeev Kumar Soni

CONTENTS

DOI: 10.1201/9781003220831-17

17.1 INTRODUCTION

We are in a state of dynamic interaction with the environment. We affect the environment and, in turn, the environment affects us. One of the basic components of our living environment, next to air, is water. Human culture and civilizations are virtually linked with water resources. Human indifferences, ignorance, and greed combine globally to waste it. Due to the continuous industrial revolution for the creation of conveniences for humans, several harmful chemicals are released in air and water (by industrial effluents) either without any treatment or with improper treatment.

Human activities such as the addition of chemical fertilizers in the soil, spraying the toxic chemicals on crops, large-scale use of fossil fuels, dumping of organic solid waste at landfill sites, and burning of agricultural waste residues are increasing the atmospheric burden of particulate matters and disturb the gaseous equilibrium. With all these activities, the ratio of carbon dioxide to oxygen in the atmosphere continues to increase at an alarming rate; the greenhouse effect is an inevitable result because of the change in this ratio. There is a need for a long-range environmentally sound approach to industrialization to greatly reduce and even reverse the rate at which the earth is surely becoming an unfit habitat for life. Though this would be ideal, it would not probably be enough. The biological revolution is expected to provide the technological tools to prevent and cure the destructive effects of pollution (Kandpal, 2021).

Today biotechnological processes are playing a significant role in the protection and rehabilitation of the environment and are advantageous alternatives or adduct to physical and chemical treatment technologies. A variety of microbes and their enzymes are employed in a wide range of pollution-control processes and also reclaim waste land (Saravanan et al., 2021). Even certain plants that remove and/or degrade toxic compounds are also being used. Table 17.1 depicts a summary of some of the major types of wastes and pollutants generated by various human activities.

Environment biotechnology is concerned both with its implications and applications in the wider context of the environment because increasing industrialization, urbanization, and other developments are constantly threatening the environment, and the natural resources are

TABLE 17.1
Major Types of Wastes and Pollutants Generated by Various Human Activities

Human Activity	Wastes	Air	Pollutants Surface Water	Soil/Ground Water
Manufacturing				
Chemical industry	Chemical refuse (solid/liquid)	Volatile organic compounds, SOx, NOx,	Organic and inorganic compounds, heavy metals, cyanides	Organic and inorganic compounds, heavy metals, cyanides
Oil refinery	Wastewater	Volatile hydrocarbons, SOx, NOx	Oil, acids, heavy metals, phenolic etc.	Oil, acids, phenolic phenolics, etc.
Fertilizer	Contaminated gypsum	SOx, NOx, NH_3	Inorganic chemicals	Inorganic chemicals
Energy production	Fly ash, heavy metals, radioactive waste, etc.	CO, CO_2, SOx NOx, aerosols	Heat, coal residues, fly ash etc.	Fly ash, coal residues
Agriculture and Dairy				
Agriculture	Crop residues, pesticides, plastics	Pesticides	Pesticides	Pesticides
Dairy	Animal refuse	NH_3, CH_4	NO_3^-, PO_4^{3-} (from manure)	NO_3^-, PO_4^{3-} (from manure), feed residue
Transport	Scrapped vehicles used up oil, grease etc.	CO_2, CO, SOx NOx, volatile hydrocarbons, lead	Oil products	Oil products
House-building and Domestic				
House-building etc.	Building and demolition refuse wood, metals, fly ash	CO_2, CO, SOx NOx, heavy metals, volatile organic compounds	NO_3^-, PO_4^{3-}, O_2 Binding compounds	Metals, wood, fly ash
Domestic	Sewage, garbage, scrapped utensils	CO_2, CO, SOx NOx, C.F.C., CH_4	Percolation water, NO_3^-, PO_4^{3-}	Solid refuse, Percolation water,

being depleted. The most essential and urgent environmental problems which need attention include:

(i) Chemical wastes,
(ii) Ground and surface waters contaminated with chemicals from the increasing use in agriculture,
(iii) Effects on wild life and flora with decreasing diversity because of agricultural use of pesticides,
(iv) Heavy metals such as mercury, cadmium, and lead spread as fertilizers, through gasoline consumption,
(v) Increasing agricultural and municipal organic wastes.

Research is being carried out to develop various processes and techniques to deal with these problems. The biggest impact of biotechnology could be in developing a new strategy for the utilization of natural resources, especially biomass. It could bring a shift away from resource strategies based on exhaustible fossils such as crude oil, coal, and natural gas, towards biomass. An increased application of biotechnology within the framework of this new resource strategy would alter the pressure on the environment and could contribute to harmonizing resource use with the production, consumption, and disposal of wastes.

Environmental biotechnology according to Prof. Moser includes:

(i) **Pollution cleanup** involving cleanup of oil spills and detoxification of contaminated soil, treatment of domestic and industrial wastewater supplies.
(ii) **Pollution control** involving the recovery of heavy toxic metals from mining water, use of enzymes rather than chlorine in pulp and paper manufacture.
(iii) **Pollution protection** involving closed cycling practice at enzyme production plants implies that raw materials are renewable, waste material is a biodegradable sludge that can be used as a local fertilizer.

17.2 MANAGEMENT OF WASTE

For managing the wastes produced by different factories, the following strategies need to be applied:

(a) Minimize the amount of waste generated
(b) Reduce the toxicity of waste that is generated
(c) Find more satisfactory ways of disposing of and degrading wastes that have been generated
(d) Find new alternatives for the treatments causing pollution
(e) Reduce the use of chemical fertilizers and pesticides

Biotechnology is playing an important role in every phase of the above-mentioned points by using the following strategies involving bioprocess technology:

(a) Point source reduction of environmental pollutants by processes like bioremediation, bioaugmentation/replacement of chemical fertilizers with organic manures and biofertilizers/replacement of chemical pesticides with biopesticides/use of enzymes in bioremediation and replacement of harsh chemical processes
(b) Off-site utility of various organic solid wastes and industrial by-products for transformation into a variety of value-added products including compost (organic manure), industrial enzymes, biofuels, biodegradable plastics, etc.
(c) Waste destruction/treatment
(d) Metagenomics and environment protection

17.2.1 Bioprocess Technologies for Waste Management

Bioprocess technologies are a set of technologies that use biocatalysts, which are living cells or their enzymes to:

- Produce a molecular product
- Change one molecule into another
- Degrade a molecule

These offer several advantages over chemical processes, and some of these include the following:

(i) Biocatalysts are the ultimate in sustainability. Microbes and cells reproduce, so they continually renew themselves and their molecular component.
(ii) Unlike many chemical processes, bioprocesses must occur under conditions that are compatible with life. They, however, also produce waste, less destructive to the environment than those of chemical processes.
(iii) Because enzyme-catalyzed reactions are very specific, they generate more undesirable by-products than a chemical process.

Many of the advantages that biocatalysts provide become limitations if the process must be carried out under extreme conditions. Now certain systems that involve biological catalysts to degrade, detoxify or accumulate contaminating chemicals, have been developed. The application of biological agents, mostly microorganisms, for the treatment of environmental chemicals has been directed mostly towards remedial activities. Modern biotechnology techniques have enhanced the performance of these natural processes in pollution control (Sharma et al., 2021). For example, mercury is a highly toxic metal that accumulates in the food chain when released in water. In the Minamata accident, inhabitants of the Japanese island of Kyushu suffered the toxic effects of fish poisoned by mercury-rich industrial effluents. Since naturally thriving mercury-tolerant bacteria are rare and cannot be easily grown in culture, the metallothionein gene has been inserted into *Escherichia coli*, at Cornell University. Such genetically modified bacteria grow well in culture. The genetically engineered bacteria are placed inside bioreactors that efficiently remove mercury from water. The bacteria are later incinerated. The accumulated mercury is then recovered. Existing techniques of mercury removal are expensive and inefficient (Wagner-Döbler, 2003).

Each year more than 2 million tons of oil enter the sea. Half of it is derived from industrial effluents, sewage, and river outflows, and the rest from non-tanker shipping and natural seepage from below the seafloor. Only about 18% of the total comes from refineries, off-shore operations, and tanker activities. Generally, most oils do have relatively low toxicity to the environment. It can, however, have catastrophic and immediate effects on bird and animal life associated with water. To deal with contaminated sites, the following approaches are made:

(a) Assessment of the nature and degree of the hazards,
(b) The choice of remedial action.

Clean-up operation of contaminated soil can involve:

(i) In-site processing,
(ii) *In situ* treatment,
(iii) Off-site processing.

17.3 POINT SOURCE REDUCTION

A large area of the earth's surface and the oceans and other waterways have been contaminated with oil-derived compounds and toxic chemicals. The contamination of soil normally results from a range of activities related to our industrialized society. Contaminated land contains substances that when present in sufficient quantities or concentrations can probably cause harm to human beings, directly or indirectly, and to the environment in general. Many xenobiotic industrially derived compounds can show high levels of recalcitrance and while in many cases, only small concentrations reach the environment, they are biomagnified, i.e., an increase in the concentration as it passes through the food chain (Marghade et al., 2021).

17.3.1 Bioremediation

Biological methods of remediation variously termed bioremediation, biorestoration, bioreclamation or biotreatment are now replacing physical and chemical methods of separation and/or removal of the pollutants. Bioremediation employs living organisms, such as microbes and plants, to extract, eliminate, and/or bind toxins in forms that are not harmful to the environment. These include biostimulation, biotransformation, biostabilization, and biofiltration. These natural processes have been employed for many years to eliminate pollutants (Adams et al., 2015). Different types of pollutants found in the environment are:

Inorganic

Metals: Cd, Hg, Ag, Co, Pb, Cu, Cr, Fe

Radionuclides

Nitrates, nitrites, phosphates, cyanides, asbestos

Organic

Biodegradable: sewage, domestic, agricultural, and process waste
Petrochemical wastes: oil, diesel, BTEX (benzene, toluene, ethylbenzene, xylenes)
Synthetic: pesticides, organohalogens, PAHs (polyaromatic hydrocarbons)

Biological

Pathogens: bacteria, viruses

Gaseous

Gases: SO_2, CO_2, NO_2, methane

Volatiles

Chlorofluorocarbons, volatile organic compounds (VOCs),

They are present as a result of industrial effluents, outfalls, mine waters, landfill run-offs, waste tips, and accidental spillages. These pollutants can be found in all regions of the environment: marine, estuaries, lakes, and soil.

17.3.1.1 Microbial Bioremediation

This can be defined as the natural or managed biological degradation of environmental pollution involving microorganisms. Microbial bioremediation is normally carried out by the indigenous microorganisms and their activity can be enhanced by the supply of nutrients or by enhancing the microbial population in a process known as bioaugmentation (Adams et al., 2015).

The basic principle of microbial bioremediation is to optimize the environmental conditions so that microbial biodegradation can occur rapidly and as completely as possible. For the biological transformation of xenobiotic compounds that are introduced into the ecosystem naturally, present microbes in soil and water environments are potential candidates. Microbial populations in natural environments exist in a dynamic equilibrium. This can, however, be altered by modifying environmental conditions such as the availability of nutrients so that microbes work efficiently (Azubuike et al., 2016).

It has been observed that not the individual strains, but consortia of microorganisms, act on pollutant molecules. The metabolic effects of microorganisms on pollutants can take forms and are not always to the environmental advantage of the ecosystems. *In situ* promotion of microbial growth during bioremediation can be achieved by the addition of nutrients. When the indigenous microbial populations are exposed to specific polluting compounds for longer periods, sub-populations develop a limited metabolic ability to utilize and, thus, degrade the offending pollutant. The growth of these particular microbes, however, will invariably be nutrient-limited. When essential growth nutrients such as nitrogen and phosphorus are added, normally the growth stimulates and there is an increase in the breakdown of pollutants. Nutrient supplementation is done for the following purposes:

(a) To enrich the useful microbes
(b) To scale up from the mixture by bioreactor cultivation
(c) To re-inoculate large quantities of the 'cocktail' of microbes into the contaminated site

This method was successfully applied to beaches along Prince William Sound and the Gulf of Alaska in 1989–1990 to clean up the oil spillage from the oil tanker Exxon Valdez. The application of fertilizers (nutrients) in various formulations to the beaches stimulated the indigenous microorganisms to degrade the oil to less harmful products that subsequently became part of the food chain. This method is used to dechlorinate recalcitrant chemicals such as polychlorinated biphenyls (PCBs) (considered highly toxic and indestructible industrial pollutants).

The extraction of the contaminants and their treatment on the surface in bioreactors, the stimulation of microbial growth *in situ* by providing extra nutrients, and the supplementation of the microbial population are the other processes that can be applied. The indigenous microbial population can be stimulated by the addition of nitrogen and phosphorus-containing substrates (Adams et al., 2015). The addition of a second carbon source to stimulate co-metabolism can be another alternative. Over the years, several *in situ* aerobic co-metabolism processes have been demonstrated. The major problem with this process is to ensure that the substrates added are distributed through the soil and not just close to the injection bore. To reduce recirculation, systems have been investigated for trichloroethylene removal where oxygen and methane, the co-metabolic substrate, were added to the subsoil. Bioremediation to environmental clean-up can be applied as discussed hereafter.

17.3.1.1.1 Microbial Bioremediation of Soils

Soils contain a very large number of microorganisms. These include several hydrocarbon utilizing bacteria and fungi representing 1% of the total population of some 10^4–10^6 cells per gram of soil. In addition, cyanobacteria and algae have also been found to degrade hydrocarbons. Hydrocarbon-contaminated soils have been found to contain more microorganisms than uncontaminated soils, but the diversity of the microorganisms was reduced (Bamforth and Singleton, 2005). The fate of organic compounds in the environment is affected by many factors which can be grouped as (i) affecting the growth and metabolism of the microorganisms and (ii) affecting the degradation of compounds.

Factors affecting the growth of microorganisms are as follows:

- Presence of other biodegradable organic material
- Presence of nitrogen- and phosphorus-containing inorganic compounds
- Oxygen levels
- Temperature
- pH
- Presence of water, soil moisture
- Number of types of microorganisms present
- Presence of heavy metals

Factors affecting the degradation of the compound are as follows:

- Bacterial growth and metabolism
- Chemical structure of the organic compound
- Availability and/or solubility
- Photochemistry

17.3.1.1.2 Microbial Degradation of Hydrocarbons

The biodegradation of the hydrocarbons is associated with microbial growth and metabolism and therefore, any of the factors affecting microbial growth will influence degradation. If the microorganisms cannot use the hydrocarbons as their sole source of energy and carbon skeletons, some other growth substrate will be needed. In some cases, if another substrate is present, the microorganisms may use this in preference to the hydrocarbons (Ławniczak et al., 2020). The microorganisms may also require supplementation with nitrogen and phosphorus-containing compounds as demonstrated in marine conditions. Aerobic degradation of hydrocarbons is considerably faster than the anaerobic process so that a supply of oxygen will be needed to maintain aerobic conditions if rapid degradation is required. Soil with an open structure will encourage oxygen transfer. Waterlogged soil will have the reverse effect. The microbial growth is affected by temperature. At low temperatures, the rate of degradation is generally slow. Nutrient addition to soils at the temperature of 4–10°C has been shown to have little effect, as the low temperature is responsible for reduced growth. The pH of the soil will affect both the growth of microorganisms and the solubility of the compound to be degraded. In the beginning, the presence of a large number of hydrocarbon-degrading microorganisms in the soil will be an advantage. As most soils contain these types of microorganisms, growth soon increases, so that seeding with specific hydrocarbon-degrading organisms is probably not needed. Hydrocarbon contamination may also be associated with high levels of heavy metals which may inhibit microbial growth depending upon the concentration and type of metals.

17.3.1.1.2.1 Pathways of Degradation of Hydrocarbons

Soil microorganisms degrade petrochemicals, PAHs, and BTEX compounds. These microorganisms use them as a source of carbon and energy for cell synthesis. Hydrocarbons are stable reduced compounds, their degradation generally proceeds by oxidation under either aerobic or anaerobic conditions. Many attempts have been made to study the microbial degradation of monocyclic and polycyclic aromatic hydrocarbons. Aliphatic hydrocarbons are converted into alcohols and then sequentially oxidized to carboxylic acids which are β-oxidized (Bamforth and Singleton, 2005).

Monocyclic aromatics are first hydroxylated by a dioxygenase enzyme to cis- 1,2-dihydroxy-1,2-dichlorobenzene, which is then converted to catechol.

One of the two pathways can be taken by the subsequent metabolism of catechol:

(a) Ortho cleavage yields cis, cis-muconate,
(b) Meta cleavage however yields 2-hydroxymuconic semialdehyde. Both pathways lead to compounds that can enter the Krebs cycle.

The first two stages of the degradation of benzene are common for the breakdown of many other monocyclic and polycyclic aromatic hydrocarbons (Figure 17.1).

In the general aromatic ring, hydroxylation is followed by ring cleavage, and the oxygenases carry out both of these reactions. The incorporation of two oxygen molecules causes the introduction of two hydroxyl groups which can undergo either meta- or ortho-cleavage. The incorporation of a single oxygen molecule is catalyzed by monooxygenases and both enzymatic systems can be used to degrade polycyclic aromatic hydrocarbons (Figure 17.2).

Many of the monocyclic aromatic hydrocarbon-degrading bacteria have chromosomal genes coding for the enzymes of the pathway for the degradation of hydrocarbons, but plasmids have been found that code for the degradation of compounds such as camphor, octane, toluene, naphthalene, and some herbicides and pesticides.

Some plasmids code only for some parts of the pathway. It is helpful to the soil to be able to transfer the degradative ability between bacteria, as this allows the rapid adaptation of the population to a particular compound. The best-studied plasmid-based pathway has been the toluene degradation by the bacterium *Pseudomonas putida* mt-2 and the Plasmid TOL. The TOL plasmid

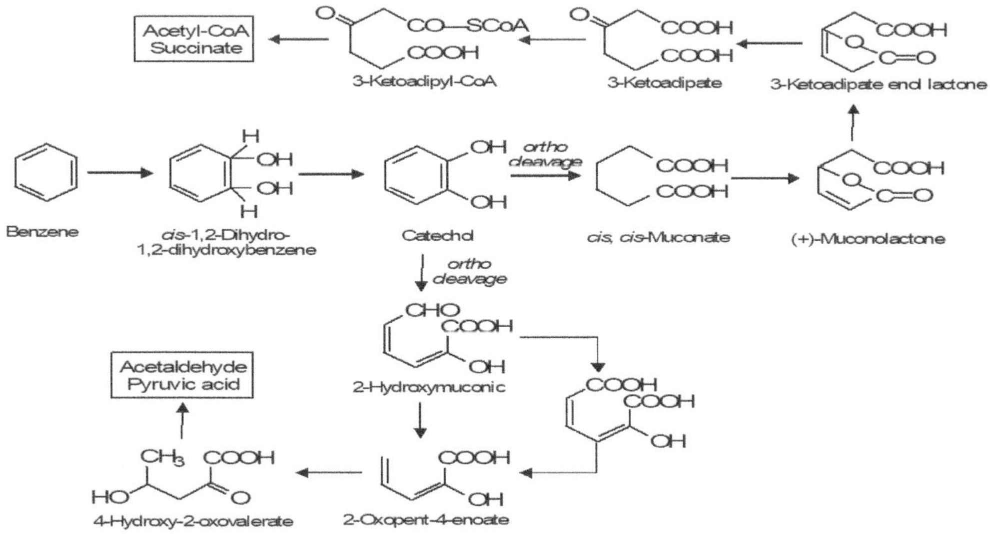

FIGURE 17.1 Catechol metabolism pathways.

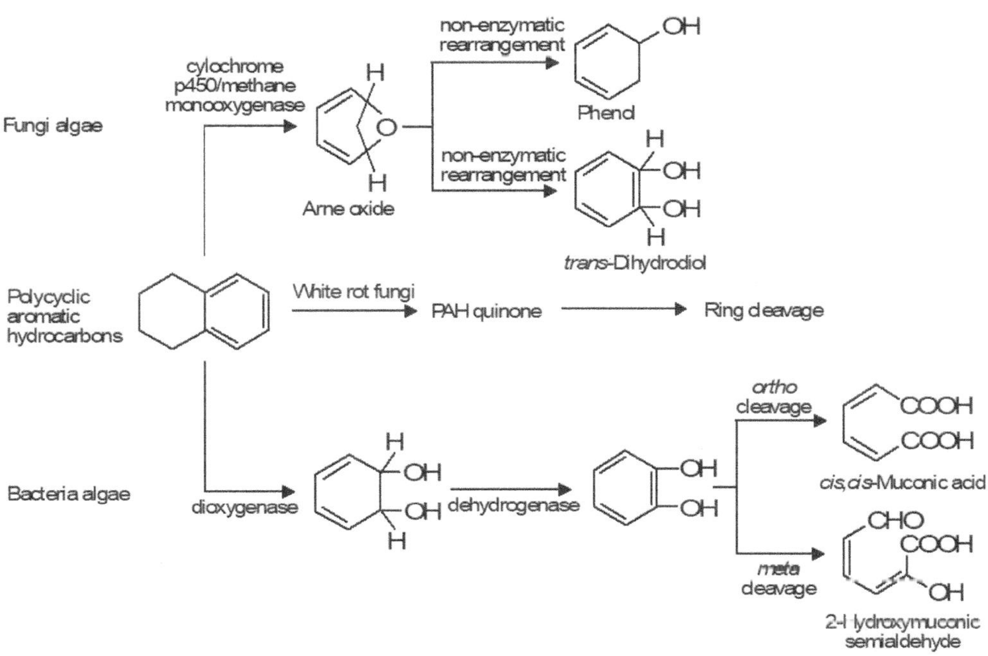

FIGURE 17.2 Ring cleavage pathway.

has been shown to confer the ability to degrade toluene and other benzene derivatives (Martínez-Lavanchy et al., 2010). The *xyl* genes are arranged into two groups or operons:

(i) The xylCAB code for the degradation to benzoic acid,
(ii) The other group xylXYZLEGFJIH code for the breakdown of benzoate into pyruvate and acetaldehyde.

The bioremediation of hydrocarbon-contaminated soil cannot always be maintained under aerobic conditions. It is because of

(a) Waterlogging,
(b) The fine particle structure of the soil and blocking of the soil pores with the biomass itself.

However, aliphatic, monocyclic, and polycyclic aromatic hydrocarbons can be anaerobically degraded provided oxygen can be obtained from water under methanogenic conditions and from sulfate under sulfur-reducing conditions. In sandy and organic soils, several moisture management systems were tested for biodegradation effectiveness. According to the findings, a combination of drying (to promote bioavailability) and traditional moisture management (to activate hydrocarbon degrading microbes) is advised (Adams et al., 2011). The hydrocarbons are converted to central metabolic intermediates by

(a) Hydration,
(b) Dehydration,
(c) Reductive dehydroxylation,
(d) Nitroreduction, and
(e) Carboxylation

The central intermediates and benzoyl CoA and sometimes resorcinol are reduced and hydrolyzed and finally transformed into compounds that can enter the Krebs cycle. The only disadvantage with anaerobic degradation is that the process is much slower than the aerobic pathway. Table 17.2 depicts some of the genes, enzymes, and their initial part of the pathway.

The degradation of both hydrocarbons and other organic molecules involves the ability of some enzymes to function with compounds other than their normal substrate. This condition, often

TABLE 17.2
Genes with Their Enzymes and Initial Part of Pathway

Initial Part of Pathway	Gene	Enzymes
	xylA	Xylene oxygenase
	xylB	Benzyl alcohol dehydrogenase
	xylC	Benzaldehyde dehydrogenase
Later part of pathway	*xylX, Y, Z*	Toluate dioxygenase
	xylE	Catechol 2,3-dioxygenase
	xylF	2-Hydroxymuconic semialdehyde hydrolase
	xylG	2-Hydroxymuconic semialdehyde dehydrogenase
	xylH	4-Oxalocrotonate tautomerase
	xylI	4-Oxalocrotonate decarboxylase
	xylJ	2-Oxopent-4-enoate hydratase
	XylK	2-Oxo-4-hydroxypentenoate aldolase
	xylL	Dihydroxycyclohexadiene carboxylate dehydrogenase
	xylR	Regulatory protein
	xylS	Regulatory protein

known as gratuitous metabolism, is probably a result of broad enzyme specificity. Another feature is known as co-metabolism, which describes the metabolism of a substrate not required for growth in which no apparent benefit is derived but the organism co-metabolism is an undefined term describing the imprecise specificity of the enzyme system and their induction system. For trichloroethylene with the co-metabolic substrates of phenol and toluene have been described *in situ* aerobic co-metabolism.

17.3.1.1.3 Microbial Degradation of Synthetic Organic Compounds

The environment is also open to contamination by several organic compounds, many of which are on the priority pollutant lists. Many of these compounds do not occur naturally. These essentially new synthetic organic compounds are often referred to as xenobiotics, from the Greek *Xenos*, meaning new. Many thousands of compounds have been synthesized and some of these find their way into the environment. Most biocides and herbicides are released into the environment by direct use. Others such as polychlorobiphenyls (PCBs), which are used as hydraulic fluids, plasticizers, adhesives, lubricants, flame retardants, and dielectric fluids in transformers, are released into the environment during production, from spillages and disposal. The chlorinated solvents such as trichloroethene, carbon tetrachloride, and tetrachloroethene are another group of compounds found frequently to contaminate ground water. Trichloroethene is a priority pollutant and contamination is found at industrial, commercial, and military sites as a result of disposal and spillage. The contaminants such as dibenzo-p-dioxins and dibenzofurans can be formed during the combustion of PAHs. The release of some of these xenobiotics into the environment can occur on a very large scale (Singh, 2017).

The problem of toxic waste disposal is enormous. Worldwide production in 1985 of just one chemical that is released into the environment pentachlorophenol was more than 50,000 tons. Incineration and chemical treatment have been used to break down toxic chemicals but these methods are costly and often create new environmental difficulties. With the discovery in the mid-1960s of several soil microorganisms that are capable of degrading xenobiotic chemicals such as herbicides, pesticides, refrigerants, solvents, and other organic compounds the notion that microbial degradation might provide an economic and effective means of disposing of toxic chemical wastes gained credence.

Members of the genus *Pseudomonas* are the most predominant group of soil microorganisms that degrade xenobiotic compounds (Mawadet al., 2021). Biochemicals assay has shown that various *Pseudomonas* strains can break down and, as a consequence, detoxify more than 100 different organic compounds. In many cases, one strain can use several different related compounds as the sole carbon source.

The biodegradation of complex organic molecules generally requires the concerted efforts of several different enzymes. The genes that code for the enzymes of these biodegradative pathways are sometimes located in the chromosomal DNA although they are more often found on large (approximately 50 to 200 kilobase [kb]) plasmids. In some organisms, the genes that contribute to the degradative pathway are found on both chromosomal and plasmid DNA.

Degradative bacteria, in most cases, enzymatically convert xenobiotic, non-halogenated aromatic compounds to either catechol or protocatechuate. Then, through a series of oxidative cleavage reactions, catechol and protocatechuate are processed to yield either acetyl coenzyme A (acetyl-CoA) and succinate or pyruvate and acetaldehyde, compounds that are readily metabolized by almost all organisms. Halogenated aromatic compounds, which are the main components of most pesticides and herbicides, are converted to catechol, protocatechuate, hydroquinone, or the corresponding halogenated derivatives by the same enzymes that degrade the non-halogenated compounds. However, for the halogenated compounds, the rate of degradation is inversely related to the number of halogen atoms that are initially present in the target compound. Dehalogenation, the removal of halogen substituent from an organic compound, is the critical requirement for detoxification and often occurs by a non-selective dioxygenase reaction that replaces the halogen on a benzene ring with a hydroxyl group. This step may occur either during or after the biodegradation of the original halogenated compound.

Many of the xenobiotic compounds released into the environment can be microbially degraded. The others however are removed only slowly and, in some cases, so slowly as to render them effectively permanent. There are a number of unwanted consequences of the persistence of organic molecules in the environment. They include protracted exposure to the compound by organisms in the environment, which can increase its toxic effect. The very low rates of degradation mean that organisms tend to accumulate the compound, in a process known as biomagnification. In this case, the first organism may be the prey of another and as a consequence, the concentration of the compound will be increased in the second organism. Thus, a very low level of the compound in the environment, under certain circumstances, can be magnified to high and toxic levels (Sun et al., 2020). An example of this is the accumulation of DDT in grebes (fish-eating water birds), which were at the top of the food chain on a lake treated with 0.01–0.02 ppm DDT to control gnats. The level found in the grebes was some 100,000 times the DDT level applied. Another problem with these persistent compounds is that in many cases there is little information on their toxicity and long-term effects. The persistence of xenobiotics in the environment is influenced by

(a) The structure and properties of the molecules,
(b) Their toxicity to microorganisms,
(c) The environmental conditions.

Natural complex polymeric molecules, such as lignin, are slow to degrade. Many of the xenobiotic compounds are also complex in structure and therefore, slow to degrade. For many of the xenobiotics, their activity and persistence are also linked to the presence of halogens in the compound. The toxicity and persistence of the organohalogen are influenced by the number of halogen molecules, their position, and the type of halogen. Due to their hydrophobic properties, organochlorines are effective biocides. It allows them to pass through or into membranes, disrupting activity; they are also effective because of their ability to inhibit oxidative phosphorylation. The solubility or organochlorines decreases as the chlorine content increases, which often increases their toxicity. Degradation of these organochlorines by microorganisms has to overcome their increased toxicity and these organochlorines by microorganisms have to overcome their increased toxicity and their low solubility in the aqueous phase. The problem may also be increased as some organochlorine compounds dare a mixture of isomers, as found with PCBs which decreases the rate of degradation.

The rate of degradation of xenobiotics in soil and water is also dependent on the presence of microorganisms with the enzymatic capability to degrade the molecule. As xenobiotics are not normally found in nature, the level of the degrading microorganism may be very low. In many cases, a period of adaptation is required before degradation occurs. The environmental conditions which affect the degradation of xenobiotics are essentially the same as those which affect the degradation of hydrocarbons.

17.3.1.1.3.1 Pathways of Degradation of Synthetic Organic Compounds (Xenobiotics)

Organisms capable of degrading xenobiotic have been found in soil and sediment, particularly from contaminated sites. They include bacteria, fungi, and algae. The degradation pathways for some of the chloroaromatics have been determined under both aerobic and anaerobic conditions. The chloroaromatics are normally cleaved by monooxygenases and dioxygenases similar to those found with the degradation of PAHs (Dhakal et al., 2017; Mishra et al., 2021). Superimposed on the degradation is the process of dehalogenation which can be by four mechanisms:

Oxidative dehalogenation: the halogen is removed and replaced by two hydroxyl ions.

Eliminative dehalogenation: the simultaneous removal of the halogen and an adjacent hydrogen ion.

Hydrolytic (substitutive) dehalogenation: the substitution of the halogen with a hydroxyl ion.
Reductive dehalogenation: the halogen is replaced by a hydrogen ion.

Following are examples of the degradation of organochlorines. These illustrate individual pathways and general processes.

Pentachlorophenol (PCP) is a herbicide and fungicide used for the preservation of wood and is a priority pollutant. Some microorganisms have been isolated which can degrade PCP under aerobic and anaerobic conditions, including *Flavobacterium*, *Arthrobacter*, *Rhodococcus*, and the white-rot fungus *Phaerochaete chryososporium*. In most of the pathways, the breakdown of chlorophenols consists of the dechlorination and hydroxylation of the aromatic ring followed by ring cleavage. The oxygenases catalyze both hydroxylation and ring cleavage. In PCP degradation where aerobic degradation starts with reductive dehalogenation the first step appears to be the rate-limiting step (Kumar et al., 2021).

The use of polychlorinated biphenyls (PCBs) has been banned for over 20 years due to their toxicity, but as their use was extensive, contaminated sites and sediments remain. PCBs can be aerobically degraded and the first step is similar to that for monocyclic aromatics involving a dioxygenase which is similar to the dioxygenases for naphthalene and toluene.

The most widely used triazine herbicide is atrazine. It is effective against broad leaved weeds. It had been employed for some 40 years and was considered to be recalcitrant. However, pure cultures of bacteria have been isolated which can degrade atrazine (Fernandes et al., 2018). Although, bacterial consortia have also been reported to degrade atrazine. Atrazine is converted to cyanuric acid in three steps and the cyanuric acid can be converted to CO_2 and NH_3. Cyanuric acid can also be metabolized by soil bacteria that cannot degrade atrazine. In coding for the enzymes that convert atrazine to cyanuric acid three genes are involved. In *Pseudomonas* spp. these genes are found on a plasmid. The isolation of these three enzymes has allowed the determination of the form of sharing of metabolism that occurs in a bacterial consortium that can degrade atrazine (Figure 17.3).

Synthetic organophosphorus compounds were used as domestic and agricultural pesticides. The usual methods of detoxification are chemical treatment, landfill, and incineration. Soil microorganisms such as *Pseudomonas diminuta* and *Flavobacterium* spp. can degrade organophosphorus using a hydrolase enzyme. Purified hydrolase enzyme has been shown to detoxify organophosphate. Hydrolase enzyme has been immobilized to various supports and has been shown to detoxify organophosphates, but the costs are too high due to the need to purify the enzyme (Mulla et al., 2020). If whole cells can be used, the need for purification would be removed. It will reduce the cost. Attempts have been made to clone the gene for the hydrolase and express it on the surface of the cell. A surface-expressed hydrolase has been developed. It will eliminate the problem of enzyme purification and the transport of the organophosphates into the cells (Figure 17.4).

Despite the ability of many naturally occurring microorganisms to degrade some different xenobiotic chemicals, there are limitations to the biological treatment of these waste materials. For example:

(1) No single microorganism can degrade all organic wastes.
(2) High concentrations of some organic compounds can inhibit the activity or growth of degradative microorganisms.
(3) Most contaminated sites contain a mixture of chemicals, and an organism that can degrade one or more of the components of the mixture may be inhibited by other components.
(4) Many non-polar compounds absorb onto the particulate matter in soils or sediments and become less available to degradative microorganisms.
(5) Microbial biodegradation of organic compounds is often quite slow.

One way to address some of these problems is to transfer by conjugation into recipient strain plasmids that carry genes for different degradative pathways. If two resident plasmids contain homologous regions of DNA, recombination can occur and a single, large "fusion" plasmid with

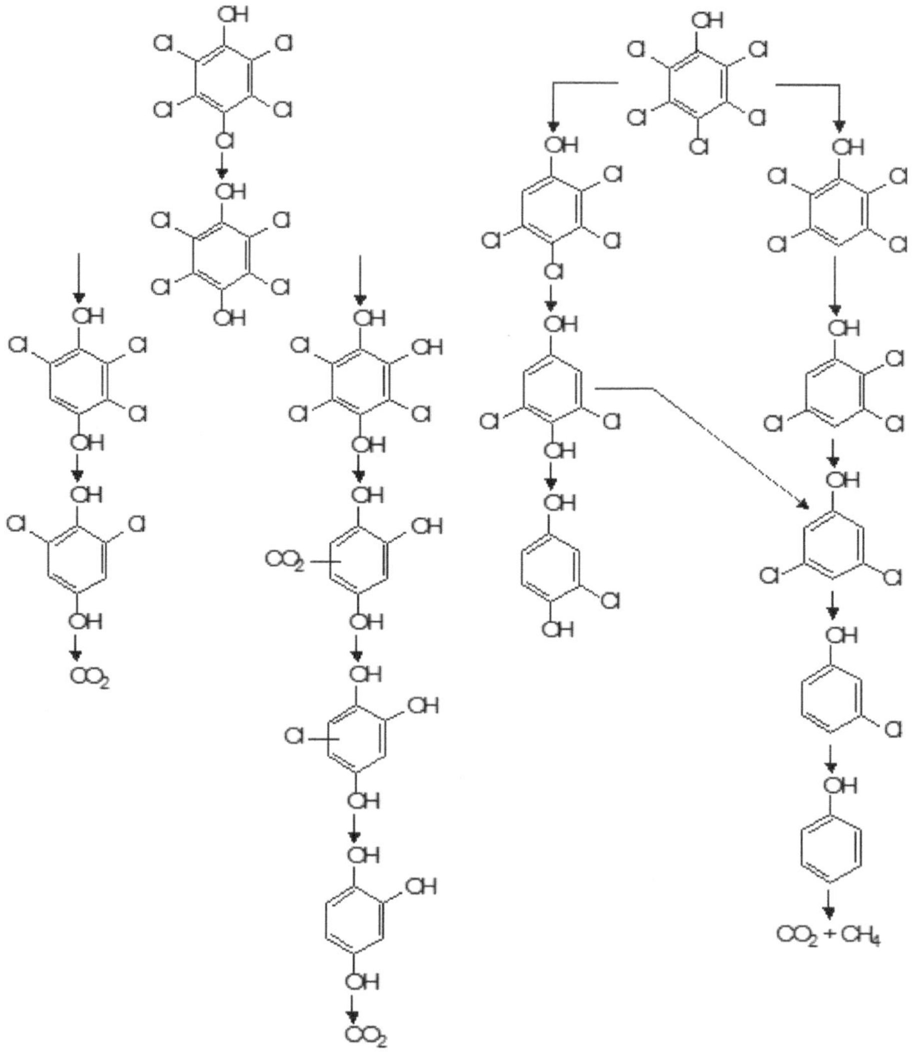

FIGURE 17.3 Pathways of PCP: (a) aerobic degradation (b) anaerobic degradation.

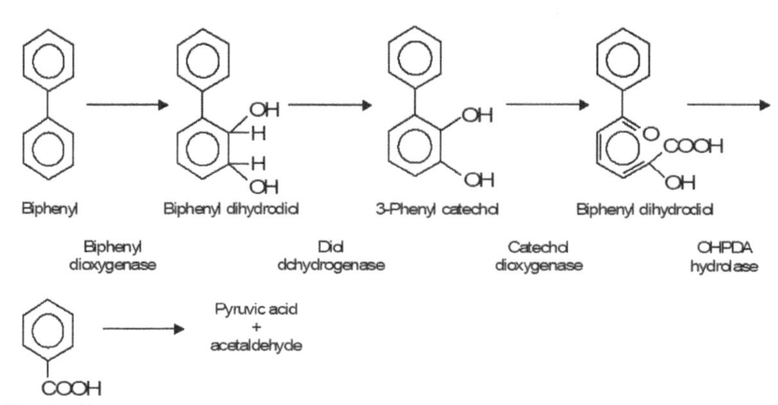

FIGURE 17.4 Degradation of atrazine.

combined functions can be created (Bhatt et al., 2021). Alternatively, if two plasmids do not contain a homologous region and in addition belong to different incompatibility groups, they can coexist within a single bacterium.

17.3.1.1.4 Microbial Degradation of Natural Organic Compounds

Cellulose is the basic component of plant material and is produced in greater quantities than any other substance. Thus, about half of photosynthetically produced biomass consists of cellulose. Plant residues in soil consist of 45 to 90% cellulose which is therefore of central importance – next to carbon dioxide – in the carbon cycle. The enzymatic cleavage of cellulose is catalyzed by cellulase. According to investigations on fungi, the cellulase system consists of at least three enzymes.

(1) Endo-β-1,4-glucanases attack the β-1,4 bonds in the center of the macromolecule and produce long-chain fragments with free ends.
(2) The Exo-β-1,4-glucanases remove disaccharide cellobiose from the ends of the cellulose chains.
(3) The β-glucosidases hydrolyze cellobiose with the formation of glucose.

The regulation of cellulase synthesis can be brought about either by catabolite repression or through substrate induction by cellobiose, whilst small amounts of cellulose are produced constitutively. The concentrations of cellobiose that have inducing or repressing effects vary in different organisms. In general, low concentrations of cellobiose act as inducers and relatively high concentrations have repressor action. In addition, cellobiose can also act as a competitive inhibitor at the enzyme level. Cellulose itself, as it is water-insoluble, does not affect the synthesis and can influence it only indirectly via the cellobiose produced by its hydrolysis. This can account for the apparent inducing action of crystalline cellulose.

Cellulose is degraded and utilized in well-aerated soils by aerobic microorganisms (Fungi, myxobacteria, and other eubacteria) and under anaerobic conditions by bacteria, a few anaerobic fungi, and protozoa (Land et al., 2015). The fungi play a significant part in the degradation of cellulose under aerobic conditions. They are more successful than bacteria in acid soils and the degradation of cellulose embedded in lignin (*i.e.*, in wood). Species of the genera *Fusarium* and *Chaetomium* are prominent. Others known to be cellulolytic are *Aspergillus fumigatus* and *A. nidulans*, *Botrytis cinerea*, *Rhizoctonia solani*, *Trichoderma viride*, *Chaetomium globosum*, and *Myrothecium verrucaria*. The last species named are used as test organisms for the demonstration of cellulolytic activity, respectively, for testing impregnation media used on cloths and coverings to prevent damage by cellulolytic microorganisms. The fungi excrete cellulases which can be isolated from the mycelium and the culture medium.

Cytophaga and *Sporocytophaga* are the most easily isolated of the aerobic cellulolytic bacteria; this is done by enrichment culture in a liquid medium. Very little is known about the utilization and initial attach of cellulose by myxobacteria. No extracellular cellulose or initial cleavage products of cellulose have ever been demonstrated. The bacterial cells are attached to the cellulose fiber with their longitudinal axis parallel to that of the fiber, they hydrolyze cellulose only when they are in close contact with it and they immediately absorb the hydrolysis products. In addition to *Cytophaga* species, myxobacteria belonging to genera *Polyangium*, *Sorangium*, and *Archangium* are also able to grow on cellulose.

The ability to grow on cellulose as a substrate is also quite common among many aerobic bacteria which could almost be considered 'omnivores'. Some of these attack cellulose only in the absence of other carbon sources. Some *Pseudomonas*-like bacteria were formerly collected in a Cellvibrio group. *Psedomonas fluorescensvar cellulose* has been described only recently. *Cellulomonas* should be mentioned among the coryneform bacteria.

Only a few cellulolytic species have been described among the actinomycetes, including *Micromonospora chalcea*, *Streptomyces cellulose*, and *S. sporangium*. Cellulose can also be degraded under anaerobic conditions by mesophilic and thermophilic eubacteria and a few fungi and protozoa. The thermophile *Clostridium thermocellum* grows in a simple synthetic medium with cellulose or cellobiose as substrate and ammonium salts as the only nitrogen source; glucose and many other sugars cannot be metabolized. The degradation of cellulose is preceded by the secretion of a yellowish, carotenoids-like substance, as has been observed in other cellulolytic bacteria. This increases the affinity of the cellulolytic enzyme for cellulose. The yellow coloration of the cellulose is a good indicator for the initiation of cellulose hydrolysis. In the case of *C. thermocellum* the multienzyme complexes described above are organized in so-called cellulosomes, which can have molecular weights of several million.

The products of this cellulose fermentation are ethanol, acetate, lactate, format, molecular hydrogen, and carbon dioxide. Extracellularly, the cellulose is probably converted to glucose. Similar fermentation of cellulose is by the mesophile, *Clostridium cellobioparum*. The long rods known as *Bacillus* disintegrates cellulose to cellulose fibers, similar to those of *Cytophaga* species, and do not excrete any cellulose into the medium.

17.3.1.1.5 Microbial Bioremediation of Water

Organic pollution from industry has increasingly affected such ecosystems as water, land, and subsurface. Many mature theories and techniques have been developed for the treatment of wastewater and surface water contaminated by organic pollutants. However, it has more recently been found difficult to successfully apply these theories and techniques to clean up the subsurface and groundwater (Srivastava et al., 2014). Almost all the engineering techniques for the treatment of subsurface and groundwater, including pump-and-treat, air sparging, and chemical flooding, are high cost, technologically difficult and otherwise fail to achieve desired clean-up targets. Since American scientists developed a theory of natural attenuation or intrinsic remediation for the remediation of subsurface and groundwater in the middle of the 1990s, intrinsic and enhanced bioremediation has been applied increasingly to the treatment of groundwater and subsurface sediments. However, the mechanism, including some microbial processes in the natural attenuation, has remained unclear although many efforts have been made to study the respiratory processes from aerobic, nitrate through iron and sulfate reduction to methanogenesis. Fermentation could probably be one of the most important processes in natural attenuation which has not yet been understood clearly.

Groundwater systems are mostly anaerobic, and particularly when the introduction of organic pollutants depletes dissolved oxygen quickly and develops the anaerobic conditions. Many microorganisms that inhabit anoxic environments can obtain their energy for growth through the fermentation of organic carbon. In fermentation, organic compounds serve not only as electron donors but also as electron acceptors. In natural environments, fermentation almost always occurs coupled with such respiration processes as nitrate, Fe (III), and sulfate-reduction to drive the biotransformation of many natural organic compounds. Since organic compounds act as both electron donors and acceptors, strictly fermentative reactions should not include any other acceptors. However, in sediments and groundwater systems fermentation usually exists together with respiration reactions, in which various inorganic electron acceptors normally participate. The conceptual basis is the methanogenic degradation of aromatic compounds since the methanogenic reactions occur without any other inorganic external electron acceptors participating. For example, benzoate acid is degraded by methanogenic microbes:

$$\text{Fermentation step: } C_7H_6O_2 + 12\ H_2O \longrightarrow 7\ CO_2 + 30\ [H]$$
$$\text{Methanogenesis: } 30\ [H] + 3.75\ CO_2 \longrightarrow 3.75\ CH_4 + 7.5\ H_2O$$
$$\text{Overall reaction: } C_7H_6O_2 + 4.5\ H_2O \longrightarrow 3.25\ CO_2 + 3.75\ CH_4$$

CO_2 is produced from the fermentative degradation of benzoic acid although it acts as the electron acceptor and carbon source of the subsequential syntrophic methanogenic reaction. The fermentation reaction goes to completion due to hydrogen being used as a favorable energy source.

River and lake waters get eutrophicated by phosphates coming from detergents in the wastewater (Glibert et al., 2005). Several microbial species which are also able to produce biosurfactants, e.g., *Nocardiaerythropolis*, are being exploited to get rid of eutrophication.

17.3.1.1.6 Microbial Bioremediation of Petroleum Oil Spills in Waters

Crude oil is an extremely complex and variable mixture of organic compounds. The majority of the compounds in crude oil are hydrocarbons which can range in molecular weight from the gas methane to the high molecular weight tars and bitumens. These hydrocarbons can also come in a wide range of molecular structures: straight and branched chains, single or condensed rings, and aromatic rings. The two major groups of aromatic hydrocarbons are:

(a) Monocyclic such as benzene, toluene, ethylbenzene and xylene (BTEX) and
(b) The polycyclic hydrocarbons (PAHs) such as naphthalene, anthracene, and phenanthrene.

Crude oil has accumulated underground as a result of the anaerobic degradation of organisms over a very long time. Under the conditions of high temperature and pressure, the organic material has been converted to natural gas, liquid crude oil, shale oil, and tars. At the underground temperatures, shale oils and tars do not flow, but crude oil is liquid and, unless contained, will escape to the surface, where the volatiles evaporate forming a tar bed. Apart from this, the release of crude oil, the main source of crude oil and oil products, such as BTEX and PAHs released into the environment comes from the leaking of storage tanks, spillages, and accidents during its transport. These BTEX compounds, although not miscible with water are mobile and can contaminate the ground water. Oil spills in the aquatic environment create PAH pollution and have an impact on marine species (Honda and Suzuki, 2020). The volatile components can be lost to the atmosphere if the leak is on the surface, but if the leak is below soil level, the mobile components can migrate down through the soil to the water table (Figure 17.5).

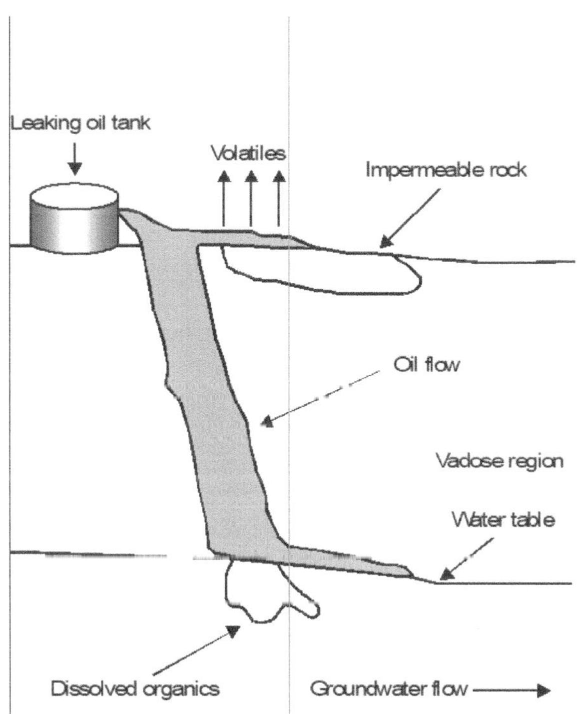

FIGURE 17.5 Underground accumulation of crude oil.

Crude oil when released at sea will not mix with sea water and will float on the surface. This allows the escape of the volatile components of those of 12 carbons and below. The floating oil, if it does not reach the shore, will be dispersed due to the action of waves. The dispersion will allow naturally occurring hydrocarbon-degrading organisms to break down the oil. Oil breakdown will occur at the interface between the oil and water and therefore, the better the oil dispersion, the greater the area, the faster the degradation. The more complex and less soluble oil components will be degraded much more slowly than the lighter oils. It is these high molecular weight components that will persist on the sand and rocks if the spill reaches shore. The first stage in the recovery and clean-up of an oil spill is to stop the release and contain the spill. The surface oil can be mechanically removed with skimmers and other machines. Chemical dispersants can be used on both floating oil and oil which has reached a rocky shore, but care has to be taken as the detergents can be as harmful to the environment as the oil. Although both physical and chemical methods are efficient, not all the oil can be removed and it is the remaining high molecular weight oil that may need to be removed by some form of bioremediation. Crude oil can be degraded by the indigenous microbial population in the sea and on the shore, as microorganisms capable of using hydrocarbons as a source of carbon compounds and energy are widely found (Chen et al., 2020). However, it is thought that the supply of utilizable phosphorus and nitrogen compounds is limited in most marine environments, to encourage degradation. Slow-release fertilizers have been added to oil slicks at nitrogen to oil ratio of 1:100.

Oil spills from oil tankers have been recognized as a major environmental hazard. These destroy the habitat of aquatic animals and fishes and cause long-term damage to the environment. Chemical dispersants were used to remove these, but they cause major pollution problems in shallow waters due to their toxic nature and persistence in the environment. In recent years, by use of genetic engineering, oil-utilizing microorganisms have been successfully produced which would rapidly grow on oil. Many oil utilizing microbes produce surface-active compounds that can emulsify oil in water and, thus, facilitate the removal of oil.

A strain of *Pseudomonas* has been developed by Dr. Ananda Mohan Chakrabarty that produces a glycolipid emulsifier capable of reducing the surface tension of the oil-water interface and, thus, helps in the removal of oil from water. This microbial emulsifier is non-toxic and biodegradable. This has been quite successful on some sites. Some companies are now marketing the microbial inoculates that are claimed to increase significantly the rate of biodegradation of oil pollutants.

17.3.1.1.7 Bioremediation Technologies

The bioremediation of hydrocarbon and xenobiotic contaminated soils can be carried out either *in situ* or *ex-situ*. The trend in the USA is for removal, as this avoids litigation over any contamination not removed by an *in situ* treatment. Whatever treatment is carried out, the basis of the process is the stimulation of the growth and metabolism of the indigenous degrading population by providing optimum conditions (Figure 17.6).

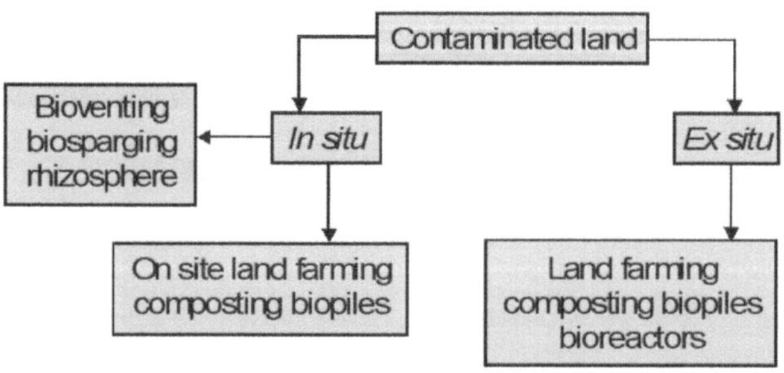

FIGURE 17.6 *In situ* and *ex-situ* processes in bioremediation of soil.

Ex-situ treatment of contaminated land involves excavation of the soil for treatment or disposal elsewhere. One bioremediation process is land treatment or landfarming, where regular tilling of the soil increases aeration and supplementation with fertilizer, encouraging the microbial population. Often the treatment area is lined and dammed to retain any contaminants that leak out. The rate of degradation depends on

(i) The microbial population,
(ii) The type and level of contamination,
(iii) The soil type.

The average half-life for degradation of diesel fuel and heavy oils is in the order of 54 days with this type of system. Another solid-phase treatment carried out after extraction is composting. Composting material such as straw, bark, and wood chips is mixed with the contaminated soil and piled into heaps. The process works in the same way as the normal compost system with a rise of temperature to 60°C and above caused by microbial activity. The higher temperature encourages the growth of thermophilic bacteria. The increased costs of this type of system restrict it to highly contaminated materials, although the process is more rapid than land farming.

In the biopile process, the soil is heaped into piles within a lined area to prevent leaching. The piles are covered with polythene and liquid nutrients applied to the surface. By applying suction to the base of the pile, as in a composting system, aeration can be improved. Any leachate formed is collected by pipes at the base and can be recycled if necessary. This type of system can be used when space is limited, and when vapor emission needs to be restricted some form of biofilter can be added to the system (Figure 17.7). Soil extracted from a contaminated site can also be treated as a solid waste (slurry) or as liquid leachate in bioreactors of various designs. The use of bioreactors gives control of parameters such as temperature, pH mixing, and oxygen supply, which can improve degradation rates. A wide range of bioreactor designs can be used and these include prepared bed reactors, slurry reactors, biofilters (fixed film), or anaerobic disasters. The processes of land farming, composting, biopiles, and bioreactors can be carried out on-site as well as on a site removed from the contaminated land (Delille et al., 2008; Emami et al., 2012).

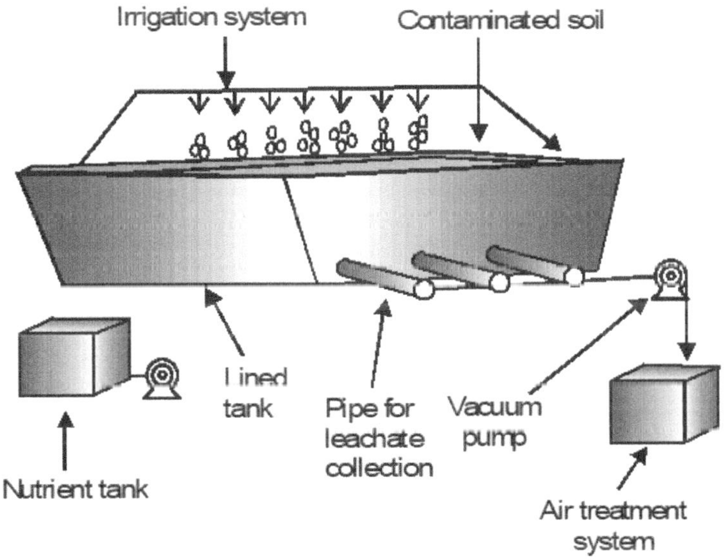

FIGURE 17.7 Biopole process for soil treatment.

17.3.1.1.7.1 Microbial Bioreactors in Bioremediation

Microbial (fungi and bacteria) bioreactors have been used in several laboratory and pilot bioremediation experiments. Many toxins in groundwater, soil, and air may be treated effectively using bioreactor methods. The system should be designed to tolerate high biomass from cell growth, as well as the provision of vital nutrients and the removal of waste components.

17.3.1.1.7.1.1 Slurry Phase Bioreactors

When a solid substrate is made into a slurry, slurry bioreactors provide an *ex-situ* environmentally benign technique to remediate primarily soils and sediments from petrochemical hydrocarbons, tars, creosotes, chlorinated solvents, herbicides, pesticides, and explosives. The slurry reactor can be operated in batch, semi-continuous, or continuous mode, with batch being the most frequent. Figure 17.8 depicts a slurry reactor in simplified form. Water is mixed in appropriate ratios with the contaminated solid matrix, enhancing contact between microorganisms, contaminants, media, and oxygen. Pollutants that have been solubilized have a higher bioavailability (Robles-González et al., 2008).

17.3.1.1.7.1.2 Stirred Tank Bioreactors

A continuous stirred tank bioreactor is made up of a cylindrical vessel with a central shaft that is powered by a motor and supports one or more agitators (impellers). For submerged cultures, stirred-tank bioreactors are the most common design. Stirred tank bioreactors are mechanically agitated, with stirrers serving as the primary gas-dispersing equipment, and offer high mass transfer rates and excellent mixing. The STR has several advantages, including effective gas transmission to growing cells, good mixing of the contents, and adjustable operating conditions, in addition to the bioreactors' commercial availability. Figure 17.9 depicts the reactor that is used as STR. The stirred tank bioreactor's fundamental flaw is its mechanical agitation, which consumes energy and can induce shear strain on microbial cells (Gargouri et al., 2011).

17.3.1.1.7.1.3 Packed Bed Bioreactors

Microbial growth on fixed-film substrata is possible in packed bed bioreactor systems. Fixed film reactors are utilized to obtain compact reactors and offer improved treatment dependability. They have the benefit of being able to remediate dilute aqueous solutions with high biomass without having to separate the biomass from the treated effluent. A simplified diagram of a laboratory-based packed bed bioreactor is shown in Figure 17.10. Organochlorine insecticides, PAHs, medicines, amines, and textile colors are all remedied using various fungi and bacteria (Tekere et al., 2007).

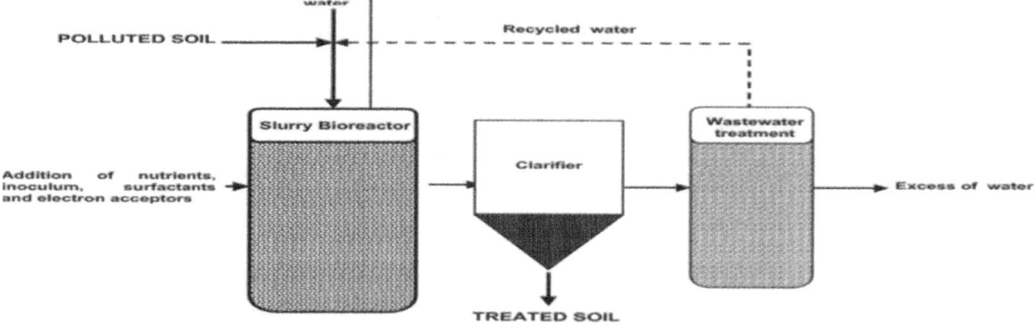

FIGURE 17.8 Slurry phase bioreactor.

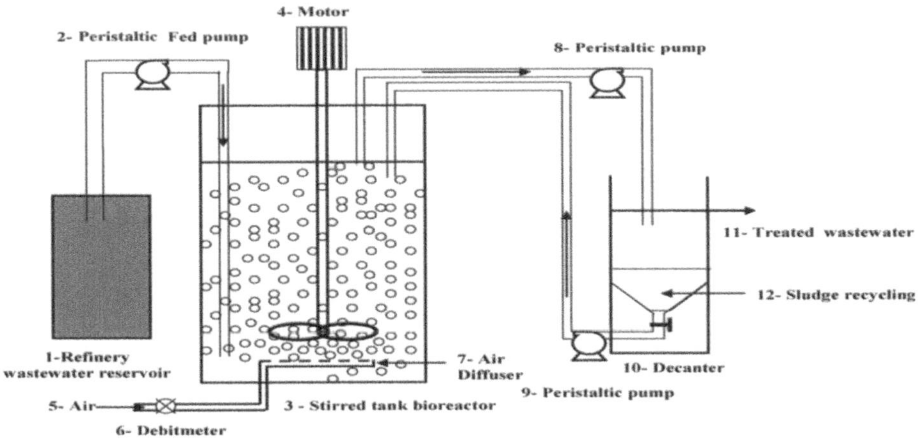

FIGURE 17.9 Stirred tank bioreactors.

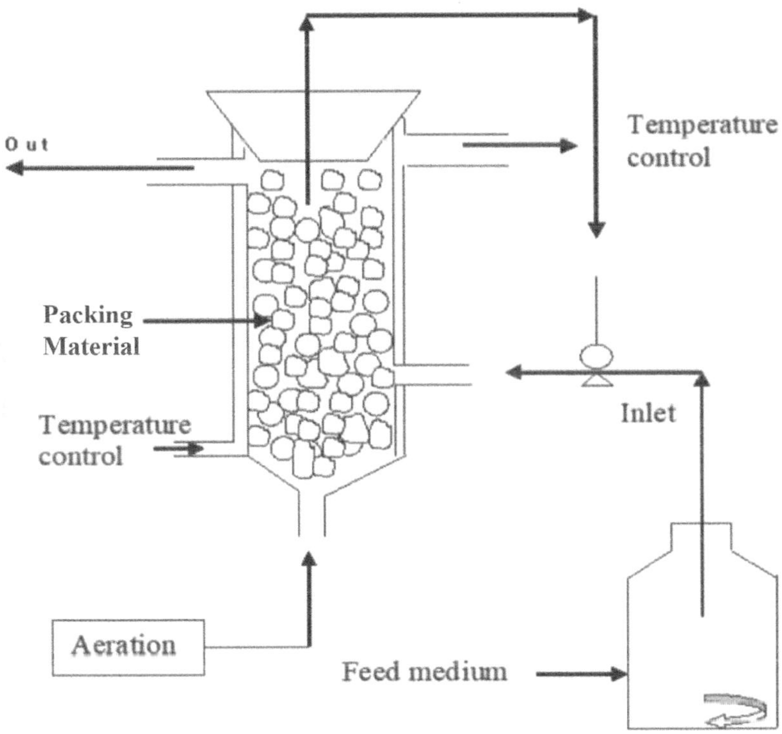

FIGURE 17.10 Packed bed bioreactors.

17.3.1.1.7.1.4 Membrane Bioreactor

Membrane bioreactors (MBR) integrate the usage of a filtration system formed by a membrane with the biological process. The membrane acts as a physical barrier between the liquid and the solid, ensuring solid retention and high-quality effluent. MBRs have the advantage of using fewer tanks and ensuring that particles are removed from treated effluent tanks to the filtration function of the membrane. Membrane fouling, on the other hand, has been identified as a key limitation in the use of membrane bioreactors in bioremediation. MBR has been tested for pentachlorophenol remediation in wastewater, textile wastewater, 1,2-dichloroethane, 1,2-dichlorobenzene, and 2-chlorophenol concentrations (Carucci et al., 2007; Hossain and Ismail, 2015).

17.3.1.1.8 Limitations of Bioremediation

There are, however, certain constraints of this technology:

(a) Indigenous degradative microbes are fully adapted to the specific environment to be treated.
(b) Introduced 'foreign' microbes must be able to survive in the new environment and be able to compete with the established indigenous microbes.
(c) There is a need for close contact of added inocula with the pollutant and in aqueous environments dilution has to be avoided.

Attempts are now being made to genetically engineered microorganisms to be able to degrade organic pollutant molecules that they are unable to attack at present. This has posed certain problems such as:

(i) Genetic stability,
(ii) Survival of the 'new' microbe in a hostile environment,
(iii) Some legislative, ethical, and perceptional problems concerned with their release into environments such as sewage systems, soils, and oceans.

17.3.1.2 Bioaugmentation

The addition of microorganisms is a process known as bioaugmentation. It has been suggested that the best approach is to stimulate the indigenous microbial populations, but this is not suitable for all occasions. Some degradative pathways can produce intermediates that trap the pathways or transform the pollutants into non-toxic compounds. Thus there may be situations that can be improved by bioaugmentation with selected or genetically manipulated organisms (Jain et al., 2011). Several fungal inoculum have been used to bioaugmented soils contaminated with PCP and this procedure removed 80–90% of PCP within four weeks. A selected strain, in a field study *Methylosinus trichosporium* was used. Here the organism was selected for its high TCE transformation rate under low copper conditions. Once the organisms had been injected, 50% of the cells attached to the soil, formed a biofilter which was efficient for the transformation of TCE.

All the soil contaminants are not easy to be removed or readily available for degradation. Many are insoluble in the aqueous phase so that many methods have been used for their extraction. One solution has been to apply biosurfactants which are surface-active molecules that solubilize, emulsify, disperse, and act as detergents. These surfactants represent a spectrum of structures and are generally more complex than chemical surfactants. The addition of biosurfactants to oil-contaminated soil has resulted in an increased rate of degradation. Various biosurfactants have been used to solubilize oils and xenobiotics such as PCBs and organophosphates. These biosurfactants can be produced *in situ* by added microorganisms. Another method of extraction has been the use of liquid carbon dioxide for the removal of compounds. The method has been successfully used for diesel removal. Soil washing has been used to remove pollutants such as pentachlorophenol. The soil is generally removed and washed by a series of scrubbing and physical separation techniques. In this technique, the contaminants are partitioned into the liquid phase, and the fine particles were collected for biotreatment or disposal. The water can be treated in several bioreactors which can be run in sequence (Schmidt and Anderson 2017).

In order to accelerate the growth of natural microbe populations and improve degradation, which preferentially feed on contaminated sites, microbes are taken from the remediation site, cultivated separately, genetically changed, and then reintroduced. For example, when soil and groundwater are contaminated with chlorinated ethenes, such as tetrachloroethylene and trichloroethylene, all important bacteria are discovered. It is employed to ensure that *in situ* microorganisms can completely eliminate and convert these pollutants to non-toxic ethylene and chloride (Niu et al., 2009). Many microbes are capable of generating ferric ions which act as oxidizing agents and transform solid metals (sulfide minerals) into water-soluble metal ions and elemental sulfur. A microorganism capable of removing a portion of organic sulfur (thiophenic sulfur) has been developed and patented in the USA. Many such organisms involving the removal of additional organic sulfur (aromatic sulfides) from coal are being developed. No cost-effective technology exists for the pre-combustion and desulfurization of coal, but biotechnology may provide a solution. Various strains of bacteria *Sulfolobus acidocaldarius*, *Pseudomonas* species (TG 232, CB1) are being developed which can remove organic sulfur and yet maintain thc fuel value of coal.

17.3.1.3 Phytoremediation with Plants

Phytoremediation is the use of plants to remove pollutants from water and soils. There are about 1.4 million polluted sites in Western Europe alone. Current techniques are costly and destroy soil structure. The use of plants that can store 10 to 500 more pollutants in their leaves and stems is cheaper; it also stabilizes the soil structure. When compared to the conventional techniques, phytoremediation techniques offer the benefit of economic feasibility (Tang et al., 2016).

The metals can be recovered from ashes and reused. There are many hyper accumulating plants. These plants accumulate lead, zinc, nickel, copper, and cobalt, among others, at levels toxic to other plants. For example, *Sebertia acuminata* can contain up to 20% of nickel in its sap (nickel is generally toxic to plants at a concentration of 0.005%) (Schaumlöffel et al., 2003). Similarly, the fern *Pteris vittata* accumulates arsenium, while conserving a very rapid growth and high biomass. The commercialization rights of the fern (now called edenfern™) for use in phytoremediation have been obtained by the firm Edenspace. The potential market for phytoremediation in the United States alone was estimated at $100 million in 2002. Table 17.3 depicts some plants and metals which they accumulate.

Genetic engineering is being used to enhance the ability of plants to accumulate heavy metals. Now the "smart plants" (plants that can detect and/or remove toxins and pollutants) have been developed by genetic engineering. They are expected to have roles in:

(i) Improving environment,
(ii) Enhancing health, and
(iii) Promoting national security.

"Smart plants" in combination with bioremediation can also be used to detect heavy metals, microorganisms, and other environmental changes. These plants could be used in phytoremediation,

TABLE 17.3
Accumulation of Metals by Some Plants

Plant	Metal
Thlaspi Caerulenscens	Cd
Lpomoea alphina	Cu
Haumanisatrum robertii	Co
T.rotundifolium	Pb
Macadamia neurophylla	Mn
Psychotria douarrei	Ni
Sebertia accuminata	Ni
T. caerulenscens	Zn

monitoring the changes in quantities or presence of toxins or pollutants (Zhu and Rosen 2009). For example, the introduction of two *Escherichia coli* genes into the genome of *Arabidopsis thaliana* which encodes an enzyme involved in the transformation of arsenate into arsenite enables the plant to accumulate three to four times more arsenium than the normal plants. Aresao Biodetection (www.aresa.dk/), a biotechnology company in Denmark, has engineered plants for the detection of landmines and/or explosives. The plant, when it comes into contact with explosive materials, changes color from green to red. The company is also developing plants for the detection of heavy metals.

17.3.2 Replacement of Chemical Fertilizers with Organic Manures and Biofertilizers

The significant rise in imbalanced use and the high cost of chemical fertilizers are major concerns for agriculture. These are extensively used for increasing the crop yield and their nutrient values. Heavy use of agrochemicals like chemical fertilizers in recent years has led to high yield. Unfortunately, it has also led to poor cropping systems due to use of these agrochemicals over the recommended levels, which has led to soil acidification, risking the contamination of underground water and the weakening of plant root systems, rendering them prone to infections and diseases. The use of other natural sources, especially biofertilizers, to improve the nutrient value of soil as well as improve the health of plants, is the necessity of time. Biofertilizers are environment-friendly inoculants of beneficial microorganisms, which not only minimize our dependency on chemically synthesized fertilizers but also directly benefit the plants by providing macronutrients, micronutrients, and plant growth-promoting hormones. Biofertilizers are widely known as materials that contain one or more species of live microorganisms that can assist in the uptake of important elements in plants by converting them from their complex non-usable form to simpler usable forms (Barman et al., 2017). It is performed through various biological processes like nitrogen fixation, phosphate or potassium solubilization, or through the secretion of plant growth-promoting hormones.

Because of the rising costs of chemical fertilizers and their hazardous effects on the environment, biofertilizers present a sustainable alternative for agricultural use (Kumar et al., 2018). Biofertilizers can be applied to seeds before sowing them, on the surface of the plants or soil. They form colonies in the rhizosphere or on the interior parts of that plant which promote the growth of the plant by enhancing the uptake of nutrients (which is also one of the main highlights of their use as fertilizers). Biofertilizers mainly include nitrogen-fixing, phosphate-solubilizing, potassium-mobilizing, and plant growth-promoting microorganisms which fulfill the major NPK requirements of the plants and secrete growth hormones. Amongst the commonly used biofertilizers benefiting the crops are *Azotobacter, Azospirillum*, blue-green algae, *Azolla*, P-solubilizing microorganisms, mycorrhizae, and *Rhizobium.*

17.3.3 Replacement of Chemical Pesticides with Biopesticides

Crop plants are damaged by a variety of pests and insects, which are generally controlled by applying chemicals known as pesticides. These are used to kill insects and other pests and to combat diseases that reduce crop yield. Chemical pesticides have not provided a complete solution to crop protection. Certain insects, due to their habitat and behavior, remain controlled. Some of the insects also develop resistance to pesticides.

The residues of these pesticides can sometimes be present in foliage, in plants, and the soil. Pesticide residues in soil interact with constituents like clay and organic matter. Adsorption of a pesticide onto a clay surface and interactions with organic matter can result in partial or even complete inactivation. The nature of soil pesticide interactions can in turn determine the extent to which a pesticide is leached into the soil. Certain persistent pesticides such as DDT and associated breakdown product 1.1-dichloro-2,2-diethane (DDE) can remain in the soil for considerable periods and accumulate in soil fauna which forms part of the food chain for higher animals. An alternative to chemical pest management can be developed by manipulation of bacteria and fungi for obtaining microbial products with fungicidal, herbicidal, and insecticidal properties (Purnomo et al., 2011).

The following microbes are used as biopesticides in the management of pests:

(i) **Bacteria as pesticides:** These belong to the families, *Pseudomoniaceae*, *Lactobacilliaceae*, *Micrococcaceae*, and *Bacilliaceae*. The majority of commercial strains belong to the genus *Bacillus* (Kachhawa 2017). The examples are:
   (a) ***Metarhizium anisopliae*:** It infects several different insects including spittle bugs in pasture and sugar cane for which it is commercially used in Brazil.
   (b) ***Beaureria bassiane*:** It attacks Colorado potato (produces cyclodepsipeptide toxin). It is used in Russia under the name Boverin.
   (c) ***Hirsutella thompsoni*:** It is used for the control of citrus mites.

Bacillus thuringiensis is a gram-positive bacterium that produces parasporal, crystalline inclusions during sporulation. These inclusions consist of proteins (insecticidal crystal proteins ICPs or delta-endotoxins) having molecular masses of 70 to 130 kDa. Most *B. thuringiensis* strains that have been characterized produce ICPs that exhibit highly specific activity against some lepidopteran (caterpillar), dipteran (mosquito and blackfly), or coleopteran (beetle) larvae. In addition, some recently discovered strains are reported to be toxic to protozoan pathogens, nematodes, and mites. *B. thuringiensis* ICPs are environmentally sound alternatives to chemical insecticides. It is expected that their use in insect control will increase in the near future. They are applied either as a spray of natural organisms or in genetically modified organisms such as plant-associated bacteria, cyanobacteria, or transgenic plants. In this context, experimental work has shifted towards developing crops transformed with ICP genes to become insect resistant. The use of these insecticidal transgenic plants will eliminate stability problems and should significantly reduce the cost of insect control. More than 50 crops, such as cotton, corn, soybean, and rice, have now been successfully transformed.

(ii) **Viruses as pesticides:** Baculoviruses are commonly used as pesticides. These are DNA viruses and can be placed in three groups:
   (a) Nuclear Polyhedrosis viruses (NPVs)
   (b) Grandulosis viruses (GVs)
   (c) Nonoccluded viruses (NOVs)

There are certain other groups of viruses that belong to this category. These are:

(a) Cytoplasmic polyhedrosis viruses
(b) Entomopox viruses
(c) Iridoviruses

Viruses have undergone extensive evaluation and have been registered for use against insects. Due to their specificity and effectiveness, viruses have the greatest potential for use in the control of integrated pest management (Jukes and Van der Merwe, 2020). Among several types of insect, viruses are nuclear polyhedro virus (NPU) and granulosis virus (GV). Insect viruses belong to the family Baculoviridae. Among viral insecticides occluded viruses are beneficial, as they are made up of proteinaceous and silicaceous covers, which are resistant to harsh environmental factors such as temperature, UV radiation, etc. Approximately 280 insect viruses have been described. Nuclear polyhedrosis viruses are rod-shaped viral particles embedded within distinct polyhedra inclusion bodies. These inclusion bodies are 0.5–3.0 micrometers and are visible under a microscope. To nucleocapsids or virions in each polyhedron is made up of double-stranded DNA that measures upto 430 mm in length. Majority of the NPV viruses have a narrow range of insecticidal activity. However, some isolates from alfalfa caterpillar *Autographa californica* have a broad spectrum of insecticidal activity, and show a target of up to 40 Lepidoptera species.

Since the virus requires living media to grow, large-scale production is still in the infant stages. The needs of larvae for NPV virus culture have been registered in the USA. Several countries like Denmark, Sweden, USA. Australia has already implemented the practice of using viral insecticides. In India, some promising works have been carried out on the use of viruses on some species. The recombinant DNA technology is being used to insert new sequences into the gene for polyhedral proteins which can be used for the synthesis of novel proteins. These new proteins could induce the protein toxins of *Bacillus thuringiensis*, by which the potential toxic effect of viruses is enhanced. Although most viruses grow in living insects, a few can be grown in insect cell culture, providing more uniformity.

17.3.4 Use of Enzymes in Bioremediation and Replacement of Harsh Chemical Processes

Tobacco plants with bacterial genes that control the synthesis of enzyme-detoxifying trinitrotoluene, TNT, have been developed. Through the introduction of genes controlling specific degradation pathways, these initiatives could improve bioremediation processes. Several attempts have been made to minimize waste at the source, especially in food processing, by applying enzymes. These include the following:

(a) **Hemicellulases and cellulases:** These are used to hydrolyze hemicelluloses in pulp to reduce consumption of bleaching chlorine in the pulping process. There is, thus, a reduction in the generation of chlorinated organic wastes in bleached pulp products (Land et al., 2015).

(b) **Invertases:** These are used in the recycling of effluents from food and dairy industries where residual starch content is 5–10%. It allows the growth of various harmful microbes: Effluents > Biofilters > Starch > Microbial invertase > Glucose > Animal feed.

(c) **Desulfurization of coal:** Coal, during combustion as a fuel, releases pollutants into the environment. SO_2 and nitrogen oxides affect air quality. Sulfurous emissions from the combustion of coal are cited as the leading contributors to acid rain. It is the largest problem in the current use of coal and the biggest constraint on its increased use. Attempts are being made to bioprocess coal by its:
   (i) Beneficiation to remove undesirable contaminants (S, N, and other trace metals),
   (ii) Conversion (by microbial liquefication, microbial solubilization, and gasification) into non-pollutant form.

Leather and textiles are among the major environmental polluting industries. Industrial discharge can be reduced by the use of enzymes through the recycling of water (Sharma et al., 2021). It will cut down the electrical and water bills and improve the quality of the final products. For this purpose, only simple adjustments are required in the plants besides replacements of 140 harsh chemicals with biological systems. Enzymes and microbes could easily be produced locally with minor additions. The industry will be set to become competitive with a reduced clean-up bill, increased earnings, and turnover.

17.4 OFF-SITE UTILITY OF VARIOUS ORGANIC SOLID WASTES FOR TRANSFORMATION INTO VALUE-ADDED PRODUCTS

A huge amount of solid waste biomass is generated every year, estimated to about more than 700 million tons by various agricultural practices and agro-based industries (Tu and Hallett, 2019). However, only 2% of the biomass is utilized by humans, and a considerable amount of biomass is sadly discarded (De Bhowmick et al., 2018). However, because of its renewable nature, food crop waste is gaining increasing research interest. Furthermore, it can potentially be converted into biofuels (Panahi et al., 2020), value-added chemicals, and cheap sources of energy for microbial fermentation as well as be exploited as a substrate for enzyme production.

Solid wastes originating from agriculture, forestry, industrial sources, and households primarily constitute the lignocellulosic biomass and it appears to be the perfect alternative for the production of various value-added products. Different lignocellulosic residues including palm wood (Sathendra et al., 2019), sugar cane bagasse (Prajapati et al., 2020), corn stover (Alavijeh et al., 2020), corn cob (Sun et al., 2018), rice straw (Roberto et al., 2020), wheat straw (Yudianto et al., 2019), sweet sorghum bagasse (Trakulvichean et al., 2019), empty palm fruit bunch (Choojit et al., 2020; Sukhang et al., 2020), and deoiled rice bran (Chugh et al., 2016) have been exploited for generation of a variety of value-added products including enzymes and biofuels. Figure 17.11 depicts the range of products that can be prepared from agricultural residues. Therefore, considering the long-term cost analysis of the products, there will always be uncertainties related to the feedstock cost curve as the demand increases. Thus, there is a need to explore new potential food crop waste residues that are cost-effective as well as readily available.

17.4.1 Composting of Food Crop Waste for the Production of Organic Manure

Composting is defined as a controlled biological activity on food crop waste residues under aerobic conditions, in which organic matter is decomposed into materials with shorter molecular chains, which are more stable, hygienic, humus-rich, and ultimately beneficial for crops and soil organic matter recycling (Sequi, 1996). Different microorganisms, such as bacteria, fungi, actinomycetes, algae,

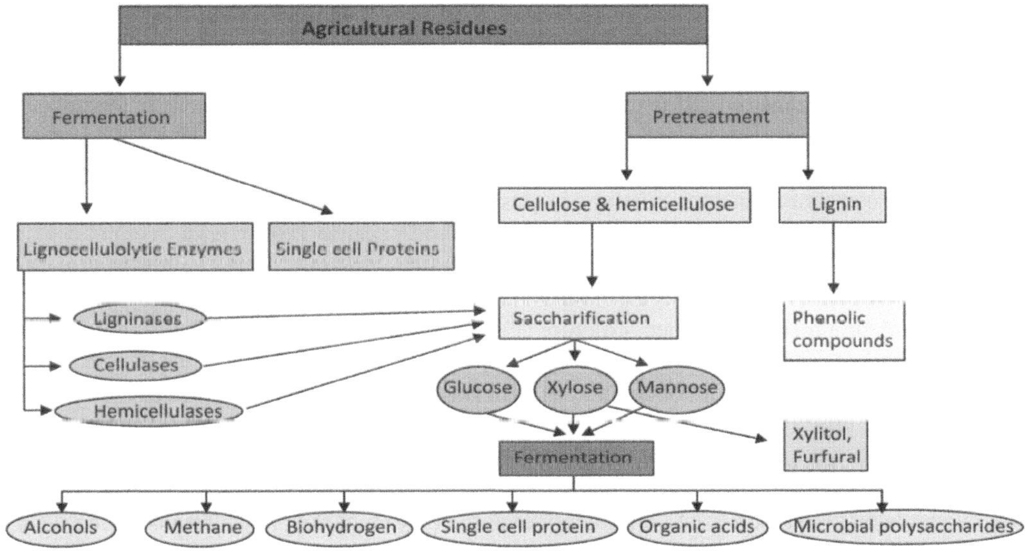

FIGURE 17.11 Range of products that can be prepared from agricultural residues.

and protozoa, mediate the process, which involves complete oxidation of organic matter, leading to the formation of a residue, known as compost, which is rich in various forms of simpler organic compounds, also known as humus, and inorganic matter including carbonates, nitrates, sulfates, phosphates, potassium, etc. and is used manure. Compost improves soil tilth and structure, increases the soil's ability to hold water and nutrients, supports living soil organisms, aids in the dissolution of mineral forms of nutrients, buffers soil from chemical imbalances, provides biological control of certain soil pests, and helps to return organic materials to the soil and keep them out of landfills, among other things. Compost can be used as a mulch, a liquid fertilizer, or as a soil amendment. Decomposition is dependent on the presence of the appropriate microorganisms, and different microbial consortia have been designed by various researchers to facilitate the quick composting of large piles of crop waste leftovers into organic-rich soil. Agricultural straw compost takes roughly 45 days to prepare on average (Kumar and Singh, 2021). Some invertebrates, such as millipedes, sow bugs, snails, and slugs, eat the organic residues of crop waste residues, shredding them to create more surface area for fungi, bacteria, and actinomycetes to operate on, which are then devoured by creatures like mites and springtails. Many kinds of worms, including earthworms, nematodes, redworms, and potworms, eat decaying vegetation and microbes and excrete organic compounds that enrich compost. The chemical and biological modification of crop wastes, as well as microbial synthesis, are all important in the production of humus. The humification is attributed to fresh microbial production or oxidative polymerization of phenolics derived by the catabolism of lignin, tannins, and polyphenols.

17.4.1.1 Technologies for Composting

Depending on the amount of food crop wastes, composting systems are divided into two categories. Composting household garbage requires only simple equipment such as a barrel, wooden box, or pile. Such technology is usually uncontrollable, and its operation is influenced by factors such as material composition, room temperature, precipitation, and so on. Despite the low productivity of household composting plants, widespread use can result in significant savings in waste collection and transportation costs, the need for expensive and space-consuming equipment is eliminated, self-contained compost is used for personal needs, soil properties are improved, and pollution and odor emissions are reduced. Composting household and small company trash is thus an appealing and socially helpful activity. The goals of industrial composting plants are to reduce waste volumes, generate high-quality compost, control composting process parameters, and reduce pollution. Sorted municipal, agricultural, and food sector waste is composted in industrial composting operations. Because these facilities process enormous amounts of heterogeneous waste, the chemical composition of the provided streams is constantly changing. This makes complex the control and management of the process. Also, mixed or industrial waste includes impurities that can be poisonous. According to composting technology and equipment design, composting facilities fall into two categories (Gajalakshmi and Abbasi, 2008): composting in open piles and in closed reactors (Figure 17.12).

Composting on open piles is the most popular method because it uses basic technology and allows trash of various compositions to be composted in huge volumes. In the composting region, waste is placed into 2-m-high and 2.5–3-m-wide long trapezoidal brick piles with perforation (Figure 17.13a), or a metal or wooden structure (Figure 17.13b). The piles are produced using specialized equipment, and their length is gradually increased. Because varied ages of composting material and microorganisms are obtained, different stages of the composting process occur along the length of the pile. The evaporation from the surface and the absence of oxygen in the deeper layers of the piles are two disadvantages of the open pile technique. To overcome the constraints of the open pile system, certain changes have been made, such as transforming the open pile into an aerated pile by laying perforated internal pipes at the bottom of the pile and transferring the air through an air compressor (Figure 17.14a).

Closed-air composting systems (Figure 17.15a) are better managed, odor emissions are minimized, the process is shortened, and higher-quality compost is created (Bari and Koenig 2012). When the volume of garbage is minimal and composting in piles is not practicable owing to

FIGURE 17.12 (a) Open pile composting system and (b) closed reactor composting system.

sensitive environmental conditions, the systems are utilized to compost mixed waste. Figure 17.15b shows how tunnel composting facilities are similar to container bins. They are made in greater sizes, with lengths of 40–70 m and widths of 2–5 m. Self-propelled machinery, conveyors, and hydraulic loaders fill them. The composting gallery can be recharged once during the composting cycle or in little increments every day or every other day. Fresh compost is extracted from the composting unit daily in the latter situation, and it might be matured in a different container or on-site. The tunnel's floor must be slanted to allow fluids to escape. In biofilters, exhaust gases are cleansed, and undesirable odors are eliminated. Anaerobic waste treatment can sometimes be done

a

b

FIGURE 17.13 (a) Perforated brick pile composting system and (b) metal cage composting system.

FIGURE 17.14 (a) Pile composting with air compressor.

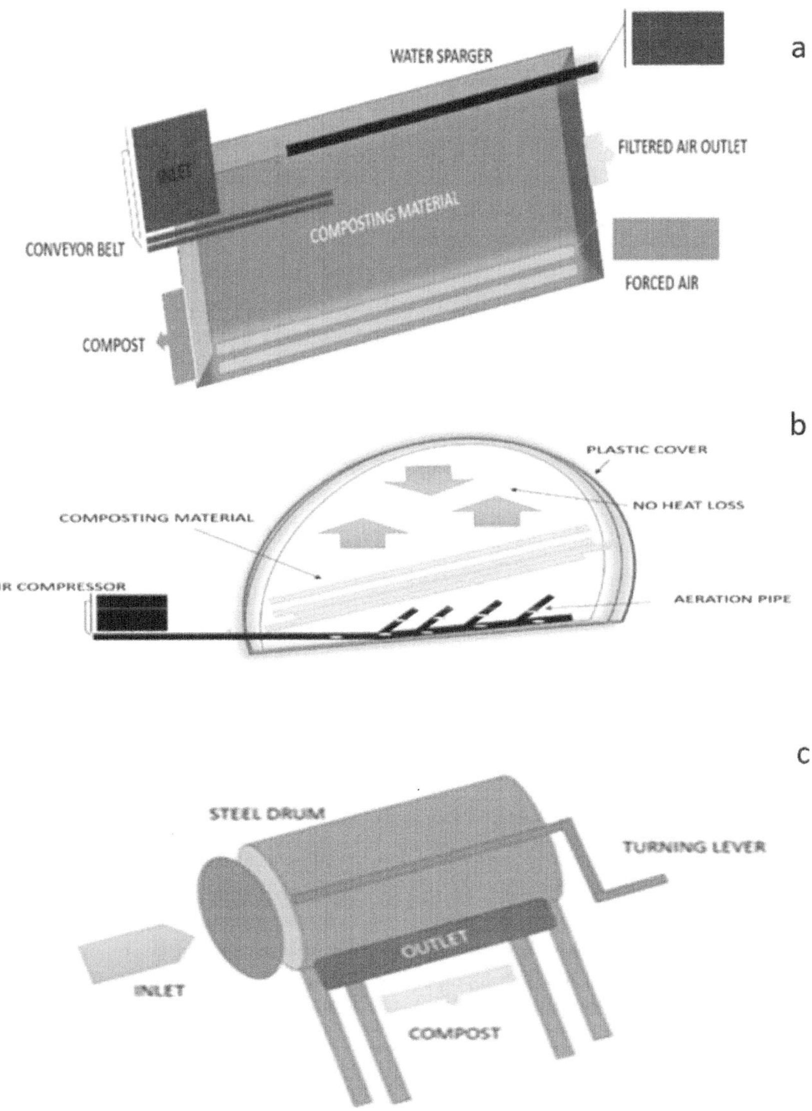

FIGURE 17.15 (a) Closed air composting system, (b) tunnel composting system, (c) rotary drum bioreactor composting system.

in sealed tunnels. The treated materials must also be hydrated in this scenario. Tunnel composting facilities are highly efficient, treating up to 15–20,000 tonnes of garbage each year (Gajalakshmi and Abbasi, 2008). Figure 17.15c shows a rotary composting machine that is similar to a closed-type concrete stirrer with thermal insulation. They are lightweight, take up little room, and can be used in biogas plants as 7 Small-Scale Composting Bioreactors.

17.4.2 Production of Industrial Enzymes from Food Crop Waste

Industrial enzymes are produced by growing suitable microorganisms including fungi and bacteria on various food crop residues by solid-state fermentation (SSF) as well as submerged state fermentation (SmF).

17.4.2.1 Solid-State Fermentation of Food Crop Waste Residues for the Production of Industrial Enzymes

SSF is a fermentation process in the absence of free-flowing water using either a natural or inert solid support. The choice of solid material used is a very crucial factor in SSF, and it should be insoluble, and act as both physical support as well as a source of nutrients for the microbes. This mimics their natural habitat and stimulates the production of enzymes as well as other industrially valuable metabolites. This fermentation is economical, as it makes the use of food crop waste as a medium or substrate for the growth of microorganisms to produce cellulases, hemicellulases, pectinases, amylases, and ligninases. SSF has the advantages of higher productivities of fermentation, high stability of the product, the higher concentration of the product, lower chances of contamination because of lower water activity demand, and growth of microbes specialized for water-insoluble substrates. Some other advantages are the use of simple instrumentation, compactness of fermenter due to lesser volume of water, no foam formation, higher fermentation capacity, lower catabolic repression, cost-effectiveness, and lesser solvent requirement for downstream processing for recovery of the product. SSF has been extensively used by various research groups for the production of hydrolytic enzymes by fungal and bacterial cultures making use of various food crop waste residues as the substrates (Table 17.4).

TABLE 17.4
Some Examples of the Production of Hydrolytic Enzymes on Various Food Crop Waste Residues by Solid-State Fermentation

Substrate	Microorganism	Enzymes	References
Wheat bran	*Thermoascus aurantiacus, Aspergillus niger*	Cellulases, glucoamylase, protease, xylanase	Negi and Banerjee (2009), Bansal et al. (2011)
Wheat bran	*Trichoderma reesei* and *Neurospora crassa*	Cellulases, mannanase, pectinase, amylases	Bansal et al. (2011)
Rice bran, papaya waste, wheat bran, black gram bran soybean, de-oiled rice bran, kitchen waste	*Aspergillus niger, Aspergillus niger, Achromobacter xylosoxidans*	α-Amylase, cellulases, xylanase, mannanase, pectinase, amylases	Kumar and Duhan (2011), Sharanappa et al. (2011), Janveja et al. (2013)
Brewer's spent grain	*Fusarium oxysporum*	Cellulases	Leite et al. (2019)
Wheat straw, paddy straw, sugar cane waste, maize straw	*Bacillus licheniformis*	Cellulase, xylanase, α-amylase	Kaur et al. (2015)
Peanut oil cake, Brewer's spent grain	*Aspergillus oryzae, Aspergillus niger*	Cellulase, xylanase	Sadh et al. (2017), Leite et al. (2019)
Orange peel, apple pomace, coffee husk, wood chips, and rice fiber	Compost	Cellulase	Marín et al. (2019a, 2019b)
Orange peels and exhausted sugarbeet cossettes	*Aspergillus awamori, Bacillus* sp.	Cellulase, xylanase, pectinase	Marzo et al. (2019)
Sugar cane bagasse	*Penicillium* sp., *Rhizomucor* sp., *Trichoderma* sp.	Cellulases	Salomão et al. (2019)
Mango peels	*Aspergillus tamarii*	Pectinase	Amande et al. (2020)
Wheat chaff	*Trichoderma reesei*	Cellulases, xylanase	Jovanović et al. (2020)
Soybean husk and flour mill waste	*Aspergillus oryzae*	Amylase	Melnichuk et al. (2020)

17.4.2.2 Liquid State Fermentation of Food Crop Waste Residues for the Production of Industrial Enzymes

Liquid state fermentation employs the cultivation of microbes in a liquid medium containing the required nutrients either under stationary or shaking conditions. Certain factors make this type of fermentation attractive for the growth of microbes and production of value-added products that include (a) uniform distribution of nutrients for the growth of microorganisms; (b) ease of control of parameters like moisture, temperature, pH, agitation, oxygen and concentration of nutrients; (c) potent technology have already been made applicable with automatic grade and availability of equipment. Liquid state fermentation has been evaluated for the production of cellulolytic, hemicellulolytic, amylolytic, ligninolytic enzymes, urease, and other bioactive metabolites.

Depending on whether the incubation is being carried out in shaking or stationary conditions, liquid type fermentation is categorized into submerged and surface culture fermentation respectively. In surface culture, fermentation microbes grow on the surface of a shallow nutrient medium and use nutrients essential for their growth thus simultaneously liberates products into it. This mode of fermentation does not require agitation in the case of fungi as they are filamentous in nature and agitation may break their mycelia, separating biomass from the liquid medium. But as compared with submerged fermentation in which there are vigorous aeration and agitation, surface culture fermentation has a slower bioreaction rate and long fermentation times. Due to this disadvantage, submerged fermentation is preferred over surface culture fermentation. Various strains of fungi, bacteria, yeast, and algae have been used in fermentation via submerged cultivation. The fermentation media used in these types of fermentation may be synthetically formulated or produced by hydrolysis of lignocellulosic biomass.

Liquid state fermentation has the advantage that the fungal hyphae is not desiccated, as it is continuously bathed in liquid medium, easy temperature control, most efficient, ease of sterilization, and economic method for the mass production of bioagents. Throughout the growth cycle of a microorganism, it is exposed to a constant temperature; oxygen availability to biomass can also be controlled at a particular level of medium saturation except in high-density cultures. Submerged culture has many advantages compared to solid substrates that are easy control of fermentation parameters, such as pH and temperature, better control of contamination, and lesser requirement of space and labor. Fermenters that provide a low-shear environment to the organism are used to understand the nature and magnitude of forces in a bioreactor. Certain factors like the cost of equipment, energy consumption, aeration breakdown, higher productivity, and yield makes surface culture fermentation preferable over submerged fermentation. Different food crop waste residues have also been used for the production of enzymes via liquid culture fermentation with different microorganisms, as revealed in Table 17.5.

17.4.3 Important Value-Added Products Obtained by Transformation of Food Crop Waste Residues

17.4.3.1 Biofuels

Biofuels are the fuels made from biomass, including food crop waste residues. They are classified into liquid biofuels, including ethanol and butanol, and gaseous biofuels, including methane and hydrogen. These are produced from renewable sources and thus may replace the dependence on fossil fuels in the near future; hence, the production of bioethanol, biobutanol, biomethane, and biohydrogen from lignocellulosic biomass is a sustainable alternative (Fatma et al., 2018).

17.4.3.1.1 Bioethanol

Conventional fossil fuels are detrimental in terms of future security, pollution, and geopolitical instability; thus the need for alternative fuels. Bioethanol can be used as a blend with gasoline or directly as fuel. The use of starchy food crops in the first generation of biofuels is in direct conflict

TABLE 17.5
Examples of the Production of Hydrolytic Enzymes on Various Food Crop Waste Residues by Liquid State Fermentation

Substrate	Microorganism	Enzymes	References
Solka-Floc cellulose	*Penicillium brasilianum*	Cellulases	Krough et al. (2004)
Tree leaves	*Pleurotus dryinus*	Cellulases, xylanase, laccase, manganese peroxidase	Elisashvili et al. (2006)
Wheat bran, starch, sugar cane bagasse	*Bacillus* sp.	Xylanase, amylase	Vidyalakshmi et al. (2009)
Partially delignified cellulignin	*Trichoderma harzianum*	Cellulases	de Castro et al. (2010)
Sugar cane bagasse, corn stover	*Acremonium* sp.	Cellulases, xylanase	De Almeida et al. (2011)
Wheat bran, rice bran	*Aspergillus niger*	Cellulase, cellulose, amylase, Pectinase	Parkash et al. (2016)
Corn cob, banana peels	*Aspergillus fumigatus*	Xylanase, pectinase	Elegbede and Lateef (2018)
Pineapple stem	*Bacillus subtilis*	Pectinase	Kavuthodi and Sebastian (2018)
Coffee waste	*Penicillium humicola*	Mannanase	Ismail et al. (2019)
Wheat bran and citrus peel waste	*Bacillus pumilus*	Xylanase, pectinase	Sharma et al. (2019)
Banana peels, coffee residues	*Bacillus subtilis*	Pectinase, mannanase	Blibech et al. (2020)
Peanut shells	*Bacillus paralichniformis*	Cellulases	Irfan et al. (2020)
Wheat chaff	*Trichoderma reesei*	Cellulases, xylanase	Jovanović et al. (2020)
Corn stover	*P. chrysosporium*	Cellulases, hemicellulases	Liu et al. (2020)
Citrus peel waste	*Bacillus safensis, Bacillus altitudinis*	Xylanase, pectinase	Thite et al. (2020)

with the food needs of the population. Second-generation biofuel uses lignocellulosic biomass, which is full of possibilities even up to the level of replacement of fossil fuels. Market share for ethanol is largest due to its use as chemical feed stock coupled with fuel use (Gamage et al., 2010).

Many of such residues like wheat straw, sugar cane bagasse, corn stover, corn cob, brewers spent grain, rice straw, and kitchen waste have been worked upon by various research groups for ethanol production using different pretreatment strategies, varying doses of cellulases, and different consortia of pentose and hexose fermenting yeasts, revealing different degrees of success. Studies on the fermentation of acid hydrolyzed de-oiled rice bran by *Pichia stipitis* NCIM3499 reported the release of 38.50 g/l of reducing sugars. The same, when subjected to fermentation, yielded 12.47 g/l of ethanol after 72 h with a fermentation efficiency of 81.74%. Other observations are ethanol concentration of 15.6 g/L from wet exploded corn stover after enzymatic hydrolysis by cellulases from *Trichoderma reesei* RUT C30 and *Aspergillus saccharolyticus* at an enzyme dose of 15FPU/g at 5% solid concentration after 72 h of hydrolysis. Other studies suggested that SHF of 2.5% H_2SO_4 pretreated wheat straw by in-house produced cellulases from *P. janthinellum* EMS-UV-8 and observed ethanol concentration and yield corresponding to 12.0 g/l and 30.2% respectively. Some studies resulted in 160 g/kg of ethanol productivity from corn stover pretreated with magnesium bisulfate after enzymatic hydrolysis with two commercial enzyme preparations (Celluclast 1.5 l and Novozyme 188) at an enzyme dose of 15 FPU/g after 48 h of hydrolysis. Some reports observed ethanol concentration of 39.5 g/l from low moisture anhydrous ammonia (LMAA) pretreated corn stover by using CTec2 and HTec2 at 7.5% solid loading after 72 h of hydrolysis. One other study reported an ethanol yield of 21.7 g/l and also obtained 83.9 g/l ethanol with Na_2CO_3 pretreated rice straw. Ethanol concentration of 6.47 g/l was also reported by simultaneous delignification, saccharification, and fermentation of rice straw. Researchers optimized bioethanol production from potato peel wastes on inputs of temperature, pH, and solid loading using simultaneous saccharification and fermentation. Subsequently, the kinetics of yeast growth and bioethanol formation under the optimized conditions were assessed using the logistic and modified Gompertz models, respectively. Maximum bioethanol concentration (22.54 g/l) and yield (0.32 g/g) were observed under optimal process conditions of 40°C (temperature), 5.78 (pH) and 12.25% w/v (solid loading). In a recent study, saccharified 0.25N NaOH pretreated rice straw using in-house produced cellulase-hemicellulase consortium from *A. niger* P-19 releasing 70 g/l of reducing sugars with 10%, solid loading and the fermentation of C6 sugars resulted in an ethanol yield of 15.6 g/l (Kaur et al., 2020).

17.4.3.1.2 Biobutanol

Biobutanol has the potential to replace even gasoline in the future. Its salient features include high energy density and miscibility with gasoline in all ratios. It is less corrosive and has low vapor pressure, allowing transportation by the existing structure. It has lesser solubility in water, reducing ground water contamination risk. Lignocellulosic biomass has been used for biobutanol production (Kumar et al., 2016). The main drawback with the use of biobutanol is its low production rate at the industrial level. Usually, the production under fermentation processes is carried out by bacteria, such as *Clostridium* spp. which may use lignocellulosic biomass.

17.4.3.1.3 Biomethane

Biogas generation has been used for meeting thermal and electrical needs at the local level. It is a mixture of methane and carbon dioxide. Lignocellulosic biomass comprising fallen leaves, stover, straws, etc. has been used for the production of biomethane (Kumar et al., 2016). Few recent studies have also explored the potential of potato peels as feedstock for the production of biogas. The performance of anaerobic digestion of potato peels in different inoculum-to-substrate ratios without and with dilute sulfuric acid pretreatment had enhanced the biogas production (Achinas et al., 2019). In addition, the impact of combined treatment with cow manure and pretreatment of potato peels was examined. Comparing the co-digestion to mono-digestion of potato peels, co-digestion in PPW/CM ratio of 60:40 increased the methane yield by 10%. In addition, grinding and acid hydrolysis applied to potato peels were positively effective in increasing the methane amount. Compared to untreated potato peels, pretreatment led to an elevation of the methane amount by 9% and 17% respectively, and alleviated the kinetics of biogas production.

17.4.3.1.4 Bio-Hydrogen

Hydrogen as a fuel has the advantage that it has a very high energy density and the only by-product is water. Wheat bran/straw, corn cob/stalk, potato peels have been used for hydrogen production with productivities ranging from 12 to 7,019 ml H_2/l (Cheng et al., 2011; Soares et al., 2020). This wide variation is due to the source of lignocellulosic, its pretreatment method, and the operational conditions of fermentation.

17.4.3.2 Organic Acids

A significant effort is being made to find bioconversion methods to utilize the bulk lignocellulosic being generated for the production of organic chemicals. Organic acids are used for the synthesis of value-added chemicals. A range of acids has been produced using lignocellulosic biomass (Prado et al., 2005).

17.4.3.2.1 Citric Acid

After ethanol, the second largest fermentation product is citric acid. It is used for medicinal purposes, heavy metal bioremediation, and nanotechnology. Different agro-industrial residues such as apple pomace, coffee husk, wheat straw, pineapple waste, mixed fruit, maosmi waste, cassava bagasse, banana, sugar beet cosset, kiwi fruit peel, cobs, husk, and cotton waste have been investigated for their potential to be used as substrates (Chukwuemeka et al., 2019; Atashnoor and Anvari, 2020). SSF gives high citric acid yield without inhibition related to the presence of certain metal ions such as Fe^{2+}, Mn^{2+}, Zn^{2+} etc.

17.4.3.2.2 Succinic Acid

Succinic acid is used in agriculture and pharma industries as well as for the production of methyl ethyl ketone, adipic acid, 1,3-butadiene, ethylene diaminedisuccinate, and 1,4-butanediol. It is conventionally produced by petro-based chemicals but with poor efficiency. *Actinobacillus succinogenes* 130Z readily converted 29.2 g/l of cellulosic glucose to 17.8 g/l of succinic acid using pretreated sweet sorghum bagasse as the substrate (Dogaris et al., 2020).

17.4.3.2.3 Lactic Acid

Lactic acid has uses in the chemical, food, textile, and pharmaceutical industries. Its demand is expected to rise as it is used in poly-lactic acid and lactate solvents. Poly-lactic acid is an environmentally safe polymer that can be used as an alternative to plastics. Currently, lactic acid is produced by the microbial route as it has a better environmental impact (Rawoof et al., 2021). Various lignocellulosic biomass sources have been used for lactic acid production via SSF using *Rhizopus oryzae*, *Lactobacillus paracasei*, and *Lactobacillus amylophilus*.

17.4.3.2.4 Acetic Acid

Acetic acid is famously known as vinegar, which is used as a preservative in the food industry, has antiseptic usage in the field of medicine, and has found pivotal roles in many industrial processes. A study using *Moorella thermoacetica* (strain ATCC 39073) was conducted for bioconversion of sugars (Iwasaki et al., 2017). Another organism, *Acetobacterium woodii,* was examined for growth on lignocellulosic substrates for the production of acetic acid; with an initial 10 g/l glucose concentration, a yield of 0.76g acetic acid per g glucose was recorded from 6.919 g/l acetate production (Karekar et al., 2020).

17.4.3.3 Microbial Flavors

Several microorganisms, including bacteria and fungi, are currently known for their ability to synthesize different aroma compounds (Cerda et al., 2019; Sharma et al., 2020). Several workers have studied the production of aroma compounds by SSF from several microorganisms such as *Neurospora* sp., *Zygosaccharomyces rouxii*, *Aspergillus* sp., *Ceratocystis fimbriata*, *Rhizopus oryzae*, *Kluyveromyces marxianus*, *B. natto*, and *B. subtilis* using a wide range of food waste residues.

17.4.3.4 Single-Cell Proteins

Single-cell proteins are extracts or whole microbes used as a dietary supplement or ingredient rich in proteins. They are used for human and animal feed. They have an advantage over agricultural practices as they have low water consumption, lesser use of land, low greenhouse gas production, and an all-weather production (Ivanovs et al., 2018). Lignocellulosic biomass can be used for single-cell protein production due to its low cost. A study using lignocellulosic hydrolysates for the production of single-cell protein by *Candida intermedia* FL023 reported high protein content. Rice straw pulp has been used as a substrate for single-cell protein production using *Trichoderma reesei* by solid-state fermentation. Also, *Chaetomium* spp. are well recognized among the group of cellulolytic fungi for their potential to produce a single-cell protein (SCP).

17.5 UTILIZATION OF VARIOUS INDUSTRIAL BY-PRODUCTS FOR TRANSFORMATION INTO VALUE-ADDED PRODUCTS

Various by-products released during usual processes can cause pollution of air, water, and soil, as they act as a potential source of growth of various harmful microorganisms. By using various biotechnological methods, these by-products of chemical, food, and dairy industries can be used for various purposes.

17.5.1 Whey

Whey is a by-product of the cheese-making industry. It is left after the separation of curd (solidified casein and butter fat) (Macwan et al., 2016). Large units release 0.5–1.5 million litres of whey per day. The solid content of whey is very little and is a good substrate of various microbes which can cause problems. Whey can be converted to various products.

(a) **Whey molasses:** Viscous liquid produced by concentrating whey to 70% solid after using reverse osmosis and evaporation.
(b) **Whey permeate:** Liquor containing ~30–35% by-product of whey protein.
(c) **Whey powder:** A powdered protein supplement.

In addition, the microbes which can hydrolyze lactose can grow and release sugar and are used in the generation of food yeast, SCP, or ethyl alcohol.

17.5.2 Single Cell Proteins (SCPs)

SCPs can be produced in single or two stages.

(a) **Single-stage process:***Kluyveromyces fragilis, Candida intermedia* are used for ethanol production.
(b) **Two-stage process:** The lactic bacteria are used to transform the lactose which is then utilized by *Candida krusei* under aerobic conditions.

In addition to this, by using various types of microflora, various products can be obtained.

17.5.3 Molasses

300–360 kg of molasses is formed for each tonne of sugar produced. Several products using microbes can be produced from molasses. These include:

(a) Alcohol: using *Saccharomyces cerevisiae.*
(b) Baker's yeast: using cane/beet molasses. Under highly aerobic conditions, strains of *Saccharomyces cerevisiae* can be grown.
(c) Citric acid: using fungus *Aspergillus niger.*
(d) Glutamic acid: using *Corynebacterium glutamic.*
(e) Acetone-Butanol: using *Clostridium acetobutylicum.*

For the production of bulk products like SCPs, various gases like CO can be used with the help of carboxydotrophic bacteria which synthesize gases (CO and H_2) from coal, lignite, etc.

$$C + H_2O \longrightarrow CO + H_2$$
$$C + \tfrac{1}{2}O_2 \longrightarrow CO$$

These carboxydotrophic bacteria can be used for the production of enzymes, insecticides, and pharmaceutical products. The anaerobic fermentation of organics, in aqueous waste to methane, is a biotechnological application that can be applied to by-product streaming of other chemicals.

17.5.4 Disposable Plastic Starch (Microbial Polysaccharides)

Bio-based polymers are used in different industries like chemical, food, petroleum, health, and bionanotechnology due to diverse functionality, rheological and physico-chemical properties.

Fermentation is used for the production of polysaccharides like schizophyllan, xanthan, dextran, pullulan, curdlan, etc. Microbial species belonging to *Xanthomonas*, *Leuconostoc*, *Sphingomonas*, and *Alcaligenes* are used industrially for the production of xanthan, dextran, gellan, and curdlan. Municipal solid waste (MSW) including sugar cane bagasse (Hilares et al., 2017), kitchen waste (Rishi et al., 2020), sugar cane baggase (Hilares et al., 2017, 2019), Asian palm kernel (Sugumaran et al., 2013), cassava bagasse (Sugumaran et al., 2014) have been exploited as cheap sources of carbon for production of pullulan by *A. pullulans* after the enzymatic hydrolysis. A study conducted using spent/damaged wheat grains as a source of sugars for growth of *Aureobasidium pullulans* 3B2 obtained the highest yield of polysaccharide production of 44.4 mg/gds. It can further be used for film casting, and 2.5% and 5% conc of EPS gave the good and properly peelable film that can be used as bioplastics. The feasibility of the production of various microbial polysaccharides using solid-state fermentation of various food crop residues has also been reported by several workers (Kalaiselvi et al., 2020; Kannah et al., 2020)

We use materials that are synthesized from petrochemicals such as polyethylene, polyvinylchloride, and polystyrene as plastic ware in our daily life. The disposal of these solid wastes becomes a major problem in many modern cities because of the synthetic and recalcitrant properties of petroleum-derived plastics. It is because the production of photodegradable plastics becomes meaningless since many cities use landfills as the means for the disposal of solid waste. Once photodegradable plastics are disposed of in a landfill and covered with soil, the photodegradable plastics cannot be exposed to sunlight and thus remain intact in landfill sites.

The alternative degradable plastics are biodegradable. The first biodegradable plastic was a plastic polymer blended with a biologically degradable polymeric compound such as starch. Once the biodegradable plastic is disposed of, microorganisms in the natural environment can degrade the starch molecules in the copolymer and disintegrate the plastic polymer. However, the physical and chemical properties of the starch-substituted biodegradable plastics are not suitable for practical usage. In addition, a large part of non-degradable plastic polymers remains, residues that can accumulate in our environment and cause pollution problems.

The production of another type of biodegradable plastic that uses PHAs, which have physical and chemical properties very similar to those of polypropylene, has been studied extensively. PHAs are microbial storage polyesters (Koller et al., 2017). In the presence of excessive carbon sources, many microorganisms form PHA granules intracellularly. There are many companies engaged in the large-scale production of PHA biodegradable plastics. One of the examples is "BiopolTM", which is a copolymer of P(3HB-Co-HV) [poly(3-hydroxybutyrate-co-hydroxyvalerate)], produced by Zeneca Bio Products (formerly Imperial Chemicals Industries) in the UK. Later, Monsanto, a leading agricultural chemical manufacturer in the US, applied genetically engineered oilseed rape plants to produce PHA biodegradable plastic polymer. However, the properties of the PHAs produced by the plant have physical and chemical properties different from those produced by microorganisms. After that, another large chemical manufacturer in the US, Union Carbide, also produced microbial biodegradable PHA plastics called "Tone Polymer", while a chemical company in Japan, Showa Denko, produced "Bionolle", which is a chemically synthesized biodegradable PHA plastic. Among these biodegradable plastics, PHB (poly-β-hydroxybutyrate) and its copolymer, P(3HB-Co-HV) [poly (3-hydroxybutyrate-co-hydroxyvalerate)], are the most widespread and thoroughly characterized PHAs (Li et al., 2011). Several bacteria, including *Alcaligenes* spp., *Pseudomonas* spp., and a number of filamentous genera such as *Nocardia* spp., produce these PHAs when they encounter unfavorable growing conditions.

In general, a specific microorganism, which occurs widely in nature, is utilized in a novel transgenic fermentation process to convert carbohydrates into PHA polymers. After the fermentation, the microorganisms have accumulated 80% of their dry weight as PHA polymer. The polymer is purified by breaking open the cells and then harvesting the PHA using an aqueous-based extraction method. Owing to their amorphous character, these biopolymers can be ideally handled as lattices. The biopolymer can be extracted with organic solvents and then purified to the usage form. A non-solvent-based process for the recovery of PHAs from bacteria was reported. The process is to solubilized the bacterial biomass by sequential treatments of heating protease digestion and detergent and leave the bacterial peptidoglycan in the cell wall intact for the ease of separation. Then the PHA granules will be purified by filtration and the final product is in a latex form. The estimated cost of production and extraction of PHA by this route will be US$5/kg, which is still higher than that of synthetic plastics.

PHA polymer can be compounded with safe, biodegradable additives to tailor processing and mechanical properties before being converted into plastic articles using standard techniques. Although stable in normal use, the PHAs can be biologically degraded when deposited in an active microbial environment such as compost. Certain microorganisms can metabolize and consume PHA polymers as their nutrients. Biodegradation of the PHAs under aerobic conditions yields carbon dioxide and water in the same amounts originally used during photosynthesis, completing the PHA biocycle (Meereboer et al., 2020). To test the biodegradability of the PHA plastics a series of tests under laboratory and field conditions should be conducted. There are standard methods published by the American Society for Testing and Materials (ASTM) for the degradation of synthetic and biodegradable plastics under laboratory conditions. A series of microorganisms, including bacteria and fungi, has been characterized as being able to degrade PHA plastics. Most of these microorganisms were isolated from an aerobic environment. Over ten different extracellular PHA depolymerase have been purified and characterized.

The critical factor limiting the use of PHA plastics is their high production cost. Alternatively, one produces novel PHAs with improved physical and chemical properties that suit the specific needs of certain usages. In Hong Kong, there are several studies on the production of biodegradable PHA plastics by microorganisms. Studies have been conducted to investigate the use of synthetic media with various amounts of organic acid (e.g., butyric acid) as the sole carbon source to produce PHA with improved physical and chemical properties by selected microorganisms such as *Alcaligenes eutrophus* and *Klebsiella pneumoniae*. By varying the concentrations of nutrient supplied, the bacteria produced PHA plastics with the ratio of HB:HV (hydroxbutyrate:Hydroxvalerate) of 97:3 and 79:21. These PHA-plastics have different thermal stabilities.

These results were due to the amount and ratio of various organic acids added to control the polymeric composition of PHAs, which in turn change the physical and chemical properties of the PHA plastics. Another approach is to produce PHA plastics with a more economical substrate, such as wastewaters from the food industry and chemical industry with xenobiotic organics. There are studies in which soya wastes from a soya milk diary and malt waste from a beer brewery were successfully used as feedstock to produce PHAs from microorganisms in activated sludge. Microorganisms, mainly bacteria, in the activated sludge fed with various amounts of wastewater from food and chemical industries produced copolymer (p(3HB-Co-HV)) of PHA plastics. Various types of PHAs were produced when the C:N ratio of the wastewater varies, with the amount of PHAs increased as the C:N ratio increased. The result was due to the wide variety of bacteria in the activated sludge that produces various types of PHA. Different bacteria have specific PHA synthetases with different substrate specificities and produce different types of PHAs. Sugars such as maltose, sucrose, and fructose are present in the industrial wastewater as carbon sources for the bacteria to produce PHAs. The use of activated sludge to convert carbon substrates into PHAs not only can produce biodegradable plastics but also serves as wastewater treatment in which the problem of disposal of activated sludge after the wastewater treatment is solved.

Recent advances in understanding the metabolism, molecular biology, and genetics of the PHA-producing bacteria and cloning of more than 20 different PHA biosynthesis genes allow the construction of various recombinant strains that can produce a much larger amount of biodegradable plastics with different types of PHA copolymers. These advances lead to the production of biodegradable plastics with costs that are comparable to or lower than that of synthetic plastics and with greatly improved physico-chemical properties (Zhang et al., 2020).

17.6 WASTE DESTRUCTION TECHNIQUES

This treatment involves the removal of waste or pollutants at the stage of release in the environment. This depends on the type of wastes encountered, whether gaseous, liquid, or solid.

17.6.1 Gaseous Waste Treatment

Many processes cause air pollution in the form of odor nuisance. Very often organic compounds are responsible for this. These odors are perceptible, and therefore become a nuisance at very low concentration. Biological treatment can control point sources of odor nuisances. In this, wastes subjected to purification are passed through biologically active filters having microbes immobilized on organic or inert material.

For this purpose, wet or dry type reactors are used:

(a) **Wet reactor or bioscrubber:** It operates as a packed bed with a counter-current of liquid (very often sewage) and gas (the contaminated air) (Barbusinski et al., 2017). Because of controlled loading, the growth rate is limited. The odorous compounds in the gas are transferred to the liquid phase and then oxidized by flora within the biofilm. This has certain benefits. These include the following:
   (i) The efficiency of scrubbing is high because the concentration of odoriferous molecules by biooxidation in the liquid phase reduces nearly to zero, thereby, mass transfer from the gaseous phase is enhanced,
   (ii) There is a great reduction of liquid volumes needed for scrubbing,
   (iii) There is no ultimate problem with effluents.

(b) **Dry reactors:** In these the beds are packed with biologically active sorptive material (compost, peat, *etc.*). The contaminated gases are blown upwards through beds.

Biofilters allow the purification (removal of toxic materials) from waste gases having phenol and its derivatives, toluene, esters like ethyl and butyl acetate (Chung et al., 2010). This process is used to purify alcohols like methanol, ethanol, butanol, and ketones like acetones from various industrial activities such as chemicals (in polymeric resin production), petrochemicals (oil refining), pharmaceuticals, metallurgical and pesticide products, *etc.* Microbes such as *Pseudomonas putida, Candida tropicalis, Fusarium flocciferium* are used as biofilters for phenol.

An important feed stock solvent is methanol. It cleans burning automotive and turbine fuel. During its manufacture, use, and disposal, there is a loss of methanol vapors to the atmosphere. Immobilized microbial cells and stationary aqueous phase in a porous medium are used in a column for recovery of methanol vapors.

Methanol is utilized as a growth substrate by methylotrophic microorganisms such as *Methylomonas, Aeromonas, Achromobacter, Pseudomonas, Flavobacterium*, etc.

The combustion process produces various gases which are colorless, odorless, and extremely toxic, *e.g.*, CO (carbon monoxide). These include blast furnaces (25–30%); automobile exhaust (0.5–12%); faulty domestic heating systems; and a small amount from a cigarette, cigar, and pipe smoking. A number of bacterial species utilize CO as an energy source. Such aerobic bacteria are collectively known as Carboxydotrophic, *e.g., Pseudomonas carboxydovorans, P. carboxydohydrogens, P. compransoris, Bacillus schlegelli.*

17.6.2 Liquid Waste Treatments

Various biotechnological wastewater treatments are available. The amounts of waste materials discharged into bodies of water around the world are steadily rising. Population increases, the dependence of agriculture on massive amounts of fertilizers and pesticides, expansion of the food industry, and growth of other industrial processes, all contribute to the volume of sewage and wastewater and the content of undesirable substances (Lombardi et al., 2017). The wastewater treatment aims to remove:

(a) Compounds with a high biochemical oxidation demand,
(b) Pathogenic organisms and viruses,
(c) A multitude of human-made chemicals.

Recycling of human, animal, and vegetable wastes has been practiced for a long time. In many cases, it provides fertilizers or fuel. It, however, has also been a source of diseases to humans and animals by residual pathogenicity of enteric (intestinal) bacteria. Concurrently, efficient waste collection and specific treatment processes for deleterious wastes have been developed in urban communities because it is not possible to discharge high volumes of waste into natural land and waters.

17.6.2.1 Wastewater Treatment

A variety of biological treatment systems has been developed. These range from aerobic cesspits, septic tanks, and sewage farms to gravel beds, percolating filters, and activated sludge processes coupled with anaerobic digestion. These systems or bioreactors are used to alleviate health hazards and reduce the amount of biologically oxidizable organic compounds. They produce a final effluent or outflow that can be discharged into the natural environment without any adverse effects.

The efficiency of such bioreactor assemblies relies on the metabolic versatility of mixed microbial populations (microbial ecology). These bioreactors contain a range of microorganisms with the overall metabolic capacity to degrade most organic compounds entering the system.

In industrialized nations, the largest biotechnological industry has been the biological treatment of domestic wastewaters and sewerage. Waterborne diseases, such as typhoid, cholera, and dysentery have been virtually eliminated by the controlled use of microorganisms in these communities. Major epidemics may quickly develop if water and sewage treatments are seriously interrupted. Biotechnology, therefore, besides producing a whole new range of useful products, also plays an important role in the reduction of infectious diseases of humans and animals through water and sewage treatment processes.

The biological disposal of organic wastes involves three stages:

(i) Primary processing,
(ii) Secondary processing,
(iii) Microbial digestion.

A stage involving chemical precipitation may also be included.

(i) **Primary processing:** It is performed to remove coarse particles and solids. This leaves the dissolved organic materials to be degraded or oxidized by microorganisms in a highly aerated, open bioreactor.
(ii) **Secondary processing:** It needs considerable energy input to drive the mechanical aerators that actively mix the whole system. This ensures regular contact of the microorganisms with the substrates and air.
(iii) **Microbial digestion:** The microorganisms multiply and form biomass or sludge that can either be removed and dumped or passed to an anaerobic digester (bioreactor) that will reduce the volume of solids, the odor, and the number of pathogenic microorganisms. Methane or biogas may be generated there. This can be used as fuel. The value of biogas, however, is marginal because of its content of CO_2 and hydrogen sulfide.

Dilute organic liquid wastes can be degraded by the percolating or trickling filter bioreactor. The liquid in this system flows over a series of surfaces, which may be stones, gravel, plastic sheets, *etc.* The microorganisms attached to them remove organic matter which is used by them for their essential growth. Excessive microbial growth can be a problem. It creates blockages and loss of biological activity. Such techniques are widely used in water purification systems. The trickling filter tank is 3–10 feet deep and consists of a bed of crushed stone, gravel, slag, or some other inert support material on which the microbial population forms a thin film. The liquid waste is evenly applied to the top of the filter and percolates downwards, depositing organic matter on the support and microbial film. Aerobic conditions are maintained by the upward flow of air through the spaces between the particles of the packed solid support. Heterotrophic organisms in the upper part of the filter obtain energy and nutrients by oxidizing the organic matter. Their number thus multiplies and a new film is formed (Daigger and Boltz, 2011). The mixed microbial population in the upper layers of trickling filters mainly consists of aerobic chemoheterotrophic bacteria and fungi like *Zoogloea, Pseudomonas, Alcaligenes, Ascoidea, Fusarium, Sepedonium*, and *Geotrichum*. Certain chemoheterotrophs degrade many organic nitrogen-containing substrates with concomitant release of ammonia. In the lower layers of the filter, the autotrophic nitrifying bacteria oxidize ammonia to nitrite (*Nitrosomonas*) and then nitrate (*Nitrobacter*). BOD is reduced to about 95% in the trickling filter tank.

In the activated sludge process, the sewage that has received primary treatment is mixed with activated sludge (an inoculum of microorganisms) and continually aerated. The organisms that degrade organic matter are the same as those in the trickling filter tank but slime-forming bacteria, primarily *Zoogloea remigera*, play a particularly important role. As these organisms grow, they form clumps called flocs to which soluble organic matter, as well as protozoa and other organisms, become attached. When the effluent from an activated sludge tank passes into a sedimentation tank, these flocs sediment out and are transferred to an anaerobic sludge digestor. Much of the reduction of BOD in this process occurs due to the floc-dependent removal of organic matter. A major field of biotechnological interest in the future is microbiological effluent treatment. For treating complex wastes, integrated systems are being developed. Combined algal/bacterial systems for waste and water treatments are developed in countries with high annual hours of sunlight. Such processes can lead to the formation of relatively pure water and algal/bacterial biomass. These may be used for animal feeding and biogas formation. Even the bulk organic chemical formation may be there.

A deep-shaft fermentation system has now been developed for wastewater treatment (Mazumder and Dikshit, 2002). The deep shaft is, in fact, a hole in the ground (up to 150 m in depth), devised to allow the cycling and mixing of wastewater, air, and microorganisms. It is most economical in land use and power and produces much less sludge than do conventional systems. Various biotechnological wastewater treatments are now available.

Anaerobic digestion is also for wastewater treatment (Park et al., 2020). This is very useful for highly polluted wastewater with greater excess of carbon in relation to nitrogen and phosphorus. The microorganisms use carbon in organic molecules as electron acceptors. Sludge digestion is the most widely used anaerobic process. The anaerobic process is used because of the following reasons:

(a) Sludge product is reduced many folds *e.g.*, 0.5–1.5 kg of sludge solids for 1 kg of BOD removed in the aerobic process as compared to 0.1–0.2 kg for each kg of BOD removed in anaerobic process.
(b) There is a drastic reduction in the energy input.
(c) There is a production of an amount of biogas (methane and CO_2) with highly loaded wastewater.

After a period of aerobic digestion, sewage sludge is digested under anaerobic conditions as well. Anaerobic digestion is also used to treat materials with a high content of insoluble organic matter, such as cellulose, and to degrade concentrated industrial wastes such as those from the food processing industry. Anaerobic degradation is a complex multisubstrate multiorganism reaction. It involves the following stages:

- **(i) Stage I:** Polymeric materials like carbohydrates, proteins, and lipids are hydrolyzed to monomers like amino acids, sugars, fatty acids by biomass (XHY), *i.e.*, hydrolyzing microbes.
- **(ii) Stage II Acidogenesis:** The monomeric compounds are fermented to organic acids like propionic butyric acetate by XAG, *i.e.*, acid-generating biomass.
- **(iii) Stage III Acetogenesis:** The organic acids with more than 3 C per molecule are converted to acetic acid and H2O.
- **(iv) Stage IV:** Conversion of acetic acid to CH4 and water along with organic acids with less than 3 carbons by methanogenic bacteria.

The major metabolic stages in the decomposition of organic wastes into methane and CO_2 have been summarized here:

Complex organic compounds
(polysaccharides, fats, proteins)

↓

Hydrolysis by extracellular
bacterial enzymes (hydrolysis)

↓

Monomeric compounds
(sugars, fatty acids, amino acids)

Acidogenesis ↓

Higher organic acids

Acetogenesis ↓

Acetic Acid, H_2, CO_2

Methanogenesis ↓

CH_4

The degradative and fermentative reactions in the anaerobic secondary treatment process can be divided into two stages:

(a) Acid forming
(b) Methane forming

Acid forming stage: In this complex organic polymers, including carbohydrates, fats, and proteins are hydrolyzed by extracellular hydrolytic enzymes (polysaccharides, lipases, and proteases) and converted to volatile short-chain fatty acids, alcohols, and ketones. The fatty acids are fermented to acetate, carbon dioxide, and hydrogen. Numerous different bacterial species participate in the acid-forming stage. These species include some that are known to be obligate aerobes, but that may grow by utilizing alternative electron acceptors, such as nitrate.

Methane forming stage: It restricts anaerobes of genera *Methanobacterium. Methanobacillus, Methanococcus*, and *Methanosarcina* convert the acetate, hydrogen, and carbon dioxide to methane. The success of anaerobic digestion depends on the cooperative interaction between microorganisms with different metabolic capabilities. An important feature of anaerobic digestion is that most of the free energy present in the substrate is conserved in the methane that is produced. The use of methane for energy generation offsets the cost of the sewage treatment process.

Retention time is significant in anaerobic stage sludge digestion. It is because of the relatively long mean generation time of methanogens and acetogens. Therefore, reactors in which biomass is retained, wash out of slower growing microbes is avoided. In the anaerobic contact process, sludge is recirculated from a settling tank back to a fully mixed digester. By increasing the temperature on the digester, reaction rates are improved. For this purpose, warm influent is used in the food industry or separated digester heating is carried out (Kim et al., 2011).

17.6.3 Biological Pulping in Paper Manufacture

Making paper from wood requires the separation of wood fibers from each other and then reforming them into a sheet. The objective of all pulp-processing operations is to efficiently remove lignin, which glues together the fibers, without damaging the valuable cellulose fibers. This is done by various mechanical and chemical processes called luping, but they suffer from disadvantages like damage to cellulose fibers, high costs, high energy use, and corrosion.

A lignin-degrading and modifying enzyme has been isolated from *Phanerochaete chrysosporium* and is found to be useful since it may reduce energy costs and corrosion thus increasing the life of the system (Singh and Chen, 2008). These also reduce environmental hazards associated with bleach plant effluents. Treatment of pulp with ligninolytic fungi such as *Trametes Versicolor, P. chrysosporium*, and *Pleurotus ostreatus* also increases the tensile strength of the paper. Research is being carried out in this area to:

(i) Develop fermentation process, including optimizing conditions for the action of the fungus,
(ii) Design a reactor that provides the optimum conditions and
(iii) Scale up the fermentation.

17.6.4 Removal of Various Toxic Materials

Xenobiotics enter the environment through different paths, some as a component of fertilizers, pesticides, and herbicides are distributed by direct applications, others, such as the polycyclic aromatic hydrocarbons, dibenzo-p-dioxins, and dibenzofelltrans, are released by the combustion process. Many kinds of xenobiotics are found in the waste effluents produced by the manufacture and consumption of all the commonly used synthetic products. Xenobiotics released to the environment on a large scale include halogenated aliphatic and aromatic compounds, nitroaromatics, phthalate esters, and polycyclic aromatic hydrocarbons. The local concentration of xenobiotics in the environment depends on the following factors:

(i) The amount of compound released,
(ii) The rate at which it is released,
(iii) The extent of its dilution in the environment,
(iv) The mobility of the compound in a particular environment,
(v) Its rate of degradation.

Many toxic xenobiotics present in the environment can become progressively more concentrated in each link of a food chain, a process called biomagnification (Maddela et al., 2021). These include

DDT, phthalate esters, and polychlorobiphenyls (PCBs). Xenobiotics may be biodegradable, persistent, or recalcitrants. Some microorganisms are metabolically able to carry out biodegradation of some of these compounds. Phenols in an aqueous solution can be removed by treatment with mushroom tyrosinase. The reduction or order of substituted phenol is catechol > *p.* cresol > *p.* chlorophenol > *p.* methoxy phenol. In the treatment of tyrosinase alone, no precipitate was formed but the color changed from colorless to dark brown. The colored compounds are removed by chitin and chitosan which are available in shellfish waste. Tyrosine immobilized on cation exchange resin can be repeatedly used. In just 2 hours, 100% of phenol can be removed by this treatment. In this enzymatic reaction phenol and other aromatic amines are polymerized to form less water-soluble compounds and then precipitates are separated by filtration.

Biodegradation in a particular environment requires the presence of suitable microorganisms. This may involve a complex microbial community. The environment must also be suitable for the growth of these organisms and for any chemical transformation reaction to proceed at a significant rate. Important factors include:

(i) The concentration of toxic chemicals,
(ii) The presence of other substrates and nutrients,
(iii) Temperature,
(iv) pH,
(v) Oxygen concentration, *etc.*

Several possible mechanisms lead to active biodegradation of xenobiotics. Gratuitous biodegradation occurs when an enzyme can transform a compound other than its natural substrate. The prerequisite for this is that the unnatural substrate should bind to the active site of an enzyme in such a manner that the enzyme can exert its catalytic activity. Co-metabolism is the process in which a substrate is modified often by stoichiometric conversion to a single product but is not utilized for growth by an organism that is grown on or metabolizing another substrate. The degradation of a range of compounds proceeds more readily with a mixed culture of organisms.

17.6.5 Heavy Metal Treatment

A major problem is the pollution of the aquatic environment by toxic metals and radionuclides. Heavy metals, even at or below currently permissible levels, depress soil and aquatic microbial biomass, ATP content, and N_2 fixation in soil.

To abate the toxic metal pollution by selectively using and enhancing the natural processes, diverse biotechnological approaches are being utilized by which microorganisms interact with toxic metals.

Detoxification of metal pollution can be achieved by.

(i) Biosorption,
(ii) Extracellular precipitation,
(iii) Uptake by purified biopolymers and other specialized molecules derived from microbial cells.

Microorganisms and microbial products can be efficient accumulators of soluble and particulate forms of metals and radionuclides, particularly from dilute external concentrations. A variety of mechanisms, ranging from physico-chemical interactions (biosorption) to processes dependent on metabolism (*e.g.*, transport or protein synthesis) may be involved.

It has been reported that a resistant phosphatase in *Citrobacter* species cleaves HPO_4^{2+} from phosphate-containing organic molecules and serves to precipitate the metals as cell-bound metal phosphates ($MHPO_4$). Metal phosphate loads can reach upto 9 g/kg biomass. This system can be applied to the treatment of recalcitrant uranium/tributyl phosphate wastes as well. Microbes can accumulate gold and silver and sequester them inside their cell membrane. Gallium can be complexed with bacterial and fungal siderophores normal for iron chelation.

17.6.5.1 Biosorption

Metal uptake takes place by whole biomass (living/dead) via physico-chemical mechanisms such as adsorption or ion exchanges. The metal toxicity may lead to the inactivation of living cells, therefore, most of the living organisms are exploited to decontaminate effluents containing metals at subtoxic levels, *e.g.*, various algal and cyanobacterial blooms encouraged by the addition of sewage effluents reduce the level of Cd, Cu, Zn, Hg, and Fe in mining effluents. Metals are removed from the water column with 99% efficiency.

The nuclear industry currently relies on physico-chemical techniques to treat radioactive effluents like ion exchange, ultrafiltration, and flocculation. Some high-activity effluents may be stored without processing. Most appropriate technologies for development and application to radionuclide removal should be based on biosorption by biomass in the short term, by modified or unmodified biopolymers in the medium term. In the longer term, biopolymers like lignin, chitin, chitosan, and cellulosic have some potential (but this may depend on the application of genetic and protein engineering, *e.g.*, for the production of appropriate metallothioneins and siderophores). The major opportunity in the short term lies with dead biomass, particularly since waste effluents from the nuclear industry are toxic (Singh et al., 2020).

Other mechanisms such as precipitation, particulate entrapment, *etc.* can also be used in such complex systems. By these, metals in the sediments are concentrated in a form that greatly reduces environment mobility and biological availability.

For biosorption, fungi including yeasts are also used. It is because waste fungal biomass arises as a by-product from several industrial fermentations (Mathew et al., 2016). Fungal biomass can remove thorium from some very acidic process streams. The biosorption role of components within fungal cell walls such as chitin and chitosan, a potentially useful biosorption agent, is based on chitin from crustacean shells. Metals containing particles, *e.g.*, Zn dust, magnetite, and other metal sulfides can be removed by fungal mass such as *Aspergillus niger*, citric acid fermentation waste by a combination of absorption and precipitation.

17.6.5.1.1 Biosorption by Immobilization of Biomass

The microbial biomass is immobilized in particles of the specified size and mechanical strength with favorable mass transport properties and a very low weight percentage of inactive additives. Bioadsorbents such as living or dead biomass, including white-rot fungi and various microbial cultures, also form a potential source for the process of bioadsorption (Gupta et al., 2016). The examples of this category are:

- Granulated *Bacillus* preparation for removing Cd, Cu, Cr, Hg, Zn at a range of concentration.
- For small volumes, a fixed bed reactor containing 20 kg of granules is used. A large fluidized pulsed bed system containing 80–90 kg of biomass is used for larger flows. After loading, metals are stripped from biomass.
- Immobilized particles of diameter 0.7–1.3 mm containing *Rhizopus arrbhizus* biomass have 12–23% added polymers. There was an improved removal of uranium at low polymer contents and low particle diameter,
- For the biosorption of Cu^{2+}, Pb^{2+}, Zn^{2+}, Au^{2+} from mixtures, fluidized bed reactor of algenated and polyacrylamide immobilized algae, *e.g., Chlorella vulgaris* and *Spirulina platensis* have been used.

Biosorption using inactive (dead microbial) biomass is useful because

(a) solution toxicity does not affect the biomass biosorption uptake,
(b) there is no biomass requirement,
(c) there is no problem of maintenance of culture purity.

17.6.5.2 Transformations of Metals by Microorganisms

Increasing human needs and anthropological activities immediately degrade the environment by releasing massive amounts of heavy metals into water bodies and food chains. Agricultural operations, mining electroplating, and copper smelting are among the anthropological activities. Heavy metals accumulating in sea water and lakes are a big concern that needs to be addressed. Several metal and metalloid species can be transformed by microbes (Dhuldhaj and Pandya, 2021). This involves oxidation, reduction, methylation, and dealkylation, *e.g.*, Hg^{2+} to Hg (0) with mercuric reductase. Au^{3+} can be reduced to elemental Au and Ag+ to elemental Ag, which are deposited on culture vessels by many bacterial, algal, fungal species. Bacteria, *Zoogloea ramigera* in activated sludge has metal binding capacity. Other bacterial polymers, *e.g.*, polysaccharides from *Klebsiella aerogenes, Arthrobacter viscous, Pseudomonas* species, *Aureobasidium pullulans* also have the metal binding capacity.

17.6.5.3 Detoxification of Metal Ions

Several microorganisms or their enzyme can be employed to detoxify various toxic metals and some examples are discussed hereafter.

17.6.5.3.1 Demethylation and Reduction of Mercury

Several hundred million tonnes of toxic mercury-containing wastes are discharged into our environment. The ionic forms of mercury such as Hg^+ and Hg^{2+} and organic mercury (R-Hg^+) such as methyl mercury and phenylmercury are extremely toxic and even mutagenic to living organisms. Many microorganisms are resistant to the toxic ionic forms of mercury. Mercury resistance found in bacteria can be categorized into two classes, narrow-spectrum resistance (resistant to inorganic mercury only) and broad-spectrum resistance (resistance to inorganic and organic mercury). Among these microorganisms, the mercury resistance of Gram-positive strains such as *Staphylococcus* spp., *Bacillus* spp., and Gram-negative bacteria such as *Pseudomonas* spp. and *Shigella* spp. are extensively studied. Organic mercury has been reported to be more toxic than its inorganic forms since the penetration of organic mercury into microbial cells is comparatively easier and more rapid than inorganic mercury. However, most of the discharged inorganic mercury can be rapidly and easily transformed biologically into organic forms in the environment. Therefore, the broad spectrum mercury-resistant microorganisms are important to detoxify mercury-contaminated sites in our environment.

What is the difference between mercury narrow-spectrum resistance and broad-spectrum resistance in microorganisms? Many studies on the biochemistry and mechanisms of the mercury resistance of Gram-negative and Gram-positive bacteria reveal that the narrow-spectrum resistance bacterium has only an enzyme called mercury reductase that is used to catalyze the reduction of Hg^{2+} to Hg0. Further studies on the mercury resistance genes of *Pseudomonas stutzeri* indicate that a plasmid pPB confers broad-spectrum mercury resistance. The plasmid is self-transmissible. Two pPB regions, separated by 25–30 Kb and sharing homology with the transposon, Tn501 *mer* (Hg detoxification) genes. Two regions were cloned separately and each was shown to carry a cluster of functional and independently regulated *mer* genes. One cluster conferred resistance only to inorganic mercury (with at least *mer*A and other mercury transporting genes such as *mer*T for narrow-spectrum mercury resistance) and other gene clusters, upstream from *mer*A have a novel *mer*B gene encoding for organomercurial lyase (Priyadarshanee et al., 2021).

The microorganisms with broad-spectrum mercury resistance have, in addition to the mercury reductase, an organomercurial lyase that removes the organic residue(s) of organic mercury. The function of organomercurial lyase is to remove the methyl group of methyl mercury and to produce a free methyl group (CH_3) and inorganic mercury (Hg^{2+}). The inorganic mercury can be further transformed into Hg0 by the reduction mediated by the mercury reductase. Thus the broad-spectrum mercury resistant microorganisms can grow in the presence of organic and/or inorganic mercury.

In addition, the elementary mercury (Hg0) is volatile and most of them leave the aquatic environment and enter the aquatic and atmospheric environment. Although the total mercury content of the aquatic and atmospheric environments remains unchanged, the availability of mercury of organisms in the aquatic environment, mainly in ionic form, will be greatly reduced. Thus the reduction of Hg^{2+} to Hg0 (also called mercury volatilization) is one of the most effective methods to detoxify mercury in the aquatic environment. Although many microorganisms contain mercury reductases and organomercurial lyases, only a few mercury reductases and quite a few organomercurial lyases have been purified from bacteria. Most of the properties of these enzymes are based on the purified enzymes from Gram-negative bacteria (*e.g., Pseudomonas* spp.).

17.6.5.3.1.1 Mercury Reductase

Mercury reductases purified from the cellular membrane of Gram-negative bacteria are either dimeric and trimeric with monomers of 58,700 daltons. The native enzymes require FAD (flavin) for reduction function and preferentially use NAD(P)H as an electron donor. A reducible active-site disulfide (cys-135, cys-14) and a C-terminal pair of cysteines (cys-558, cys-559) were identified. All four cysteines are required for efficient Hg^{2+} reduction. These cysteine residues at its active site and the electrons donated by NAD(P)H are transferred via FAD to reduce the disulfur bond (-S-S-) of two adjacent cysteine residues at the active site; and converted them into two cysteine residues (with -SH groups). The electrons captured by the enzyme are then used to reduce Hg^{2+} to Hg0 (Sone et al., 2010).

17.6.5.3.1.2 Organomercurical Lyase

The enzyme is inducible in the presence of organic mercury. It has been reported to be located in the cytoplasm of the bacterial cell. The enzyme is 19,000–20,000 daltons (19–20 KD). Its function is the cleavage of the C-Hg bond of organic mercury by protonolysis. For instance, methylmercury is detoxified by demethylation to mercuric ion by a bacterial enzyme.

Other than mercury reductase and organomercurial lyase, the enzymes involved in the transportation of Hg^{2+} such as *mer*T and *mer*P are also important to the mercury resistance of bacteria (Priyadarshanee et al., 2021).

*Mer*T encodes a transmembrane protein. The molecular biology of *mer*T and *mer*P proteins of Tn501 isolated from *Escherichia coli* carrying the pPB plasmid was studied. Similar to the biochemistry of *mer*A protein, an active site with cys-24 and cys-25 in the first transmembrane region of *mer*T protein was essential for the transport of Hg^{2+} through the cytoplasmic membrane. A cys-33-ser mutation in *mer*P appears to block transport to Hg^{2+} by *mer*T protein, while deletion of whole *mer*P only slightly reduced Hg^{2+} resistance of the bacterium and suggests that a functional *mer*T protein is sufficient for Hg^{2+} transport across the cytoplasmic membrane.

17.6.5.3.2 Reduction of Chromium

Hexavalent chromium ion (Cr^{6+}) is a common and toxic pollutant in soils and waters. Trivalent chromium ion (Cr^{3+}) is less toxic as compared with its oxidized form (*i.e.*, Cr^{6+}). Chemical reduction of Cr^{6+} is feasible and some operations have been practiced. However, the addition of reduced compounds such as NAD(P)H that are potential toxicants for living organisms renders the large-scale application of the chemical method to detoxify Cr^{6+}. Biological detoxification of Cr^{6+}becomes a promising approach to solve the pollution problem of toxic chromium ions. In some cases, chromate (*i.e.*, Cr^{6+}) is used as a terminal electron acceptor during anaerobic respiration by microorganisms. Once Cr^{6+} in chromate is reduced to Cr^{3+} Equation (1), the toxicity of the chromium is greatly reduced. In addition, since Cr^{3+} compounds are insoluble and tend to precipitate out from the aqueous phase, the reduced chromium ion (Cr^{2+}) will not be available for uptake by microorganisms as well as other organisms in the contaminated environments.

$$Cr_2^{6+}\,O_7^{2-} \longrightarrow Cr^{3+}\,O_4^{2-} \quad (1)$$

There are many cases using bioreactors with bacteria to detoxify Cr^{6+} in an aqueous solution. Many bacteria such as *Bacillus* spp., *Pseudomonas* spp., and *Thiobacillus ferrooxidans* are able to reduce Cr^{6+}. Both monoculture of Cr^{6+}-reducing bacteria or bacterial consortium have been used in detoxification of Cr^{6+} in an aqueous solution (Baillet et al., 1998). Bacterial cells immobilized by appropriate matrices (*e.g.*, DuPont Bio-Sep beads) in either packed-bed or fixed-film (biofilm) reactors efficiently reduced Cr^{6+} in the presence of reducing compounds such as sulfite or thiosulfate. Under selected conditions, up to 200 mg/l of Cr^{6+} can be completely reduced to Cr^{6+} within 24 hours. Further study indicates that a chromate reductase is associated with the Cr^{6+}-reducing bacteria such as *Pseudomonas putida* PRS2000. Both cell suspension and cell-free extract can reduce Cr^{6+}. The enzyme is a soluble protein and the crude enzyme activity is heat-labile and with a Km of 40 M of chromate and is NAD(P)H dependent. Neither sulfate nor nitrate affects chromate (Cr^{6+}) reduction either *in vitro* or with intact cells. In the reduction of Cr^{6+} by *Pseudomonas ambigua* G-1 an intermediate, Cr^{5+}, was identified by electron spin resonance during the enzymatic reduction of Cr^{6+}. This suggested that the chromate reductase in this bacterium reduced Cr^{6+} to Cr^{3+} with at least two reaction steps via Cr^{5+} as an intermediate (Rahman et al., 2007).

17.6.5.3.3 Precipitation of Phosphate

Citrobacter freundii, a facultative anaerobic bacterium isolated from marine sediments, was employed in the laboratory to induce the precipitation of carbonate and phosphate minerals under anaerobic conditions. The ability of *C. freundii* ZW123 to precipitate carbonate and phosphate minerals has been proven for the first time. The bacteria's production of carbonic anhydrase, alkaline phosphatase, and ammonium not only supported an elevation in pH but also drove the super saturation and precipitation of carbonate and phosphate minerals during the trials (Sun et al., 2020).

17.6.5.3.4 Control of N_2 in Environment

Due to the excessive chemical synthesis of nitrogen fertilizers, use of internal combustion engines, and xenobiotic compounds, there is an increased N_2 in the wastewater. As a result, there is an increase in the concentration of ammonia (NH_3), which has an offensive smell and also causes eutrophication of water resources.

The bacteria with the appropriate enzyme complement use ammonia as an electron donor for denitrification in the absence of O_2.

$$5NH_4^- + 3NO_3^- \longrightarrow 4N_2 + 9H_2O + 2H^+$$
$$DG0 = -1483.5\ KJ$$

Here chemolithotrophic bacteria are involved. They are used in the Anamox TM process. In this, O_2 is not essential for nitrification, therefore, it is a total anaerobic wastewater treatment process that is capable of achieving nitrification and denitrification simultaneously. Several various bacteria have been found to degrade nitrogen-containing heterocyclic pesticides like dixuron, metamitron, *etc.* Metamitron is degraded by organisms like *Rhodococcus* species (Long et al., 2021).

17.6.5.3.5 Removal of Cyanide

Cyanide-rich wastes are yielded during the production of electrophoretic objects, plastic, and steel. They often get mixed to various degrees with phenol, thiocyanate, heavy metals, and organic nutrients. Different types of cyanogenic glycosides are produced by >200 plant species, *e.g., Falk, Sorghum, Cassava, etc. Saccharomyces cerevisiae* was identified as the cyanide-degrading microbe in Baijiu fermentation using metatranscriptomic analysis, and an indigenous strain *S. cerevisiae* MT-1 was isolated from Baijiu fermentation (Shen et al., 2021). Certain microbes have been found to be able to degrade cyanide to non-toxic products. Three enzymes able to degrade cyanide have been identified in *Chromobactrium violacerum.* Most important is cytoplasmic B cyanoalanine synthetase. Other cyanides degrading enzymes are rhodanase, cyanide hydratase, nitrogenases, and cyanide oxidase. Cyanide hydratase is secreted by *Glieocercopora sorghii* (leaf spot of *Sorghum*).

17.6.6 Role of Microbes in Geological Processes

In certain geological processes, microbes act as catalytic agents. These include mineral formation, mineral degradation, sedimentation, weathering, and geochemical cycling. When bituminous coal seams are exposed to air and moisture during mining, microbial pyrite oxidation takes place. Huge volumes of sulfuric acid produced in this way have created pollution on an unprecedented scale. Microbial weathering of building stone such as limestone, leading to defacement or structural changes, are other detrimental acts of microbes.

17.6.6.1 Bioleaching

Microbes are increasingly used to extract commercially important elements by solubilization (bioleaching). Using microbial agents, metals such as cobalt, copper, zinc, lead, or uranium can be more easily separated from low-grade ores. In extractive metal leaching, the biological reactions are usually concerned with the oxidation of mineral sulfides. These specific reactions can be carried out by bacteria, fungi, yeasts, algae, and even protozoa (Mishra et al., 2005). Many minerals exist in close association with other substances such as sulfur, *e.g.*, an iron sulfide that must be oxidized to free the valuable metal. A bacterium *Thiobacillus ferroxidans* can oxidize both sulfur and iron; the sulfur in the ore wastes is converted by the bacteria to sulfuric acid. Simultaneously, this enhances the oxidation of iron sulfide to iron sulfate. The crushed ore (normally in large heaps) is repeatedly washed with a bioleaching solution containing live microorganisms and some essential nutrients (phosphate/ammonia) to the metal that can easily be separated (downstream processing) from the sulfuric acid into which it has been extracted.

Canada, India, the USA, and the countries of the former Soviet Union are using bacterial leaching to recover uranium from low-grade ore (0.01 to 0.5% U3O8). Four thousand tons of uranium is extracted per year by the USA alone in this manner. It is primarily used as fuel in nuclear power generation. The microbial recovery of uranium from otherwise useless low-grade ores is an important contribution to energy production. Bioleaching of uranium ores is an important contribution to the economics of nuclear power stations. Bioleaching also helps in the recovery of uranium from low-grade nuclear waste.

Bioleaching technology is, thus, of significant importance for extracting the increasingly scarce metals necessary for modern industry. However, the process of bioleaching is relatively slow. As much as 40% of organic sulfur has been removed in the bench process. The process is bacterial leaching, *i.e.*, extraction or solubilization of metals from minerals contained either in rock matrix or in the form of concentrated mineral products from processing plants. Since most of the metals occur mainly as sulfides, acidophilic thiobacilli from family Thiobacteriaceae, *e.g., Thiobacillusferrooxidans, T. thiooxidans, T. acidophilus. Thiobacillus* substrates are used for desulfurization. For example: oxidation of pyrite (FeS_2 minerals form of ferrous sulfide) to ferro sulfide.

Pyrites are oxidized as follows:

(a) Autooxidation of ferrous sulfide in the presence of O2, H2O, and H2SO4.

$$2FeS_2 + 7O_2 + 2H_2 \longrightarrow 2FeSO_4 + 2H_2SO_4 \quad \text{(i)}$$
$$2FeS_2 + 2H_2SO_4 + O_2 \longrightarrow 2FeSO_4 + 2H_2O + 4S \quad \text{(ii)}$$

(b) Oxidation of pyrite in the presence of bacteria.

$$4FeSO_4 + 2H_2SO_4 + O_2 \longrightarrow 2Fe_2(SO_4)_3 + 2H_2O \quad \text{(iii)}$$
$$2S + 3O_2 + 2H_2O \longrightarrow 2H_2O + 2H_2SO_2 \quad \text{(iv)}$$
$$FeS_2 + Fe_2(SO_4)_3 \longrightarrow 3FeSO + 2S \quad \text{(v)}$$

Overall Reaction

$$4FeS_2 + 5O_2 + 2H_2O \longrightarrow 2Fe_2(SO_4)_3 + 2H_2SO_4$$

As in these reactions, S is not produced as an end product, thus, production of toxic gases like SO_2 can be prevented.

Ferric sulfate is a strong oxidizing agent and can dissolve several economically important copper sulfide minerals by the following reactions:

$$Cu\,FeS_2 + 2Fe_2(SO_4)_3 \longrightarrow CuSO_4 + 5FeSO_4 + 2S^0$$

Chalocopyrite

$$Cu_2S + 2Fe_2(SO_4)_3 \longrightarrow 2CuSO_4 + 4FeSO_4 + S^0$$

Chalcocite

$$Cu_5FeS_4 + 6Fe_2(SO_4)_3 \longrightarrow 5CuSO_4 + 13FeSO_4 + 4S^0$$

Bomite

Leaching by $Fe_2(SO_4)_3$ is called indirect because it is independent of the presence of oxygen or microbial action. The rate of metal extraction from sulfide ores depends on the concentration of ferric ion (Fe^{3+}). In the absence of catalysis, the rate of oxidation of Fe^{2+} is very slow and leaching would likewise be very slow.

T. ferrooxidans increases the rate of Fe^{2+} oxidation by tenfold. Indirect leaching depends on the ability of microbes to supply the necessary $Fe_2(SO_4)_3$ by oxidizing Fe^{2+} to Fe^{3+}.

T. ferrooxidans also derives energy from the oxidation of elemental sulfur (S0) to sulfuric acid.

$$2S^0 + 3O_2 + 2H_2O \longrightarrow 2H_2SO_4$$

The sulfuric acid maintains the low pH that is optimal for the acidophilic microbe and suppresses the loss of ferric sulfate by hydrolysis.

$$Fe_2(SO_4)_3 + H_2O_2 \longrightarrow 2Fe(OH)SO_4 + H_2SO_4$$

The sulfuric acid also leaches various copper oxide minerals.

$$Cu_3(OH)_2(CO_3)_2 + 3H_2SO_4 \longrightarrow 3CuSO_4 + 2CO_2 + 4H_2O$$

Azurite

$$CuSiO_3.\ 2H_2O + H_2SO_4 \longrightarrow CuSO_4 + SiO_2 + 3H_2O$$

Chrysocolla

In the direct leaching process *T. ferrooxidans* becomes attached to the mineral particles, and then enzymes associated with the cell membrane catalyze oxidative attack on the crystal lattice of the metal sulfide. Oxidation of minerals proceeds in two steps:

$$CuS + 0.5O_2 + 2H^+ \longrightarrow Cu^{2+}\ S^0 + H_2O$$
$$S^0 + 1.5\ O_2 + H_2O \longrightarrow H_2SO_4$$

Other acidophilic chemolithotrophs believed to be important to the leaching process include *T. thiooxidans, T. acidophilus, T. organoparus, Leptospirilum ferroxidans*, and species belonging to the genus *Sulfolobus*.

Metal can also be concentrated from dilute solution using microorganisms (bioaccumulators). The microorganisms, bacteria, yeasts, and molds can actively take up the metals in various ways and such processes have potential use in extracting rare metals from dilute solutions (Irawati et al., 2017). Similarly, microorganisms are being used to extract toxic metals from industrial effluents and reduce subsequent environmental poisoning. As mentioned earlier, certain plants called smart plants are also helping us in this direction. Bacterial bioleaching also helps in the removal of sulfur-containing pyrite from high sulfur coal. As high sulfur coal produces sulfur dioxide pollution after burning, it is not in use. But it is not feasible to ignore the high sulfur coals because of their increasing availability. Here, the bacterial removal of pyrite (which contains most of the sulfur) from high sulfur coal is bound to help and the process is expected to be economically viable.

For prospecting for petroleum deposits, aliphatic hydrocarbon-utilizing bacteria are also being used. Microbes are used to release petroleum products from oil shelves and tar sands. In all these systems, the natural geological site acts as a bioreactor. It allows water and microorganisms to flow over the ore and be collected after natural seepage and outflow. Recycling by mechanical pumping can also be used. In all these activities, multidisciplinary approaches are required and new biotechnological techniques such as designing an organism for a specific function could yield further benefits. The overall picture of this area of biotechnology is one of the rapid and exciting developments. There is a growing awareness of the value of an unpolluted environment.

17.6.7 Biofeedstock Derived from Wastes

Most chemical feedstocks are also derived from petroleum. Many petrochemicals, such as the molecular feedstocks for polyethylene and plastic are small organic (carbon-containing) molecules that are hooked together to create polymers. The lactic acid molecules are hooked together to create the polymer, polylactic acid, which is marketed under the name Natureworks. Natureworks is a biodegradable polymer that can be used as a starting material for manufacturing packaging materials and fibers for clothing, pillows, and comforters. The bedding company Pacificcourt feather plans to replace 80% of its polyester products with polylactic acid over the next five years.

17.7 METAGENOMICS AND ENVIRONMENT PROTECTION

The growth of biorefineries for the production of biofuels and value-added products is aided by the widespread use of biomass. But the road to use the biomass as feedstock for the production of biofuels, novel biocatalysts, antimicrobial compounds, and value-added products is extremely arduous. This path is long and winding, as the current natural system is incapable of handling all of the tasks involved in decomposing complex biomass (Hug et al., 2018). The reason behind it is that traditional microbiology and molecular biology are still not capable of cultivating approximately 99% of microorganisms. The introduction of next-generation sequencing, such as metagenomics, has revolutionized the study of the "unseen majority" in a variety of habitats. The term "metagenomics" refers to the collection of DNA from all microorganisms in a community without necessarily identifying all of the species involved (Attwood et al., 2019). The current metagenomics research of novel enzymes, novel biocatalysts, and host systems opens a new window for improving the bioenergy industry, which plays a significant part in the bioeconomy (Pandey and Singhal, 2021). Metagenomics profiling is based on 16S RNA for bacteria and archaea, which is a phylogenetic marker and which relies on PCR-based method. The examination of microbes having the capability to produce industrial enzymes in biofuel production, such as microbial glycoside hydrolase, novel biocatalysts is referred to as metagenomics (Alves et al., 2018).

Traditional methods that rely on the ability of microbial cells to proliferate in an artificial environment given by differential and selective media have been culture-dependent procedures. But these commonly used methods are highly laborious and; in assessing the microbial diversity, the methods which are preferred are the sole source of carbon utilization, plate counts, morphology, microscopy, gram staining, analysis of fatty acids. The advancement in next-generation sequencing which comprising of metagenomics approaches enabled researchers for in-depth insight into microbial diversity. The notion of metagenomics has been thoroughly investigated for elucidating the microbiome of adverse habitats, but rising energy demand need the identification of novel enzymes and producer cells from diverse biomass, which can only be performed through metagenomics techniques (Biswas and Sarkar, 2018). Currently, two common approaches have been explored to understand various aspects of the microbiome: amplicon-based metagenomics and shotgun metagenomics. Figure 17.16 shows a topographic depiction of in silico techniques to metagenomics (Kumar et al., 2015; Gracia and Sanchez, 2016).

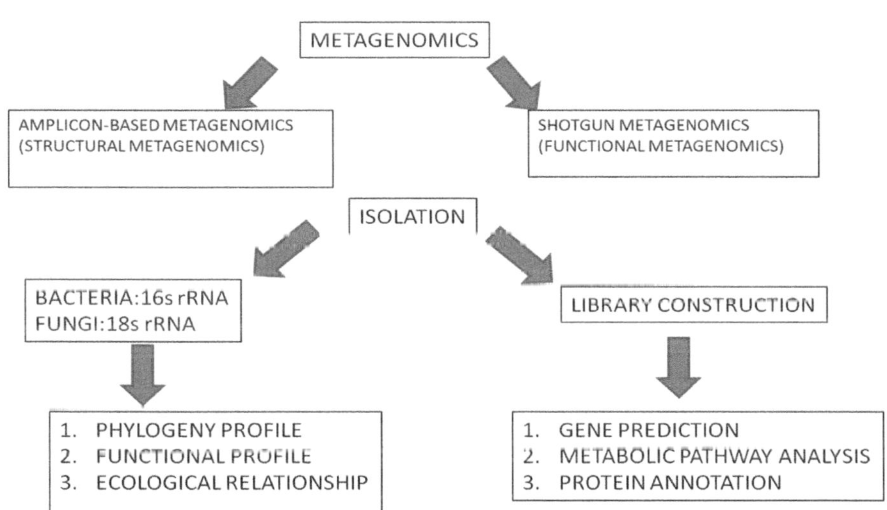

FIGURE 17.16 Topographic depiction of in silico techniques to metagenomics.

Source: Adapted from Kumar et al. (2015); Gracia and Sanchez (2016)

17.7.1 Amplicon-Based Metagenomics (Structural Metagenomics)

Amplicon-based metagenomics is also known as structural metagenomics. In amplicon-based metagenomics, the bacterial diversity is looked upon by analyzing the 16S RNA sequence, whereas internal transcribed spacer (ITS) sections of the 18S rRNA gene have been utilized to identify fungal and eukaryotic diversity. However, a drawback of this structural metagenomics method is that, even if two organisms belong to distinct species, they may have identical 16S rRNA sequences, causing the true investigation to be misled (Jovel et al., 2016).

17.7.2 Shotgun Metagenomics (Functional Metagenomics)

The identification of gene function that expresses the intended function and identifies the phylogenetic analysis of cloned DNA is known as function-driven analysis. In function-based metagenomics, novel proteins are identified based on their functions in medicine, industry, agriculture. Furthermore, post-sequencing steps such as metagenomic sequence assembly, sequence binning, contig formation, Open Reading Function prediction, functional annotation, variant analysis, metabolic reconstruction, community taxonomic profile, and data simulation for correcting sequencing errors play critical roles in decoding the real information from a variety of environmental samples (Kumar et al., 2015).

17.7.3 Metagenomics in Biorefineries and Biofuel Industry

From the past two or three decades, the uncontrolled usage of non-renewable fuels headed us to environmental pollution as well as fuel deficiency. Ultimately, we need to come up with a different approach that doesn't harm the environment and diminishes our dependency on fossil fuels. The best alternate for replacing these fuels is biofuels. The use of lignocellulose biomass as the feedstock for biofuel production is being more studied as the biorefinery industry evolves. The various genes encoding cellulase, hemicellulase, xylanases, ligninases, esterases, lipases are for the depolymerization of various lignocellulosic substrates that has a broader working range at varied temperature, pH, and oxygen level are explored by the use of metagenomics analysis (Chukwuma et al., 2020). Various prokaryotic systems, such as *Thermus thermophilus, Burkholderia graminis, Ralstonia metallidurans, Sulfolobus solfataricus, Caulobacter vibrioides, Pseudomonas putida, Bacillus subtilis, Streptomyces*, and *Proteobacteria* spp., have been used as suitable hosts for compatible enzyme assays (such as esterase, cellulase, lignin peroxidase, and lipase) by heterologous gene expression for metagenomic studies (Craig et al., 2010). Table 17.6 outlines the function of metagenomics in identifying enzymes from diverse microorganisms that can be employed in the development of biorefineries in harsh environments.

TABLE 17.6
Function of Metagenomics in Identifying Enzymes and Their Industrial Application

Depolymerizing Enzymes	Environmental Sources of Microorganisms	Bacterial Sources	Fungal Sources	Industrial Application of Enzymes	References
Cellulases	Bamboo fiber soaking pit & bamboo pulp pit, oil reservoir	*Bacillus brevis, B. thuringeinsis, Cellulomonas fimi, Clostridium* sp., *Prevotella ruminicola, B. ceresus, B. subtilis, Thermotoga maritime, Rhodothermus marinus*	*Aspergillus* sp., *Agaricus bisporus, Coprinus truncorum, Penicillium chrysogenum, Paecilomyces thermophile, Neocallimastix patriciarum, Piromyces* sp.	Paper & pulp industry, Textile industry, health & personal care, food & brewing industry, green biotechnology	Obeng et al. (2017)

Depolymerizing Enzymes	Environmental Sources of Microorganisms	Bacterial Sources	Fungal Sources	Industrial Application of Enzymes	References
Hemicellulase (Xylanases and Mannanase)	Sugar cane bagasse, Norway Hot springs, Sea bottom, Compost	*Thermotoga* sp., *Rhodothermus marinus, Pedicoccus acidilactici, Bacillus brevis, Bacillus thuringeinsis, Arthrobacter* sp., *Clostridium* sp., *Kitasatospora* sp.	*Trichoderma* sp., *Aspergillus* sp., *Rhizomucor pusillus*	Biorefineries, paper & pulp industries, textile industries, chemicals & pharmaceuticals	Bhardwaj et al. (2019)
Laccases (De-lignifying enzyme)	Marine sediment from Qiongdongnan basin, South China Sea	*Azospirillium lipoferum, Bacillus subtilis, S. maltophilia, Streptomyces coelicolor, M. mediterranea*	*Stereum Ostrea, Fomitella fraxinea, Lentinus tigrinus, Ganoderma* sp	Paper & pulp industry, textile industry, health & personal care, food & brewing industry	Desai and Nityanand (2011)
Amylases	Soil, Western Ghats, cow dung	*Bacillus amyloliquefaciens, Bacillus lichenoformis, Bacillus coagulans, Chromohalobacter* sp.	*Aspergillus oryzea, Penicillium fellutanum, Thermomyces lanuginosus, Aspergillus niger*	Biofuel, baking industries, starch processing, brewing industry	Sundarram and Murthy (2014)
Lipases	Anaerobic lagoon of the wastewater treatment plant, German forest soil	*S. aureus, S. hyicus, S. epidermis, Steptomyces* sp., *Pseudomonas* sp.	*Penicillium aurantiogriseum, Rhizopus rhizopodiformis, Rhizopus pusillus, Penicillium restricted, Penicillium simplicissimum, Rhizopus oligosporus*	Biofuel, baking industries, starch processing, brewing	Mehta et al. (2017)
Esterases	Sediment, Compost, Hot Springs	*Bacillus subtilis, Pseudomonas fluorescens, Bacillus pumilis, Escherichia coli, Streptomyces lividans*	*Aspergillus awamori, Monascus* sp., *Rhodosporidium toruloides, Aureobasidium pullulans*	Biofuel industry, paper & pulp industry, food industry, pharmaceutical industries, cosmetic industries	Kulkarni et al. (2013)
Proteases	Surface sand from Gobi and Death valley deserts, Marine sediment soil	*Bacillus amyloliquefaciens, Teredinibacter tumerae, Bacillus pumilus, Pseudomonas aeruginosa, Streptomyces*	*Aspergillus flavus, Rhizopus oligosporus, Phanerochaete radicata, Penicillium* sp., *Aspergillus parasiticus, Beauveria feline*	Detergent industries, dairy industries, leather industries, pharmaceutical industries, baking industries	Dhillon et al. (2016)

TABLE 17.7
Range of Metagenomics in Sustainable Agriculture

Target	Advantages	References
Diseases Epidemiology	Identification of unculturable virus and bacterial plant pathogens, e.g. phytoplasmas and *Candidatus liberibacter*	Padhi et al. (2019)
Plant Growth-Promoting Metabolites	Plant-assisted beneficial microbiomes produce Indole Acetic Acid	Ufarté et al. (2015)
Bioremediation	For the remediation of heavy metal ions and radionucleotides (uranium) from metagenomic approaches, numerous species of denitrifying bacteria and *Geobacter* have been found.	Ufarté et al. (2015)
Hydrolyzing Enzymes	Successful identification of novel enzymes such as monooxygenases, trichlorophenol degrading enzymes, polyaromatic hydrocarbon-degrading enzymes, laccases, phenol and alkane degrading enzymes, carboxylesterases	Ufarté et al. (2015)

17.7.4 Metagenomics in Agriculture

Intensive agricultural methods and urbanization have a catastrophic influence on soil biodiversity, as well as making soils more susceptible to pathogenic diseases. Because of the loss of soil fertility and the overuse of chemical fertilizers, soil health is compromised. Thus, metagenomic methods may be utilized as a guideline to generate fresh creative agricultural solutions for a sustainable agri-economy, reducing crop loss and maintaining soil health. Plant viral pathogen diversity study offers hope for early disease detection while also aiding our understanding of viral ecology and disease epidemiology. Table 17.7 depicts the range of metagenomics in sustainable agriculture.

17.8 SUMMARY

The techniques of biotechnology are providing us with novel methods for diagnosing environmental problems and assessing normal environmental conditions so that we can be better-informed environmental stewards. Methods for detecting harmful organic pollutants in the soil using monoclonal antibodies and the polymerase chain reaction have been developed. The scientists in government labs have produced antibody-based biosensors that detect explosives at old ammunition sites. These methods are not only cheaper and faster than laboratory methods that require large and expensive instruments, they are portable too.

The following industries are to gain from the contributions of biotechnology:

- ***The chemical industry.*** These can use biocatalysts to produce novel compounds. They will reduce waste by-products and improve the purity of chemicals.
- ***The plastics industry.*** There will be a decrease in the use of petroleum for plastic production by making "green plastics" from renewable crops such as corn or soybeans.
- ***The paper industry.*** There will be an improvement in manufacturing processes, including the use of enzymes to lower toxic by-products from pulp processes.
- ***The textiles industry.*** There will be less-toxic by-products of fabric dyeing and finishing processes. With the addition of enzymes to their active ingredients fabric detergents are becoming more effective.

- ***The food industry.*** Baking processes will be improved. There will also be an improvement in fermentation-derived preservatives and analysis techniques for food safety.
- ***The livestock industry.*** To increase nutrient uptake and decrease phosphate by-products enzymes can be used.
- ***The energy industry.*** Enzymes can be used to manufacture cleaner biofuels from agricultural wastes.
- ***Environmental health.*** Different solid waste management approaches also including composting, biofertilizer, and biopesticide can improve the overall health of our environment.

BIBLIOGRAPHY

Achinas S, Li Y, Achinas V, Euverink GJW (2019). Biogas potential from the anaerobic digestion of potato peels: Process performance and kinetics evaluation. *Energies*, 12:2311.

Adams RH, Kanga-Leyva K, Guzmán-Osorio FJ, Escalante-Espinosa E (2011). Comparison of moisture management methods for the bioremediation of hydrocarbon contaminated soil. *African Journal of Biotechnology*, 10:394–404.

Adams GO, Fufeyin PT, Okoro SE, Ehinomen I (2015). Bioremediation, biostimulation and bioaugmention: a review. *International Journal of Environmental Bioremediation & Biodegradation*, 3:28–39.

Alavijeh RS, Karimi K, van den Berg C (2020). An integrated and optimized process for cleaner production of ethanol and biodiesel from corn stover by *Mucor indicus*. *Journal of Cleaner Production*, 249:119321.

Alves LF, Westmann CA, Lovate GL, Siqueira GMV, Borelli TC, Guazzaroni ME (2018). Metagenomic approaches for understanding new concepts in microbial science. *International Journal of Genomics*, 2018:1–15.

Amande T, Adebayo-Tayo B, Ndubuisi-Nnaji U, Ado B (2020). Production and partial characterization of pectinases from mango peels by *Aspergillus tamarii*. *Journal of Microbiology, Biotechnology and Food Sciences*, 9:59–62.

Atashnoor F, Anvari M (2020). Citric acid production by agricultural wastes with *Aspergillus niger* using solid state fermentation and process optimization. *Cellular and Molecular Researches (Iranian Journal of Biology)*, 34: In press.

Attwood GT, Wakelin SA, Leahy SC, Rowe S, Clarke S, Chapman DF, Muirhead R, Jacobs JME (2019). Applications of the soil, plant and rumen microbiomes in pastoral agriculture. *Frontiers in Nutrition*, 6:1–17.

Azubuike CC, Chikere CB, Okpokwasili GC (2016). Bioremediation techniques – classification based on site of application: principles, advantages, limitations and prospects. *World Journal of Microbiology and Biotechnology*, 32(11): 1–8.

Baillet F, Magnin JP, Cheruy A, Ozil P (1998). Chromium precipitation by the acidophilic bacterium *Thiobacillus ferrooxidans*. *Biotechnology Letters*, 20:95–99.

Bamforth SM, Singleton I (2005). Bioremediation of polycyclic aromatic hydrocarbons: current knowledge and future directions. *Journal of Chemical Technology & Biotechnology*, 80:723–736.

Bansal N, Tewari R, Gupta JK, Soni R, Soni SK (2011). A novel strain of *Aspergillus niger* producing a cocktail of hydrolytic depolymerising enzymes for the production of second generation biofuels. *BioResources*, 6:552–569.

Barbusinski K, Kalemba K, Kasperczyk D, Urbaniec K, Kozik V (2017). Biological methods for odor treatment – a review. *Journal of Cleaner Production*, 152:223–241.

Bari QH, Koenig A (2012). Application of a simplified mathematical model to estimate the effect of forced aeration on composting in a closed system. *Waste Management*, 32:2037–2045.

Barman M, Paul S, Choudhury AG, Roy P, Sen J (2017). Biofertilizer as prospective input for sustainable agriculture in India. *International Journal of Current Microbiology and Applied Sciences*, 6(11):1–77.

Bhardwaj N, Kumar B, Verma P (2019). A detailed overview of xylanases: an emerging biomolecule for current and future prospective. *Bioresources and Bioprocessing*, 6:1–36.

Bhatt P, Bhandari G, Bhatt K, Maithani D, Mishra S, Gangola S, Bhatt R, Huang Y, Chen S (2021). Plasmid-mediated catabolism for the removal of xenobiotics from the environment. *Journal of Hazardous Materials*. 420:126618.

Blibech M, Farhat-Khemakhem A, Kriaa M, Aslouj R, Boukhris I, Alghamdi OA, Chouayekh H (2020). Optimization of β-mannanase production by *Bacillus subtilis* US191 using economical agricultural substrates. *Biotechnology Progress*, 36(4):e2989.

Biswas R, Sarkar A (2018) "Omics" tools in soil microbiology: the state of the art. In: Adhya T, Lal B, Mohapotra B, Paul D, Das S (eds) *Advances in soil microbiology: recent trends and future prospects. Microorganisms for sustainability*. Springer, Singapore, pp. 35–64.

Carucci A, Manconi I, Manigas L (2007). Use of membrane bioreactors for the bioremediation of chlorinated compounds polluted groundwater. *Water Science and Technology*, 55:209–216.

Cerda A, Artola A, Barrena R, Font X, Gea T, Sánchez A (2019). Innovative production of bioproducts from organic waste through solid-state fermentation. *Frontiers in Sustainable Food Systems*, 3:63. https://doi.org/10.3389/fsufs.2019.00063

Chen Q, Bao B, Li Y, Liu M, Zhu B, Mu J, Chen Z (2020). Effects of marine oil pollution on microbial diversity in coastal waters and stimulating indigenous microorganism bioremediation with nutrients. *Regional Studies in Marine Science*, 39:101395.

Cheng CL, Lo YC, Lee KS, Lee DJ, Lin CY, Chang JS (2011). Biohydrogen production from lignocellulosic feedstock. *Bioresource Technology*, 102:8514–8523.

Choojit S, Chansiriphota S, Phiboon K (2020). The optimization of ethanol production from oil palm empty fruit bunches hydrolysate by using a statistical experimental design. *International Journal of Integrated Engineering*, 12:261–270.

Chugh P, Soni R, Soni SK (2016). Deoiled rice bran: a substrate for co-production of a consortium of hydrolytic enzymes by *Aspergillus niger* P-19. *Waste and Biomass Valorization*, 7:513–525.

Chukwuemeka IC, Ethel OC, Kalu AD, Chigozie NC (2019). Citric acid production by *Aspergillus niger* using banana and plantain peels. *GSC Biological and Pharmaceutical Sciences*, 8(2). https://doi.org/10.30574/gscbps.2019.8.2.0111

Chung YC, Cheng CY, Chen TY, Hsu JS, Kui CC (2010). Structure of the bacterial community in a biofilter during dimethyl sulfide (DMS) removal processes. *Bioresource Technology*. 101:7165–7168.

Chukwuma OB, Rfatullah M, Tajarudin HA, Ismail N (2020). Lignocellulolytic enzymes in biotechnological and industrial processes: A review. *Sustainability*, 12:1–31.

Craig JW, Chang FY, Kim JH, Obiajulu SC, Brady SF (2010). Expanding small-molecule functional metagenomics through parallel screening of broad-host-range cosmid environmental DNA libraries in diverse Proteobacteria. *Applied and Environmental Microbiology*, 76:1633–1641.

Daigger GT, Boltz JP (2011). Trickling filter and trickling filter-suspended growth process design and operation: A state-of-the-art review. *Water Environment Research*, 83:388–404.

de Almeida MN, Guimarães VM, Bischoff KM, Falkoski DL, Pereira OL, Gonçalves DS, de Rezende ST (2011). Cellulases and hemicellulases from endophytic *Acremonium* species and its application on sugarcane bagasse hydrolysis. *Biotechnology and Applied Biochemistry*, 165:594–610.

De Bhowmick G, Sarmah AK, Sen R (2018). Lignocellulosic biorefinery as a model for sustainable development of biofuels and value added products. *Bioresource Technology*, 247:1144–54.

de Castro AM, Pedro KCNR, da Cruz JC, Ferreira MC, Leite SGF, Pereira N (2010). *Trichoderma harzianum* IOC-4038: a promising strain for the production of a cellulolytic complex with significant β-glucosidase activity from sugarcane bagasse cellulignin. *Biotechnology and Applied Biochemistry*, 162:2111–2122.

Delille D, Duval A, Pelletier E (2008). Highly efficient pilot biopiles for on-site fertilization treatment of diesel oil-contaminated sub-Antarctic soil. *Cold Regions Science and Technology*, 54:7–18.

Desai SS, Nityanand C (2011). Microbial laccases and their applications: a review. *Asian Journal of Biotechnology*, 3:98–124.

Dhakal K, Gadupadi G S, Robertson L W (2017). Sources and toxicities of phenolic polychlorinated biphenyls (OH-PCBs). *Environmental Science and Pollution Research*. 25:16277–16290.

Dhillon A, Sharma K, Rajulpati V, Goyal A (2016) Proteolytiv enzymes. In: *Current developments in biotechnology and bioengineering production, isolation and purification of industrial products*, 1st edn. Elsevier, Amsterdam, pp. 149–173.

Dhuldhaj U, Pandya U (2021). Combinatorial study of heavy metal and microbe interactions and resistance mechanism consort to microbial system. *Geomicrobiology Journal*, 38:181–189.

Dogaris I, Lo E, Brabo-Catala L, Ammar EM, Philippidis GP (2020). Biochemical conversion of sweet sorghum bagasse to succinic acid. *Journal of Bioscience and Bioengineering*, 129:104–109.

Elegbede JA, Lateef A (2018). Valorization of corn-cob by fungal isolates for production of xylanase in submerged and solid state fermentation media and potential biotechnological applications. *Waste and Biomass Valorization*, 9(8):1273–1287.

Elisashvili V, Penninckx M, Kachlishvili E, Asatiani M, Kvesitadze G (2006). Use of *Pleurotus dryinus* for lignocellulolytic enzymes production in submerged fermentation of mandarin peels and tree leaves. *Enzyme and Microbial Technology*, 38:998–1004.

Emami S, Pourbabaee AA, Alikhani HA (2012). Bioremediation principles and techniques on petroleum hydrocarbon contaminated soil. *Technical Journal of Engineering and Applied Sciences*, 2:320–323.

Fatma S, Hameed A, Noman M, Ahmed T, Shahid M, Tariq M, Tabassum R (2018). Lignocellulosic biomass: a sustainable bioenergy source for the future. *Protein and Peptide Letters*, *25*:148–163.

Fernandes AF, Braz VS, Bauermeister A, Paschoal JA, Lopes NP, Stehling EG (2018). Degradation of atrazine by *Pseudomonas* sp. and *Achromobacter* sp. isolated from Brazilian agricultural soil. *International Biodeterioration & Biodegradation*, 130:17–22.

Gajalakshmi S, Abbasi SA (2008). Solid waste management by composting: state of the art. *Critical Reviews in Environmental Science and Technology*. 38:311–400.

Gamage J, Lam H, Zhang Z (2010). Bioethanol production from lignocellulosic biomass, a review. *Journal of Biobased Materials and Bioenergy*, 4:3–11.

Garcia RAB, Sanchez MDR (2016). Lignocellulosic metagenomes to lignocellulolytic genes: trends, challenges and future prospects. *Biofuels, Bioproducts and Biorefining*, 10:864–882.

Gargouri B, Karray F, Mhiri N, Aloui F, Sayadi S (2011). Application of a continuously stirred tank bioreactor (CSTR) for bioremediation of hydrocarbon-rich industrial wastewater effluents. *Journal of Hazardous Materials*, 189:427–434.

Glibert PM, Seitzinger S, Heil CA, Burkholder JM, Parrow MW, Codispoti LA, Kelly V (2005). Eutrophication. *Oceanography*. 18(2):198.

Gupta A, Joia J, Sood A, Sood R, Sidhu C, Kaur G (2016). Microbes as potential tool for remediation of heavy metals: a review. *Journal of Microbial & Biochemical Technology*, 8:364–372.

Hilares RT, Orsi CA, Ahmed MA, Marcelino PF, Menegatti CR, da Silva SS, dos Santos JC (2017). Low-melanin containing pullulan production from sugarcane bagasse hydrolysate by *Aureobasidium pullulans* in fermentations assisted by light-emitting diode. *Bioresource Technology*, 230:76–81.

Hilares RT, Resende J, Orsi CA, Ahmed MA, Lacerda TM, da Silva SS, Santos JC (2019). Exopolysaccharide (pullulan) production from sugarcane bagasse hydrolysate aiming to favor the development of biorefineries. *International Journal of Biological Macromolecules*, 127:169–177.

Honda M, Suzuki N (2020). Toxicities of polycyclic aromatic hydrocarbons for aquatic animals. *International Journal of Environmental Research and Public Health*, 17(4):1363.

Hossain K, Ismail N (2015). Bioremediation and detoxification of pulp and paper mill effluent: A review. *Research Journal of Environmental Toxicology*, 9(3):113.

Hug JJ, Bader CD, Remškar M, Cirnski K, Müller R (2018). Concepts and methods to access novel antibiotics from actinomycetes. *Antibiotics*, 7:1–47.

Irawati W, Parhusip AJ, Christian S, Yuwono T (2017). The potential capability of bacteria and yeast strains isolated from Rungkut Industrial Sewage in Indonesia as a bioaccumulators and biosorbents of copper. *Biodiversitas Journal of Biological Diversity*, 18:971–977.

Irfan M, Bakhtawar J, Shakir HA, Khan M, Ali S (2020). Utilization of peanut shells as substrate for cellulase production in submerged fermentation through Box-Behnken Design. *International Journal of Biological Chemistry*, 12:28–39.

Ismail SA, Khattab OKH, Nour SA, Awad GE, Abo-Elnasr AA, Hashem AM (2019). A Thermodynamic Study of Partially-Purified *Penicillium humicola* β-mannanase Produced by Statistical Optimization. *Jordan Journal of Biological Sciences*, 12:209–217.

Ivanovs K, Spalvins K, Blumberga D (2018). Single cell protein production from waste biomass: review of various agricultural by-products. *Agronomy Research*, 16:1493–1508.

Iwasaki Y, Kita A, Yoshida K, Tajima T, Yano S, Shou T, Saito M, Kato J, Murakami K, Nakashimada Y (2017). Homolactic acid fermentation by the genetically engineered thermophilic homoacetogen *Moorella thermoacetica* ATCC 39073. *Applied and environmental microbiology*. 83:e00247–17.

Jain PK, Gupta VK, Bajpai V, Lowry M, Jaroli DP (2011). GMO's: Perspective of bioremediation. In: *Recent advances in environmental biotechnology*, Jain PK, Gupta VK and Bajpai V (Eds.). LAP Lambert Academic Publishing AG and Co. KG, Germany, pp. 6–23.

Janveja C, Rana SS, Soni SK (2013). Environmentally acceptable management of kitchen waste residues by using them as substrates for the co-production of a cocktail of fungal carbohydrases. *International Journal of Chemical and Environmental Engineering*, 4:20–29.

Jovanović M, Vučurović D, Bajić B, Dodić S, Vlajkov V, Jevtić-Mučibabić R (2020). Optimization of the simultaneous production of cellulase and xylanase by submerged and solid-state fermentation of wheat chaff. *Journal of the Serbian Chemical Society*, 85:177–189.

Jovel J, Patterson J, Wang W, Hotte N, O'Keefe S, Mitchel T, Perry T, Kao D, Mason AL, Madsen KL, Wong GKS (2016). Characterization of the gut microbiome using 16s or shotgun metagenomics. *Frontiers in Microbiology*, 7:1–17.

Jukes M, Van der Merwe M (2020). Viruses as biopesticides. *Quest*, 16:14–5.

Kachhawa D (2017). Microorganisms as a biopesticides. *Journal of Entomology and Zoology Studies*, 5:468–473.

Kalaiselvi P, Parameswari E, Sebastian SP, Davamani V, Suganya K (2020). Potential for economic utilization of cassava starch factory wastes. *Biotica Research Today*. 12:1325–1327.

Kandpal G (2021). Review on impact of chemical fertilizers on environment. *International Journal of Modern Agriculture*, 10:758–763.

Kannah RY, Merrylin J, Devi TP, Kavitha S, Sivashanmugam P, Kumar G, Banu JR (2020). Food waste valorization: biofuels and value added product recovery. *Bioresource Technology Reports*. 11:100524.

Karekar SC, Srinivas K, Ahring BK (2020). Continuous in-situ extraction of acetic acid produced by Acetobacterium woodii during fermentation of hydrogen and carbon dioxide using Amberlite FPA53 ion exchange resins. *Bioresource Technology Reports*. 12:100568.

Kaur PS, Kaur S, Kaur H, Sharma A, Raj P, Panwar S (2015). Solid substrate fermentation using agro industrial waste: new approach for amylase production by *Bacillus licheniformis*. *International Journal of Current Microbiology and Applied Sciences*, 4:712–717.

Kaur A, Rishi V, Soni SK, Rishi P (2020). A novel multi-enzyme preparation produced from *Aspergillus niger* using biodegradable waste: a possible option to combat heterogeneous biofilms. *AMB Express*, 10:1–16.

Kavuthodi B, Sebastian D (2018). Biotechnological valorization of pineapple stem for pectinase production by *Bacillus subtilis* BKDS1: Media formulation and statistical optimization for submerged fermentation. *Biocatalysis and Agricultural Biotechnology*, 16:715–722.

Kim J, Novak JT, Higgins MJ (2011). Multistaged anaerobic sludge digestion processes. *Journal of Environmental Engineering*, 137:746–753.

Koller M, Maršálek L, de Sousa Dias MM, Braunegg G (2017). Producing microbial polyhydroxyalkanoate (PHA) biopolyesters in a sustainable manner. *New Biotechnology*, 37:24–38.

Krough KBR, Morkeberg A, Jorgensen H, Frisvad JC, Olsson L (2004). Screening genus *Penicillium* for producers of cellulolytic and xylanolytic enzymes. *Applied Biochemistry and Biotechnology*, 114:389–401.

Kulkarni S, Patil S, Satpute S (2013). Microbial esterases: an overview. *International Journal of Current Microbiology and Applied Sciences*, 2:135–146.

Kumar A, Duhan JS (2011). Production and characterization of amylase enzyme isolated from *Aspergillus niger* MTCC-104 employing solid state fermentation. *International Journal of Pharma and Bio Sciences*, 2:250–258.

Kumar, A., Gautam, A., and Dutt, D. (2016). Biotechnological transformation of lignocellulosic biomass in to industrial products: an overview. *Advances in Bioscience and Biotechnology*, 7:149–168.

Kumar S, Krishnani KK, Bhushan B, Brahmane MP (2015). Metagenomics: retrospects and prospects in high throughput age. *Biotechnology Research International*, 1–13. https://doi.org/10.1007/s13399-021-01585-9

Kumar SM, Reddy GC, Phogat M, Korav S (2018). Role of bio-fertilizers towards sustainable agricultural development: A Review. *Journal of Pharmacognosy* and *Phytochemistry*, 7:1915–1921.

Kumar M, Gupta A, Srivastava S (2021). Degradation and biotransformation of pentachlorophenol by microorganisms. In: *Climate resilience and environmental sustainability approaches*, Kaushik A, Kaushik CP, Attri SD (Eds.), Springer Nature, Singapore Pte Ltd, pp. 299–318.

Kumar P, Singh R K (2021). Selection of sustainable solutions for crop residue burning: an environmental issue in northwestern states of India. *Environment, Development and Sustainability*, 23(3), 3696–3730.

Land M, Hauser L, Jun SR, Nookaew I, Leuze MR, Ahn TH, Karpinets T, Lund O, Kora G, Wassenaar T, Poudel S (2015). Insights from 20 years of bacterial genome sequencing. *Functional & Integrative Genomics*, 15:141–161.

Ławniczak Ł, Woźniak-Karczewska M, Loibner AP, Heipieper HJ, Chrzanowski Ł (2020). Microbial degradation of hydrocarbons – basic principles for bioremediation: a review. *Molecules*, 25:856.

Leite P, Silva C, Salgado JM, Belo I (2019). Simultaneous production of lignocellulolytic enzymes and extraction of antioxidant compounds by solid-state fermentation of agro-industrial wastes. *Industrial Crops and Products*, 137:315–322.

Li SY, Dong CL, Wang SY, Ye HM, Chen GQ (2011.) Microbial production of polyhydroxyalkanoate block copolymer by recombinant *Pseudomonas putida*. *Applied Microbiology and Biotechnology*, 90:659–969.

Liu J, Yang J, Wang R, Liu L, Zhang Y, Bao H, Jang JM, Wang E, Yuan H (2020). Comparative characterization of extracellular enzymes secreted by *Phanerochaete chrysosporium* during solid-state and submerged fermentation. *International Journal of Biological Macromolecules*, 152:288–294.

Lombardi L, Nocita C, Bettazzi E, Fibbi D, Carnevale E (2017). Environmental comparison of alternative treatments for sewage sludge: An Italian case study. *Waste Management*, 69:365–376.

Long Z, Wang X, Wang Y, Dai H, Li C, Xue Y, Deng Y, Zhang H, Yu Y, Fang H (2021). Characterization of a novel carbendazim-degrading strain *Rhodococcus* sp. CX-1 revealed by genome and transcriptome analyses. *Science of The Total Environment*, 754:142137.

Macwan SR, Dabhi BK, Parmar SC, Aparnathi KD (2016). Whey and its utilization. *International Journal of Current Microbiology and Applied Sciences*, 5:134–155.

Maddela NR, Pinos MR, Ezugwu CI, Reddy KV, Ranjit P (2021). Recent advances in microbial remediation techniques for xenobiotics-polluted soil. *Microbial Products for Health, Environment and Agriculture*, 259–294. https://doi.org/10.1007/978-981-16-1947-2_12

Marghade DT, Chahande AD, Tiwari MS, Patil PD (2021). Microbial degradation of xenobiotic compounds. In: *Recent Advances in Microbial Degradation. Environmental and Microbial Biotechnology*, Inamuddin, Ahamed MI, Prasad R (Eds.). Springer, Singapore, pp. 173–217. https://doi.org/10.1007/978-981-16-0518-5_7.

Marín M, Sánchez A, Artola A (2019a). Production and recovery of cellulases through solid-state fermentation of selected lignocellulosic wastes. *Journal of Cleaner Production*, 209:937–946.

Marín M, Artola A, Sánchez A (2019b). Optimization of down-stream for cellulases produced under solid-state fermentation of coffee husk. *Waste and Biomass Valorization*, 10:2761–2772.

Martínez-Lavanchy PM, Müller C, Nijenhuis I, Kappelmeyer U, Buffing M, McPherson K, Heipieper HJ (2010). High stability and fast recovery of expression of the TOL plasmid-carried toluene catabolism genes of *Pseudomonas putida* MT 2 under conditions of oxygen limitation and oscillation. *Applied and environmental microbiology*, 76:6715–6723.

Marzo C, Díaz AB, Caro I, Blandino A (2019). Valorization of agro-industrial wastes to produce hydrolytic enzymes by fungal solid-state fermentation. *Waste Management & Research*. 37:149–56.

Mathew BB, Jaishankar M, Biju VG, Beeregowda KN (2016). Role of bioadsorbents in reducing toxic metals. *International Journal of Toxicology*, 1–13. https://doi.org/10.1155/2016/4369604

Mawad A, Albasri H, Shalkami AG, Alamri S, Hashem M (2021). Synergistic degradation of phenanthrene by constructed *Pseudomonas* spp. consortium compared with pure strains. *Environmental Technology & Innovation*, 11:101942.

Mazumder D, Dikshit AK (2002). Applications of the deep-shaft activated sludge process in wastewater treatment. *International journal of environment and pollution*, 17:266–272.

Meereboer KW, Misra M, Mohanty AK (2020). Review of recent advances in the biodegradability of polyhydroxyalkanoate (PHA) bioplastics and their composites. *Green Chemistry*, 22:5519–5558.

Mehta A, Bodh U, Gupta R (2017) Fungal lipases: a review. *Journal of Biotech Research*, 8:58–77.

Melnichuk N, Braia MJ, Anselmi PA, Meini MR, Romanini D (2020). Valorization of two agroindustrial wastes to produce alpha-amylase enzyme from *Aspergillus oryzae* by solid-state fermentation. *Waste Management*, 106:155–161.

Mishra D, Kim DJ, Ahn JG, Rhee YH (2005). Bioleaching: a microbial process of metal recovery; a review. *Metals and Materials International*, 11:249–256.

Mishra S, Lin Z, Pang S, Zhang W, Bhatt P, Chen S (2021). Recent advanced technologies for the characterization of xenobiotic degrading microorganisms and microbial communities. *Frontiers in Bioengineering and Biotechnology*, 9:31. https://doi.org/10.3389/fbioe.2021.632059

Mulla SI, Ameen F, Talwar MP, Eqani SA, Bharagava RN, Saxena G, Tallur PN, Ninnekar HZ (2020). Organophosphate pesticides: Impact on environment, toxicity, and their degradation. In: *Bioremediation of industrial waste for environmental safety*. Springer, Singapore, pp. 265–290.

Niu GL, Zhang JJ, Zhao S, Liu H, Boon N, et al. (2009). Bioaugmentation of a 4- chloronitrobenzene contaminated soil with *Pseudomonas putida* ZWL73. *Environmental Pollution*, 57:763–771.

Negi S, Banerjee R (2009). Optimization of extraction and purification of glucoamylase produced by A. awamori in solid state fermentation. *Biotechnology and Bioprocess Engineering*, 14:60–66.

Obeng EM, Adam SNN, Budiman C, Ongkudon CN, Maas R, Jose J (2017) Lignocellulases: A review of emerging and developing enzymes, systems, and practices. *Bioresources and Bioprocessing*, 4:1–22.

Padhi EMT, Maharaj N, Lin SY, Mishchuk DO, Chin E, Godfrey K, Foster E, Polek M, Leveau JHJ, Slupsky CM (2019). Metabolome and microbiome signatures in the roots of citrus affected by Huanglongbing. *Phytopathology*, 109:2022–2032.

Panahi SHK, Dehhaghi M, Aghbashlo M, Karimi K, Tabatabaei M (2020). Conversion of residues from agro-food industry into bioethanol in Iran: An under-valued biofuel additive to phase out MTBE in gasoline. *Renewable energy*, 145:699–710.

Pandey M, Singhal B (2021). Metagenomics: adding new dimensions in bioeconomy. *Biomass Conversion and Biorefinery*. 123.

Park JH, Park JH, Lee SH, Jung SP, Kim SH (2020). Enhancing anaerobic digestion for rural wastewater treatment with granular activated carbon (GAC) supplementation. *Bioresource Technology*, 315:123890.

Parkash O, Tewari R, Sharma A, Soni SK (2016). Cellulase production by *Aspergillus niger* using lignocellulosic substrates and standardization of fermentation process and parameters. *Journal of Multidisciplinary Engineering Science and Technology*, 3:5033–39.

Prado F, Vandenberghe L, Woiciechowski A, Rodrigues-Leon J, Soccol C (2005). Citric acid production by solid-state fermentation on a semi-pilot scale using different percentages of treated cassava bagasse. *Brazilian Journal of Chemical Engineering*, 22: 547–555.

Prajapati B P., Jana U K, Suryawanshi R K, Kango N (2020). Sugarcane bagasse saccharification using *Aspergillus tubingensis* enzymatic cocktail for 2G bio-ethanol production. Renewable Energy, 152, 653–663.

Priyadarshanee M, Chatterjee S, Rath S, Dash HR, Das S (2021). Cellular and genetic mechanism of bacterial mercury resistance and their role in biogeochemistry and bioremediation. *Journal of Hazardous Materials*, 126985.

Purnomo AS, Mori T, Takagi K, Kondo R (2011). Bioremediation of DDT contaminated soil using brown-rot fungi. *International Biodeterioration & Biodegradation*, 65:691–695.

Rahman MU, Gul S, Haq MZ (2007). Reduction of chromium (VI) by locally isolated *Pseudomonas* sp. C-171. *Turkish Journal of Biology*, 31:161–166.

Rawoof SA, Kumar PS, Vo DV, Devaraj K, Mani Y, Devaraj T, Subramanian S (2021). Production of optically pure lactic acid by microbial fermentation: a review. *Environmental Chemistry Letters*, 19:539–556.

Rishi V, Sandhu AR, Kaur A, Kaur J, Sharma S, Soni SK (2020). Utilization of kitchen waste for production of pullulan to develop biodegradable plastic. *Applied Microbiology and Biotechnology*, 104:1307–1317.

Roberto IC, Castro RC, Silva JP, Mussatto SI (2020). Ethanol production from high solid loading of rice straw by simultaneous saccharification and fermentation in a non-conventional reactor. *Energies*, 13:2090.

Robles-González IV, Fava F, Poggi-Varaldo HM (2008). A review on slurry bioreactors for bioremediation of soils and sediments. *Microbial Cell Factories*, 7:5. https://doi.org/10.1186/1475-2859-7-5

Sadh PK, Chawla P, Bhandari L, Duhan JS (2017). Bio-enrichment of functional properties of peanut oil cakes by solid state fermentation using *Aspergillus oryzae*. *Journal of Food Measurement and Characterization*, 12:622–633.

Sequi P (1996). The role of composting in sustainable agriculture, In: *The Science of Composting*, M. Bertoldi, P. Sequi, B. Lemmens, and T. Papi, eds., Blackie Academic & Professional, London, pp. 23–29.

Saravanan A, Senthil Kumar P, Dai-Viet N. Vo, Jeevanantham S, Karishma S, Yaashikaa PR (2021). A review on catalytic-enzyme degradation of toxic environmental pollutants: Microbial enzymes. *Journal of Hazardous Materials*, 419:126451.

Sathendra ER, Baskar G, Praveenkumar R, Gnansounou E (2019). Bioethanol production from palm wood using *Trichoderma reesei* and *Kluveromyces marxianus*. *Bioresource Technology*, 271:345–352.

Salomão GS, Agnezi JC, Paulino LB, Hencker LB, de Lira TS, Tardioli PW, Pinotti LM (2019). Production of cellulases by solid state fermentation using natural and pretreated sugarcane bagasse with different fungi. *Biocatalysis and Agricultural Biotechnology*, 17:1–6.

Schaumlöffel D, Ouerdane L, Bouyssiere B, Łobiński R (2003). Speciation analysis of nickel in the latex of a hyperaccumulating tree *Sebertia acuminata* by HPLC and CZE with ICP MS and electrospray MS-MS detection. *Journal of Analytical Atomic Spectrometry*, 18:120–127.

Schmidt T, Anderson WA (2017). Biotrickling filtration of air contaminated with 1-butanol. *Environments*, 4:57. https://doi.org/10.3390/environments4030057

Sharma A, Sharma P, Singh J, Singh S, Nain L (2020). Prospecting the potential of agroresidues as substrate for microbial flavor production. *Frontiers in Sustainable Food Systems*. 4:18. https://doi.org/10.3389/fsufs.2020.00018

Sharma D, Sharma G, Mahajan R (2019). Development of strategy for simultaneous enhanced production of alkaline xylanase-pectinase enzymes by a bacterial isolate in short submerged fermentation cycle. *Enzyme and Microbial Technology,* 122, 90–100.

Sharma P, Sirohi R, Tong YW, Kim SH, Pandey A (2021). Metal and metal(loids) removal efficiency using genetically engineered microbes: Applications and challenges. *Journal of Hazardous Materials*, 416: 125855.

Sharanappa A, Wani KS, Pallavi P (2011). Bioprocessing of food industrial waste for α-amylase production by solid state fermentation. *International Journal of Advanced Biotechnology and Research*, 2:473–480.

Shen T, Wu Q, Xu Y (2021). Biodegradation of cyanide with *Saccharomyces cerevisiae* in Baijiu fermentation. *Food Control*, 127:108107.

Singh D, Chen S (2008). The white-rot fungus *Phanerochaete chrysosporium*: conditions for the production of lignin-degrading enzymes. *Applied Microbiology and Biotechnology*, 81:399–417.

Singh R (2017). Biodegradation of xenobiotics-a way for environmental detoxification. *International Journal* of *Development* Research, 7:14082–14087.

Singh S, Kumar V, Datta S, Dhanjal DS, Sharma K, Samuel J, Singh J (2020). Current advancement and future prospect of biosorbents for bioremediation. *Science of The Total Environment*, 709:135895.

Soares JF, Confortin TC, Todero I, Mayer FD, Mazutti MA (2020). Dark fermentative biohydrogen production from lignocellulosic biomass: Technological challenges and future prospects. *Renewable and Sustainable Energy Reviews*, 117:109484.

Sone Y, Pan-Hou H, Nakamura R, Sakabe K, Kiyono M (2010). Roles played by MerE and MerT in the transport of inorganic and organic mercury compounds in Gram-negative bacteria. *Journal of Health Science*, 56:123–127.

Srivastava J, Naraian R, Kalra SJ, Chandra H (2014). Advances in microbial bioremediation and the factors influencing the process. *International Journal of Environmental Science and Technology*, 11:1787–1800.

Sugumaran KR, Gowthami E, Swathi B, Elakkiya S, Srivastava SN, Ravikumar R, Gowdhaman D, Ponnusami V (2013). Production of pullulan by *Aureobasidium pullulans* from Asian palm kernel: a novel substrate. *Carbohydrate Polymers*, 92:697–703.

Sugumaran KR, Jothi P, Ponnusami V (2014). Bioconversion of industrial solid waste – cassava bagasse for pullulan production in solid state fermentation. *Carbohydrate Polymers*, 99:22–30.

Sukhang S, Choojit S, Reungpeerakul T, Sangwichien C (2020). Bioethanol production from oil palm empty fruit bunch with SSF and SHF processes using *Kluyveromyces marxianus* yeast. *Cellulose*, 27:301–314.

Sun B, Zhao H, Zhao Y, Tucker ME, Han Z, Yan H (2020). Bio-precipitation of carbonate and phosphate minerals induced by the bacterium *Citrobacter freundii* ZW123 in an anaerobic environment. *Minerals*, 10(1):65.

Sun J, Wang J, Tian K, Dong Z, Liu X, Permaul K, Singh S, Prior BA, Wang Z (2018). A novel strategy for production of ethanol and recovery of xylose from simulated corncob hydrolysate. *Biotechnology Letters*, 40:781–788.

Sun RX, Sun Y, Xie XD, Yang BZ, Cao LY, Luo S, Wang YY, Mai BX (2020.) Bioaccumulation and human health risk assessment of DDT and its metabolites (DDTs) in yellowfin tuna (*Thunnus albacares*) and their prey from the South China Sea. *Marine Pollution Bulletin*. 158:111396.

Sundarram A, Murthy T (2014). α-Amylase production and applications: a review. *Applied and Environmental Microbiology*, 2:166–175.

Tang X, Li Q, Wu M, Lin L, Scholz M (2016). Review of remediation practices regarding cadmium-enriched farmland soil with particular reference to China. *Journal of environmental management*, 181:646–662.

Tekere M, Read JS, Mattiasson B (2007). Polycyclic aromatic hydrocarbon biodegradation by a subtropical white rot fungus in packed bed and suspended carrier bioreactor systems. *Environmental technology*, 28:683–691.

Thite VS, Nerurkar AS, Baxi NN (2020). Optimization of concurrent production of xylanolytic and pectinolytic enzymes by *Bacillus safensis* M35 and *Bacillus altitudinis* J208 using agro-industrial biomass through Response Surface Methodology. *Scientific Reports*, 10:1–2.

Trakulvichean S, Chaiprasert P, Otmakhova J, Songkasiri W (2019). Integrated economic and environmental assessment of biogas and bioethanol production from cassava cellulosic waste. *Waste & Biomass Valorization*, 10:691–700.

Tu WC, Hallett JP (2019). Recent advances in the pretreatment of lignocellulosic biomass. *Current Opinion in Green and Sustainable Chemistry*, 20:11–17.

Ufarté L, Laville É, Duquesne S, Veronese G (2015). Metagenomics for the discovery of pollutant degrading enzymes. *Biotechnology Advances*, 33:1845–1854.

Vidyalakshmi R, Paranthaman R, Indhumathi J (2009). Amylase production on submerged fermentation by *Bacillus* spp. *World Journal* of *Che*mical Education, 4:89–91.

Wagner-Döbler I (2003). Pilot plant for bioremediation of mercury-containing industrial wastewater. *Applied Microbiology and Biotechnology*, 62(2), 124–133.

Yudianto D, Nainggolan EA, Millati R, Hidayat C, Lennartsson P, Taherzadeh MJ, Niklasson C (2019). Bioconversion of pretreated wheat straw to ethanol by *Monascus purpureus* CBS 109.07 and *Fusarium venenatum* ATCC 20334 using simultaneous saccharification and fermentation. *Biodiversitas Journal of Biological Diversity*, 20(8). https://doi.org/10.13057/biodiv/d200817

Zhang X, Lin Y, Wu Q, Wang Y, Chen GQ (2020). Synthetic biology and genome-editing tools for improving PHA metabolic engineering. *Trends in Biotechnology*, 38:689–700.

Zhu YG, Rosen BP (2009). Perspectives for genetic engineering for the phytoremediation of arsenic-contaminated environments: from imagination to reality? *Current Opinion in Biotechnology*, 20:220–224.

18 Anti-Biofouling of Heterogeneous Surfaces
A Green Approach Using Microbial Enzymes

Praveen Rishi, Arashdeep Kaur, Ritika Sahi, Harmanpreet Kaur and Sanjeev Kumar Soni

CONTENTS

DOI: 10.1201/9781003220831-18

18.1 INTRODUCTION

Bacteria live either as independent planktonic cells or as consortium of surface-attached communities called biofilms, which are ubiquitous in nature. Biofilms, also referred to as "city of microbes", are an assemblage/aggregation of microbial cells (bacteria, algae, fungi and protozoa) enclosed in self-produced EPS matrix (Watnick and Kolter, 2000; Meliani and Bensoltane, 2015; Payne and Boles 2016). Biofilm formation is one of the most remarkable survival strategies adopted by bacteria, in natural as well as under clinical settings (Flemming *et al.*, 2016). From human teeth to industrial sites, biofilms can form wherever nomadic planktonic bacteria sense that their current surroundings can provide sufficient nutrients, water and a compatible surface (Costerton *et al.*, 1995). EPS is the defining feature of biofilm which plays many important functions in the establishment and persistence of biofilms, including structural stability, adhesion and aggregation of microbial cells, desiccation tolerance, and sorption of organic and inorganic compounds. It also provides protection to the cells against antimicrobials as well as the host immune system (Costerton *et al.*, 1995; Wang *et al.*, 2016b; Fang *et al.*, 2020).

Though greatly beneficial for bacterial survival, biofilms can have detrimental effects on other organisms, including humans. The protective environment provided by the biofilm architecture makes the cells much more resistant to different antimicrobial agents. Up to 1000-fold increased antibiotic tolerance has been reported in biofilm cells as compared to their planktonic counterparts (Rogers *et al.*, 2010; Johani *et al.*, 2018). Also, higher doses of antibiotics for a longer period of time are needed to combat biofilms, which can induce antibiotic resistance in pathogens. The reasons for enhanced antimicrobial resistance include physical impedance, enzymatic inactivation of the drugs and lowered metabolic rates in biofilm-associated cells (Lewis, 2012). As a result, biofilms are the primary cause of nosocomial infections and infections related to indwelling medical devices such as intravascular catheters, cerebrospinal fluid shunts, peritoneal dialysis catheters, intraocular lenses and cardiac pacemakers (Kaplan *et al.*, 2003).

The resistance of biofilms towards antimicrobial agents is a major challenge to the food industry also particularly the meat and poultry industry, breweries, dairy industry and food processing units, resulting in outbreaks of contamination (Alvarez-Ordóñez *et al.*, 2019). Additionally, biofilm development on equipments and pipelines often leads to their biofouling, resulting in spoilage of food, economic losses and hygiene problems (Simoes *et al.*, 2010; Galie *et al.*, 2018). Many a times, biofilms form thick and viscous layers on heat exchangers and pipelines, resulting in reduced heat exchange, which in turn compromises the efficiency of the equipments.

Looking at the challenges posed by biofilms, it becomes imperative to understand about the major factors involved in the biofilm formation and to develop various strategies for their management.

18.2 COMPONENTS OF BIOFILMS

Broadly, all the biofilms are comprised of three major components: water, the microbial cells and the EPS. The details of each are discussed in the following subsections.

18.2.1 Water

Water is the largest component and accounts for 97% of any biofilm matrix. It provides a hydrated environment in the biofilm's EPS matrix. Water acts as buffer and protects the cells present in the biofilm against the fluctuations in the water potential. Due to its hygroscopic nature, EPS matrix does not allow the fast drying of the surroundings and therefore provides protection against desiccation (Meliani and Bensoltane, 2015).

18.2.2 Biofilm Cells

Biofilms are considered to be established communities, and their basic structural unit is a microcolony which may be comprised of multiple or mono species. In a mono-species bacterial biofilm, the cells are genetically identical but they are functionally different, i.e. the cells are differentiated into various cell types adapted to perform different functions, e.g. some cells might help in the establishment of biofilms while others are responsible for the maintenance and recalcitrance while others help in the dispersal of a biofilm. The cells within biofilm are not uniformly distributed; they are often aggregated to form microcolonies and in between these microcolonies lie the zones of EPS and water channels (Costerton *et al.*, 1999). The organisms within biofilms not only adapt themselves to extreme and harsh environment/conditions but also communicate with each other (via quorum sensing) (Annous *et al.*, 2009; Satpathy *et al.*, 2016). The proximity of cells within the microcolonies of biofilms also known as cohesion provides a favourable environment for exchange of genes, nutrients and quorum sensing (Donlan, 2002). However, considerable phenotypic variations among bacteria can be observed in a mono-species biofilm. These variations depend on the position and requirements of the organisms in the biofilm. For example, there are differences in the levels of various factors like oxygen, nutrients, antimicrobial agents etc. for bacteria present on the surface as compared to the cells which are deep-seated in the biofilm matrix. Thus, depending on these factors, the cells within the same biofilm may express completely different set of genes. The cells in the heterogeneous population belong to different species. The association between the heterogeneous cells within biofilms helps to metabolize the substrates that are otherwise difficult to use by a single species (Azeredo *et al.*, 2017).

18.2.3 Exopolymeric Substances (EPS)

The EPS of biofilms also known as the "the house of biofilms" (Flemming *et al.*, 2007) is synthesized by the microbes present in a biofilm. The EPS mainly consist of different biopolymers like carbohydrates, proteins, glycopeptides, nucleic acids, lipids, glycolipids, levan and humic acid substances (Cortes *et al.*, 2011; Hobley *et al.*, 2015). These biopolymers act as cementing material that holds the biofilm cells together (Flemming and Wingender, 2010), for this reason the EPS is also known as the 'cement' of biofilms. EPS of the biofilms play a crucial role in the establishment, protection and stability of biofilm structure. It provides mechanical cohesive stability to the biofilms and also endows a protective environment for the microorganisms encased within the biofilms. It is also responsible for the extremely high recalcitrance of biofilm cells to antimicrobials, host defences and provides protection against harsh environmental conditions (Del Pozo and Patel, 2007). For example, EPS matrix impairs or slows down the penetration of antibiotics or other antimicrobials into the biofilms. It also provides protection to the biofilm cells from phagocytosis. Besides playing the protective role, the matrix also serves as a scavenging system as it helps in trapping and concentrating essential minerals and nutrients from the surrounding environment (Flemming *et al.*, 2007).

The exopolysaccharides present in the biofilms matrix are responsible for cohesion and adhesion interactions, the proteins are used as carbon and energy source, while eDNA serves as a genetic pool that aids in the transmission of resistance gene (Limoli *et al.*, 2015). The EPS matrix also interacts with the surrounding environment, aids in the attachment of the biofilms to surfaces, helps in the sequestration of dissolved/particulate substances from the environment through sorption and therefore provides nutrients for the organisms in a biofilm. It is formed during the attachment stage i.e. when the biofilms attach to a biotic or abiotic surface (Flemming and Wingender, 2001a, 2001b).

18.2.3.1 Polysaccharides Present in Biofilms

Biofilm formation is a prerequisite and an intelligent survival strategy adopted by microorganisms for the successful existence (Sutherland, 2001; Flemming et al., 2016). As mentioned earlier, EPS forms the integral and fundamental component of bacterial biofilms and a broad range of microorganisms have been reported to produce EPS. Over 99% of microorganisms on earth have been reported to live encased within these biopolymers. However, the proportion of EPS in a biofilm has been found to vary between 50–90% (approximately) of the total organic matter found in the matrix (Flemming and Wingender, 2001a; Donlan, 2002; Flemming and Wingender, 2010). Important reasons for these variations are age of the biofilms and type of organisms present in the biofilm as well as the interactions between the organisms in heterogeneous biofilms. These polymers can be composed of single or multiple structural units and the arrangements of these units also vary with the kind of EPS in different microorganisms (Roca *et al.*, 2015).

Among the various polymers associated with the biofilms EPS, polysaccharides are the most studied components of biofilm, but due to various complexities, the composition of the biofilm polysaccharides has not been completely deciphered yet. However, studies suggest that certain monomers found in biofilm EPS are identical to those found in the cell-wall material of plants (Sutherland, 1990). Various environmental conditions such as temperature, pH, nutrients, oxygen and nitrogen levels, properties of the surface (biotic/abiotic) onto which they attach, desiccation etc. affect the amount and composition of EPS (Mayer *et al.*, 1999; Ahimou *et al.*, 2007).

The polysaccharides present in biofilm matrix vary in chemical composition as well as in physical properties. For example, some polysaccharides are neutral in nature while others are polyanionic or polycationic. The nature of the polysaccharides is often influenced by the environment as well as the association of the polysaccharides with other molecules like proteins, lipids, eDNA and lectins. The bacterial extracellular polysaccharides are composed of homo- and hetero-polysaccharides, particularly of glucose, fucose, mannose, galactose, fructose, mannuronic acid, glucoronic acid or pyruvate-based complexes. Further, different bonds between the saccharides give rise to a multitude of different polysaccharides including cellulose, dextrans, amylopectin, levans, polymannans and alginate (Limoli *et al.*, 2015). Another factor that influences the nature of the matrix polysaccharides is the heterogeneous/multispecies nature of the biofilms wherein the biofilms include different bacteria or fungal species. Due to interspecies interactions among the microorganisms an array of different polysaccharides interact with each other, resulting in generation of an array of permutations and unique biofilm architecture (Jenkinson and Lamont, 1997).

18.2.3.2 Proteins

Proteins are another major component of biofilm EPS (Beenken *et al.*, 2010). Many cell-cell and cell-host tissue contacts within a biofilm are mediated by surface proteins. Different surface proteins are responsible for cell-cell as well as cell-host tissue interactions, such as the fibronectin binding proteins (O'Neill *et al.*, 2008), protein A (Merino *et al.*, 2009) and biofilm associated protein (BAP) (Trotonda *et al.*, 2005), and have been reported to be important for cell-cell and cell-surface interactions occurring during the development of biofilms. Apart from surface proteins, the cytoplasmic proteins also play a vital role in the biofilm matrix by associating themselves with the cells, whenever the cells sense a fall in the pH. Another example of proteins are the amyloids, first recognized in human neurodegenerative diseases and are often associated with the biofilms. Amyloids have emerged as an important proteinaceous component of many microbial biofilms, including *S. aureus*.

18.2.3.3 eDNA

Extracellular genomic DNA (eDNA) is another important component, identified in many bacterial biofilms (Okshevsky and Meyer, 2015; Vorkapic *et al.*, 2016). Treatment of growing or mature biofilms with DNase results in inhibition or biofilm formation/or disruption of established biofilms of various pathogens. The autolytic activity of cells often results in the release of genomic DNA from the cells which aids in the cell adhesion. The chemically charged nature of DNA modulates the cell surface properties and promotes adhesion of cells (Rice *et al.*, 2007; Mann *et al.*, 2009).

18.3 DISTRIBUTION AND UBIQUITY OF BIOFILMS

Biofilms have a profound effect on living organisms, including humans. Practically, biofilms can develop wherever there is sufficient supply of water and nutrients (Costerton *et al.*, 1995). Biofilms may develop on a variety of surfaces, including living tissues, in-dwelling medical devices, industrial or potable water system piping or natural aquatic systems. Biofilm formation has an adverse effect on the surfaces wherever they are formed, e.g. biofilms on medical implants often result in chronic infections, development of biofilms on pipelines results in reduced flow rate, development and adherence of biofilms to ship hulls results in increased drag and higher fuel consumption, in turn, biofilms in the food industry often lead to product contamination sometimes resulting in outbreak of infections. Some examples that exhibit the ubiquitous nature of biofilms are described next.

18.3.1 Biofilm Formation on Medical Devices

Association between biofilm formation and medical device-related infections was reported for the first time by Johanson *et al.* (1972). Due to their inherent 'resistant' nature, biofilms are of utmost concern as they develop on a wide range of medical devices and on prosthetics such as cardiac pacemakers and catheters (Marrie *et al.*, 1982; Peters *et al.*, 1982; Hall-Stoodley *et al.*, 2004).

The emergence of biofilm-related infections associated with the widespread use of medical devices in healthcare has given rise to the term 'polymer-associated infection' (Percival *et al.*, 2015). The polymer-associated infections are associated with high rates of morbidity and mortality (Donlan, 2008). In addition, such infections are huge financial burdens on any country's healthcare services. The most commonly reported HCAIs include lower respiratory tract infections and ventilator-associated pneumonia (VAP), catheter-associated urinary tract infections (CAUTIs) and surgical-site infections (SSIs). These infections are most commonly associated with Gram-positive bacteria, such as *S. aureus*, *S. epidermidis* and *E. faecalis*. Gram-negative bacteria, including *K. pneumonia*, *E. coli*, *P. aeruginosa* and *P. mirabilis* and amongst yeasts, *Candida* species are reported (Donlan, 2001). The treatment of these infections poses a major challenge due to the ability of these pathogens to form biofilms, which ultimately makes them more resistant to antimicrobial agents including antibiotics. Since biofilm formation requires attachment to some surface, either biotic or abiotic, medical devices are easy targets for their development; this ability has been implicated in the medical device-related infections (Vinh and Embil, 2005; Percival *et al.*, 2012; Seth *et al.*, 2012). Various factors are responsible for the initial contamination of medical devices. Some of the common reasons are contamination by the patient's or healthcare workers skin, environmental factors, peri-operative or post-operative contamination or contaminated water (von Eiff *et al.*, 1999, 2005).

18.3.2 Biofilm Formation on Wounds

The development of biofilms in wounds is a major hindrance to wound healing (James *et al.*, 2008), particularly in individuals who have reduced immunity. Biofilm formation results in increased resistance to antimicrobials as a result of obstruction in penetration of these agents. In addition, increased drug resistance in MDR organisms is also challenging. Other factors such as reduced perfusion limits the delivery of antibiotics (particularly systemically to the wound), resulting in increased complications in healing of these wounds (Lipsky, 2004; Guo and Dipietro, 2010).

18.3.3 Biofilms in Food Industry

Another major concern associated with the biofilms is the biofouling of surfaces and machinery in the food industries, which not only results in spoilage of food but also a source of outbreak of diseases caused by a range of pathogens. Biofilms pose serious problems in the various sectors of food industry. The resistance of biofilms towards antimicrobials agents is a major hazard to industries, particularly the meat and poultry industry, breweries, dairy industry and food-processing units. Biofilm development on equipments, pipelines etc. leads to major outbreaks of contaminations, resulting in spoilage of food, economic losses and hygiene problems (Simões *et al.*, 2006).An additional problem is metal corrosion. Many a times, biofilms form thick and viscous layers on heat exchangers and pipelines resulting in reduced heat exchange and in turn compromising the efficiency of the equipments (Lequette *et al.*, 2010).

18.4 ANTIMICROBIAL RESISTANCE AND RECALCITRANCE OF BIOFILMS

The protection provided to the microbial cells within the biofilm lifestyle often results in severe biofouling and dangerous or persistant infections (Johani *et al.*, 2018). Combating the biofilm associated infections requires higher antibiotic doses for a prolonged period of time. According to certain studies, 65% of nosocomial infections in medical settings are reported to be biofilm associated, costing the healthcare system billions of dollars (Costerton *et al.*, 1999; Potera, 1999; Mah and O'toole, 2001). These biofilm associated infections are 10 to 1000 times more resistant to the effects of antimicrobial agents as compared to their planktonic counterparts (Fleming *et al.*, 2017).

The type of resistance observed in these adherent cells is unique to biofilm-associated organisms and is markedly distinct from conventional antimicrobial resistance (Anderl *et al.*, 2000; Del Pozo and Patel, 2007; Lebeaux *et al.*, 2014). The resistance observed in biofilms is not due to "classic" genetic mechanisms (i.e. gene mutation, genetic exchange), but is determined by certain attributes of biofilm lifestyle (Orazi and O'Toole, 2019). Some of the potential mechanisms implicated in biofilm resistance towards antimicrobial agents are

- Restricted penetration of the antimicrobial agents
- Antimicrobial inactivating or modifying enzymes
- Horizontal gene transfer
- Quorum sensing
- Altered growth rate (presence of persister cells)
- Stress response to hostile environmental conditions and overexpression of genes

18.5 VARIOUS APPROACHES USED FOR BIOFILM CONTROL

Since the majority (nearly 80%) of the infectious diseases involve biofilm formation by the associated pathogens, efforts to curb biofilm formation becomes imperative. Treatment of infections involving biofilm formation poses a difficult and complicated challenge for microbiologists as well as clinicians. Combating biofilms effectively is a huge challenge, both under the clinical as well as environmental settings. Various obstacles that are being faced to combat biofilms are as follows:

- Biofilms have complex structural and biological properties, such as the presence of EPS, physical and chemical heterogeneity of the biofilm cells, which lead to remarkable therapeutic challenges.
- Complex physico-chemical and biological attributes with multiple genetic and molecular factors, polymicrobial interactions within biofilms further increase the recalcitrance of biofilms.
- The existing antimicrobial and mono-therapeutic approaches to combat biofilm associated infections, need to be replaced by multi-targeted or combinatorial therapies.

In the majority of the cases, antibiotic treatment alone is often inadequate to take care of biofilm infections. Many antibiotics were originally developed to target individual bacterial cells. Currently, the strategies used to combat biofilms use analogues of quorum-sensing molecules, metagenomics, targeting virulence factors and use of various agents which disperse biofilms. Combining ethnopharmacology with modern synthetic biology is another approach used to combat biofilm associated antimicrobial resistance. Another alternate approach to treat the established biofilms is the use of enzymes that degrade the biofilm matrix.

18.5.1 Agents that Destroy Persister Cells

The study of antimicrobial activity against persisters is crucial to delineate whether current antimicrobials may be used in this scenario. Development of new compounds to destroy this latent population is a potential approach. Such approaches require utilization of compounds which can enter the cells without the need of active transport and kill the persister cells without requiring any cell machinery (as there is little or no metabolism) (Wood, 2017). Examples of such an approach include utilizing the DNA cross linking agents mitomycin C (Kwan *et al.*, 2015) and cisplatin (Chowdhury *et al.*, 2016). An alternative approach is to wake persister cells and then treat them with traditional antibiotics because adding sugars and glycolysis intermediates (e.g. mannitol, glucose, fructose, pyruvate) rapidly wakes persister cells (Allison *et al.*, 2011).

18.5.2 Quorum-Quenching Enzymes

A number of bacterial enzymes have been identified which are capable of enzymatic inactivation of N-acylhomoserine lactones. These enzymes show considerable promise as "quenchers" of quorum sensing (Roche *et al.*, 2004). Over the years, a range of quorum-quenching enzymes and inhibitors have been identified from different sources. Several potential quorum-sensing targets for the inhibition of *P. aeruginosa* virulence have been described (Dong and Zhang, 2005).

18.5.3 Agents that Cause Biofilm Self-Destruction

Another interesting strategy for biofilm eradication arises from the observation wherein the biofilms are self-destructed, e.g. *Pseudomonas fluorescens* has been reported to form biofilm under well-oxygenated environments, once the oxygen is depleted by the growing biofilm mass, a specific exopolysaccharide lyase is induced that digests the biofilm matrix, resulting in the liberation of the cells from biofilms (Allison *et al.*, 1998).

18.5.4 Approaches to Enhance the Activity of Conventional Antimicrobial Agents

Several physical approaches to biofilm eradication are being evaluated including the use of an electrical current (Jass *et al.*, 1995), radiofrequency electrical current (Caubet *et al.*, 2004), electromagnetic fields (McLeod *et al.*, 1999) and ultrasound (Rediske *et al.*, 2000), in combination with antimicrobial therapy.

18.5.5 Quorum-Sensing Inhibition

Cell-to-cell communication, also known as quorum sensing (QS), helps in co-ordination among the bacteria. This communication depends on the accumulation, release and simultaneous detection of special signalling molecules called auto-inducers (AIs) that are released extracellularly AIs have been found to regulate other processes also like sporulation, virulence-factor secretion, bioluminescence and biofilm maturation (Rutherford and Bassler, 2012; Verma and Miyashiro, 2013). Since, resistance of biofilms towards antimicrobials is primarily due to the QS system, so inhibition of this communication can be a promising strategy for biofilm control/inhibition (Lazar, 2011). Various methods are available for the disruption of the QS system, e.g. inhibition/direct degradation of the signal molecules. Another approach is the mimicking of signalling molecules by synthetic compounds (analogues), so that the signal molecules are unable to bind to their receptors

(Kalia, 2013). Three classes of quorum-quenching enzymes, namely (i) acetyl homoserine lactone (AHL) acylase, (ii) AHL lactonases and (iii) oxidoreductases have been reported to degrade QS signals (Dong and Zhang, 2005).

18.5.6 Agents that Destroy or Alter the Biofilm Matrix

Due to their importance in establishing and maintaining the biofilm architecture, EPS disruption provides an interesting alternative to tackle biofilms. Substances able to depolymerize, dissolve, or impede the synthesis of the matrix would leave biofilm cells exposed to environmental conditions. Various agents including enzymes, antibiofilm peptides, and dispersal molecules that target the biofilm architecture are promising methods for biofilm control (Fleming and Rumbaug, 2017).

18.6 CURRENT STRATEGIES TO TREAT BIOFILMS

For the effective control of undesirable biofilms, it is important to have a complete understanding of the microorganism, the nature of the EPS formed as well as the type/nature of the surface where the biofilm establishes is necessary. Steps should be taken for the prevention of biofilm formation rather than treating it. However, there is no known technique for the complete control or prevention of biofilm formation without causing any adverse side effects. Traditionally, various physical and chemical methods are employed for the control and removal of biofilms. Chemicals such as surfactants, chelating agents, hypochlorite, monochloramine and concentrated urea are employed for control of biofilms. However, the use of chemicals is non-ecofriendly, expensive and also leads to corrosion of surfaces.

Other attempts to combat biofilms include modification of the surface materials in order to inhibit biofilm formation, e.g. incorporation of antimicrobial agents onto the surface of materials (Weng *et al.*, 1999; Park *et al.*, 2004), coating of the surfaces with antimicrobial products (Gottenbos *et al.*, 2001), silver (Hashimoto, 2001) or modification of the physico-chemical properties of the surfaces (Thouvenin *et al.*, 2003). Use of silicone rubber implants (covalently coupled with quaternary ammonium coatings) also reduced the biofilm associated infection rate (Gottenbos *et al.*, 2001). Although these studies involve biomedical applications, their use on an industrial scale is restricted, possibly due to the carryover of antimicrobial agents into food products in case the surface coatings release these agents (Simoes *et al.*, 2010). Various anionic and non-ionic surfactants also possess anti-biofilm activity and inhibit bacterial adhesion to the surfaces (Sharahi *et al.*, 2019).

Bacteriophages are also used to treat various infectious diseases in humans. Like biofilms, phages are also ubiquitous in nature and may provide a natural, non-toxic and highly specific alternative for controlling pathogens involved in biofilm formation (Kudva *et al.*,1999; Harper *et al.*, 2014; Huang 2019; Milho *et al.*, 2019). A number of herbs and plants are documented in literature that possess antimicrobial activities. Anti-biofilm properties have been attributed to certain natural compounds like honey, plant extracts and essential oils. Honey is known for its antibacterial, antioxidant, anti-inflammatory and wound-healing properties. It has been reported to possess antimicrobial activity against many bacteria and fungi (Nolan *et al.*, 2019; Al-kafaween *et al.*, 2020). Numerous plant extracts and their active compounds have been extensively evaluated for their antimicrobial as well as antibiofilm activities (Cox and Balick, 1994; Howell *et al.*, 1998; Teanpaisan *et al.*, 2017). Biosurfactants (BSs) produced by microorganisms have the ability to accumulate at the liquid-liquid/air-liquid interfaces due to varying degrees of hydrogen bonding and polarity (Sharahi *et al.*, 2019). They have diverse structures and may be phospholipids, glycolipids, lipoproteins, lipopeptides, fatty acids, neutral lipids or particulate in nature (Makkar and Rockne, 2003; Banat, *et al.*, 2010; De Rienzo *et al.*, 2015). BSs have antimicrobial, antiadhesive properties with strong dispersal properties which make them a hopeful candidate for dispersing and eradicating biofilms (Merghni *et al.*, 2017).

In spite of these known strategies, increasing antibiotic resistance among pathogens and the slow rate of newer antibiotic discovery have forced the scientific community to look for alternatives.

Also, various chemical disinfectants are widely used to take care of microbial contamination in healthcare as well as industrial manufacturing units, have adverse effects in the equipments since they leave toxic products behind (Hansen *et al.*, 2004).

To overcome these problems, the efforts have now been shifted to use bio-based antimicrobial agents. In this regard, enzymes are now being exploited against pathogens and biofilms. Different classes of enzymes can take care of the biofilms at different levels, e.g. during biofilms attachment, formation, EPS disruption, quorum quenching to name a few (Thallinger *et al.*, 2013).

18.7 ENZYMES AS ANTIBIOFILM AGENTS

Enzymatic treatment is also called the wave of the future. The application of enzymes for the control of biofilms provides an attractive alternative, especially in cases where the classical treatments involving antibiotics and chemical agents do not give satisfactory results. Although individual enzymes have been used for degradation of biofilms (Loiselle and Anderson, 2003; Kalpana *et al.*, 2012; Lamppa and Griswold, 2013), due to heterogeneity of the extracellular polysaccharides in the biofilm EPS, a mixture of carbohydrases might prove to be more effective in degradation of these multi-component biofilms. Enzymes can be incorporated/adsorbed/immobilized/grafted into and onto various surfaces to prevent microbial colonization. Figure 18.1 shows the different classes of enzymes which can attack biofilms.

18.7.1 Proteolytic Enzymes

18.7.1.1 Subtilisins

Subtilisins are serine proteases produced by *Bacillus* sp. and are widely used in the food industry to control biofilms (Longhi et al., 2008). These proteases hydrolyze adhesions, bacterial proteins that play role in attachment, thereby preventing co-aggregation of microorganisms, which is essential for biofilms formation.

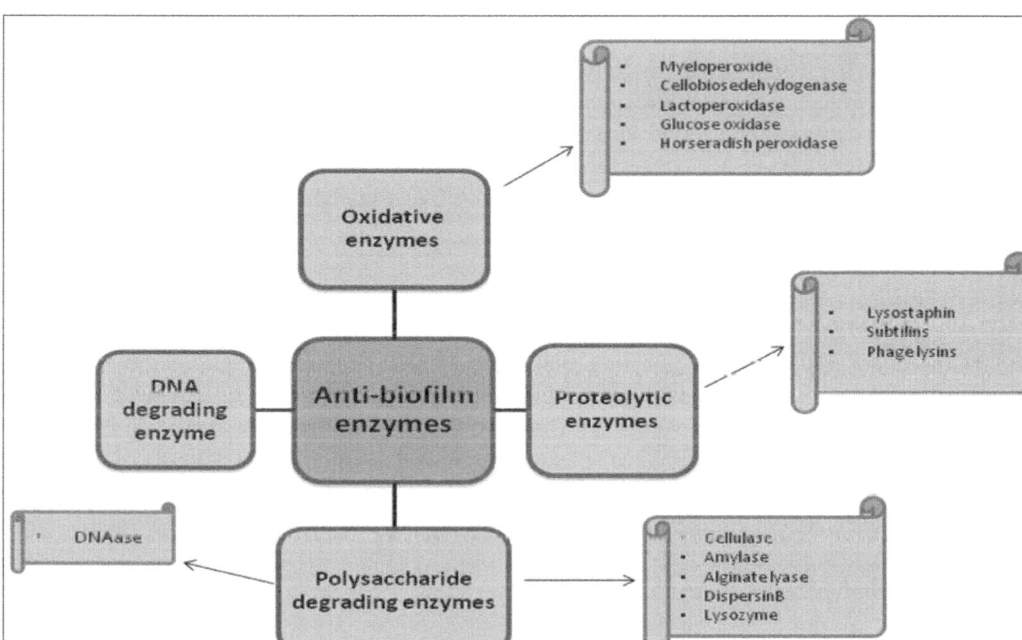

FIGURE 18.1 Different classes of enzymes used as anti-biofilm enzymes.

18.7.1.2 Lysostaphin

Lysostaphin, a metalloendopeptidase, was isolated in 1964 from *Staphylococcus simulans* by Schindler and Schuhardt. It cleaves the Staphylococcal cell wall at the glycine residues (Kumar *et al.*, 2008). Lysostaphin has been reported to have three different catalytic activities: glycylglycineendopeptidase, endo-beta-Nacetylglucosamidase and N-acetyl muramyl-L-alanine amidase activity (Recsei *et al.*, 1987).

18.7.1.3 Bacteriophage Lysins/Endolysins

Endolysins include bacteriophage muralytic enzymes which include enzymes like N-acetylmuramidases, lytic transglycosylases, endopeptidases, endo-β-N-acetylglucosaminidases, lytic transglycosylases, and N-acetylmuramoyl-l-alanine amidases (Navarre *et al.*, 1999; Borysowski *et al.*, 2006).

18.7.1.4 Papain

Papain, obtained from latex of *Carica papaya* is a complex mixture of proteolytic enzymes and peroxidases. Papain acts upon various functional groups, such as hydroxyl, carbonyl, amino and phosphoryl and reduces biofilm formation. Several studies suggest that papain has the ability to not only inhibit the biofilm formation by Methicillin-resistant pathogens like *S. epidermidis* (MRSE) and *S. haemolyticus* (MRSHa) but also disrupts the established biofilms of MRSE (Oliveira *et al.*, 2014).

18.7.2 Polysaccharide-Degrading Enzymes

Polysaccharide hydrolyzing/degrading enzymes including dispersin B, amylases, alginate lyase and lysozyme are most commonly used for the dispersion of biofilms.

18.7.2.1 Dispersin B (DspB)

This enzyme is a soluble glycoside hydrolyase produced by a Gram-negative periodonto pathogen *Actinobacillus actinomycetemcomitans* and has been reported to possess antimicrobial as well as anti-biofilm activity. This enzyme hydrolyzes β-1,6-N-acetyl-d-glucosamine, a polymer which is required for the attachment of biofilms onto various surfaces (Chaignon *et al.*, 2007). Engineered DspB has also been used and found to be effective against *Staphylococcus* and *E. coli* biofilms. Coating of catheters with DspB, triclosan1 and chlorhexidine silver sulfadiazine has also found to be effective against *S. epidermidis*, *S. aureus* and *E. coli* (Darouiche *et al.*, 2009).

18.7.2.2 Deoxyribonuclease 1

Deoxyribonuclease 1 (DNasel) is the enzyme that degrades extracellular DNA, also called e-DNA, present in the biofilms matrix. This degradation of the e-DNA renders the biofilm matrix/EPS weak and therefore more susceptible to antimicrobial agents. DNasel reduces initial biofilm attachment as well as formation in *Pseudomonas aeruginosa*. Deoxyribonucleases can detach pre-formed biofilms, sensitize biofilm bacteria to biocide killing and finally inhibit biofilm formation (Tetz *et al.*, 2009). DNase-coated nanoparticles have been recommended to enhance the delivery of antibiotics in case of infections involving biofilms (Baelo *et al.*, 2015). Such nanoparticles prevent biofilms formation and also had the ability to disperse the preformed biofilms.

18.7.2.3 Carbohydrases as Antibiofilm Agents

Carbohydrases are the hydrolytic enzymes that find a wide application in industrial processes and products. In the broadest sense, the term carbohydrase is considered to encompass diverse enzymes involved in hydrolysis and synthesis of carbohydrates (Contesini *et al.*, 2013). The most important carbohydrases to accomplish these goals are of microbial origin and include amylases (α-amylases and glucoamylases), invertases, inulinases, galactosidases, glucosidases, fructosyl transferases, pectinases and glucosyltransferases.

Most of the carbohydrases applied in the industry are obtained from microbial sources, such as bacteria, yeasts and fungi, since this approach allows for maximum consistency. Microorganisms grow faster and easier than mammalian and plant cells, and the enzyme production is neither influenced by climatic conditions or seasonal changes, nor by regulatory or ethical issues related to animal slaughter or tree or plant felling. Extracellular enzyme-producing microorganisms are preferred, since this simplifies downstream processing, hence further lowering costs (Tufvesson *et al.*, 2010; Simpson *et al.*, 2012). Apart from these applications, the use of some carbohydrases has also been explored for the dispersal of biofilms. The application of enzymes for the control of biofilms would provide an interesting alternative especially in the cases when the classical treatments involving chemical agents do not give satisfactory results.

A number of enzymes, including proteases, amylase, alginate lyase, lysostaphin, lysozymes, oxidative enzymes, dispersin B and DNAase, are used for degradation or disruption of biofilms (Loiselle and Anderson, 2003, Darouiche *et al.*, 2009; Tetz *et al.*, 2009; Nijland *et al.*, 2010; Torres *et al.*, 2011; Kalpana *et al.*, 2012; Lamppa and Griswold, 2013; Thallinger *et al.*, 2013; Ye *et al.*, 2017). However looking at the heterogeneity of the extracellular polysaccharides in the EPS the complex nature of the biofilms formed as a result of intra and interspecies interactions of the microorganisms, the application of multiple enzymes/carbohydrases is more meaningful.

Keeping in view the heterogeneous nature of polysaccharides present in the EPS of biofilms, recently a cocktail of nine different carbohydrases, including cellulases, pectinase, hemicellulases (xylanase and mannanase), amylases and alginate lyase, has been evaluated to combat heterogeneous biofilms from different surfaces (Kaur *et al.*, 2020). This study primarily focused on the carbohydrases, since a large portion of biofilm EPS is composed of heterogeneous polysaccharides. The enzymes cocktail was obtained from a soil isolate, *Aspergillus niger* and enzymes were produced using economically viable media containing kitchen waste as its composition. The in-house-produced enzyme cocktail was evaluated for its potential to eradicate the biofilms of the selected pathogens. The cocktail was found to effectively disperse the biofilms of the selected pathogens wherein 89.75 ± 5.0%, 95.26 ± 4.0%, 97.20 ± 2.0%, 67.42 ± 5.0%, 86.32 ± 5.0% and 72.75 ± 7.0% eradication of *E. coli*, *S.* Typhi, *S.* Typhimurium, *P. aeruginosa, S. aureus* and MRSA biofilms respectively was observed. Further, the in-house-produced enzyme cocktail was found to possess noteworthy potential to eradicate/disperse biofilms of *S. aureus* from contact lens and dentin discs (Figure 18.2 and Figure 18.3) and biofilms of *E. coli* and *P. aeruginosa* from Foley's catheter. The efficacy of the multiple carbohydrases was also tested to degrade/disperse thick slime layer (commonly known as 'black gunk') developed on the inner surface of kitchen drainage pipe. Black gunk formation, which occurs as a result of biofouling, is a focal point for formation of heterogeneous

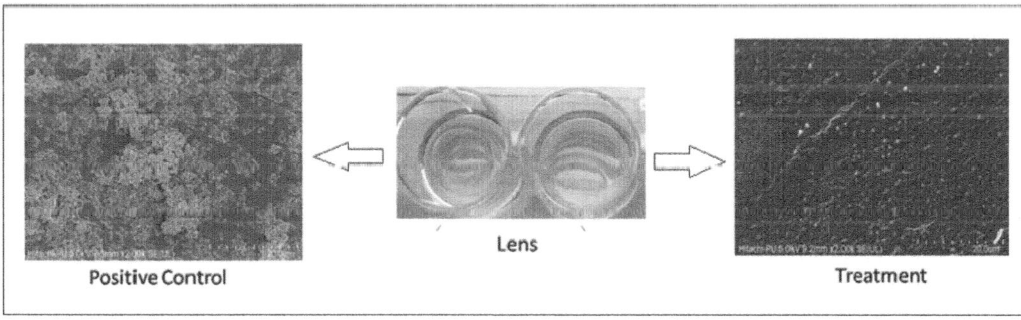

FIGURE 18.2 Effect of enzyme cocktail obtained from *A. niger* on *S. aureus* biofilm developed on contact lens.

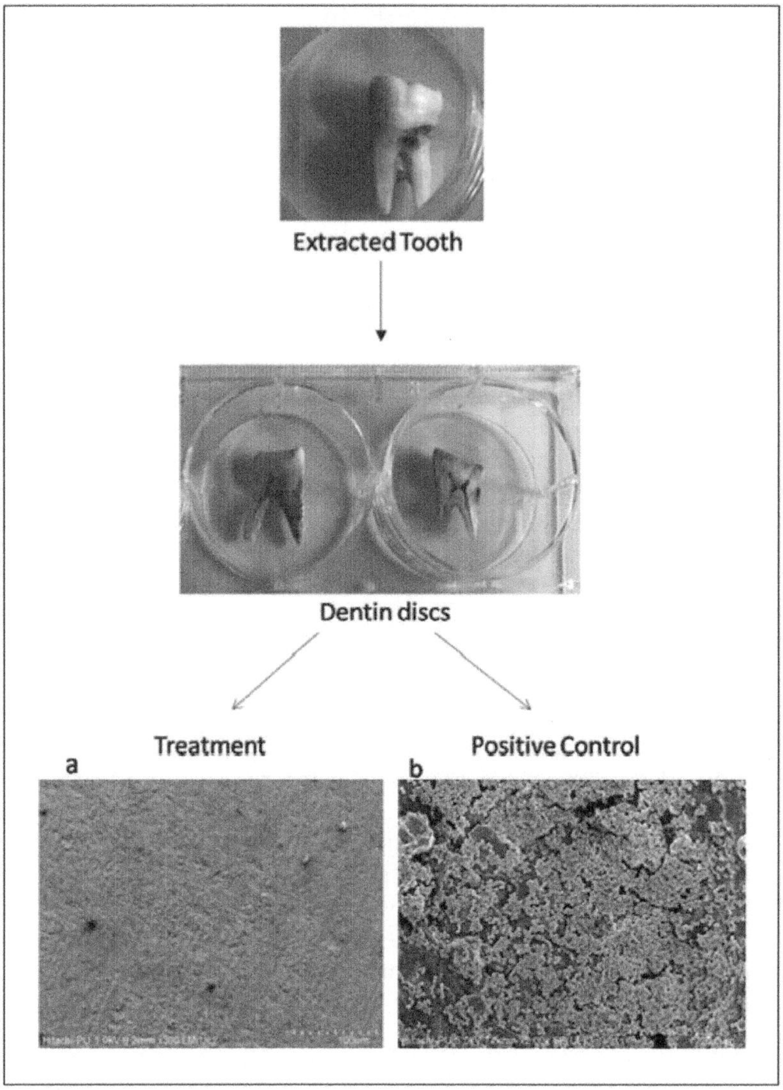

FIGURE 18.3 Effect of in-house-produced enzyme cocktail from *A. niger* on *S. aureus* biofilm developed on dentin discs.

biofilms, formed as a result of intra and inter-species interactions among the microorganisms. The multiple carbohydrases were found to possess noteworthy potential to eradicate/disperse the 'black gunk' and biofilms of certain pathogens from kitchen drainage pipes and various biopolymer surfaces including contact lenses, catheter and dentin discs respectively, by hydrolyzing the specific bonds present in the polysaccharides of the biofilm EPS.

In the study, the potential of multiple enzymes in conjunction with antibiotics was also assessed. Minimum inhibitory concentrations (MICs) and biofilm bactericidal concentrations (BBCs) of three antibiotics, viz. ampicillin, ciprofoxacin and chloramphenicol, were found to be significantly reduced when used in conjunction with the enzyme cocktail. This effect might be attributed to the breakdown or hydrolysis of the biofilm EPS by enzymatic action which act as a physical barrier for the diffusion of the antibiotics into the biofilm. The reduced effective concentration of antibiotics may not only help in minimizing the chances of emergence of resistance but may also reduce the level of toxic effects associated with the antibiotics.

18.8 CONCLUDING REMARKS

Application of enzymes for the dispersal of biofilms and cleaning of slime would provide an interesting anti-biofouling alternative under natural, industrial as well as clinical settings, especially in situations when the classical treatments involving chemical agents do not give satisfactory results. In view of the application of multiple enzymes as anti-biofouling agent for heterogeneous surfaces, it appears to be a promising green approach to further translate the cocktail of enzymes obtained from a single strain into a commercially viable as well as cost-effective option. Moreover, the reduced effective concentration of antibiotics may not only help in minimizing the chances of emergence of resistance but also provide the opportunity to add carbohydrases to the existing armamentarium against biofilm-mediated infections.

BIBLIOGRAPHY

Ahimou, F., Semmens, M. J., Haugstad, G., and Novak, P. J. 2007. Effect of protein, polysaccharide, and oxygen concentration profiles on biofilm cohesiveness. *Applied and Environmental Microbiology* 73:2905–2910.

Al-kafaween, M. A., Hilmi, A. B. M., Jaffar, N. et al. 2020. Antibacterial and antibiofilm activities of Malaysian Trigona honey against *Pseudomonas aeruginosa* ATCC 10145 and *Streptococcus pyogenes* ATCC 19615. *Jordan Journal of Biological* 13:69–76.

Allison, D. G., Ruiz, B., SanJose, C., Jaspe, A., and Gilbert, P. 1998. Extracellular products as mediators of the formation and detachment of *Pseudomonas fluorescens* biofilms. *FEMS Microbiology Letters* 167:179–184.

Allison, K. R., Brynildsen, M. P., and Collins, J. J. 2011. Metabolite-enabled eradication of bacterial persisters by aminoglycosides. *Nature* 473:216–220.

Alvarez-Ordóñez, A., Coughlan, L. M., Briandet, R., and Cotter, P. D. 2019. Biofilms in food processing environments: challenges and opportunities. *Annual Review of Food Science and Technology* 10:173–195.

Anderl, J. N., Franklin, M. J., and Stewart, P. S. 2000. Role of antibiotic penetration limitation in *Klebsiella pneumoniae* biofilm resistance to ampicillin and ciprofloxacin. *Antimicrobial Agents and Chemotherapy* 44:1818–1824.

Annous, B. A., Fratamico, P. M., and Smith, J. L. 2009. Scientific status summary: quorum sensing in biofilms: why bacteria behave the way they do. *Journal of Food Science* 74:R24–R37.

Azeredo, J., Azevedo, N. F., Briandet, R. et al. 2017. Critical review on biofilm methods. *Critical Reviews in Microbiology* 43:313–393.

Baelo, A., Levato, R., Julián, E. et al. 2015. Disassembling bacterial extracellular matrix with Dnase-coated nanoparticles to enhance antibiotic delivery in biofilm infections. *Journal of Controlled Release* 209:150–158.

Banat, I. M., Franzetti, A., Gandolfi, I. et al. 2010. Microbial biosurfactants production, applications and future potential. *Applied Microbiology and Biotechnology* 87:427–444.

Beenken, K. E., Mrak, L. N., Griffin, L. M. et al. 2010. Epistatic relationships between *sarA* and *agr* in *Staphylococcus aureus* biofilm formation. *PLoS One* 5, no. 5 (May):10790. https://journals.plos.org/plosone/article?id=10.1371/journal.pone.0010790

Borysowski, J., Weber-Dąbrowska, B., and Górski, A. 2006. Bacteriophage endolysins as a novel class of antibacterial agents. *Experimental Biology and Medicine* 231:366–377.

Caubet, R., Pedarros-Caubet, F., Chu, M. et al. 2004. A radio frequency electric current enhances antibiotic efficacy against bacterial biofilms. *Antimicrobial Agents and Chemotherapy* 48:4662–4664.

Chaignon, P., Sadovskaya, I., Ragunah, C. et al. 2007. Susceptibility of staphylococcal biofilms to enzymatic treatments depends on their chemical composition. *Applied Microbiology and Biotechnology* 75:125–132.

Chowdhury, N., Wood, T. L., Martinez-Vázquez, M., Gracia-Contreras, R., and Wood, T. K. 2016. DNA-crosslinker cisplatin eradicates bacterial persister cells. *Biotechnol Bioeng* 113:1984–1992.

Contesini, F., de Alencar Figueira, J., Kawaguti, H. et al. 2013. Potential applications of carbohydrases immobilization in the food industry. *International Journal of Molecular Sciences* 14:1335–1369.

Cortes, M. E., Consuegra, J., and Sinisterra, R. D. 2011. Biofilm formation, control and novel strategies for eradication. *Science against Microbial Pathogens: Communicating Current Research and Technological Advances* 2:896–905

Costerton, J. W., Lewandowski, Z., Caldwell, D. E., Korber, D. R., and Lappin-Scott, H. M. 1995. Microbial biofilms. *Annual Review of Microbiology* 49:711–745.

Costerton, J. W., Stewart, P. S., and Greenberg, E. P. 1999. Bacterial biofilms: a common cause of persistent infections. *Science* 284:1318–1322.

Cox, P. A., and Balick, M. J. 1994. The ethnobotanical approach to drug discovery. *Scientific American* 270:82–87.

Darouiche, R. O., Mansouri, M. D., Gawande, P. V., and Madhyastha, S. 2009. Antimicrobial and antibiofilm efficacy of triclosan and DispersinB® combination. *Journal of Antimicrobial Chemotherapy* 64:88–93.

De Rienzo, M. A. D., Banat, I. M., Dolman, B., Winterburn, J., and Martin, P. J. 2015. Sophorolipid biosurfactants: possible uses as antibacterial and antibiofilm agent. *New Biotechnology* 32:720–726.

Del Pozo, J. L., and Patel, R. 2007. The challenge of treating biofilm-associated bacterial infections. *Clinical Pharmacology and Therapeutics* 82:204–209.

Dong, Y. H., and Zhang, L. H. 2005. Quorum sensing and quorum-quenching enzymes. *The Journal of Microbiology* 43:101–109.

Donlan, R. M. 2001. Biofilms and device-associated infections. *Emerging Infectious Diseases* 7:277–281.

Donlan, R. M. 2002. Biofilms: microbial life on surfaces. *Emerging Infectious Diseases* 8:881–890.

Donlan, R. M. 2008. Biofilms on central venous catheters: is eradication possible? *Current Topics in Microbiology and Immunology* 322:133–161.

Fang, K., Park, O. J., and Hong, S. H. 2020. Controlling biofilms using synthetic biology approaches. *Biotechnology Advances* 40:107518.

Fleming, D., Chahin, L., and Rumbaugh, K. 2017. Glycoside hydrolases degrade polymicrobial bacterial biofilms in wounds. *Antimicrobial Agents and Chemotherapy* 61, no. 2 (February):01998–16. https://journals.asm.org/doi/epub/10.1128/AAC.01998-16

Flemming, H. C., and Wingender, J. 2001a. Relevance of microbial extracellular polymeric substances (EPSs)-Part II: technical aspects. *Water Science and Technology* 43:9–16.

Flemming, H. C., and Wingender, J. 2001b. Relevance of microbial extracellular polymeric substances (EPSs)-Part I: structural and ecological aspects. *Water Science and Technology* 43:1–8.

Flemming, H. C., and Wingender, J. 2010. The biofilm matrix. *Nature Reviews Microbiology* 8:623–633.

Flemming, H. C., Neu, T. R., and Wozniak, D. J. 2007. The EPS matrix: the "house of biofilm cells". *Journal of Bacteriology* 189:7945–7947.

Flemming, H. C., Wingender, J., Szewzyk, U., Steinberg, P., Rice, S. A., and Kjelleberg, S. 2016. Biofilms: an emergent form of bacterial life. *Nature Reviews Microbiology* 14:563–575.

Galie, S., Garcia-Gutierrez, C., Miguelez, E. M., Villar, C. J., and Lombo, F. 2018. Biofilms in the food industry: health aspects and control methods. *Frontiers in microbiology* 9:898.

Gottenbos, B., van der Mei, H. C., Klatter, F., Nieuwenhuis, P., and Busscher, H. J. 2001. In vitro and in vivo antimicrobial activity of covalently coupled quaternary ammonium silane coatings on silicone rubber. *Biomaterials* 23:1417–1423.

Guo, S. A., and DiPietro, L. A. 2010. Factors affecting wound healing. *Journal of Dental Research* 89:219–229.

Gupta, P., Sarkar, S., Das, B., Bhattacharjee, S., and Tribedi, P. 2016. Biofilm, pathogenesis and prevention – a journey to break the wall: a review. *Archives of Microbiology* 198:1–15.

Hall-Stoodley, L., Costerton, J. W., and Stoodley, P. 2004. Bacterial biofilms: from the natural environment to infectious diseases. *Nature Reviews Microbiology* 2:95–108.

Hansen, E. H., Schembri, M. A., Klemm, P., Schäfer, T., Molin, S., and Gram, L. 2004. Elucidation of the antibacterial mechanism of the *Curvularia haloperoxidase* system by DNA microarray profiling. *Applied and Environmental Microbiology* 70:1749–1757.

Harper, D. R., Parracho, H. M., Walker, J. et al. 2014. Bacteriophages and biofilms. *Antibiotics* 3:270–284.

Hashimoto, H. 2001. Evaluation of the anti-biofilm effect of a new antibacterial silver citrate/lecithin coating in an *in-vitro* experimental system using a modified Robbins device. *Journal of the Japanese Association for Infectious Diseases* 75:678–685.

Hobley, L., Harkins, C., MacPhee, C. E., and Stanley-Wall, N. R. 2015. Giving structure to the biofilm matrix: an overview of individual strategies and emerging common themes. *FEMS Microbiology Reviews* 39:649–669.

Howell, A. B., Vorsa, N., Marderosian, A. D., and Foo, L. Y. 1998. Inhibition of the adherence of P-fimbriated *Escherichia coli* to uroepithelial-cell surfaces by proanthocyanidin extracts from cranberries. *New England Journal of Medicine* 339:1085–1086.

Huang, J. Y. 2019. Bacteriophage to combat biofilms in hospital drains. *bioRxiv* 522227

James, G. A., Swogger, E., Wolcott, R. et al. 2008. Biofilms in chronic wounds. *Wound Repair and Regeneration* 16:37–44.

Jass, J., Costerton, J. W., and Lappin-Scott, H. M. 1995. The effect of electrical currents and tobramycin on *Pseudomonas aeruginosa* biofilms. *Journal of Industrial Microbiology* 15:234–242.

Jenkinson, H. F., and Lamont, R. J. 1997. Streptococcal adhesion and colonization. *Critical Reviews in Oral Biology & Medicine* 8:175–200.

Johani, K., Abualsaud, D., Costa, D. M. et al. 2018. Characterization of microbial community composition, antimicrobial resistance and biofilm on intensive care surfaces. *Journal of Infection and Public Health* 11:418–424.

Johanson, W. G., Jr Pierce, A. K., Sanford, J. P., and Thomas, G. D. 1972. Nosocomial respiratory infections with Gram-negative bacilli The significance of colonization of the respiratory tract. *Annals of Internal Medicine* 77:701–706.

Kalia, V. C. 2013. Quorum sensing inhibitors: an overview. *Biotechnology Advances* 31:224–245.

Kalpana, B. J., Aarthy, S., and Pandian, S. K. 2012. Antibiofilm activity of α-amylase from *Bacillus subtilis* S8–18 against biofilm forming human bacterial pathogens. *Applied Biochemistry and Biotechnology* 167:1778–1794.

Kaplan, J. B., Ragunath, C., Ramasubbu, N., and Fine, D. H. 2003. Detachment of *Actinobacillus actinomycetemcomitans* biofilm cells by an endogenous β-hexosaminidase activity. *Journal of Bacteriology* 185:4693–4698.

Kaur. A., Rishi, V., Soni, S. K., and Rishi, P. 2020. A novel multi-enzyme preparation produced from *Aspergillus niger* using biodegradable waste: a possible option to combat heterogeneous biofilms. *AMB Express* 10:1–16.

Kudva, I. T., Jelacic, S., Tarr, P. I., Youderian, P., and Hovde, C. J. 1999. Biocontrol of *Escherichia coli* O157 with O157-specific bacteriophages. *Applied Environmental Microbiology* 65:3767–3773.

Kumar, A., Vemula, P. K., Ajayan, P. M., and John, G. 2008. Silver-nanoparticle-embedded antimicrobial paints based on vegetable oil. *Nature Materials* 7:236.

Kwan, B. W., Chowdhury, N., and Wood, T. K. 2015. Combatting bacterial infections by killing persister cells with mitomycin C. *Environmental Microbiology* 17:4406–4414.

Lamppa, J. W., and Griswold, K. E. 2013. Alginate lyase exhibits catalysis independent biofilm dispersion and antibiotic synergy. *Antimicrobial Agents and Chemotherapy* 57:137–145.

Lazar, V. 2011. Quorum sensing in biofilms – how to destroy the bacterial citadels or their cohesion/power? *Anaerobe* 17:280–285.

Lebeaux, D., Ghigo, J. M., and Beloin, C. 2014. Biofilm-related infections: bridging the gap between clinical management and fundamental aspects of recalcitrance toward antibiotics. *Microbiology and Molecular Biology Reviews* 78:510–543.

Lequette, Y., Boels, G., Clarisse, M., and Faille, C. 2010. Using enzymes to remove biofilms of bacterial isolates sampled in the food-industry. *Biofouling* 26:421–431

Lewis, K. 2012. Persister cells: molecular mechanisms related to antibiotic tolerance. *Handbook of Experimental Pharmacology* 211:121–133.

Limoli, D. H., Jones, C. J., and Wozniak, D. J. 2015. Bacterial extracellular polysaccharides in biofilm formation and function. *Microbial Biofilms*:223–247.

Lipsky, B. A. 2004. Medical treatment of diabetic foot infections. *Clinical Infectious Disease* 39:S104–S114.

Loiselle, M., and Anderson, K. W. 2003. The use of cellulase in inhibiting biofilm formation from organisms commonly found on medical implants. *Biofouling* 19:77–85.

Longhi, C., Scoarughi, G. L., Poggiali, F. et al. 2008. Protease treatment affects both invasion ability and biofilm formation in *Listeria monocytogenes. Microbial Pathogenesis* 45:45–52.

Mah, T. F. C., and O'toole, G. A. 2001. Mechanisms of biofilm resistance to antimicrobial agents. *Trends in Microbiology* 9:34–39.

Makkar, R. S., and Rockne, K. J. 2003. Comparison of synthetic surfactants and biosurfactants in enhancing biodegradation of polycyclic aromatic hydrocarbons. *Environmental Toxicology and Chemistry* 22:2280–2292.

Mann, E. E., Rice, K. C., Boles, B. R. et al. 2009. Modulation of eDNA release and degradation affects *Staphylococcus aureus* biofilm maturation. *PLoS One* 4, no. 6 (June):5822. https://journals.plos.org/plosone/article?id=10.1371/journal.pone.0005822

Marrie, T. J., Nelligan, J., and Costerton, J. W. 1982. A scanning and transmission electron microscopic study of an infected endocardial pacemaker lead. *Circulation* 66:1339–1341.

Mayer, C., Moritz, R., Kirschner, C. et al. 1999. The role of intermolecular interactions: studies on model systems for bacterial biofilms. *International Journal of Biological Macromolecules* 26:3–16.

McLeod, B. R., Fortun, S., Costerton, J. W., and Stewart, P. S. 1999. Enhanced bacterial biofilm control using electromagnetic fields in combination with antibiotics. *In Methods in Enzymology* 310:656–670.

Meliani, A., and Bensoltane, A. 2015. Review of *Pseudomonas* attachment and biofilm formation in food industry. *Poultry, Fisheries and Wildlife Sciences* 3:2–7.

Merghni, A., Dallel, I., Noumi, E. et al. 2017. Antioxidant and antiproliferative potential of biosurfactants isolated from *Lactobacillus casei* and their anti-biofilm effect in oral *Staphylococcus aureus* strains. *Microbial Pathogenesis* 104:84–89.

Merino, N., Toledo-Arana, A., Vergara-Irigaray, M. et al. 2009. Protein A-mediated multicellular behavior in *Staphylococcus aureus*. *Journal of Bacteriology* 191:832–843.

Milho, C., Andrade, M., Boas, D. V., Alves, D., and Sillankorva, S. 2019. Antimicrobial assessment of phage therapy using a porcine model of biofilm infection. *International Journal of Pharmaceutics* 557:112–123.

Navarre, W. W., Ton-That, H., Faull, K. F., and Schneewind, O. 1999. Multiple enzymatic activities of the murein hydrolase from staphylococcal phage φ11 identification of a d-alanyl-glycine endopeptidase activity. *Journal of Biological Chemistry* 274:15847–15856.

Nijland, R., Hall, M. J., and Burgess, J. G. 2010. Dispersal of biofilms by secreted, matrix degrading, bacterial DNase. *PLoS One* 5, no. 12 (December):15668. https://journals.plos.org/plosone/article?id=10.1371/journal.pone.0015668

Nolan, V. C., Harrison, J., and Cox, J. A. 2019. Dissecting the antimicrobial composition of Honey. *Antibiotics* 8:251.

Okshevsky, M., and Meyer, R. L. 2015. The role of extracellular DNA in the establishment, maintenance and perpetuation of bacterial biofilms. *Critical Reviews in Microbiology* 41:341–352.

Oliveira, H. L. D. C. D., Fleming, M. E. C. K., Silva, P. V. et al. 2014. Influence of papain in biofilm formed by methicillin-resistant *Staphylococcus epidermidis* and methicillin-resistant *Staphylococcus haemolyticus* isolates. *Brazilian Journal of Pharmaceutical Sciences* 50:261–267.

O'Neill, E., Pozzi, C., Houston, P. et al. 2008. A novel *Staphylococcus aureus* biofilm phenotype mediated by the fibronectin-binding proteins, FnBPA and FnBPB. *Journal of Bacteriology* 190:3835–3850.

Orazi, G., and O'Toole, G. A. 2019. "It takes a village": mechanisms underlying antimicrobial recalcitrance of polymicrobial biofilms. *Journal of Bacteriology* 202, no. 1 (December):00530–19. https://journals.asm.org/doi/epub/10.1128/JB.00530-19

Park, S. I., Daeschel, M. A., and Zhao, Y. 2004. Functional properties of antimicrobial lysozyme – chitosan composite films. *Journal of Food Safety* 69:M215–M221.

Payne, D. E., and Boles, B. R. 2016. Emerging interactions between matrix components during biofilm development. *Current Genetics* 62:137–141.

Percival, S. L., Suleman, L., Vuotto, C., and Donelli, G. 2015. Healthcare-associated infections, medical devices and biofilms: risk, tolerance and control. *Journal of Medical Microbiology* 64:323–334.

Percival, S. L., Hill, K. E., Williams, D. W., Hooper, S. J., Thomas, D. W., and Costerton, J. W. 2012. A review of the scientific evidence for biofilms in wounds. *Wound Repair and Regeneration* 20:647–657.

Peters, G., Locci, R., and Pulverer, G. 1982. Adherence and growth of coagulase-negative staphylococci on surfaces of intravenous catheters. *Journal of Infectious Disease* 146:479–482.

Recsei, P. A., Gruss, A. D., and Novick, R. P. 1987. Cloning, sequence, and expression of the lysostaphin gene from *Staphylococcus simulans. Proceedings of the National Academy of Sciences* 84:1127–1131.

Rediske, A. M., Roeder, B. L., Nelson, J. L. et al. 2000. Pulsed ultrasound enhances the killing of *Escherichia coli* biofilms by aminoglycoside antibiotics *in vivo. Antimicrobial Agents and Chemotherapy* 44:771–772.

Rice, K. C., Mann, E. E., Endres, J. L. et al. 2007. The *cidA* murein hydrolase regulator contributes to DNA release and biofilm development in *Staphylococcus aureus*. *Proceedings of the National Academy of Sciences* 104:8113–8118.

Roca, C., Alves, V. D., Freitas, F., and Reis, M. A. 2015. Exopolysaccharides enriched in rare sugars: bacterial sources, production, and applications. *Frontiers in Microbiology* 6:288.

Roche, D. M., Byers, J. T., Smith, D. S., Glansdorp, F. G., Spring, D. R., and Welch, M. 2004. Communications blackout? Do N-acylhomoserine-lactone-degrading enzymes have any role in quorum sensing? *Microbiology* 150:2023–2028.

Rogers, S. A., Huigens, R. W., Cavanagh, J., and Melander, C. 2010. Synergistic effects between conventional antibiotics and 2-aminoimidazole-derived antibiofilm agents. *Antimicrobial Agents and Chemotherapy* 54:2112–2118.

Rutherford, S. T., and Bassler, B. L. 2012. Bacterial quorum sensing: its role in virulence and possibilities for its control. *Cold Spring Harbor Perspectives in Medicine* 2:a012427.

Satpathy, S., Sen, S. K., Pattanaik, S., and Raut, S. 2016. Review on bacterial biofilm: an universal cause of contamination. *Biocatalysis and Agricultural Biotechnology* 7:56–66.

Seth, A. K., Geringer, M. R., Hong, S. J., Leung, K. P., Mustoe, T. A., and Galiano, R. D. 2012. *In vivo* modeling of biofilm-infected wounds: a review. *Journal of Surgical Research* 178:330–338.

Sharahi, J. Y., Azimi, T., Shariati, A., Safari, H., Tehrani, M. K., and Hashemi, A. 2019. Advanced strategies for combating bacterial biofilms. *Journal of Cellular Physiology* 234:14689–14708.

Simoes, M., Simoes, L., and Vieira, M. J. 2010. A review of current and emergent biofilm control strategies. *LWT Food Science and Technology* 43:573–583.

Simões, M., Simões, L. C., Machado, I., Pereira, M. O., and Vieira, M. J. 2006. Control of flow-generated biofilms with surfactants: evidence of resistance and recovery. *Food and Bioproducts Processing* 84:338–345.

Simpson, B. K., Rui, X., and Klomklao, S. 2012. Enzymes in food processing. *Food Biochemistry and Food Processing*:181–206.

Sutherland, I. W. 1990. *Biotechnology of microbial exopolysaccharides*. Cambridge University Press.

Sutherland, I. W. 2001. Biofilm exopolysaccharides: a strong and sticky framework. *Microbiology* 147:3–9.

Teanpaisan, R., Kawsud, P., Pahumunto, N., and Puripattanavong, J. 2017. Screening for antibacterial and antibiofilm activity in Thai medicinal plant extracts against oral microorganisms. *Journal of Traditional and Complementary Medicine* 7:172–177.

Tetz, G. V., Artemenko, N. K., and Tetz, V. V. 2009. Effect of DNase and antibiotics on biofilm characteristics. *Antimicrobial Agents and Chemotherapy* 53:1204–1209.

Thallinger, B., Prasetyo, E. N., Nyanhongo, G. S., and Guebitz, G. M. 2013. Antimicrobial enzymes: an emerging strategy to fight microbes and microbial biofilms. *Biotechnology Journal* 8:97–109.

Thouvenin, M., Langlois, V., Briandet, R. et al. 2003. Study of erodable paint properties involved in antifouling activity. *Biofouling* 19:177–186.

Torres, C. E., Lenon, G., Craperi, D., Wilting, R., and Blanco, Á. 2011. Enzymatic treatment for preventing biofilm formation in the paper industry. *Applied Microbiology and Biotechnology* 92:95–103.

Trotonda, M. P., Manna, A. C., Cheung, A. L., Lasa, I., and Penadés, J. R. 2005. SarA positively controls bap-dependent biofilm formation in *Staphylococcus aureus*. *Journal of Bacteriology* 187:5790–5798.

Tufvesson, P., Lima-Ramos, J., Nordblad, M., and Woodley, J. M. 2010. Guidelines and cost analysis for catalyst production in biocatalytic processes. *Organic Process Research and Development* 15:266–274.

Verma, S., and Miyashiro, T. 2013. Quorum sensing in the squid-*Vibrio* symbiosis. *International Journal of Molecular Sciences* 14:16386–16401.

Vinh, D. C., and Embil, J. M. 2005. Device-related infections: a review. *Journal of Long-term Effects of Medical Implants* 15:467–488.

von Eiff, C., Heilmann, C., and Peters, G. 1999. New aspects in the molecular basis of polymer-associated infections due to staphylococci. *European Journal of Clinical Microbiology & Infectious Diseases* 18:843–846.

von Eiff, C., Jansen, B., Kohnen, W., and Becker, K. 2005. Infections associated with medical devices. *Drugs* 65:179–214.

Vorkapic, D., Pressler, K., and Schild, S. 2016. Multifaceted roles of extracellular DNA in bacterial physiology. *Current Genetics* 62:71–79.

Wang, H., Wang, H., Xing, T., Wu, N., Xu, X., and Zhou, G. 2016b. Removal of *Salmonella* biofilm formed under meat processing environment by surfactant in combination with bio-enzyme. *LWT-Food Science and Technology* 66:298–304.

Watnick, P., and Kolter, R. 2000. Biofilm, city of microbes. *Journal of Bacteriology* 182:2675–2679.

Weng, Y. M., Chen, M. J., and Chen, W. 1999. Antimicrobial food packaging materials from poly (ethylene-co-methacrylic acid). *LWT-Food Science and Technology* 32:191–195.

Wood, T. K. 2017. Strategies for combating persister cell and biofilm infections. *Microbial Biotechnology* 10:1054–1056.

Ye, J., Shao, C., Zhang, X. et al. 2017. Effects of DNase I coating of titanium on bacteria adhesion and biofilm formation. *Materials Science and Engineering* C78:738–747.

19 Biofertilizers for Sustainable Agriculture

Current Trends and Future Perspective

Sanjeev Kumar Soni, Ritika Manhas, Yudhvir Jakhar, Apurav Sharma and Raman Soni

CONTENTS

DOI: 10.1201/9781003220831-19

19.1 INTRODUCTION

Agricultural work has been in practice since time immemorial. People have depended on agriculture to feed the ever-rising population, currently at approximately 7.5 billion, as per data of 2018, and projected to be 9.2 billion by 2050. The increased demand for food has put pressure on traditional agriculture methods to increase its production. Intensification and extensification methods have been reported to deal with these global environmental issues of concern by increasing the use of external off-farm inputs to increase yields as well as by increasing the land used for the cultivation of food (Matson and Vitousek, 2006). Agriculture intensification led to the Green Revolution Technology, which was based on three principles viz., people need to eat, land resources are limited and therefore one last resort to increase yield is by external inputs. Although green revolution technology multiplied the agriculture produce by many folds, a decline in the soil quality was observed widely, making it a necessity to explore options like the Green Revolution or sustainable agriculture.

The Green Revolution led to heavy use of synthetic fertilizers, pesticides and the withdrawal of water. All this resulted in the promotion of monoculture crops. Increased production, mechanization and improved livelihoods were the positive impacts, whereas the negatives included loss of biodiversity, declining soil fertility, environmental contamination and degradation, pest resistance amongst others. The Green Revolution gave boomerang effects due to loss of efficiency, productivity, stability and sustainability, which called for reviewing of the technology being used to explore more holistic and sustainable options. India contains 2.2% of the world's geographical area, with 46% cultivable land used to support 18% of the human population and 15% of the livestock population of the world. Agriculture is the predominant means of livelihood for more than 50% of the Indian population. The agriculture and allied sector contribute to more than 15% of India's gross domestic product (GDP). Nearly 60% of the cultivable land is prominently monsoon dependent, which means any adversity in agriculture would be affecting the Indian economy significantly (Srivastava et al., 2016), and, therefore, sustainable agriculture is the need of the hour to tackle the present environmental problems of an inefficient biosphere brought into existence due to short-sighted mismanagement and underestimation of the interplays of biotic and abiotic components in sustainable living. It is estimated that by 2050, to meet the nine billion population, the world's food grain production has to increase by approximately 70% (Neupane et al., 2022). The deficit of food grain production will put pressure and lead to an energy crisis, thereby increasing the costs of chemical fertilizers for the farmers, rendering them unaffordable to small and marginal farmers, thus, widening the gap between nutrient removal from the soil and replenishment, leading to poor levels of soil fertility (Kumar et al., 2017). Sustainable agriculture can help in stabilizing and further improving yield over time by increasing the buffering capacity of agriculture fields concerning environmental fluctuations, pest problems, soil organic matter (SOM) and nutrients (like N and P). Sustainable agriculture talks about bringing traditional agriculture practices into use again combined with organic farming and the use of green manure, which may prove to be an inexpensive, environment-friendly and sustainable alternative for mineral fertilization using agrochemicals. It puts major emphasis on the use of biofertilizers to improve soil quality by increasing SOM, water retention, nutrient cycling, soil enzymatic activity, and decreasing soil erosion and greenhouse gas (GHG) emissions (Meki et al., 2013). Recently, Honorable Prime Minister of India Mr. Narendra Modi has also advocated for the adoption of organic farming practices to achieve increased and sustainable agricultural productivity for sustainable development, quoting Sikkim state (India) as a model example (Das, 2020).

Nutrient supplementation along with the inoculation of plant growth-promoting rhizobacteria (PGPR) strains as biofertilizers is expected to act synergistically to be more effective in increasing crop growth, yield and soil fertility. Biofertilizer, an environmentally friendly fertilizer is now used in most countries. Biofertilizers are organisms that enrich the nutrient quality of the soil.

The main sources of biofertilizers are bacteria, fungi and cyanobacteria (blue-green algae). The most prominent feature of these partners is their symbiotic association with plants, which delivers several benefits to plants, including nutrition, disease resistance and tolerance to unfavourable climatic conditions. Biofertilizers will help to solve such problems as increased salinity of the soil and chemical run-offs from the agricultural fields. Thus, biofertilizers are important if we are to ensure a healthy future for the generations to come. These microorganisms supply, in addition to nitrogen, a considerable amount of organic matter, enriching the structure of the soil. Inoculants of these microorganisms have proved their technical feasibility, economic viability and social acceptability. They are therefore called a 'biofertilizer'. Hence the term 'biofertilizer' or microbial inoculants may be defined as preparations containing strains of nitrogen-fixing, phosphate solubilizing, potassium mobilizing or cellulolytic microorganism (Barman et al., 2017).

The application of biofertilizers may help to improve the degraded cultivable lands by promoting the flourishing of microbial communities and diversity (Zhen et al., 2014). The contributions of nitrogen-fixing and phosphate solubilizing bacteria like cyanobacteria (*Anabaena*) and bacteria (*Azospirillum*, *Azotobacter*, *Pontibacter*, *Pseudomonas*, *Burkholderia*, and *Rhizobium*), fungi (*Trichoderma*), etc., have been known for improved crop nutrition and soil health. Biofertilizers hence are considered safe and sustainable for the development of the agriculture sector.

19.2 BIOFERTILIZERS

Biofertilizers are widely known as materials that contain one or more species of live microorganisms that can assist in the uptake of important elements in plants by converting them from their complex non-usable form to simpler usable form. It is performed through various biological processes like nitrogen fixation, phosphate or potassium solublization or through secretion of plant growth-promoting hormones. They also help in cellulose and biodegradation of soil, compost and other environments. They have no toxic effects on the plants or soil. In other words, they are microbial inoculants of bacteria, fungus and algae that work by increasing the availability of nutrients to plants in combination or alone. Because of the rising costs of chemical fertilizers and their hazardous effects on the environment, biofertilizers present a sustainable alternative for agricultural use (Kumar et al., 2017). Biofertilizers can be applied to the seeds before sowing them, or on the surface of the plants or soil. They form colonies in the rhizosphere or on the interior parts of that plant which promotes the growth of the plant by enhancing the uptake of nutrients (which is also one of the main highlights of their use as fertilizers). The use of biofertilizers was reported to increase crop yield by 10–40% due to increasing contents of proteins, essential amino acids, vitamins and nitrogen fixation (Bhardwaj et al., 2014). Biofertilizers are a cheap source of nutrients. The use of live microorganisms provides an excellent supply of macro as well as micronutrients. Their use can increase the organic matter in the soil and counteract the negative impact of the chemical fertilizers in the soil (Gaur, 2010). There are numerous microorganisms in the rhizosphere of plants that especially have a beneficial effect on their growth, as considerable amounts of these have a functional relationship and are part of a holistic system with plants. Varying amounts of microbes in the soil form a vital component in biotic activities of the soil ecosystem, which makes them essential in nutrient mobilization and sustainability for the production of crops (Ahemad and Kibret, 2014). On the contrary, chemical fertilizers are not only expensive, they also disturb the equilibrium of agricultural ecosystems by causing a tremendous amount of pollution leading to an ecological imbalance in the environment. All of these problems can be avoided by the use of biofertilizers (Al-Khiat, 2006). The utilization of biofertilizers has immense advantages over the use of conventional chemical fertilizers in agriculture. Biofertilizers are safer than most of the chemical fertilizers currently in use. They are nontoxic and microorganisms are not accumulated in the food chain. They do not need repeated application in most of the known cases. In the case of biofertilizers as pest control or biocontrol agents, development of resistance against the microorganism is seldom noticed (Wu et al., 2005).

Biofertilizers mainly include nitrogen-fixing, phosphate solubilizing, potassium mobilizing and plant growth-promoting microorganisms that fulfill the major NPK requirements of the plants and secrete growth hormones. Amongst the commonly used biofertilizers benefiting the crops are *Azotobacter, Azospirillum*, blue-green algae, *Azolla*, P-solubilizing microorganisms, mycorrhizae, and *Rhizobium* (Selvakumar et al., 2009).

19.3 ESSENTIAL SOIL NUTRIENTS FOR SUSTAINABLE AGRICULTURE

The combined effect of physical factors such as the air, water, light, pH, micronutrients and macronutrients determines the overall yield of crops and growth of plants. The quality of soil progressively decreases due to continuously being used for crop production season after season and needs replenishment of the nutrients at the same or higher rate. The major factors contributing to the loss of soil fertility and quality deterioration, other than the consumption of nutrients by the plants, are leaching, over-tillage, improper irrigation, bush burning and so on, and these are leading to poor crop yield, which can be addressed by the use of a biofertilizer (Wander, 2010).

Heavy use of agrochemicals like chemical fertilizers and pesticides in recent years has led to high yield. Unfortunately, it has also led to poor cropping systems due to use of these agrochemicals over the recommended levels, which has led to soil acidification, risking the contamination of underground water and the weakening of plant root systems, rendering them prone to infections and diseases (Chun-Li et al., 2014).

The essential nutrients include both macronutrients required in larger quantities and micronutrients required in smaller quantities (Table 19.1). Nitrogen, phosphorus, potassium, calcium, magnesium and sulfur are the macronutrients, while others are the micronutrients. Macronutrients are further divided into primary and secondary nutrients that are always present in limited quantities and those that are rarely present in limiting quantities respectively. NPK thus forms the primary nutrients in soil required for plant growth.

19.3.1 Nitrogen

Seventy-eight percent of the earth's atmosphere comprises nitrogen gas. Nitrogen is the major constituent of proteins and nucleic acids that are present among all living organisms. Nitrogen is abundant in the soil but in an insoluble form that plants cannot use. Nitrogen is required for the synthesis of enzymes, proteins, chlorophyll, DNA and RNA and transport in xylem and phloem, e.g., as osmoregulation, alkaloids, and is therefore important in plant growth and production of food and feed. Plants can only uptake nitrogen in the form of nitrates and ammonia by the process of biological nitrogen fixation, thunder, chemical fertilizer, and decomposition of organic matter. Slow growth, low protein levels, poor yield, low-quality produce, and inefficient water use are all symptoms of an insufficient supply of accessible nitrogen to plants. A large number of prokaryotes

TABLE 19.1
Major Macronutrients and Their Role in Agriculture (Modified from Devi and Sumathy, 2017)

Element	Application	Deficiency
Nitrogen	Promotes leaf and stem growth	Yellowish coloring of leaves and short growth of the stalk
Phosphorus	Promotes plant growth and fruit ripening	Lack of flowering
Potassium	Stomata turgor; growth and enzyme catalyst	Stunted growth; leaf necrosis (death); reduced gas exchange.

and archaebacteria can biologically fix nitrogen and make them available to plants (Itelima et al., 2018).

19.3.2 Phosphorus

Phosphorus is a primary macronutrient required in large amounts, relatively less than nitrogen and potassium; it is not possible to achieve normal plant growth without it. Unlike nitrogen, there is no large atmospheric source of phosphorus known that can be made biologically available. Phosphorus is known to promote growth in legumes by increasing yield, nodule number and nodule mass. It plays a vital role in seed formation and forms a main component of seeds as well as fruits (Scalenghe et al., 2012). Root development, stalk and stem strength, crop maturity and production, crop quality, and resistance to plant diseases are associated with phosphorus nutrition (Mohammadi and Sohrabi, 2012). Soluble phosphorus is usually found in very low micromolar quantities or less in soil. Although ordinary soils have a phosphate concentration of roughly 0.05% (w/w), only 0.1% of total P is accessible to plants in the form of phosphates (Fankem et al., 2006). It can be added to the soil through fertilizers, bone meal, and superphosphates. Deficiency of soil P is often associated with stunted growth and late maturity of plants. Some microorganisms are involved in the solubilization of insoluble phosphate known as phosphate solubilizers. Excess phosphorus cannot be properly assimilated by plants and leads to early maturation of plants (Salhia, 2010).

19.3.3 Potassium

Potassium is the second-highest element required by the plant after nitrogen. Its functions include enhancing cold endurance in plants, resistance against fungal and bacterial diseases, and drought tolerance in the plants by affecting stomatal regulation (Mfilinge et al., 2014). Potassium is absorbed by the plants in the form of ions through fertilizer, wood ash, or decaying organic matter. It is known to influence crop growth and water economy due to its effects on water uptake, maintenance of turgor pressure, root growth, and transpiration. It increases the production of carbohydrates, resulting in thicker cell walls and straw of cereal plants. Potassium deficiency in a plant inhibits enzyme activities along with disruption in the plant's protein and carbohydrate metabolism. The deficiency of potassium reduces the viability of seeds making them difficult to preserve in marketable conditions. It is an activator of multiple enzymes participating in photosynthesis, it builds cellulose and forms a chlorophyll precursor, and produces enzymes for respiration. Potassium deficiency may cause necrosis, interveinal chlorosis, wilting or brown spotting (Itelima et al., 2018).

19.4 CLASSIFICATION OF BIOFERTILIZERS

Microorganisms have been grouped in different categories based on their functions and asssociation as describes in Table 19.2.

19.4.1 Nitrogen-Fixing Microorganisms as Biofertilizers

Biological nitrogen-fixing bacteria play a vital role in improving soil quality for plants by assisting in the production of plant growth hormones along with nitrogen fixation by converting elemental nitrogen into the usable form. They help protect the plant against pathogens. The deficiency of nitrogen can lead to stunted growth amongst other physiological changes. Numerous free-living and symbiotic nitrogen-fixing bacteria play a significant role in enhancing nitrogen fixation in soil.

The mechanism of nitrogen fixation by plant growth-promoting bacteria has been pictorially represented in Figure 19.1. The N-fixing process is carried out by the enzyme complex called nitrogenase, which consists of dinitrogenase reductase and dinitrogenase. Dinitrogenase reductase

TABLE 19.2
Components of Various Biofertilizers (Modified from Kumar et al., 2017)

S. No.	Components of Biofertilizers	Examples
Nitrogen-fixing microorganisms		
1	Free-living N-fixers	*Azospirillum, Azotobacter, Beijerinkia, Clostridium, Klebsiella, Anabaena* and *Nostoc*
2	Symbiotic N-fixers	*Azospirillum, Rhizobium, Frankia, Anabaena and Azolla*
Phophate-solubilizing/P-mobilizing microorganisms		
3	Phosphate solubilizers	*Achromobacter sp, Aereobacter sp., Erwinia sp., Bacilluis subtilis, Pseudomonas striata, Rhizobium, Burkholderia, Agrobacterium, Micrococcus, Acetobacter, Flavobacterium and Bacillus megaterium var phosphaticum, Penicillium, Aspergillus awamori, A. niger*
4	P-mobilizers	*Amanita, Boletus, Gigaspora, Glomus, Laccaria, Sclerocystis*
Potassium-solubilizing/mobilizing microorganisms		
5	K-solubilizers	*Pseudomonas, Burkholderia, Acidothiobacillus ferrooxidans, Bacillus mucilaginosus, Bacillus edaphicus, Bacillus circulans, Paenibacillus* spp., and *Aspergillus* sp.
6	K-mobilizers	*Bacillus mucilaginosus, Azotobacter chroococcum,* and *Rhizobium* sp.
Zinc-solubilizing microorganisms		
7	Zn-solubilizers	*Bacillus subtilis, Thiobacillus thioxidans, Saccharomyces*
Sulfur-oxidizing microorganisms		
8	S-oxidizers	*Thiobacillus*
Plant growth-promoting microorgniams		
9	Plant growth-promoting rhizobacteria	*Pseudomonas, Azospirillum, Azotobacter, Klebsiella, Enterobacter, Alcaligenes, Arthrobacter, Burkholderia, Bacillus,* and *Serratia*

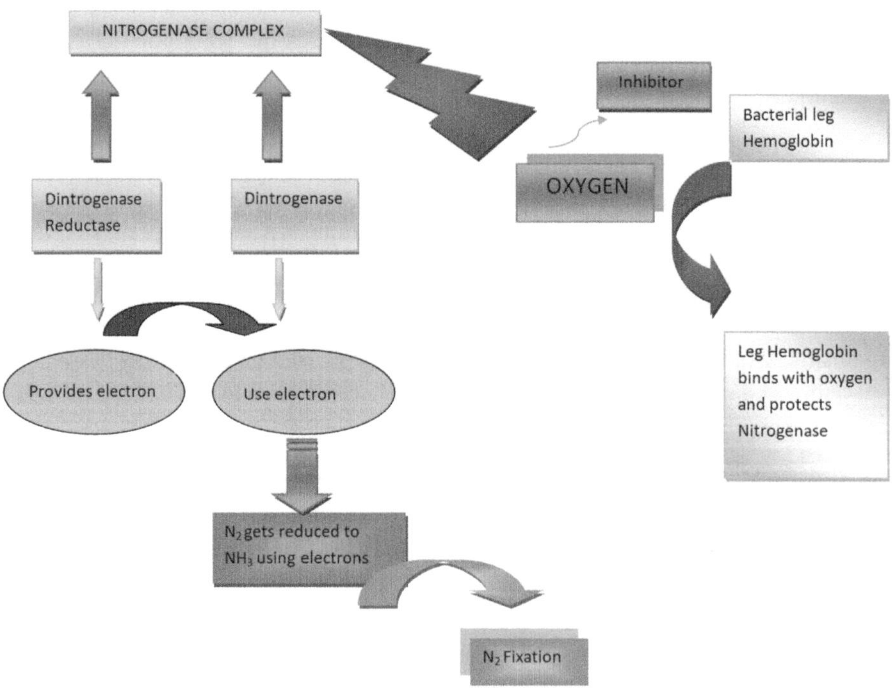

FIGURE 19.1 Molecular N_2 fixation mechanism by plant growth-promoting rhizobacteria.

Source: Adapted from Mahanty et al. (2017)

provides electrons and using these electrons, dinitrogenase reduces N_2 to NH_3. Oxygen is a potent inhibitor for the enzyme complex since the enzyme can bind to O_2 and becomes inactive. But bacterial leghemoglobin has a stronger affinity for oxygen and thus binds to free oxygen more strongly and effectively. Thus, the presence of leghemoglobin keeps the nitrogenase enzyme complex active by protecting it from oxygen.

19.4.1.1 *Rhizobium*

Rhizobium is an N_2-fixing symbiotic bacteria belonging to the Rhizobiaceae family. They establish a symbiotic relationship with plant roots by infecting them and forming tumor-like growths called root nodule structures. These nodules are ammonia-producing factories. They vary morphologically and physiologically from free-living conditions to the bacteroids of nodules. Seven genera are *Rhizobium*, *Bradyrhizobium*, *Sinorhizobium*, *Azorhizobium* and *Mesorhizobium* are collectively called rhizobia. Rhizobacteria that fix nitrogen in on-leguminous plants are call diazotrophs; they form nonobligate interactions with hosts (Verma et al., 2010). Nif genes responsible for nitrogen fixation are found in both symbiotic and free-living forms (Black et al., 2012). *Azorhizobium*, another strain of *Rhizobium*, forms stem-nodules and fixes nitrogen in that area (Gourion et al., 2015). *Rhizobium* is reported to produce large amounts of IAA to promote plant growth. In a study, *Bradyrhizobium*, also a well-documented nitrogen-fixing strain, inoculated in mucuna seeds enhanced the total organic carbon, N_2, potassium, and phosphorus contents of soil, thereby significantly increasing plant growth, plant biomass also resulting in the reduction of weed population (Youssef and Eissa, 2014). The use of *Rhizobium* as biofertilizers can lead to fixing of up to 100–300 kg N/ha in a given season and leave the soils substantially rich in nitrogen for following crops (Bhattacharyya and Jha, 2012). In a study, inoculating *Rhizobium* on French beans resulted in an increased yield, plant height, high pod numbers (Wani et al., 2013). They have a significant influence on legume vegetable crops thereby recording an increase in yield ranging from 4–13% (Ruiz-Sánchez et al., 2010).

19.4.1.2 *Azotobacter*

Azotobacter represents a diverse group of free-living diazotrophic, aerobic, photoautotrophic and nonsymbiotic bacteria which belongs to the azotobacteriaceae family and is commonly occurring in soil. They have been reported to significantly improve soil fertility by harnessing atmospheric nitrogen and its fixation in the roots. The genus *Azotobacter* includes six species amongst which *A. chroococcum* occurs most commonly in soils all over the world. The other reported species include *Azotobacter vinelandii, Azotobacter beijerinckii, Azotobacter insignis* and *Azotobacter macrocytogenes* (Mishra et al., 2013). They are usually present in neutral and alkaline soils (Moraditochaee et al., 2014). They secrete vitamin B complex; different phytohormones such as gibberellins, naphthalene acetic acid (NAA); and other substances, which along with inhibiting the growth of pathogens, promote plant root growth and minerals uptake (Mathivanan et al., 2015). In a study, it was reported that *Azotobacter* secretes substances that are reported to inhibit root pathogens affecting plants and to considerably improve both the root growth and uptake of nutrients essential for the plants (Babalola, 2010). *Azotobacter* was reported to add up to 15–93 kg N/ha on *Paspslum notatum* roots (Youssef and Eissa, 2014). Another strain of *Azotobacter indicum* is known to produce several antifungal substances, which are used to inhibit the growth of multiple fungal pathogens in the root system of plants, thereby reducing the seedling mortality to a certain extent (Martin et al., 2011). *A. chroococcum* is the dominant inhabitant in arable soils to fix N_2 in the amounts of about 2–15 mg N_2/g of carbon source in culture media. An abundant amount of slime produced by it plays a key role in soil aggregation. Its number in soils rarely exceeds 10^5/g due to the lack of organic matter in it and the presence of antagonistic microorganisms in the soil (Kumar et al., 2017). The population of *Azotobacter* in the rhizosphere of crop plants has been reported to be quite low. Rice, maize, sugar cane, bajra, vegetables, and plantation crops have all been found to have the organism in their rhizospheres (Wani et al., 2013).

19.4.1.3 *Azospirillum*

Azospirillum is a gram-negative, non-nodule forming aerobic nitrogen-fixing which is an endosymbiotic bacterium belonging to the family Spirilaceae (Mehnaz, 2015). Although there are many species under this genus, like *Azospirillum amazonense, Azospirillum halopraeferens,* and *Azospirillum brasilense,* the mainly beneficial species that are primary inhabitants of soil, rhizosphere and also the intercellular spaces of root cortex of graminaceous plants which include *Azospirillum lipoferum* and *A. brasilense* (Mishra et al., 2013). Plants that use the C4 dicarboxylic route for photosynthesis create an associative symbiosis with *Azospirillum* due to nitrogen fixation on organic malic and aspartic acids (Mishra and Dash, 2014). Hence it is mainly recommended for use in crops of maize, pearl millet, sugar cane, etc. Application of *Azospirillum* improving plant growth and yield in commercially important crops like tomato, brinjal, chilies, etc. is well documented (Mohandas, 1988). The use of *Azospirillum* has appeared amongst the newer technologies used in plant implantations. Apart from nitrogen fixation, inoculation with *Azospirillum* has been reported to produce plant growth promoters like IAA, gibberellins, cytokinin. They enhance the development of roots to increase the uptake of plant nutrients (Trabelsi and Mhamdi, 2013). They also help to develop disease resistance and drought tolerance in the plants. *Azospirillum* is known to have met the nitrogen requirements of non-leguminous plants because of their associative activities that showed an increase in yield and nitrogen economy of vegetables (Sundaravelu and Muthukrishnan, 1993). In a study, it was reported that the inoculated strain *A. brasilense* sp. 245 resulted in the production of different phytohormones that considerably enhanced the maize growth substantially (Perrig et al., 2007). Another study showed that *Azospirillum* and *Pseudomonas* inoculated in the rhizosphere of wheat modified the cultivable bacterial community and changed the carbon source utilization profiles of the soil microflora at tillering and grain filling stages (Naiman et al., 2009).

19.4.1.4 Cyanobacteria

Cyanobacteria are the most abundant group of organisms on the earth. They are autotrophic and found in a variety of environments, especially in the marine and freshwater. Marine water is the richest nutrient source for the cultivation of cyanobacteria (Chittora et al., 2020). They are generally unicellular and often grow in large colonies. Cyanobacteria comprise a broad range of bacteria with different shapes and sizes. They can cover 150 genera that have been identified so far. Classification of cyanobacteria was proposed in 1985, in which four orders of the bacteria have been identified as *Chroococcales, Nostocales, Oscillatoriales,* and *Stigonematales.* The most efficient nitrogen-fixing cyanobacteria are *Nostoclinkia, Anabaena variabilis, Aulosira fertilisima, Calothrix* sp., *Tolypothrix* sp., *and Scytonema* sp. They are present in the rice crop cultivation area. Cyanobacteria generally form complex associations with bacteria and green algae and structures known as cyanobacterial mats. Cyanobacteria consist of two important cells: (i) heterocysts (responsible for nitrogen fixation and ammonia synthesis) and (ii) vegetative cells (exhibit normal photosynthesis and reproductive growth). Besides fixing atmospheric nitrogen, they induce the growth of plants by producing auxins and gibberellins. Reports have suggested that rice production increased from 15% to 38% with the supplementation of BGA inoculants (Kumar et al., 2018). In India, free-living and symbiotic cyanobacteria have been applied in the cultivation of rice (Thajuddin and Subramanian, 2005).

19.4.2 Phosphate Solubilizing Microorganisms as Biofertilizers

Phosphate is present in the soil in an insoluble form, which makes it unavailable for the plants for utilization, as plants are only capable of utilizing it in two soluble forms i.e., monobasic and diabasic forms. The PSMs consist of different bacteria and fungi which are capable of fulfilling about 20–25% of phosphorus necessity of the plants and hence decreases the burden of manufacturing of phosphate fertilizers (chemical) as well as the burden on farmers to purchase to increase the output (Verma et al., 2017). Phosphate solubilizing bacteria (PSBs) make up about 1–50%, while phosphate solubilizing fungi (PSFs) are only 0.1–0.5% among the P-solubilizers (Chen et al., 2006). These PSMs employ different mechanisms for solubilization of phosphate by converting insoluble phosphate in soil into soluble forms HPO_{2-} and HPO_4^{2-} through the process of organic acid production, chelation and ion exchange reactions, proton extrusion and make them available to plants (Nahas, 1996). Phosphate solubilizing microorganisms secret organic acids (gluconic acids), inorganic acid (hydrochloric acid) and enzymes that act on insoluble phosphates and convert it into a soluble form (chelate cations like Al, Fe, Ca) thus providing phosphorus to the plants. But the solubilizing capacity of hydrochloric acid is less effective compared to organic acids. Actinomycetes are also considered as potential phosphate solubilizing organisms which can tolerate extreme environments like drought, and antibiotics. Reports also suggested that 20% of actinomycetes can solubilize phosphorus (Hamdali et al., 2008.)

19.4.2.1 Phosphate Solubilizing Fungi (PSF)

Aspergillus sp. and *Penicillium* sp. are the two important phosphate solubilizing fungi. It has been reported that *Aspergillus niger* improves the quantity of dry biomass of chick pea plants by 22–33% concerning the non-inoculated control (Kapri and Tewari, 2010). Phosphate solubilizing activity has also been shown by symbiotic nitrogenous *Rhizobia* and nemato fungus *Arthrobotrys oligospora* (Thakur et al., 2014).

19.4.2.2 Phosphate Solubilizing Bacteria (PSBs)

Phosphorus becomes immobilized and gets washed away from the soil thereby becoming unavailable to the plants; PSB microbes solubilize inorganic phosphorus and make it available to plants in adequate quantities even from poor sources. Higher concentrations of PSB are found in rhizospheric soils than the nonrhizospheric soils (Youssef and Eissa, 2014). PSB also increases the efficiency of nitrogen-fixing bacteria by augmenting the process of biological nitrogen fixation (Mohammadi and Sohrabi, 2012). PSB produce organic acid and lower the pH of the soil around them in order to dissolve bound phosphates (Bhattacharyya and Jain, 2000). *Bacillus polymyxa* and *Pseudomonas striata* increases the yield when used in combination with peat. Similar results were reported in tomato (Shukla et al., 2009), cauliflower (Kachari and Korla, 2009), turmeric (Padmapriya and Chezhiyan, 2009) and peas (Chaykovskaya et al., 2001). The crops of French beans are adversely affected by the unavailability of phosphorus in the soil. It requires phosphorus in high amounts for optimum growth and yield. For effective utilization of nutrients, inoculation with both PSB and VAM is considered effective. It is established through research that PSB increased the phosphorus uptake in low fertility soils and resulted in improved plant growth (Ramana et al., 2010, 2011). Quality of tomatoes yield after treatment with biofertilizer consisting of phosphobacteria and *Azospirilium* in parameters of TSS, ascorbic acid and lycopene contents was recorded to be higher (Kumaran et al., 1998).

The mechanism of inorganic phosphorus solubilization by phosphate-solubilizing rhizobacteria has been pictorially represented in Figure 19.2. Low molecular weight organic acids such as gluconic and citric acids are produced by solublization of phosphorus by bacterium represented in Figure 19.2. The hydroxyl (OH) and carboxyl (COOH) group of gluconic and citric acids chelate the cations bound to phosphate and thus converts insoluble phosphorus into a soluble organic form. The mineralization of soluble phosphorus occurs by synthesizing different phosphatases which catalyze the hydrolysis process. When plants incorporate these solubilized and mineralized phosphorus molecules, and, eventually, overall plant growth and crop yield significantly increase.

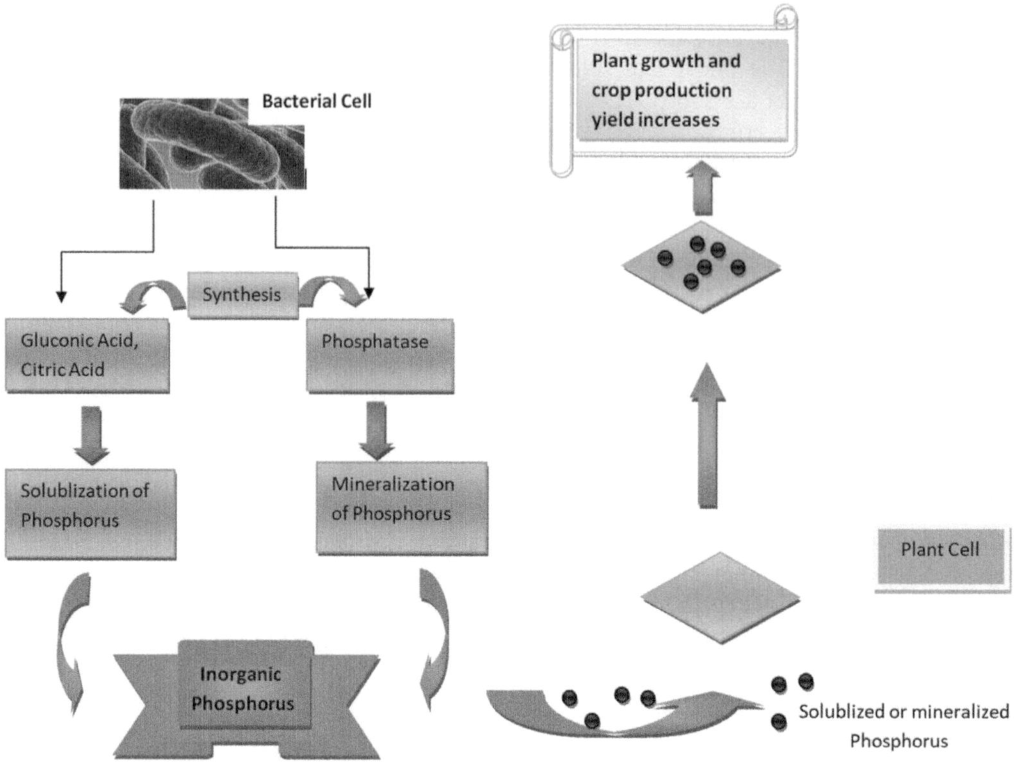

FIGURE 19.2 Inorganic phosphorus solubilization by phosphate-solubilizing rhizobacteria.

Source: Adapted from Mahanty et al. (2017)

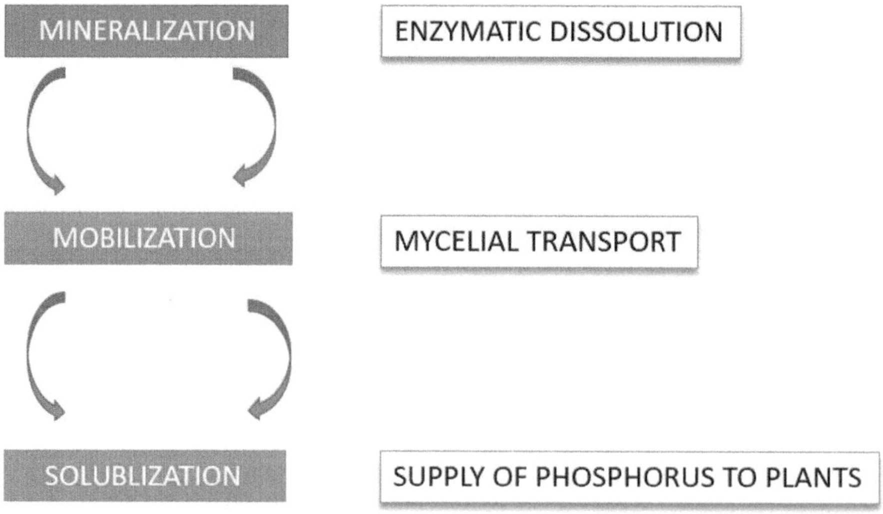

FIGURE 19.3 Phosphorus mobilization by phosphate-mobilizing mycorrhizal fungi.

19.4.3 Phosphorus Mobilizing Microorganisms as Biofertilizers

Microorganisms mobilize phosphorus by scavenging it from a phosphorus-enriched environment and then channelizing it to the plants. They can even solubilize the organic phosphorus by using the phosphatase enzyme present in their hyphae (Ezawa et al., 1995). The generalized mechanism of phosphorus mobilization to plants is illustrated in Figure 19.3. Arbuscular mycorrhizal fungi (AMF) are known to be significant contributors to phosphate mobilization. The potential of AMF can be fully perceived when the soil is deficient in phosphorus (Ghorbanian et al., 2012). These AMF have special structures known as Vesicles and Arbuscules, which remain symbiotically associated with the root of the plants in the rhizosphere and enhance phosphorus uptake by the extensive external fine absorbing hyphae (Karandashov et al., 2004).

The substantial size of the hyphae can exploit a larger soil volume (Jakobsen et al., 1992). The fungi transports nutrients and water to the plant roots, and these, in turn, receive the carbon requirements from the host plants. Studies by Yao et al. (2001) reveal that the AMF not only mobilizes soil phosphates when its availability is low but also provides a greater surface area, which permits increased contact with the phosphates concerning the non-mycorrhizal root systems. The supply of phosphorus to the host plants depends upon the amounts of external mycelium produced by the fungi. The kinetics of nutrient uptake by the mycorrhizal roots are quite different from that of the non-mycorrhizal ones (Cress et al., 1979). Colonization of the roots by AMF thus helps in enhanced growth of the plants in low fertility soils or soils with uneven distribution of nutrients. Hence the phosphate mobilizing microorganisms can be considered as broad-spectrum fertilizers (Itelima et al., 2018).

19.4.4 Potassium Solubilizing Microorganisms as Biofertilizers

Potassium solubilizers consist of a wide range of saprophytic bacterial and fungal strains (Ahmad et al., 2016). The presence of salt in the soil adversely affects nutrition acquisition in plants which results in salt stress-induced osmotic stress thereby enhancing K deficiency in plants (Römheld and Kirkby, 2010). The use of potassium-solubilizing bacteria may prove useful in facilitating plants growing in low potassium saline soils as the insoluble potassium can be converted into soluble forms by acidification, exchange reactions, chelation, and formation of polymeric substances (Bhattacharyya and Jha, 2012). Under greenhouse and field circumstances, inoculating seeds with potassium-solubilizing bacteria from various plants results in a considerable increase in germination percentage, plant growth, and potassium uptake by plants (Anjanadevi et al., 2016). Their use not only mobilizes the insoluble potassium present in the soil but also replaces the manufactured potassium fertilizer that incurs huge costs. Inoculation with potassium solubilizers promotes the growth of their populations and relieves the stress of salt on plant growth leading to higher nutrient uptake, increased accumulation of ions, regardless of salinity levels. Due to their excellent capacity to produce a wide variety of enzymes and metabolites, *Pseudomonas* has been of particular interest in the category of potassium solubilizing bacteria as they impart plants an ability to withstand biotic and abiotic stress conditions. In a study, *B. subtilis* along with *B. cereus* strain has been recorded to alleviate the adverse effects of salinity in soyabean (Chakraborty et al., 2011). Multiple studies have also shown beneficial effects to the growth of eggplant, pepper and cucmber on treatment with potassium-solubilizing bacteria (Han and Lee, 2006).

19.4.5 Potassium-Mobilizing Microorganisms as Biofertilizers

Bacillus sp. is an example of potassium-mobilizing biofertilizers, which mobilizes inaccessible forms of potassium in the soil and makes it accessible to the plants. *Bacillus* sp. and *Aspergillus* sp. are established phosphorus solubilizers and are also known to mobilize potassium as well, along with solubilizing phosphate (Itelima et al., 2018). Hence, these microbes are performing a dual function. A generalized flow sheet indicating the microbial solubilization of precipitated potassium and mobilization of exchangeable potassium to plants is depicted in Figure 19.4.

19.4.6 Zinc-Solubilizing Microorganisms as Biofertlizers

Zinc is found up to 0.008% in the earth's crust, however more than half of Indian soils are zinc deficient, with levels below the essential level of 1.5 ppm accessible zinc (Katyal and Rattan, 1993). Farmers compensate this lag by adding soluble zinc sulfate ($ZnSO_4$) from external sources of which the crop utilizes only 1–4%, and the rest 75% gets converted to different mineral fractions, and hence not absorbed by the plants. Zinc-solubilizing microorganisms solubilize zinc through various mechanisms (Figure 19.5), one of which is acidification. These microbes produce organic acids in soil which sequester the zinc cations and decrease the pH of the nearby soil thereby involving

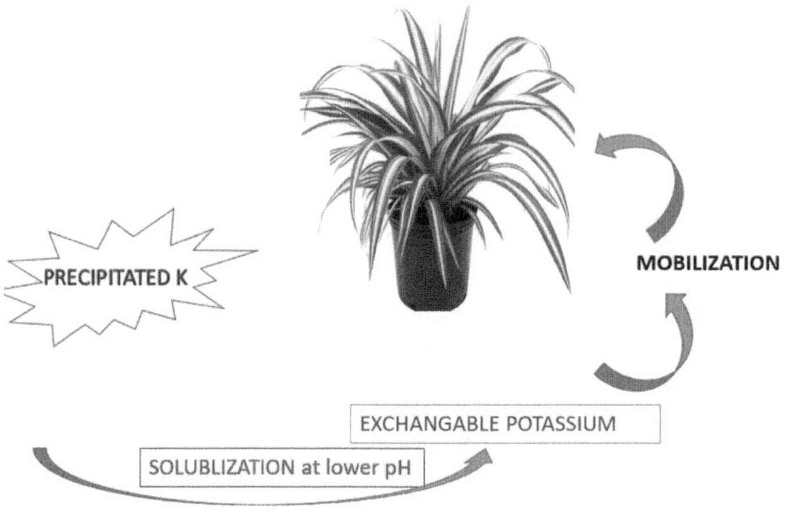

FIGURE 19.4 Microbial solubilization and mobilization of potassium to plants.

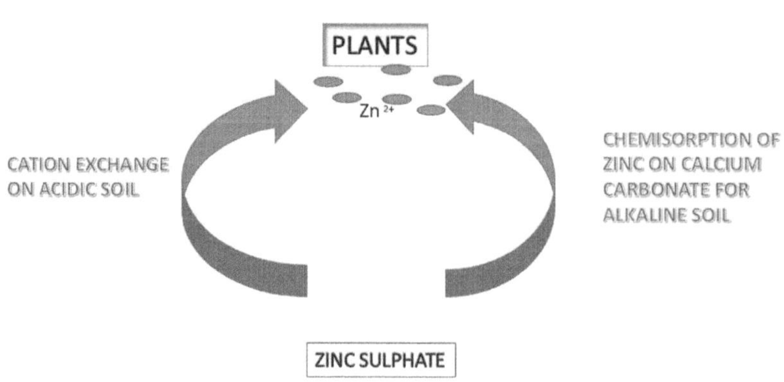

FIGURE 19.5 Generalized mechanisms of microbial solubilization of zinc.

cation exchange. The other mechanism involves chemisorptions of zinc on calcium carbonate with simultaneous complexation by organic ligands under alkaline conditions (Kumar et al., 2018). *Bacillus* sp. has been conjugated with cheap, insoluble zinc compounds such as zinc carbonate, and zinc sulfide in place of costly zinc sulfate and successfully applied for zinc solubilization (Yadav and Saxena, 2018).

19.4.7 Sulfuroxidizers as Biofertilizers

These microorganisms oxidize sulfur to sulfates so that the plants can utilize it. *Thiobacillus* sp. is one of the sulfur-oxidizing biofertilizers (Itelima et al., 2018). This grows and mobilizes the insoluble sulfur present in the soil, ultimately increasing the uptake of sulfur and making it readily available to the plant. The generalized mechanism of microbial oxidation of sulfur and its mobilization to plants is depicted in Figure 19.6.

19.4.8 Plant Growth-Promoting Rhizobacteria as Biofertilizers

Groups of bacteria that colonize the rhizosphere or roots of plants and impart beneficial effects to them are known as plant growth-promoting rhizobacteria. In Figure 19.7 various plant growth-enhancing mechanisms of plant growth-promoting rhizobacteria (PGPR) have been listed. Plant growth-promoting rhizobacteria (PGPR) are known to act as biofertilizers, biostimulants and bio-protectants as well. PGPRs promote plant growth by the production of phytohormones as well as other important metabolites. Production of indole acetic acid (IAA), gibberellic acid and cytokinins (Kloepper et al., 2007) are well documented in several studies along with the production of metabolites such as HCN, antibiotics and siderophores of multiple kinds. Amongst auxins, IAA is the most physiologically active. One of the prominent mechanisms of plant growth promoters is, due to the high amounts of phytohormones secreted there is an alteration in root architecture and a great number of fine roots grow out of the plants that increase its surface area for absorption leading to an increase in uptake of nutrients (Han et al., 2006). Several studies are exploring the potential of fungi as PGPRs.

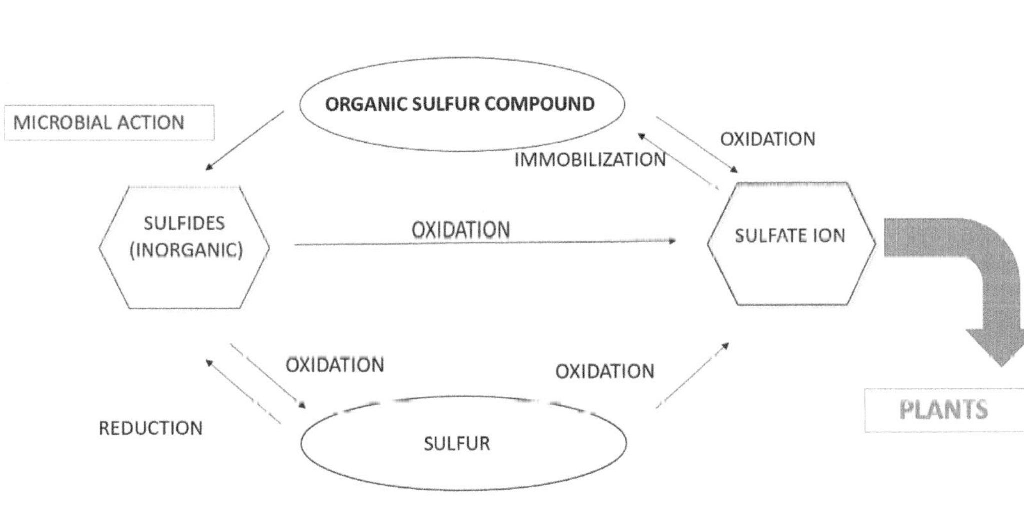

FIGURE 19.6 Generalized mechanism of microbial oxidation and mobilization of sulfur.

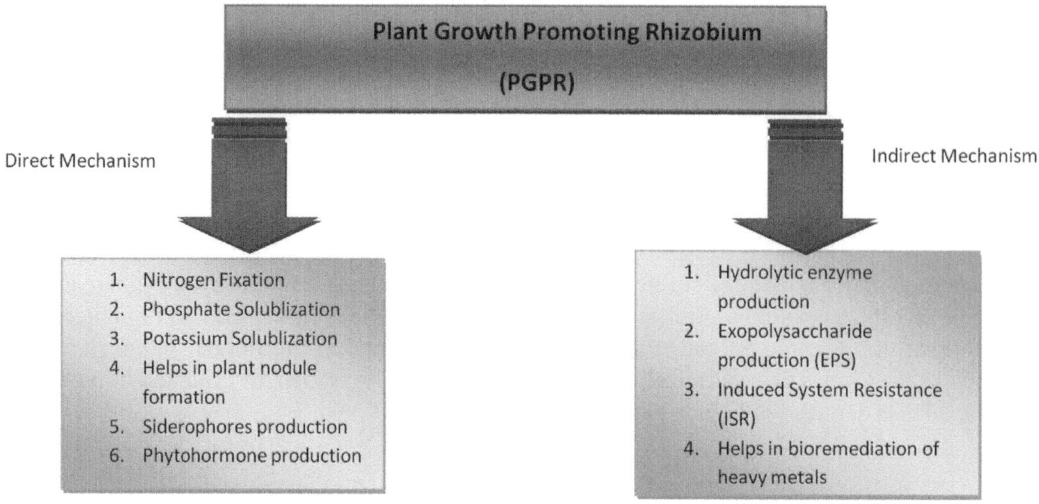

FIGURE 19.7 Plant growth-enhancing mechanisms of plant growth-promoting rhizobacteria (PGPR).
Source: Adapted from Mahanty et al. (2017)

19.5 ADVANTAGES OF USING BIOFERTILIZERS

One of the biggest benefits of using biofertilizers is that they are eco-friendly. Moreover, they are inexpensive, fix nitrogen biologically and directly make them available to the plants. On the contrary, chemical fertilizers usually get immobilized, as in the case of phosphorus, and are rendered unavailable to the plants. They lead to soil enrichment and their use over time gives spectacular results whereas the use of chemical fertilizers over time may lead to reduced fertility of the soil in addition to various kinds of polluting effects given by it. The plant growth promotion effect leads to improved root proliferation in the plants. Biofertilizer microbes have the ability to break down complex nutrients into simple forms making them available to the plants as in the case of breaking down organic matter that is abundantly present in the soil. They ensure an adequate supply of nutrients to the host plants regulating their development and physiology positively. Also help to increase the crop yield many folds, protect host plants against many of the plant pathogens may be exposed to, enhance the growth of beneficial microorganisms in the soil and promote the development of mycorrhizal associations.

19.6 DISADVANTAGES OF USING BIOFERTILIZERS

Undoubtedly biofertilizers offer multiple benefits, but they certainly have a few downsides to them, which act as constraints in their implementation and at times offer disadvantages too. At times, the implementation costs for biofertilizers are higher than their chemically synthesized counterparts. The long-term usage of biofertilizers may result in the accumulation of heavy metals, nutrients, salts, etc, over time, which may have adverse effects on the development of the plants and surrounding ecosystems. Due to the concentrations being low, large amounts of biofertilizers are required for application in fields. The use of biofertilizers doesn't show immediate effects. The unavailability of good carrier material is the biggest hindrance as it directly affects the shelf life of the biofertilizer. Trained technical staff, if unavailable, may lead to faulty production amongst other infrastructural constraints. Lack of awareness amongst the farmers and marketing constraints pose a big problem in India. There are environmental constraints present like limited cropping seasons.

19.7 PRODUCTION OF BIOFERTILIZERS

Production of biofertilizers is a multi-step process including isolation of desired microbes and studying their growth profiles, formulation of inoculum of a single type of microbes or a combination of different desired microbes, method of propagation of these organisms, selection of a suitable carrier material, studying the most effective method of application and determining its shelf life in suitable storage conditions. All of these steps are critical to the optimum performance of the biofertilizer (Bhattacharjee and Dey, 2014).

To start with the selection of the desired microorganism has to be made, whether to use a nitrogen fixer, a phosphate solubilizer, or a combination of multiple types of biofertilizers. Then the desired bacterium is isolated and propagated in large quantities after screening it for plant growth-promoting traits. The production of biofertilizer can be done in a flask to start with and later in a fermenter. The biofertilizer can be produced in liquid form or can be produced in a powder form which will require a suitable carrier to increase its shelf life as well as efficiency for long-term use. The most commonly used carriers are peat, lignite, tapioca flour, etc., the effects of such made biofertilizer can then be studied to analyze its effectiveness and limitations, which, on being satisfactory, can then be moved onto mass production (Khosro and Yousef, 2012).

Carrier-based biofertilizers are usually adsorbed onto a carrier material with mainly the intent to increase its shelf life for longer periods and the time between their manufacturing in the factory up till their delivery to the field. They help in delivering bacteria in good physiological conditions and viable numbers for them to give effective results. They provide the time for the bacteria to grow in new soil conditions by giving them a short-term niche-like atmosphere. But choosing a suitable carrier material that is cheap and readily available is an area that is still being explored as there is no one universal carrier for all types of biofertilizers (Arora et al., 2010). A good carrier must have a few properties that stand non-negotiable for increasing the efficiency of the biofertilizers. A good carrier must be easy to handle, cheap, sterilizable without degradation of its physical and chemical properties, very high water retention capacity of more than 50% is desired, should be nontoxic in nature and adhesive to the seeds but without the formation of lumps within itself with desirably a high organic content (Ritika and Utpal, 2014). All of these properties play a key role in the easy handling of the biofertilizers, increase their efficiency and provide it with water retention for longer viability of the organisms Very commonly used carrier materials in the case of biofertilizers include the use of sawdust, rice bran, lignite, talcum powder, peat, wood charcoal, manure, etc. (Bhattacharjee and Dey, 2014). In a study, sludge ash generated as a by-product of the combustion of sludge from sewage treatment plants was used as a carrier for the production of biofertilizer of *Rhizobium* sp. and compared its shelf life to that of a lignite-based carrier. There was a significant enhancement of five months in the shelf life and 10–50% enhancement in the lentil crop yields (Paliya et al., 2019).

To date, no single carrier has been found to possess all the qualities just listed, and researches are still going on to develop one satisfying most of the criteria. Carrier-linked biofertilizers have enhanced effectiveness and water ration capacity. Integrating microorganisms into carrier materials makes them easy to handle and store long term (Khosro and Yousef, 2012). They become cost-effective and easy to produce. Carriers like clay particles, peat, charcoal, sawdust, talcum dust, vermiculite, bentonite, diatomaceous earth, manure, rice or wheatbran, zeolite, rock phosphate pellets, charcoal, compost, and polymer-based carriers are considered to be suitable ones. Polymer-based carriers are alginate beads, in which the bacteria are immobilized in its matrix, subsequently releasing the microorganism in the soil environment. *Azospirillum brasilense* (>10^{11} cfu/g inoculant) has been successfully encapsulated using alginate beads (with a diameter in the range 100–200 μm) and this biofertilizer enhanced the production of wheat and tomato (Bashan et al., 2002). While 10–40 μm particle size of carrier material has been used for seed coating, the carrier with particle size 500–1500 μm is used for soil implantation.

To overcome these problems, liquid biofertilizers came into the picture, which is quite a quantum leap in their manufacturing industry history. Liquid biofertilizers are microbial suspensions with special cell protectants that stimulate the formation of resting spores for the longer shelf life and tolerance to adverse conditions. These biofertilizers involve the formation of liquid inoculants to enhance their effectiveness and lifespan. The latent forms germinate and produce active cells once they reach the soil. These active cells utilizing the carbon sources already present in the soil and multiply. Some of the frequently used cell protectants are glycerol, molasses, and sugars like glucose, maltose, sucrose, trehalose, etc. (Brar et al., 2012). The liquid biofertilizers deliver many advantages over the carrier-based biofertilizers in terms of shelf life, thermostability, maintenance of high cfu count (>10^9 cfu/ml, which is constant for 24 months), better survival on seeds and in soil, and easy handling and storage by farmers. Another significant benefit is the dosage, or the amount of the formulation required is also ten times lesser than that of carrier form. Because of suspension form, it can be packed in variable volumes and hence save the extra quantity utilized in the case of carrier-based biofertilizers (Borkar, 2015). The application of 1 ml of liquid biofertilizer is equivalent to that of 1 kg of 5 months old carrier-based counterparts (1000 times). Its easy administration procedure in irrigation systems involves hand sprayers, power sprayers, or fertigation tanks (Vandergheynst et al., 2007). It can also be as basal manure and mixed along with farmyard manure. But it is also the expensive one and hence the less popular (Mahdi et al., 2010). All the suitable biofertilizers (carrier-based or liquid-based) are initially studied to analyze their effectiveness and limitations. On being satisfactory, these are moved onto mass production (Khosro andYousef, 2012).

19.8 METHODS OF APPLICATION OF BIOFERTILIZERS

The method of application of a biofertilizer depends upon whether it is carrier-based or liquid-based. Since it is to be used by farmers who may not have the desired skill if required. So, the scientists have to devise simple application methods that can encourage the farmers to use biofertilizers. Ideally, farmers should apply these to their fields with their setup so that no extra need for infrastructure shall be required. Irrigation or sprayer methods could be utilized in case of applications of liquid formulations to the soil. For peach orchards and plants like citrus, mango, vines, guava, apple, etc., spraying is usually endorsed. Spreader centrifuges can be used to spread powdered or encapsulated products over the fields. In other cases, the seeds may be coated with inoculants.

19.8.1 Seed Treatment

It is the most common method for the application of biofertilizers. One packet of the inoculant having approximately 200 g is sufficient to treat 10 kg of seeds, or we can say for every 1 kg of seeds, 20 gm fertilizer is applied. So it turns out that the quantity of the seed to be sown (which indirectly depends on the crop or plant for which seeds are required) becomes the deciding factor behind the amount of fertilizer needed to be applied in a particular piece of agricultural land.

The simple process involves mixing the biofertilizer and water in the ratio of 1:2 to form slurry poured in the container along with the seeds before its application. The combination is subjected to a tender and thorough mixing using an adhesive generally like a 10% jaggery solution as a sticker or like gum acacia etc., such that the seeds are uniformly coated with the biofertilizer (Kumar et al., 2017). The seeds are then spread on a clean sheet or cloth in the shade for 30 min to dry, and direct sunlight exposure is avoided. Then they are sown immediately. This method is usually preferred for crops like pulses, oil seeds, and fodder (García-Fraile et al., 2015).

19.8.2 Seedling Root Dip

The treatment involves the utilization of diluted formulation of biofertilizer, where a ratio of 1:10 is maintained for the dilution. In general, this method is recommended for transplanted crops. In rice production, a bed is prepared in the field. The field is then supplied with water and the appropriate biofertilizers. *Azospirillum* is used for seedling root dip for about 8–10 hours, particularly for rice, and then transplanted. This method is suitable for crops such as paddy, tomato, potato, cabbage, onion, brinjal, chilli, etc., which are replanted at the seedling stage. It can also be applied for treating ornamental plants like rose, jasmine, dahlia, marigold, chrysanthemum, and many more (García-Fraile et al., 2015).

19.8.3 Soil Application or Main Field Treatment

This treatment involves the utilization of four packets of the recommended biofertilizers, which are mixed with a specific quantity of compost or dried powdered manure from the farmyard (20 kg) and kept overnight. This produced combination is then disseminated in the soil either at the time of seeding or just before transplanting (Rana and Ramesh, 2013). Broadcasting biofertilizers in the soil may require five to ten times the amount of biofertilizers used in sowing.

19.8.4 Set Treatment or Tuber Inoculation

This method is preferred for processing sugar cane, sliced potato bits, and the base of banana suckers. For such a treatment, the biofertilizer and water are mixed in the ratio of 1:50. The explants are dipped into the suspension containing the biofertilizer for 30 min. Culture suspension is generally prepared by mixing 1 kg (five packets, each of 200 g) of biofertilizer in 40–50 liters of water. After 30 mins of dipping the cut, pieces need to be subsequently dried under the shade. After drying, seeds are planted in the field (Rana and Ramesh, 2013). In another direct approach, the biofertilizer could be mixed with different carriers like soil, compost, manure, rice husks, etc., in the ratio of 1:25 (1 kg per 25 kg of carrier) and further administered in the soil.

19.9 CURRENT TRENDS AND FUTURE PERSPECTIVE FOR USING BIODEGRADABLE MUNICIPAL SOLID WASTE AS A FEEDSTOCK FOR THE PRODUCTION OF BIOFERTILIZERS

Every year, the globe produces 2.01 billion tons of municipal solid waste (MSW), with at least 33% of that not being managed in an environmentally sound manner. USA produces the highest amount of 258 million tons, followed by 220 million tons by China and 168 million tons by India (Figure 19.8). One major hurdle experienced is the disposal of the same, the source segregation of the MSW into biodegradable and non-biodegradable fractions. Segregation of collected waste after it is mixed is nearly impossible considering the huge amount of waste generated and the manpower required for its segregation, along with the funds that will have to be directed for this make it an impractical approach in many countries including India.

Various technologies employed for the management of MSW include composting of waste, incineration, pyrolysis, biogas production, production of refuse derived fuel (RDF), and biofuels such as bioethanol which are part of waste to energy projects. Since waste to energy plants is still a relatively new technology in India they are yet to be as profitable as their counterparts in developed nations. Research is still going strong in these areas to make them a profitable alternative (Patel and Ahluwalia, 2018).

Composting is a better established low-cost alternative process for minimizing the volume of municipal solid waste disposed of, with several plants operating all over the world. With a high organic content present, there is potential for economic gain through resource recovery.

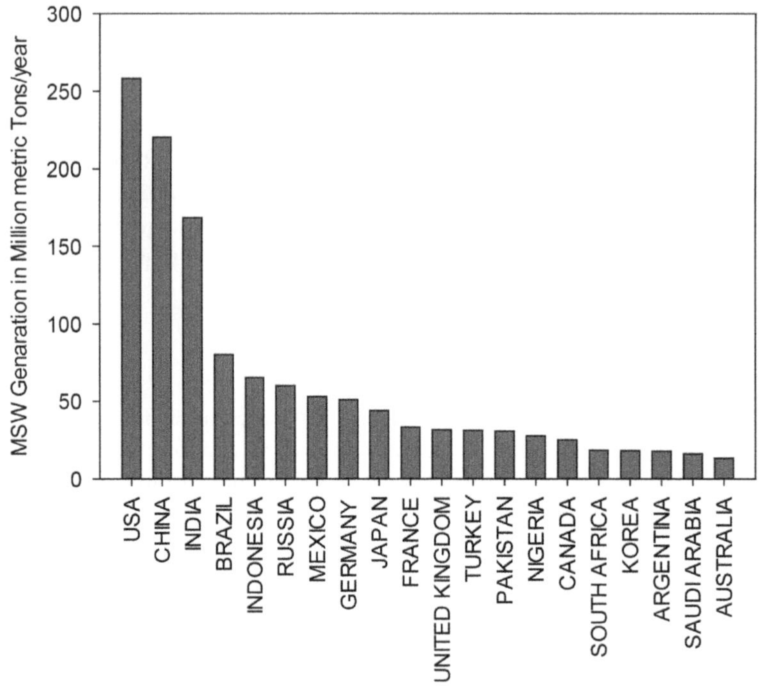

FIGURE 19.8 Municipal solid waste generation in 2017 by select countries.

Source: Adapted from Statistica (2021)

Transforming these high amounts of organic wastes into biofertilizers can give an alternative for harnessing all the nutrients and use them for conditioning and amendment of soils safely and beneficially creating a zero-waste process. In this process, the biodegradable fraction of the municipal solid waste can be used for the propagation of PGPRs as well as a carrier for the production of biofertilizers (Talwar, 2013). The abundant amounts of nutrients and high moisture content present in the biodegradable municipal solid waste (BMSW) give optimum conditions for the growth of PGPRs, which can feed on it and multiply in numbers high enough to be used as biofertilizers.

BMSW is rich in carbohydrate content, which can be converted into simpler sugars that act as a source of nutrition for the growth of PGPRs. The waste status of BMSW makes it an eco-friendly zero-cost substrate for the growth of PGPRs, such factors contribute towards making BMSW a resource rather than treating it as waste due to the presence of abundant nutrition, high organic and moisture content that renders it deteriorate. Consequently, utilization of these for the growth of N-fixing, phosphate-solubilizing, potassium-mobilizers, and PGPRs for the formation of biofertilizers is a promising approach for MSW management resulting in less MSW being dumped into the landfills.

19.10 RECENT DEVELOPMENTS IN THE USE OF MSW FOR BIOFERTILIZER PRODUCTION

The expanding population is putting tremendous pressure on cultivable lands around the world. To meet such challenges, a continuous expansion of the food-producing ecosystem is required. In this context, chemical fertilizers have to provide nutrients to support plant growth and boost crop productivity. In every production system, crops must be provided with major nutrients for better growth and substantial yields (Velmourougane et al., 2017). Fertilizers are nutrient-rich sources that fulfill the major nutrient requirements of plants like nitrogen, phosphorus, phosphate, zinc, and others. Undoubtedly, chemical fertilizers have offered benefits to fulfill the need of a rapidly growing population, but the excessive and uneven use of these chemical fertilizers resulted in negative effects on the health of agricultural soils leading to both lower production yield and lower efficiency of fertilizers.

Sustainable agriculture can help in meeting society's present food and fiber needs, sustain the economic viability of farm operations without compromising the ability of current or future generations to meet their needs. The key principles associated with sustainability in agriculture are to decrease the amounts of non-renewable and unsustainable inputs, particularly environmentally harmful ones, and promoting traditional agriculture practices like organic farming, use of green manure, a sustainable alternative for chemical fertilizers. Sustainable agriculture puts major emphasis on the use of biofertilizers to improve soil quality by increasing soil organic matter, water retention, nutrient cycling, and decreasing soil erosion. The application of biofertilizers provides an economical and environmentally promising strategy and beneficial in replenishing and maintaining long-term soil property and plant health (Kumar et al., 2015).

Developing countries like India are dealing with two major issues at the same time, one is to fulfill the food requirement of expanding population, and the other is the safe disposal of biodegradable solid waste. Outrageous waste-stream generation gleaned from various sectors of the current linear economic model make up one of the major societal challenges to be confronted regarding management and disposal of municipal solid waste and agricultural residues. The organic fraction of MSW accounts (OFMSW) approximately for 46 weight % of the total municipal waste streams with large content of food waste, kitchen waste and leftovers from residences, restaurants and markets (Campuzano and González-Martínez, 2016).

Developing countries like India have to adopt a scientific and eco-friendly strategy to deal with both of these issues. The widely used approaches are composting and fermentation, which turned organic waste into valuable products (Rada, 2015). Composting is the major approach for the degradation of organic fraction of biodegradable solid waste in several countries (Ofori-Boateng et al., 2013). Vermicomposting is another approach for handling organic waste, which is a time-consuming process.

The second-largest populated country in the world, India is facing solid waste management to be the major concern to its economy. The circular economy in distinction to linear economy insinuated all 'waste' should become 'food' for another agricultural/industrial process. The traditional linear economy implies the 'take, make, dispose off' model of production. In comparison to linear economy, the circular economy implies either a by-product or recovered resource for another agricultural/industrial process or as regenerative resources for nature (Diacono et al., 2019).

Conventional waste disposal methods such as open landfill, dumping and open burning are low-cost treatment methods that are themselves an environmentally threatening approach due to several environmental concerns such as leachate production and gas emissions (Barati et al., 2017). Waste management and global resource depletion can be solved together, for an eco-innovation focused on a 'zero waste approach' society and economy. Possible uses of residual organic materials include the production of biofertilizers, energy recovery, production of chemicals (volatile organic acids, alcohols, etc.), and utilization in farm animal nutrition (feedstock).

Composting is the method most commonly applied worldwide. The process aims at stabilizing the organic matter under aerobic conditions to produce a material with soil improver properties, referred to as compost (Cesaro et al., 2019). Municipal organic waste, which can be subjected to biodegradation, consists of food waste (up to 46%), cardboard and paper packaging materials, wood residues after the sanitary pruning of plants, litter, and sewage sludge and are unconventional raw materials for compost production. Composting still represents the most applied technique for the biological stabilization of organic solid waste, as it is considered easier to operate than anaerobic processes, generates a sanitized product, with soil improver properties that can be used in soil conditioning and nutrient supply (Cesaro et al., 2019).

The increasing environmental awareness among the common masses has led the significance on the development of new strategies for the effective management of organic waste with zero-waste production. Compost can be enriched with different nitrogen-fixing, phosphate-solubilizing, and potassium-solubilizing bacteria or archaea (Velmourougane et al., 2017). Microorganisms used in the biofertilizer formulation also show significant tolerance towards chemical fertilizers used by masses for increasing productivity which not only inhibits the growth of pests in the field but on another side toxic to the ecosystem and ultimately resulting in their bioaccumulation (Nathiya et al., 2020).

19.11 CURRENT STATUS OF MSW MANAGEMENT AND BIOFERTILIZER PRODUCTION IN INDIA

India is the second-biggest consumer of fertilizer in the world, next only to China. The Indian fertilizer companies produced around 32.4 million tons of fertilizer in the year 2012–13. However, the total availability was short of demand and was through imports. The indiscriminate use of chemical-based fertilizers has resulted in soil pollution and contamination, as well as the destruction of beneficial microorganisms and insects, making crops more susceptible to disease and lowering soil fertility. As the demand is higher than production, increasing the fertility of soil my making use of biofertilizers is one of the alternatives. The required biofertilizers in India is roughly 550,000 metric tonnes based on crop area; however, overall output is far less than the requirement, indicating the likelihood of a rise in biofertilizer production (Barman et al., 2017).

Furthermore, because plants are unable to absorb or utilise all of the nutrients provided through synthetic fertilizers, a significant proportion of nutrients are either fixed in the soil or leached out, polluting water bodies. To make agriculture sustainable, it is now necessary to execute a balanced and acceptable use of nutrients that is both cost-effective and environmentally friendly. The government of India has also taken a great interest to harness the full potential of the biofertilizers by introducing them alone or in combination with chemical fertilizers to the farmers.

Composting has been well documented, which results in the generation of organic fertilizer, soil conditioner-like product; reduction of waste volume; reduction in presence of pathogens, control germination of weeds in agricultural fields, and elimination of undesirable odorous compounds. In composting of food waste, microorganisms, including bacteria, fungi, mold and actinomycetes use the organic components in food waste, degrading them into short-chain chemicals, e.g. humic acid. Compost can be carried out in a vessel, aerobic windrow or aerobic pile (Pandey et al., 2016). During the composting process, the temperature often rises to a high level, as microbes release heat (55°C for 5–7 days or 75°C for 2–3 days). This increase in temperature is responsible for the deactivation of pathogens and weed seeds.

Biofertilizer is an environmentally friendly fertilizer now used in most countries. Biofertilizers are organisms that enrich the nutrient quality of the soil. Bacteria, fungus and cyanobacteria are the main sources of biofertilizers, which are known to provide benefits such as plant nutrition, disease resistance and adverse climate tolerance. These techniques have proved to be successful biofertilizers that form a healthy relationship with the roots. Thus, biofertilizers are important if we have to ensure a healthy future for the generations to come. These microorganisms supply in addition to nitrogen, a considerable amount of organic matter enriching the structure of the soil. Inoculants of these microorganisms have proved their technical feasibility, economic viability and social acceptability. They are therefore called a 'biofertilizer' (Barman et al., 2017).

The biggest downside in attaining agricultural sustainability has been the malfunctioning of beneficial rhizospheric microorganisms, such as efficient commercialization of PGPR, PGPB due to a host of factors, viz. environmental, climatic and species-specific or niche-specific tripartite soil-plant-microbe interactions. The additional facet which has decreased the estimated industrial recognition of utilizing PGPR as biofertilizers is a lack of appropriate and economically viable carrier. The advantageous effect of these bacterial populations can be optimally deliberated only when carrier-based formulations are acquiescent for field-level appliances and made accessible to the customers. Beneficial bacteria are currently applied in both forms of solid and liquid carrier-based inoculants preparations in a range of agro-economically vital plants (Dutta and Thakur, 2017). However, the triumph of carrier-based formulations primarily relies on the bacterial strain used for inoculant production. Liquid formulations possess their advantages for smaller-scale agricultural systems, such as easier to handle, working faster, being uniform and easy to store; solid-based formulations are inexpensive and easier to manufacture. Historically these carriers have been classified as inert materials (alginate, talcum powder, vermiculite, and other inorganic minerals), coal, clays, peat and plant waste materials (compost, farmyard manure and soybean meal) (Klayraung et al., 2009). In comparison with biofuel and biochemical production, the importance of generating biofertilizer from OFMSW, food waste has been underestimated. Replacing synthetic chemical fertilizer with biofertilizer derived from OFMSW, food waste would reduce the requirement for synthetic fertilizers, reducing the environmental impact of OFMSW, food waste and directly benefiting food production (Chenyu et al., 2018).

Agricultural residues are generated during crop cultivation activities. They are rich in lignocellulose material and have a relatively low moisture content of 10–15% for air-dried wheat straw (Pensupa et al., 2013). Although agriculture residues have a low economic value, they are an important renewable carbon and mineral resource for the soil. Agricultural residues can be degraded either in situ or collected, taken off the farm, and converted into biofertilizer before being added back to the soil. Returning agriculture residues as a biofertilizer in a correct manner has been shown to improve the organic content of the soil, modify the soil particle structure, reduce water evaporation, improve niche microorganism activities and decrease fertilizer loss (Jordan et al., 2010).

Key market players in the biofertilizer industry are Novozymes (Denmark), Kiwa Biotech (China), Symborg (Spain), Lallemand Inc. (Canada), Rizobacter Argentina S.A (Argentina), Mapleton Agribiotech (Australia), T. Stanes & Company Ltd (India), SOM Phytopharma (India), Ltd International Panaacea Limited (India), Madras Fertilizers Limited (India), National Fertilizers Limited (India), Rashtriya Chemicals & Fertilizers Ltd (India), Varsha Bioscience and Technology India Pvt. Limited (India), IFFCO-Indian Farmers Fertiliser Cooperative (India).

19.12 CONCLUSION

The increasing environmental awareness among the common masses has led the significance on the development of new strategies for the effective management of organic waste with zero-waste production. Because of these concerns, compost is being enriched with biofertilizer formulation. Looking at the future perspective, an improvement in the quality of biofertilizers being supplied is required both in liquid- and carrier-based forms. Further, there is a need to enhance the production and plant growth-promoting properties of already available microbial inoculants. Spreading of awareness amongst the masses should be spread about the use of biofertilizers and their beneficial effects on soil health.

BIBLIOGRAPHY

Ahemad, M., and Kibret, M. 2014. Mechanisms and applications of plant growth-promoting rhizobacteria: current perspective. *Journal of King Saud University-Science*, 26: 1–20.

Ahmad, M., Nadeem, S.M., Naveed, M., and Zahir, Z.A. 2016. Potassium-solubilizing bacteria and their application in agriculture. In: *Potassium solubilizing microorganisms for sustainable agriculture*, Springer, 293–313.

Al-Khiat, S.H.A. 2006. *Effect of cyanobacteria as conditioner and biofertilizer on growth and some biochemical characteristics of tomato.* Faculty of Science, King Saud University, Riyadh, 190.

Anjanadevi, I.P., John, N.S., John, K.S., Jeeva, M.L., and Misra, R.S. 2016. Rock inhabiting potassium solubilizing bacteria from Kerala, India: characterization and possibility in chemical K fertilizer substitution. *Journal of Basic Microbiology*, 56: 67–77.

Arora, N.K., Khare, E., and Maheshwari, D.K. 2010. Plant growth-promoting rhizobacteria: Constraints in bioformulation, commercialization, and future strategies. In: *Plant growth and health-promoting bacteria*, Springer, 97–116.

Babalola, O.O. 2010. Beneficial bacteria of agricultural importance. *Biotechnology Letters*, 32: 1559–1570.

Barati, M.R., Aghbashlo, M., Ghanavati, H., Tabatabaei, M., Sharifi, M., Javadirad, G., Dadak, A., and Soufivan, M.M. 2017. Comprehensive exergy analysis of a gas engine-equipped anaerobic digestion plant producing electricity and biofertilizer from the organic fraction of municipal solid waste. *Energy Conversion and Management*, 151: 753–763.

Barman, M., Paul, S., Choudhury, A.G., Roy, P., Sen, J. 2017. Biofertilizer as prospective input for sustainable agriculture in India. *International Journal of Current Microbiology and Applied Sciences*, 6: 1177–1186.

Bashan, Y., Hernandez, J.P., Leyva, L.A. 2002. Alginate microbeads as inoculant carriers for plant growth-promoting bacteria. *Biology and Fertility of Soils*, 35: 359–368.

Bhardwaj, D., Ansari, M.W., Sahoo, R.K., and Tuteja, N. 2014. Biofertilizers function as key player in sustainable agriculture by improving soil fertility, plant tolerance and crop productivity. *Microbial Cell Factories*, 13: 66.

Bhattacharjee, R., and Dey, U. 2014. Biofertilizer, a way towards organic agriculture: a review. *African Journal of Microbiological Research*, 8: 2332–2343.

Bhattacharyya, P.N., and Jha, D.K. 2012. Plant growth-promoting rhizobacteria (PGPR): Emergence in agriculture. *World Journal of Microbiology and Biotechnology*, 28: 1327–1350.

Bhattacharyya, P., and Jain, R.K. 2000. Phosphorus solubilizing biofertilisers in the whirlpool of rock phosphate-challenges and opportunities. *Fertiliser News*, 45: 45–52.

Black, M., Moolhuijzen, P., Chapman, B., Barrero, R., Howieson, J., Hungria, M., Bellgard, M. 2012. The genetics of symbiotic nitrogen fixation: comparative genomics of 14 rhizobia strains by resolution of protein clusters. *Genes*, 3: 138–166.

Borkar, S.G. 2015. *Microbes as biofertilizers and their production technology*. Wood head Publishing India Pvt. Ltd., 7–153.

Brar, S.K., Sarma, S.J., and Chaabouni, E. 2012. Shelf-life of biofertilizers: an accord between formulation-sandgenetics. *Journal of Biofertilizers and Biopesticides*, 3: 5–6.

Campuzano, R., and González-Martínez, S. 2016. Characteristics of the organic fraction of municipal solid waste and methane production: a review. *Waste Management*, 54: 3–12.

Cesaro, A., Conte, A., Belgiorno, V., Siciliano, A., and Guida, M. 2019. The evolution of compost stability and maturity during the full-scale treatment of the organic fraction of municipal solid waste. *Journal of Environmental Management*, 232: 264–270.

Chakraborty, U., Roy, S., Chakraborty, A.P., Dey, P., and Chakraborty, B. 2011. Plant growth promotion and amelioration of salinity stress in crop plants by a salt-tolerant bacterium. *Recent Research in Science and Technology*, 3: 61–70.

Chaykovskaya, L.A., Patyka, V.P., and Melnychuk, T.M. 2001. Phosphate mobilising microorganisms and their influence on the productivity of plants. In: *Plant nutrition*, Springer, 668–669.

Chen, Y.P., Rekha, P.D., Arunshen, A.B., Lai, W.A., and Young, C.C. 2006. Phosphate solubilizing bacteria from subtropical soil and their tricalcium phosphate solubilizing abilities. *Applied Soil Ecology*, 34: 33–41.

Chenyu, D.U., Jwan, J.A., Darren, G., Danni, F., Mengyuan, Y., Liwei, R., Shuang, L., and Diannan, L. 2018. Valorization of food waste into biofertiliser and its field application. *Journal of Cleaner Production*, 187: 273–284.

Chittora, D., Meena, M., Barupal, T., Swapnil, P., and Sharma, K. 2020. Cyanobacteria as a source of biofertilizers for sustainable agriculture. *Biochemistry and Biophysics Reports*, 22: 100737

Chun-Li, W., Shiuan-Yuh, C., and Chiu-Chung, Y. 2014. Present situation and future perspective of biofertilizer for environmentally friendly agriculture. *Annual Reports*, 1–5.

Cress, W.A., Throneberry, G.O., and Lindsey, D.L. 1979. Kinetics of phosphorus absorption by mycorrhizal and nonmycorrhizal tomato roots. *Plant Physiology*, 64: 484–487

Das, J. 2020. An enquiry into the challenges of organic farming in Sikkim-a study of agriculture and agribusiness and rural change in India. *Our Heritage*, 68: 93–113.

Devi, V., and Sumathy, V.J.H. 2017. Production of biofertilizer from fruit wastes. *European Journal of Pharmaceutical and Medical Research*, 4: 436–443.

Diacono, M., Persiani, A., Testani, E., Montemurro, F., and Ciaccia, C. 2019. Recycling agricultural wastes and by-products in organic farming: biofertilizer production, yield performance and carbon footprint analysis. *Sustainability*, 11: 3824.

Dutta, J., and Thakur, D. 2017. Evaluation of multifarious plant growth-promoting traits, antagonistic potential and phylogenetic affiliation of rhizobacteria associated with commercial tea plants grown in Darjeeling, India. *PloS One*, 12: 1–24.

Ezawa, T., Saito, M., and Yoshida, T. 1995. Polyphosphates in intraradical and extraradical hyphae of an arbuscular mycorrhizal fungus, *Gigaspora margarita*. *Plant Soil*, 176: 57–63

Fankem, H., Nwaga, D., Deubel, A., Dieng, L., Merbach, W., and Etoa, F.X. 2006. Occurrence and functioning of phosphate solubilizing microorganisms from oil palm tree (*Elaeis guineensis*) rhizosphere in Cameroon. *African Journal of Biotechnology*, 5: 2450–2460.

García-Fraile, P., Menéndez, E., and Rivas, R. 2015. Role of bacterial biofertilizers in agriculture and forestry. *Bioengineering*, 2: 183–205.

Gaur, V. 2010. Biofertilizer – necessity for sustainability. *Journal of Advanced Development*, 1: 7–8.

Ghorbanian, D., Harutyunyan, S., Mazaheri, D., Rasoli, V., and Moheb, I.A. 2012. Influence of arbuscular mycorrhizal fungi and different levels of phosphorus on the growth of corn in water stress conditions. *African Journal of Agricutural Research*, 7: 2575–2580.

Gourion, B., Berrabah, F., Ratet, P., and Stacey, G. 2015. *Rhizobium* – legume symbioses: the crucial role of plant immunity. *Trends in Plant Science*, 20: 186–194.

Hamdali, H., Bouizgarne, B., Hafidl, M., Lebrihl, A., Virolle, M.J., and Ouhdouch, Y. 2008. Screening for rock phosphate solubilizing actinomycetes from moroccan phosphate mines. *Applied Soil Ecology*, 38: 12–19.

Han, H.S., and Lee, K.D. 2006. Effect of co-inoculation with phosphate and potassium solubilizing bacteria on mineral uptake and growth of pepper and cucumber. *Plant Soil and Environment*, 52: 130–136.

Itelima, J.U., Bang, W.J., and Onyimba, I.A. 2018. A review: biofertilizer - a key player in enhancing soil fertility and crop productivity. *Journal of Microbiology and Biotechnology Reports*, 2: 22–28.

Jakobsen, I., Abbott, L.K., and Robson, A.D. 1992. External hyphae of vesicular – arbuscular mycorrhizal fungi associated with *Trifolium subterraneum* L. I. Spread of hyphae and phosphorus inflow into roots. *New Phytologist*, 120: 371–380.

Jordan, A., Zavalam L.M., and Gil, J. 2010. Effects of mulching on soil physical properties and runoff under semi-arid conditions in southern Spain. *Catena*, 81: 77–85.

Kachari, M., and Korla, B.N. 2009. Effect of biofertilizers on growth and yield of cauliflower cv. PSB K-1. *Indian Journal of Horticulture*, 66: 496–501.

Kapri, A., and Tewari, L. 2010. Phosphate solubilization potential and phosphatase activity of rhizospheric *Trichoderma* spp. *Brazilian Journal of Microbiology*, 41: 787–879.

Karandashov, V., Nagy, R., Wegmuller, S., Amrhein, N., and Bucher M. 2004. Evolutionary conservation of a phosphate transporter in the arbuscular mycorrhizal symbiosis. *Proceedings of National Academy of Sciences, USA*, 101: 6285–6290.

Katyal, J.C., and Rattan, R.K. 1993. Distribution of zinc in Indian soils. *Fertilizer News*, 38: 15–26.

Khosro, M., and Yousef, S. 2012. Bacterial bio-fertilizers for sustainable crop production: a Review. *APRN Journal of Agricultural and Biological Science*, 7: 237–308.

Klayraung, S., Viernstein, H., and Okonogi, S. 2009. Development of tablets containing probiotics: effects of formulation and processing parameters on bacterial liability. *International Journal of Pharmaceutics*, 370: 54–60.

Kloepper, J.W., Gutierrez-Estrada, A., and McInroy, J.A. 2007. Photoperiod regulates elicitation of growth promotion but not induced resistance by plant growth-promoting rhizobacteria. *Canadian Journal of Microbiology*, 53: 159–167.

Kumar, A., Bahadur, I., Maurya, B.R., Raghuwanshi, R., Meena, V.S., Singh, D.K., and Dixit, J. 2015. Does plant growth-promoting rhizobacteria enhance agricultural sustainability. *Journal of Pure and Applied Microbiology*, 9: 715–724.

Kumar, S.M., Reddy, G.C., Phogat, M., and Korav, S. 2018. Role of bio-fertilizers towards sustainable agricultural development: A review. *Journal of Pharmacognosy and Phytochemistry*, 7: 1915–1921.

Kumar, R., Kumawat, N., and Sahu, Y.K. 2017. Role of biofertilizers in agriculture. *Popular Kheti*, 5: 63–66.

Kumaran, S.S., Natarajan, S., and Thamburaj, S. 1998. Effect of organic and inorganic fertilizers on growth, yield and quality of tomato. *South Indian Horticulture*, 46: 203–205.

Mahanty, T., Bhattacharjee, S., Goswami, M., Bhattacharyya, P., Das, B., Ghosh, A., and Tribedi, P. 2017. Biofertilizers: a potential approach for sustainable agriculture development. *Environmental Science and Pollution Research*, 24: 3315–3335.

Mahdi, S.S., Hassan, G.I., Samoon, S.A., Rather, H.A., Dar, S.A., and Zehra, B. 2010. Biofertilizers in organic agriculture. *Journal of Phytology*, 2: 42–54.

Martin, X.M., Sumathi, C.S., and Kannan, V.R. 2011. Influence of agrochemicals and *Azotobacter* sp. application on soil fertility in relation to maize growth under nursery conditions. *Eurasian Journal of Biosciences*, 5: 19–28.

Mathivanan, R., Umavathi, S., Ramasamy, P.K., and Thangam, Y. 2015. Influence of vermicompost on the activity of the plant growth regulators in the leaves of the Indian butter bean plant, Dolichos lablab L. *International Journal of Advanced Research in Biological Sciences*, 2: 84–89.

Matson, P.A., and Vitousek, P.M. 2006. Agricultural intensification: will land spared from farming be land spared for nature? *Conservation Biology*, 20: 709–710.

Mehnaz, S. 2015. *Azospirillum*: a biofertilizer for every crop. In: *Plant microbes symbiosis: Applied facets*, Springer, 297–314.

Meki, M.N., Kemanian, A.R., Potter, S.R., Blumenthal, J.M., Williams, J.R., and Gerik, T.J. 2013. Cropping system effects on sorghum grain yield, soil organic carbon, and global warming potential in central and south Texas. *Agricultural Systems*, 117: 19–29.

Mfilinge, A., Mtei, K., Ndakidemi, P.A. 2014. Effects of *Rhizobium* inoculation and supplementation with P and K, on growth, leaf chlorophyll content and nitrogen fixation of bush bean varieties. *American Journal of Research Communication*, 2: 49–87.

Mishra, D., Rajvir, S., Mishra, U., and Kumar, S.S. 2013. Role of bio-fertilizer in organic agriculture: A review. *Research Journal of Recent Sciences*, 2: 39–41.

Mishra, P., and Dash, D. 2014. Rejuvenation of biofertilizer for sustainable agriculture and economic development. *Consilience*, 11: 41–61.

Mohammadi, K., and Sohrabi, Y. 2012. Bacterial biofertilizers for sustainable crop production: A review. *ARPN Journal of Agriculrural and Biological Science*, 7: 307–316.

Mohandas, S. 1988. Nitrogen fixation in tomato (*Lycopersicon esculentum* Mill 'Pusa Ruby'). *Plant and Soil*, 107: 219–225.

Moraditochaee, M., Azarpour, E., and Bozorgi, H.R. 2014. Study effects of bio-fertilizers, nitrogen fertilizer and farmyard manure on yield and physiochemical properties of soil in lentil farming. *International Journal of Biosciences*, 4: 41–48.

Nahas, E. 1996. Factors determining rock phosphate solubilization by microrganisms isolated from soil. *World Journal of Microbiology and Biotechnology*, 12: 18–23.

Naiman, A.D., Latrónico, A., and de Salamone, I.E.G. 2009. Inoculation of wheat with *Azospirillum brasilense* and *Pseudomonas fluorescens*: impact on the production and culturable rhizosphere microflora. *European Journal of Soil Biology*, 45: 44–51.

Nathiya, S., Janani, R., and Kannan T.V. 2020. Potential of plant growth-promoting Rhizobacteria to overcome the exposure of pesticide in *Trigonella foenum – graecum* (fenugreek leaves). *Biocatalysis and Agricultural Biotechnology*, 23: 101493.

Neupane, D., Adhikari, P., Bhattarai, D., Rana, B., Ahmed, Z., Sharma, U., and Adhikari, D. 2022. Does climate change affect the yield of the top three cereals and food security in the world? *Earth*, 3(1): 45–71.

Ofori-Boateng, C., Lee, K.T., and Mensah, M. 2013. The prospectus of electricity generation from municipal solid waste in Ghana: a better waste management option. *Fuel Process Technology*, 110: 94–102.

Padmapriya, S., and Chezhiyan, N. 2009. Effect of shade, organic, inorganic and biofertilizers on morphology, yield and quality of turmeric. *Indian Journal of Horticulture*, 66: 333–339.

Paliya, S., Mandpe, A., Kumar, S., and Kumar, M.S. 2019. Enhanced nodulation and higher germination using sludge ash as a carrier for biofertilizer production. *Journal of Environmental Management*, 250: 109523.

Pandey, P.K., Vaddella, V., Cao, W., Biswas, S., Chiu, C., and Hunter, S. 2016. In-vessel composting system for converting food and green wastes into pathogen-free soil amendment for sustainable agriculture. *Journal of Cleaner Production*, 139: 407–415.

Patel, U., and Ahluwalia, I. 2018. *Solid waste management in India: an assessment of resource recovery and environmental impact.* http://hdl.handle.net/11540/8143

Pensupa, N., Jin, M., Kokolski, M., Archer, D.B., Du, C. 2013. A solid-state fungal fermentation-based strategy for the hydrolysis of wheat straw. *Bioresource Technology*, 149: 261–267.

Perrig, D., Boiero, M.L., Masciarelli, O.A., Penna, C., Ruiz, O.A., Cassán, F.D., Luna, M.V. 2007. Plant-growth-promoting compounds produced by two agronomically important strains of *Azospirillum brasilense*, and implications for inoculant formulation. *Applied Microbiology and Biotechnology*, 75: 1143–1150.

Rada, E.C. 2015. *Biological treatment of solid waste: enhancing sustainability*. Apple Academic Press, 310.

Ramana, V., Ramakrishna, M., Purushotham, K., and Reddy, K.B. 2010. Effect of bio-fertilizers on growth, yield attributes and yield of french bean (*Phaseolus vulgaris* L.). *Legume Research-An International Journal*, 33: 178–183.

Ramana, V., Ramakrishna, M., Purushotham, K., and Reddy, K.B. 2011. Effect of bio-fertilizers on growth, yield and quality of french bean (*Phaseolus vulgaris* L.). *Vegetable Science*, 38: 35–38.

Rana, R., and Ramesh, K.P. 2013. Biofertilizers and their role in agriculture. *Popular Kheti*, 1: 56–61.

Ritika, B., and Utpal, D. 2014. Biofertilizer, a way towards organic agriculture: A review. *African Journal of Microbiology Research*, 8: 2332–2343.

Römheld, V., and Kirkby, E.A. 2010. Research on potassium in agriculture: needs and prospects. *Plant and Soil*, 335: 155–180.

Ruiz-Sánchez, M., Aroca, R., Muñoz, Y., Polón, R., and Ruiz-Lozano, J.M. 2010. The arbuscular mycorrhizal symbiosis enhances the photosynthetic efficiency and the antioxidative response of rice plants subjected to drought stress. *Journal of Plant Physiology*, 167: 862–869.

Salhia, B.M. 2010. *The Effect of Azotobacter chrococcum as nitrogen biofertilizer on the growth and yield of Cucumis sativus*. M.Sc Thesis, Faculty of Science, The Islamic University, Gaza.

Scalenghe, R., Edwards, A.C., Barberis, E., and Ajmone-Marsan, F. 2012. Are agricultural soils under a continental temperate climate susceptible to episodic reducing conditions and increased leaching of phosphorus? *Journal of Environmental Management*, 97: 141–147.

Selvakumar, G., Lenin, M., Thamizhiniyan, P., and Ravimycin, T. 2009. Response of biofertilizers, on the growth and yield of blackgram (*Vigna mungo* L.). *Recent Research in Science and Technology*, 1: 169–175.

Shukla, Y.R., Thakur, A.K., and Joshi, A. 2009. Effect of inorganic and bio-fertilizers on yield and horticultural traits in tomato. *Indian Journal of Horticulture*, 66: 285–287.

Srivastava, P., Singh, R., Tripathi, S., and Raghubanshi, A.S. 2016. An urgent need for sustainable thinking in agriculture – An Indian scenario. *Ecological Indicators*, 67: 611–622.

Statistica. 2021. Cited from: www.statista.com/statistics/916749/global-generation-of-municipal-solid-waste-by-country/

Sundaravelu, S., and Muthukrishnan, T. 1993. Effect of seed treatment with *Azospirillum* and gibberellic acid on the growth and yield of radish (*Raphanus sativus* L. var. Japanese White). *South Indian Horticulture*, 41: 212–212.

Talwar, S.P. 2013. *Formulation and testing of locally isolated effective microbes for the development of biofertilizers*. Ph.D. Thesis, University Malaysia, Sarawak.

Thajuddin, N., and Subramanian, G. 2005. Cyanobacterial biodiversity and potential applications in biotechnology. *Current Science*, 89: 47–57.

Thakur, D., Kaushal, R., and Shyam, V. 2014. Phosphate solubilising microorganisms: role in phosphorus nutrition of crop plants- a review. *Agricultural Reviews*, 35: 159–171.

Trabelsi, D., and Mhamdi, R. 2013. Microbial inoculants and their impact on soil microbial communities: A review. *BioMed Research International*, 1–11.

Vandergheynst, J., Scher, H., Guo Hy, and Schultz, D. 2007. Water-inoil emulsions that improve the storage and delivery of the biolarvacide *Lagenidium giganteum*. *Bio Control*, 52: 207–229.

Velmourougane, K., Prasanna, R., and Saxena, A.K. 2017. Agriculturally important microbial biofilms: Present status and future prospects. *Journal of Basic Microbiology*, 57: 548–573.

Verma, P., Yadav, A.N., Kumar, V., Singh, D.P., and Saxena, A.K. 2017. Beneficial plant-microbes interactions: biodiversity of microbes from diverse extreme environments and its impact for crop improvement. In: Singh DP, Singh HB, Prabha R (eds) *Plant-microbe interactions inagro- ecological perspectives: volume2: microbial interactionsand agro-ecological impacts*. Springer Singapore, 543–580.

Verma, J.P., Yadav, J., Tiwari, K.N., and Lavakush, S.V. 2010. Impact of plant growth promoting rhizobacteria on crop production. *International Journal of Agricultural Research*, 5: 954–983.

Wander, M. 2010. *Much more that plant nutrition*. Nature publishers.
Wani, S.A., Chand, S., and Ali, T. 2013. Potential use of *Azotobacter chroococcum* in crop production: An overview. *Current Agriculture Research Journal*, 1: 35–38.
Wu, S.C., Cao, Z.H., Li, Z.G., Cheung, K.C., Wong, M.H. 2005. Effects of biofertilizer containing N-fixer, P and K solubilizers and AM fungi on maize growth: A greenhouse trial. *Geoderma*, 125: 155–166.
Yadav, A.N., and Saxena, A.K. 2018. Biodiversity and biotechnological applications of halophilic microbes for sustainable agriculture. *Journal of Applied Biology and Biotechnology*, 6: 48–55.
Yao, Q., Li, X., Feng, G., and Christie, P. 2001. Mobilization of sparingly soluble inorganic phosphates by the external mycelium of an abuscular mycorrhizal fungus. *Plant Soil*, 230: 279–285.
Youssef, M.M.A., and Eissa, M.F.M. 2014. Biofertilizers and their role in management of plant parasitic nematodes. A review. *Journal of Biotechnology and Pharmaceutical Research*, 5: 1–6.
Zhen, Z., Liu, H., Wang, N., Guo, L., Meng, J., Ding, N., Wu, G., and Jiang, G. 2014. Effects of manure compost application on soil microbial community diversity and soil microenvironments in a temperate cropland in China. *PloS One*, 9: 1–12.

20 Microbiofuels
The Sustainable Energy Source for the Future

Sanjeev Kumar Soni, Nisha Phour Dhull, Raman Soni and Apurav Sharma

CONTENTS

20.1 INTRODUCTION

Energy is the lifeline of society and the available energy on earth is generally characterized in two groups, namely non-renewable energy and renewable energy (Vakulchuk et al., 2020). The increasing human population and industrialization continuously increase the demand for non-renewable energy resources (United Nations, 2011). It has been estimated that 87% of the world's energy consumption is fulfilled from conventional fossil fuels (Vakulchuk et al., 2020). Consequently, the sources of non-renewable energies are progressively exhausting due to large demand. The major areas responsible for the consumption of energy are the industrial, transport, and electricity production sectors, which lead to many negative effects on the climate, including global warming, melting of glaciers, rise in sea level, and rapid loss of biodiversity (Lopez-Linares et al., 2014). Therefore, the need of the hour is to search for alternative, renewable, and sustainable fuel resources with lesser greenhouse gas (GHG) emissions. The currently available, renewable energy sources are

DOI: 10.1201/9781003220831-20

categorized into two groups, namely, clean energy and bioenergy, as represented in Figure 20.1. Clean energy is the energy derived from the wave, wind, hydroelectric, solar and rain energy, etc., and bioenergy is derived from biomass by biochemical processes. Bioenergy resources, also known as biofuels, include alcohols, biodiesel, biogas, hydrogen, etc. (Hossain et al., 2020). Biofuels have been studied for many years and are regarded as reliable and cost-effective energy sources (Khan et al., 2020). Due to the increasing consumption of biofuel over the world in recent past years, biofuel research and development has significantly moved toward efficient and nonconventional biomass feedstocks. In the view of the increasing global biofuel consumption trend, major research attention has been given towards feasible and low-cost resources of biofuel as reported by the research publications in the last 20 years around the world, especially from Asia, Europe, and the USA (Subramaniam et al., 2020). Depending upon feedstocks used for biofuels, they can be classified into four major categories, including 1G, 2G, 3G, and 4G generation fuels (Azad et al., 2012). The majority of exploration work has been carried out from the last decade on 1G and 2G biofuels. However, due to the major drawbacks of these two, the focus of research has shifted towards the 3G and 4G biofuels.

20.2 CLASSIFICATION OF BIOFUELS

Fuels derived from plant biomass can be categorized into primary and secondary biofuels. Biofuels produced directly from biomass are known as primary biofuels, whereas fuels produced from the biomass that is subjected to some kind of transformation are known as secondary biofuels. Secondary biofuels can further be categorized into four different generations, namely first-, second-, third-, and fourth-generation biofuels based upon the kind of feedstock used. 1G, 2G, 3G, and 4G biofuels are produced from edible food crops, lignocellulosic waste, microalgal biomass, and genetically engineered microorganisms, respectively.

First-generation biofuels are primarily formed from the edible part of the food and oil crops, which have accomplished a cost-effective level of commercial production (Nigam and Singh, 2010). However, the first-generation biofuels from edibles, including sugar cane, grains, and vegetable oils, have been facing the challenge of food-fiber production for human usage, which can harm environmental sustainability and profitability (Schenk et al., 2008). The limitation of first-generation biofuels has caused greater concern, and they are required to be replaced by second-generation biofuels. The second-generation biofuels are normally formed from non-edible lignocellulosic biomass, including food crops or whole plant biomass, organic parts of municipal solid wastes, by-products (sugar cane bagasse, cereal straw, forest residues), and feedstock that would include short-rotation forests, vegetative grasses, and further energy crops (Zhang et al., 2011). Third-generation biofuels, also known as algal fuels or microbiofuels, obtained from microalgae are suitable substitute energy resources that answer the significant disadvantages related to 1G and 2G biofuels (Chisti, 2007). Therefore, microalgal biofuels are the cutting-edge biofuels, which are today's intriguing issue in the renewable energy industry and are foreseen to be the substitutes of fossil fuels. Nowadays, algae have been recognized as an unconventional biofuel feedstock due to their higher photosynthetic rate and rapid development as contrasted with any plant. It can be grown on wasteland by using wastewater as a culture medium and, therefore, microalgae can sustain the level of water sources (Kim et al., 2014). Algae can operate three kinds of mechanisms for their biomass production in a cultivation system, i.e. autotrophic, heterotrophic, and mixotrophic (Brennan and Owende, 2010), thus producing proteins, lipids, and carbohydrates in massive quantities from biomass (Widjaja et al., 2009). Further, these cellular components can be converted into valuable co-products and biofuel leading to the concept of biorefinery (Brennan and Owende, 2010). It can yield more biofuel per acre than conventional crops. Furthermore, algal biomass can also be used for the production of biodiesel and bioethanol. Many algal species, including *Botryococcus braunni*, *Chaetoceros calcitrans*, *Chlorella* spp., *Isochrysis galbana*, *Nanochloropsis* sp., *Schizochytrium limacinum*, and *Scenedesmus* spp., have been examined as potential sources of biofuel (Rodolfi et al., 2008).

Fourth-generation biofuels are obtained from genetically engineered microalgae and thus are also known as algal fuels or microbiofuels (Zinoviev et al., 2010; Ganesh, 2014) and are considered sustainable and promising substitutes to fossil fuels. Their possible future prospects are shown in Figure 20.2. A comparison of the advantages and disadvantages associated with the different generations of biofuels is listed in Table 20.1.

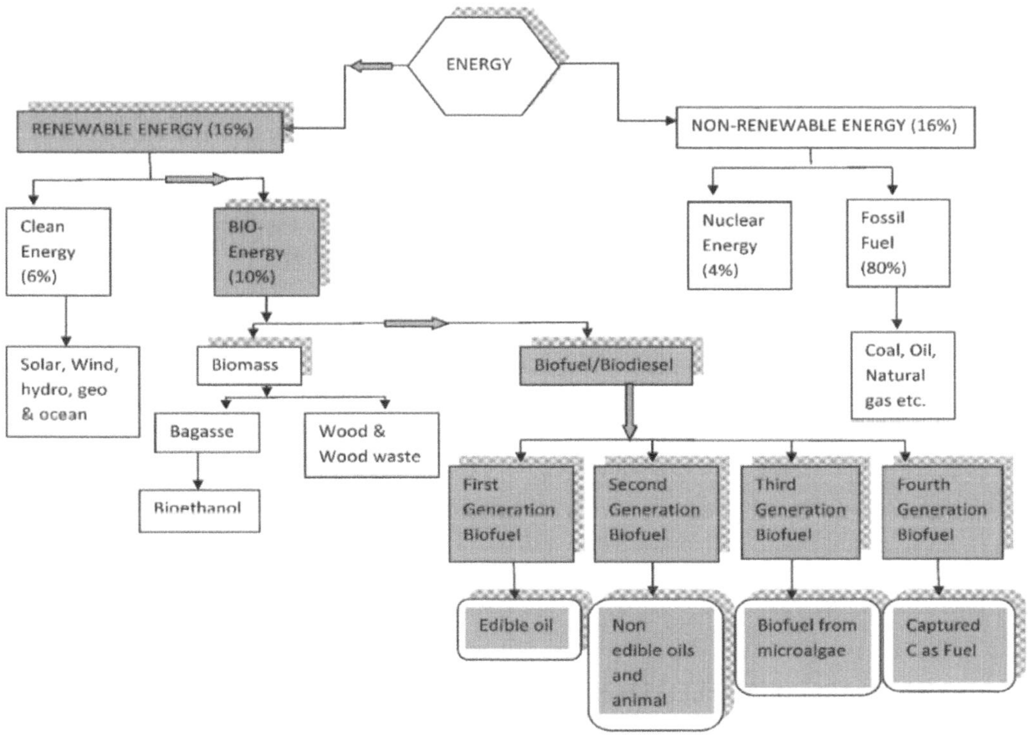

FIGURE 20.1 Types of energy resources depending upon percentage of involvement toward world energy consumption.

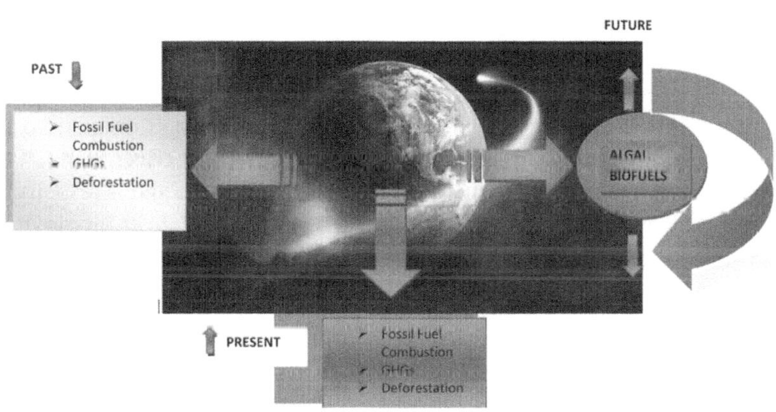

FIGURE 20.2 The schematic representation of the earth's climate over time. The less energy-related GHGs release and deforestation in past (left). The present stage of the earth (middle) represents high energy-related GHGs release and deforestation. The future (right) represents if algal biofuels are foreseen to be the substitute of fossil fuels, which would lead to a greener earth.

TABLE 20.1
The Advantages and Disadvantages of Different Generations of Biofuels

Generations Based upon Biofuels Feedstock	Advantages	Disadvantages
First	Easily available feedstock, simple manufacturing technology, low GHGs emissions	Required large area to grow feedstock, low yield, low quality of biofuels
Second	Non-arable land required, low GHGs emissions, utilize agricultural wastes as feedstock	Cost-effective strategies required for pretreatment of highly integrated feedstock and conversion of lignocellulosic biomass into biofuel
Third	A substitute of renewable source for conventional-liquid fuels with carbon-neutral combustion, CO_2 mitigation, high energy yield, bioremediation of wastewater, nonarable land can be used, can be scaled up, industries are being incorporated, exploring studies have been in progress since past years	Required controlled environment to grow, needs huge area and fertilizers, high cultivation cost, uncertain downstream processing
Fourth	High productivity of biomass with a high yield of biodiesel and bioethanol, enhanced power of CO_2 absorption	High risk of associated environmental toxicity issues with algal biofuel, only a few studies are reported using genetic modification to build up most favorable algae strains, relatively high upfront capital costs, not clear so far what the final outlay of cost per gallon will be, which is presently too high.

20.3 MICROALGAE AS FEEDSTOCK FOR MICROBIOFUELS

Microbiofuels are the 3G and 4G biofuels produced by microorganisms like cyanobacteria, microalgae, bacteria, fungi, etc. The idea of using algae for fuel production is not new. In the 1970s, algae were initially examined as a potential replacement for fossil fuels. Due to high production costs and limitations, commercial algal fuels could not be developed initially (Chisti et al., 2007). With increasing petroleum prices and an increased concern about global warming, microalgae research has shifted from the discovery and production of fine chemicals such as beta carotene and healthy foods to biofuel production. Various fuels can be produced by microalgae, such as biodiesel, bioethanol, bio-oil, syngas, and biogas. The critical first step of all biofuel synthetic processes is the conversion of light to chemical energy via photosynthesis, transforming sunlight into algal biomass accumulating carbohydrates and lipids as represented in Figure 20.3 (Beer et al., 2009). In green algae, the Calvin cycle is a vital piece of the photosynthetic procedure and is responsible for fixing CO_2. NADPH and ATP are utilized as a part of the Calvin cycle to deliver the sugars, starch, and oils that are put up to create biodiesel, bioethanol, and biomethane (Schenk et al., 2008).

20.3.1 Bioethanol

The use of microalgal biomass as a feedstock source has great potential for the production of bioethanol. Microalgae are grown mostly anywhere without soil (Widjaja et al., 2009), and also have an extremely short harvesting cycle of approximately one to ten days, compared with other crop feedstocks, which can be harvested only a handful of times per year (Schenk et al., 2008). An additional benefit is that microalgae require carbon dioxide for growth, creating the potential to absorb some of the greenhouse gas emissions created through the process (Avagyan, 2008).

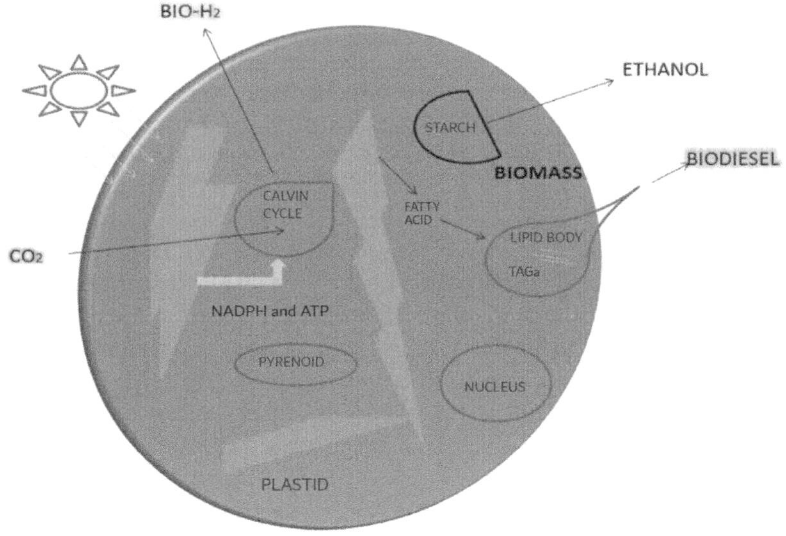

FIGURE 20.3 Metabolic complex pathways in microalgae for the synthesis of nucleic acids, proteins, lipids, carbohydrates, and H_2 related to biofuel production.

Source: Modified from Beer et al. (2009)

Microalgae are light-driven cell factories that can transform light into chemical energy by using CO_2 (Chisti, 2007). Under optimal conditions, they can double their number within hours, permitting a shorter harvesting cycle than terrestrial plants. In addition, the benefit of using microalgal-based carbohydrates is absence of lignin content in microalgae cells. Therefore microalgal biomass is much easier to convert into simpler sugars (Harun et al., 2010; John et al., 2011). Microalgae like *Scenedesmus*, *Dunaliella*, *Chlamydomonas*, *Chlorella*, and *Tetraselmis* accumulate carbohydrates in large amounts (John et al., 2011). Some researchers have studied that some species of *Chlorella* are a great source of carbohydrates, especially *C. pyrenoidosa* which have carbohydrate contents of 37–55% on dry weight basis (Ho et al., 2013; Dhull, 2016), consisting of starch, cellulose, and hemicellulose components. Microalgae provide an interesting consideration as these lack the lignin component, therefore, do not require a pretreatment step (Choi et al., 2010). This step usually increases the costs and unnecessary steps for ethanol production (Lin and Tanaka, 2006). However, a mild pretreatment step is required for the breakdown of the cell wall. In general, the following steps occur during bioethanol production: pretreatment of the algal biomass to break the cell walls, enzymatic hydrolysis to form simple sugars, and the fermentation of the resulting sugar to form bioethanol. At present, the optimum method for the pretreatment of microalgae has not yet been determined for the production of bioethanol (Li et al., 2014). Bioethanol production processes can be categorized into the following types (Girio et al., 2010):

1. Separate hydrolysis and fermentation (SHF) is a process involving the hydrolysis of the polysaccharides, which occurs separately from the fermentation.
2. Simultaneous saccharification and fermentation (SSF) is a process involving the hydrolysis of the polysaccharides and fermentation of released sugars, which occurs together.
3. Consolidated bioprocessing (CBP) is a process involving enzyme production, hydrolysis of polysaccharides, and fermentation of all sugars, which occurs in one step.

The following processing steps are involved in bioethanol production from algal biomass:

20.3.1.1 Pretreatment

Due to the simple structure of microalgae and the absence of lignin, it is possible to perform the pretreatment and hydrolysis in a single step. Although the use of a one-step process may reduce the yield of fermentable sugars, it reduces the costs involved, such as handling time and equipment. There are several key factors to be considered when analyzing an effective pretreatment method such as the chemicals required for pretreatment should be minimal and inexpensive, the pretreatment method should preferably limit the size reduction required, to avoid the costly and energy-intensive process of milling the biomass, a high yield of fermentable sugars should be achieved and heat and energy consumption should be low (Galbe and Zacchi, 2007). This is generally followed by thermal or thermochemical preteatments (Oncel et al., 2013; Dhull, 2016). Dhull (2016) employed pretreatments of autotrophically and mixotrophically cultivated biomass of *Chlorella pyrenoidosa*, which involved steaming at 121°C for 15 min. alone and in combination with 1% H_2SO_4 for the disintegration of cell wall structure to release polysaccharides for the hydrolytic enzymes to act upon. In the study, the thermal treatment alone was found to be at par with thermochemical treatment.

20.3.1.2 Hydrolysis

Enzymatic hydrolysis and acid hydrolysis are the common methods of saccharification. Cellulases and hemicellulases are used to depolymerise cellulose, xylan, and mannan of algal cell wall while amylases are employed to breakdown starch, the reserve food of a cell into simple sugars, including glucose, xylose, and mannose, which can then be used for fermentation. Due to the mild conditions under which this process operates, the utility costs are low compared to acid treatment. However, the actual cellulase and hemicellulase enzymes cost is very high as compared with the cost of the amylases. Cellulases are also a lot slower than amylases and a higher enzyme loading is required due to the complex nature of the cellulose (Lin and Tanaka, 2006). In microalgae, the released sugars after the enzymatic process are reported to have the potential in improving the production of bioethanol (Choi et al., 2010). A study by Puspawati et al. (2015) revealed that enzymatic hydrolysis of *Eucheuma cottonii* macroalgae by cellulases at 50°C produced 7.9 g/l of reducing sugar yield. Wang et al. (2014) reported that microalgae biomass when treated with acid (3% H_2SO_4) hydrolyzed at 121°C for 45 min. produced 14.5 g/l glucose in hydrolysate. Tan and Lee (2014) observed that enzymatic hydrolysis of seaweed by cellulases produced maximum glucose concentration of 8 g/l. In another study, Dhull (2016) carried out the separate enzymatic hydrolysis of autotrophically and heterotrophically cultivated *C. pyrenoidosa* biomass with in-house-produced enzyme preparation containing CMCase, FPase, β-glucosidase, α-amylase, and glucoamylase in both thermochemical and thermal pretreated residues. The hydrolysis attempted at 50°C in a water bath shaker at 150 rpm for 72 h. The autotrophic algal biomass revealed the maximum yields of total reducing sugars of 77.5 mg/g with glucose of 69.7 mg/g was obtained for thermochemical pretreatment, while the maximum yields of total reducing sugars of 76.2 mg/g and glucose of 68.1 mg/g were obtained for thermal pretreatment alone. The carbohydrate conversion of 77.5% and 76.2% were observed for thermochemical and thermal pretreatment, respectively. On the other hand, the mixotrophically cultivated biomass on pulp and paper industrial waste revealed the maximum yields of total reducing sugars of 178.1 mg/g with glucose content of of 164.7 mg/g was obtained for thermochemical pretreatment, while the maximum yields of total reducing sugars of 175.3 mg/g and glucose of 158.1 mg/g was obtained for thermal pretreatment alone.

Few reports are also available in literature where optimization of hydrolysis conditions has been carried out by varying one factor at a time and statistical tools to obtain the maximum sugars from microalgae (Kim et al., 2014; Cheng et al., 2013; Castro et al., 2015) and macroalgae (Tan and Lee, 2014; Ho et al., 2013). In a study by Trivedi et al. (2015), the enzymatic hydrolysis of *U. fasciata* feedstock for 24 h at 40°C and pH 4.0 gave maximum yield of sugar amounting to 112mg/g. Dhull (2016) used the statistical tool of response surface methodology to standardize the factors, including steam pressure, pretreatment time, enzyme dose, hydrolysis time, and solid concentration for mixotrophically cultivated *C. pyrenoidosa* biomass on pulp and paper industry wastewater and obtained the released of 410.89 mg and 318.99 mg of total reducing sugars and glucose per gram of substrate, respectively, with steam pretreatment at 14.1 psi for 56 min. and followed by hydrolysis with 3 ml of enzyme cocktail of cellulases and amylases/100 mg of biomass using solid concentration of 23 mg/ml at 50°C for 48.5 h.

20.3.1.3 Fermentation

After the pretreatment and the conversion of cellulose and starch to simple sugars, fermentation is done. Fermentation converts these compounds into bioethanol and carbon dioxide by the use of microorganisms. Many types of microorganisms are capable of bioethanol fermentation, including bacteria, yeast, and fungi. In the past, the most commonly used type has been yeasts, and more specifically, *Saccharomyces cerevisiae*, also known as baker's yeast. It is generally regarded as a safe microbe for human consumption, and as such is used in food processing and producing alcoholic beverages. *S. cerevisiae* can give bioethanol concentrations of up to 18% of the fermentation broth. It can also tolerate a wide range of inhibitors and elevated osmotic pressure (Banerjee et al., 2010). However, one of the limitations of *S. cerevisiae* is that it can only ferment hexose sugars and not the simple sugars produced from the breakdown of microalgal cell walls. Bacteria, particularly *Zymomonas mobilis*, have also been investigated. This microbe uses another pathway to digest the glucose (Entner-Doudoroff as opposed to Embden-Meyerhof) and results in higher bioethanol yield and specific bioethanol activity than *S. cerevisiae* (Lin and Tanaka, 2006). *Z. mobilis* is also generally regarded as safe and has simple nutritional needs. However, despite these advantages, it has its limitations, including its restrictive ability to ferment only glucose, fructose, and sucrose. Also, it is not as hardy as *S. cerevisiae*, therefore *S. cerevisiae* continues to be the preferred choice.

There have been few studies reported of saccharification and fermentation of microalgae (Kim et al., 2014; Cheng et al., 2013; Castro et al., 2015) and macroalgae (Tan and Lee. 2014; Doan et al., 2012). In a study involving separate hydrolysis and fermentation by Dhull (2016), the enzymatic hydrolysates obtained after the bioconversion of autotrophically and mixotrophically cultivated *C. pyrenoidosa* were subjected to fermentation by *Saccharomyces cerevisiae.* Maximum alcohol yields of autotrophically cultivated biomass amounting to 0.023 g/g was obtained with a fermentation efficiency of 70.95%, while the maximum alcohol yields of mixotrophically grown biomass on pulp and paper industry wastewater amounting to 0.132 g/g was obtained with a fermentation efficiency of 92.61%. The study was also extended to simultaneous enzymatic hydrolysis of mixotrophically cultivated algal biomass and fermentation of sugars by suitable yeast strain of *Saccahromyces cerevisae.* The same exhibited the alcohol yield of 0.128 g/g with alcohol concentration of 4.32 g/l.

20.3.2 Biodiesel

Competition between feedstocks for food supplies and biodiesel production has prompted researchers to focus on microalgal biodiesel production. The oil content of several microalgae has been observed to be over 30% (w/w), such as *Scendesmus obliquus* (35–55%), *Chlorella vulgaris* (14–40%), *Neochloris oleoabundans* (35–65%), and *Chlorella minutissima* (57%) (Gouveia, 2011). Microalgae can potentially replace current biodiesel feedstocks, which include vegetable oils. However, the current high production cost limits the commercial feasibility of producing biodiesel from microalgae. More research is needed for reducing the production costs of microalgal biodiesel production. The primary raw materials used in the production of microbiodiesel are algal oils, which contain triglycerides, free fatty acids, and other contaminants depending on the degree of pretreatment (Soni, 2007a). Since biodiesel is a mono-alkyl fatty acid ester, the primary alcohol used to form the ester is the other major feedstock. Most processes for making biodiesel use a catalyst to initiate the esterification reaction. The catalyst is required because the alcohol is sparingly soluble in the oil phase. The catalyst promotes an increase in solubility to allow the reaction to proceed at a reasonable rate. The most common catalysts used are strong mineral bases such as sodium hydroxide and potassium hydroxide (Soni, 2007a). After the reaction, the base catalyst must be neutralized with a strong mineral acid. Typical proportions for the chemicals used to make biodiesel are:

Reactants: Fat or oil (100 kg), primary alcohol (10 kg methanol)
Catalyst: Mineral base (0.3 kg sodium hydroxide)
Neutralizer: Mineral acid (0.25 kg sulfuric acid)

Methanol is the preferred alcohol for obtaining biodiesel because it is the cheapest (and most available) alcohol. Other alcohols, such as ethanol, isopropanol, and butyl, can also be used. A key quality factor for the primary alcohol is the water content. Water interferes with transesterification reactions and can result in poor yields and high levels of soap, free fatty acids, and triglycerides in the final fuel. Unfortunately, all the lower alcohols are hygroscopic and are capable of absorbing water from the air. It takes three moles of alcohol to react completely with one mole of triglyceride. The base catalyzed process typically uses an operating mole ratio of 6:1 mole of alcohol rather than the 3:1 ratio required by the reaction. The reason for using extra alcohol is that it "drives" the reaction closer to the 99.7% yield (Soni, 2007a). The unused alcohol must be recovered and recycled back into the process to minimize operating costs. Methanol is considerably easier to recover than the ethanol. Ethanol forms an azeotrope with water so it is expensive to purify the ethanol during recovery.

In biodiesel production, the oils after extraction from algal cells are transesterified. During transesterification, methanol and oils (triglycerides) are mixed inside a reactor at temperatures around 60°C with base or acid as catalysts called as transesterification pathway (Schubert, 2006). Methanol has been commonly used due to its low costs even though other alcohols, such as ethanol or 2-propanol, can be used. In the reaction, which has two reversible reactions with diglycerides and monoglycerides as intermediates, triglycerides can be converted into alkyl esters (biodiesel) with glycerol as a by-product. Transesterification consists of several consecutive, reversible reactions (Schwab et al., 1988). The triglycerides are converted step-wise to triglycerides, monoglycerides, and finally glycerol. A mole of the ester is liberated at each step. The typical reaction in the process is indicated here:

$$\text{100 liters of algal oil} + \text{24.65 liters of methanol} \rightarrow \text{103.3 liters of monoalkyl esters} \\ + \text{7.42 liters glycerol} + \text{12.33 liters unreacted methanol}$$

20.3.3 Biogas

Biogas is a colorless, odorless, inflammable gas, produced by organic waste and algal biomass decomposition (fermentation). Biogas is a renewable energy source produced by the conversion of sunlight energy to chemical energy by photosynthesis and anaerobic digestion. The composition of biogas varies among various processes. Biogas from landfills contains 30–40% carbon dioxide, 45–55% methane, and 5–15% nitrogen. Biogas from sewage digesters usually contains 35–45% carbon dioxide, 55–65% methane, and less than 1% nitrogen. Biogas from organic waste digesters usually contains 30–40% carbon dioxide, 60–70% methane, and less than 1% nitrogen (Oslaj et al., 2010). Biogas may also contain a trace amount of sulfur compounds, such as hydrogen sulfide. Methane in biogas can be further processed, such as in steam generation and the production of electricity. Production of biogas from various sources has been studied, such as from corn, animal manures, and corn stalks (Elmore, 1987; Oslaj et al., 2010; Chen et al., 2010). The methane yield from different microalgae strains is shown in Table 20.2.

TABLE 20.2
Biomethane Yield from Various Microalgae (Adapted from Harun et al., 2010)

Biomass	Methane Yield (m^3/kg)
Laminaria sp.	0. 26–0. 28
Gracitara sp.	0. 28–0. 4
Macrocystis sp.	0. 39–0. 41
I. digitata	0. 5
Ulva sp.	0. 20

Even though microalgae can be used for biogas production and grows much faster than other crops, using microalgae as a feedstock for methane production is not competitive for other feedstocks, including plant wastes, animal manures, human waste, garbage and municipal solid waste, etc., since growth costs of microalgae is more expensive than other feedstocks (Schenk et al., 2008). However, combinations of biogas production and wastewater treatment by microalgae may be feasible for fuel production. The biogas is produced under both natural and artificial conditions. The production of biogas is a biological process that takes place in the absence of air, through which the organic material is converted into, essentially, methane (CH_4) and carbon dioxide (CO_2). The process gives excellent organic fertilizer as the second by-product. The one essential requirement is an airtight container. Biogas is generated only when the decomposition of biomass takes place under anaerobic conditions. The airtight container used for the biogas production under artificial conditions is known as a digester or reactor. A variety of microorganisms harmoniously grow and produce reduced end-products. These include fermentative microbes (acidogens); hydrogen-producing, acetate-forming microbes (acetogens); and methane-producing microbes (methanogens). Methane fermentation involves a series of metabolic interactions among various groups of microorganisms. The first group of microorganisms, commonly known as non-methanogens, depolymerize the organics into monomers such as glucose and amino acids, which are subsequently converted to higher volatile fatty acids, H_2, and acetic acid. In the second stage, hydrogen-producing acetogenic bacteria, also called as non-methanogens, convert the higher volatile fatty acids to H_2, CO_2, and acetic acid. Finally, the third group, methanogenic bacteria convert H_2, CO_2, and acetate, to CH_4 and CO_2. Non-methanogens resemble the bacteria of human intestinal tract and include *E. coli*, *Micrococcus varians*, *Pseudomonas reptilivora*, *Microcollus luteus*, species belonging to *Clostridium*, *Peptococcus*, *Eubacterium*, *Lactobacillus*, *Actinomyces*, *Bacillus*, *Vibrio*, and *Sprillium*. On the other hand, methanogens belong to *Methanocoocus vennelli*, *M. omelianskii*, *M. sarcina*, *Methanobacterium ruminantum*, *M. mobilis*, and *Methanospirillum* spp. The complete digestion takes place in 30–40 days of incubation at 30°C and pH 6.4–7.5. Compounds containing sodium, magnesium, or calcium are preferred to control the pH of the digester. Presence of acid-forming bacterial population enhances CH_4 formation by producing a combination of compounds acting as substrates – methanol, acetate, formate, H_2, and other acids for methanogenic bacteria. Addition of Fe and S compounds in traces is beneficial. C, N, and P are required in ratio of 100:15:1 for optimal gas production (Soni, 2007b)

20.3.4 Synthesis Gas

Production of synthesis gas (or syngas) involves a gasification process that converts solid or liquid carbon materials, such as biomass, coal, or oil into gas with air, oxygen, and steam under thermochemical treatment (Schubert, 2006). The gas composition predominantly contains carbon monoxide (CO), hydrogen (H_2), carbon dioxide (CO_2), methane (CH_4), and nitrogen (N_2). This technology is not new and biomass gasification is already commercialized for producing heat and electricity. Also, syngas can be used for synthesizing commercial products, such as methanol and alkanes. Gasification of microalgae species *Cladophora fracta* and *Chlorella protothecoid* by steam has been studied (Demirbas, 2009). This study showed the technical feasibility of the gasification of algae by steam gasification. Gasification of *Chlorella vulgaris* with supercritical water was also studied (Chakinala et al., 2010). The feedstock for this process can be homogeneous liquid or slurry. This may give an advantage for using wet microalgae or microalgal liquid as a feedstock for reducing harvesting costs. From Chakinala et al. (2010) work, the syngas from *C. vulgaris* contained 7% H_2, 22% CO, 25% CH_4, 26% CO_2, and 20% C_2-C_3 organic chemicals. The gasification efficiency or carbon conversion was reported to be 53%. The author concluded microalgae can be a promising feedstock for supercritical water gasification.

20.3.5 Bio-Oil

The form of products obtained from thermochemical treatment of biomass can be varied, such as gases, bio-oil, and char, and depends on heating rate, temperature, and reaction time. For example, in flash pyrolysis, biomass can be converted into bio-oil when the temperature is maintained at 400–650°C. However, when the temperature is maintained at 650–900°C, a bio-oil and gas mixture was obtained (Schubert, 2006). Bio-oil can be used for power generation through internal and external combustion by mixing with diesel or natural gas. Upgrading the bio-oil is needed for use as a transportation fuel since bio-oil has a high oxygen content, low heat value, and high viscosity at ambient temperatures typical of fuel storage. Most bio-oil studies focus on conventional biomass sources, such as agricultural wastes or forestry products. The production of bio-oil from microalgae has been investigated by several workers. Pyrolysis of *Chlorella protothecoides* in absence of oxygen at 200, 300, 400, 500, and 600°C for 5, 20, 60, and 120 min. were studied by Peng et al. (2000). The maximum yield of bio-oil from *C. protothecoides* was 52% (w/w) at 500°C after heating for 5 min. The direct pyrolysis and catalytic pyrolysis of *Nannochloropsis* sp. residue after lipid extraction with HZSM-5 as a catalyst was studied (Pan et al., 2010). For direct pyrolysis of *Nannochloropsis* sp., the maximum yield of bio-oil was found to be 30% at 400°C, with 24% (w/w) of biomass converted to syngas. The liquid product yield was 45.3% (w/w). From these results, the authors suggested the direct pyrolysis temperature for *Nannochloropsis* sp. residues of 350–450°C. Liquefaction of *Chlorella vulgaris* and *Spirulina* at 300 and 350°C using KOH, Na_2CO_3, acetic, and formic acid as catalysts has also been reported (Ross et al., 2010). Acetic acid catalyst gives the best bio-crude yield at 22% (w/w). From these observations, the production of bio-oil from microalgae appears to be technically feasible.

20.3.6 Integrative Approach of Microbiofuel Production and Bioremediation

Various studies have been carried out in the biofuel production and simultaneous wastewater treatment of various effluents from industries (Menger-Krug et al., 2012; Li et al., 2011; Osundeko et al., 2013). Algal biomass can remove phosphate, nitrate, ammonium, and dyes, such as Azo dye, Remazol Black B, Remazol Red, Remazol Golden Yellow, Acid blue 290, Acid blue 324, Acid red 274, and Malachite green (Daneshvar et al., 2007). Furthermore, algae are capable of removing and recovering metal from wastewater by metal accumulation or biosorption. Biosorption is usually referred to as metal adsorption on dead biomass; however, it can also mean metal accumulation by live biomass as well. Several studies have shown that algae are capable of removing Al, Au, Co, Cd, Cr, Cu, Fe, Hg, Ni, Pb, Zn, Ur from aqueous solution (Jacinto et al., 2009; Gupta and Rastogi, 2008). Precious metals such as silver or gold can be recovered from the wastewater by the adsorption and desorption process. Removal of heavy metals by algae involves two processes: adsorption and transportation (Bates et al., 1982). During the adsorption, metals are bound into the cell wall. This process is fast and metabolism-independent. However, during the transportation process, metal ions are transported through the cell membrane and bound intracellularly. The equilibrium and kinetics of this process are observed to be affected by several factors, such as initial metal concentration, temperature, metabolic stage of algae, pH, presence of competing ions, and amount of biomass. In the biosorption process, metal accumulation can occur on the cell surface or inside the cell. The algal wall or surface contains several functional groups, such as hydroxyl, amino, carboxyl, sulphydryl, and phosphoryl. Metal ions, usually positively charged, attract and bind to those functional groups on the cell surface, since those functional groups act as a negative charge on the cell surface (Gupta and Rastogi, 2008). Those functional groups are parts of cell wall components, such as protein and polysaccharides.

Since the cell wall composition is varied among different algal species, the adsorption capability is varied among algae. However, even though the cell wall composition varies among algae, most metals are bound with proteins and polysaccharides (Kuyucak and Volesky, 1989). Once metals are inside the cell, they may precipitate or bind to intracellular components, such as proteins, cellular polyphosphates, polysaccharides, or enzymes (Wong et al., 1984). Also, metals can be absorbed as nutrients for living cells (Hammouda et al., 1995). Biosorption is not the only mechanism for polishing wastewater. Pollutants, such as ammonium and phosphate can be removed by microalgae as nutrient uptake. The removal rates of both ammonium and phosphate are reported as high as 99%. Removal of phosphates and ammonium has been reported by incubating a consortium in diluted dairy wastewater or primary clarifier effluent. Initial phosphate concentrations of wastewater are changed from 1.8 mg/l to 2.6 mg/l. The ammonium concentration of primary effluent, 10% diluted and 25% diluted dairy wastewater are 39, 30.5, and 16.3 mg/l (Woertz et al., 2009). *Scenedesmus* sp. grown in wastewater derived from remains of fish removed 94.44% ammonia nitrogen, 77.54% phosphate, and 35.59% organic matter from the wastewater (Andrade et al., 2009).

The photosynthetic rate can affect ammonium lost to the atmosphere due to the usage of carbonate, which changes the pH. Likewise, not all phosphates are consumed by microalgae, since phosphate can precipitate with a range of metal ions present in wastewater, such as calcium and magnesium (Larsdotter et al., 2007). The amount of phosphate removed by precipitation instead of consumption by microalgae depends on temperature, pH, and initial concentrations.

NASA has developed an OMEGA system, an innovative method for cultivating algae, cleaning wastewater, capturing carbon dioxide, for the production of biofuel without competing with agriculture for water, fertilizer, or land. The system has photobioreactors consisting of large plastic tubes that are flexible and allowed to float on seawater. The plastic tubes contain wastewater in which microalgae is cultivated (Harris et al., 2013).

20.3.6.1 Cultivation of Algae for Microbiofuel Production

It is important to develop a process that is viable and profitable at the commercial level. Therefore, the key step in the development of microalgae for biofuel needs the outcome of scientific findings into a marketable product. This can be achieved by outlining, designing operations, and optimization of photobioreactors. Algal biomass can be produced in the following ways.

20.3.6.1.1 Open Pond

Various specific cultivation conditions are required for the cultivation of algae in an open pond. These are the ideal climatic circumstances and sufficient supplements for optimum growth of microalgae and the requirement of typical water characteristics (e.g. high saltiness, high pH, high supplements). The most frequently used artificial system is raceway ponds. These are made of closed, oval-shaped circulation canals, counterfeit loop, 0.2–0.6 meters deep, and equipped with a paddlewheel for mixing as shown in Figure 20.4.

This maintains a continuous production cycle and avoids sedimentation (Prussi et al., 2014). However, open ponds are facing major drawbacks. They are most frequently contaminated by unknown species of algae and a high rate of water loss by evaporation, similar to land crops resulting in reduced yields (Schenk et al., 2008). Raceway ponds are less economic due to the requirement

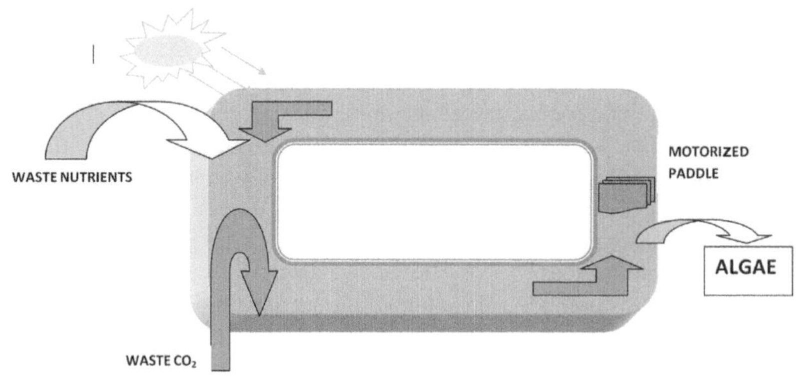

FIGURE 20.4 Diagrammatic representation of raceway pond.

of specific infrastructures like oval shape, paddlewheel, etc., and maintenance of pond integrity to maintain the faster flow rates.

20.3.6.1.2 Photobioreactors

Photobioreactors (PBRs) are described as bioreactors that have controlled and regulated biological factors. Besides, PBR has low contamination possibility, reproducible development conditions, controllable hydrodynamics and temperature, no loss of CO_2, and flexible technical design (Pulz et al., 2010).

The apparatus system is made up of clear walls or light fibers that can collect maximum sunlight. As various algal high-value products must be developed free of contamination, microalgal mass culture requires closed systems. There are different designs, with categories including tubular (helical, serpentine, manifold, and α-shaped), column (bubble columns and airlift), and flat (alveolar panels and glass plates). Research has been done over the past years to improve the designing of a photobioreactor for algal biomass production (Tredici, 2004). Regardless of numerous practicable applications, an insignificant quantity of microalgae is cultivated due to inefficient improvement in microalgal bioreactors. The various cultivation system is listed in Table 20.3.

20.3.6.1.3 Open Tanks

Open tanks with different sizes of transparent rectangular chambers have been used for the cultivation of microalgae (Hsieh and Wu, 2009) and the important physical (light; temperature; mixing intensity), chemical (dissolved CO_2 and oxygen; pH; and nitrogen), and biological (biomass concentration and biomass composition) parameters have been monitored and regulated. *C. pyrenoidosa* grew in a 100 L photobioreactor (Figure 20.5) with a working volume of 70 L illuminated with 6 CFL and mixing intensity of 150 rpm also revealed a biomass yield of 1.0 g/l after 12 days of incubation with 12:12 circadian cycles (Dhull et al., 2014).

20.3.6.2 Harvesting of Algal Biomass

The most frequently used harvesting processes include flocculation, centrifugation, filtration, sedimentation systems, and flotation by ultrasound. Some of these strategies are currently being utilized by a couple of associations, while some are still under examination. Flocculants are essential for the aggregation of algal cells. A few inorganic flocculants (syntheses of metals and salt, for example, ferric chloride or alum) and organic (chitosan-polyacrylamide) flocculants are broadly used and easily available. The suitable flocculent is chosen depending upon the microalgal cell size and the selected process used for the separation (centrifugation or filtration). Guar gum is the suitable flocculating operator for small and negative charged algae (*Chlorella* sp. Cb4 and *Chlamydomonas*

TABLE 20.3
Limitations and Advantages of Various Cultivation Systems

Culture Systems	Advantages	Limitations
Open systems 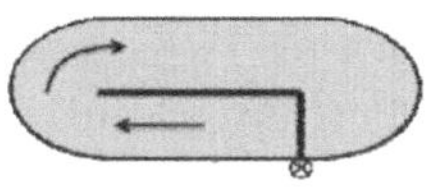	Relatively cost-effective, easy to clean up, easy maintenance, utilization of non-agricultural land, low vitality inputs	Difficult to maintain culture conditions, poor blending, low light penetration, and CO_2 use, difficult to grow algal cultures for long periods, low biomass production, limited to few strains, high contamination risk
Tubular PBR 	Large light surface, relatively economical, good biomass production, good for outdoor culture	Gradients of pH, dissolved oxygen, and CO_2 along the tubes, smell fouling, algal wall growth, need large land area, photoinhibition
Flat PBR 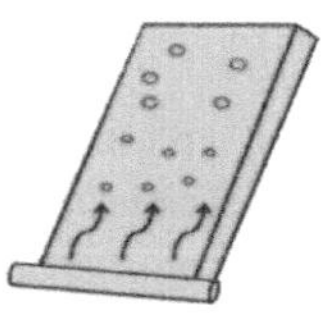	Comparatively cheap, easy to clean up, huge light surface area, good for outdoor cultures, low power utilization, relatively high biomass production, good light way, shortest oxygen path, low oxygen buildup	Hard to scale up, hard to control temperature, algal wall growth, hydrodynamic stress to some algal strains, low photosynthetic efficiency, high shear stress
Column PBR 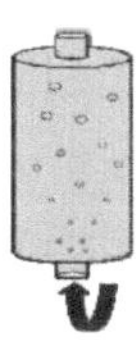	Low energy utilization, high mass transfer, high blending, the best acquaintance of photoperiods, low shear stress, easy to maintain hygiene, low photoinhibition, low photooxidation, high photosynthetic efficiency	Small light surface area, sophisticated building materials, high shear stress, the small light surface area upon scale-up, relatively costly to open ponds

FIGURE 20.5 Open glass chamber for large scale production of biomass.

Source: Cited from Dhull et al. (2014)

sp. Crp7) (Banerjee et al., 2013). Flocculants such as Flocudex CS/5000, Drewfloc 447, Magnafloc LT-25, Chemifloc CV/300, Flocusol CM/78, and Chitosan have been utilized by several experts for harvesting algae (De Godos et al., 2011; Schlesinger et al., 2012; Lin and Lu, 2021).

20.3.6.2.1 Algal Biomass Transformation

After collecting the biomass, various biochemical or thermochemical transformation procedures can be applied for the transformation of microalgae. Syngas and bio-oil are extracted through the thermochemical transformation process, whereas bioethanol, biogas, and biodiesel are extracted through the biochemical transformation process (Maity et al., 2014). *Schizochytrium* sp. and *Haematococcus pluvialis* have high lipid content, which makes them more suitable for biodiesel production. Furthermore, the potential methane producers are *Laminaria* sp., *Gracilaria* sp., *Macrocystis*, and *Ulva* sp. Additionally, the potential ethanol makers are *Chlamydomonas perigranulata* (Molina Grima, 1999; Eriksen et al., 1998) and *Clostridium saccharoperbutylacetonicum* N14 (Poulsen and Iversen, 1999).

20.4 FACTORS AFFECTING THE ALGAL CULTIVATION

It is very important to identify the physical parameters like light, temperature, pH, and carbon sources that significantly affect growth, nutrient uptake, and carbohydrate/lipid accumulation properties. The growth of microalgae depends on essential nutrients, major ions, and numerous micronutrients (carbon, nitrogen, phosphorus, Mg^{2+}, Ca^{2+}, Cl^{-}, SO_4^{2-}, molybdenum, manganese, copper, iron, zinc, and cobalt) (Sunda et al., 2005).

20.4.1 Light

Phototrophic algae depend upon the intensity of light and its accessibility is the base foundation of energy. It is the most important factor. The growth of the algae in the culture medium is zero at very low light intensities, and this is called the compensation point (Lee, 1997). With the increasing light intensity, photosynthesis increases until a point is reached where the maximum development rate is achievable. It is known as the light saturation point (Richmond, 2000). Further increase in light intensity beyond saturation level does not affect the growth of algae, resulting in photooxidation of cells. This leads to decreased photosynthesis rate by harming the light receptors of the algae, called photoinhibition (Seo et al., 2014). The is due to the highly efficient light-harvesting antennae of the cells, which absorb all the light that hits them. Therefore, all the light is absorbed by the uppermost layer of the culture medium, leaving the rest shady. However, most algae get saturated by light intensity of 20% of the sunlight-based light intensities (Torzillo et al., 2003). These issues can be resolved by minimizing the mutual shading of algal cells by proper mixing and shortening of the light path length of the algae culture system. This is additionally encouraged by proper mixing (Tredici, 2004).

20.4.2 Temperature

In general, algae development increases exponentially with rising temperature until an optimum level is reached, after which development falls. The algal growth rate increases with the increasing temperature until an optimum point is reached, after which the growth rate slows down (Borowitzka, 2016). In the open pond, the algae culture temperature can be 10–15°C below optimal during the night (Borowitzka, 2016). The optimum culture temperature is only achieved at some hours of the day (Richmond, 2000). While ideal culture temperature may not be accomplished until after early afternoon, sunlight intensity rises very quickly in the morning (Lu and Vonshak, 1999). This variation of temperature results in a comparatively low biomass production rate than closed photobioreactors. Therefore, closed photobioreactors are preferred over open ponds for maximum biomass production.

20.4.3 Carbon Source

Carbon is an important component of algal biomass, accounting for roughly 45–50% of algal (Chae et al., 2006); therefore, supplementary carbon is required in the medium, as a low concentration of CO_2 in the air (0.033%) limits the growth of algae in the culture medium. For autotrophic microorganisms, carbon dioxide is the primary carbon source and is utilized for the production of sugar during photosynthesis. This process contains two cycles: a light cycle and a dark cycle. The photosynthetic process has been included in the light cycle. In the dark cycle, every mole of carbon dioxide taken up by the algal cell is converted to one mole of carbohydrates with three moles ATP and two moles of $NADPH_2$ (Richmond, 2000). Carbon dioxide is considered to be a greenhouse gas, contributing significantly to global warming. At present, the carbon dioxide concentration in the atmosphere is estimated to be 0.036% (Barsanti and Gualtieri, 2006). Excess carbon dioxide is usually supplied into open algae ponds. The presence of carbonate ions is dependent on pH. A bicarbonate-carbonate buffer system is created as shown in the following reaction:

$$2HCO_3^- \leftrightarrow CO_3^{2-} + H_2O + CO_2$$
$$HCO_3^- \leftrightarrow CO_{2\,+}OH-$$
$$CO_3^{2-} + H_2O \leftrightarrow CO_{2\,+}2OH-$$

A carefully monitored supply of carbon dioxide is essential for growing algae. Oversupply of carbon dioxide (in comparison to the metabolic uptake rate) can cause a significant drop in pH and subsequent cessation of algal growth. Likewise, an inadequate supply of CO_2 can adversely affect algal growth. For example, the maximum biomass of *Chlorella vulgaris* has been observed with 6 to 7% (v/v) of carbon dioxide in the air and decreased when CO_2 was increased from 7% to 12% (v/v). After increasing above 12% (v/v) carbon dioxide, no significant growth was observed (Chinnasamy et al., 2009). Carbon dioxide concentration in the feed gas not only affects the maximum biomass concentration but also affects the composition of products, such as lipid. Influence of carbon dioxide on four microalgae strains including *Nannochloropsis* sp., *Nannochloropsis oculata*, *Nannochloris atomus*, and *Isochrysis* sp. were studied by (Roncarati et al., 2004). A carbon dioxide addition of 2% (v/v) in air exerted a significant and more favorable influence than a lower carbon dioxide concentration (i.e. 1% v/v in air). Desaturation of fatty acids was observed to be lower at higher CO_2 concentrations with *Chlorella vulgaris*. A similar result has been reported for *Botryococcus braunii* (Rangarao et al., 2007). As carbon dioxide concentration was increased to 2.0% (v/v), the percentage of high chain carbohydrates (> 30 carbons) increased from 34.41% (w/w) to 57.58% (w/w). The maximum biomass concentration reached 2.0 g/l on day 25. For a carbon dioxide concentration of 1% (v/v), the biomass concentration reached ~1g/l at day 25. Increasing carbon dioxide concentration had a positive influence on the total lipid composition of C16:0, C18:1, and C24:0, but have a negative influence on C18:0, C18:2, and C22:1.

20.4.4 Nitrogen Source

Nitrogen is one of the abundant elements on earth and is used for synthesizing important biological components, such as NADP, ATP, DNA chlorophyll, and protein. The composition of air contains around 79% nitrogen, which can be converted into ammonia by nitrogen fixation under a low nitrogen environment (Barsanti and Gualtieri, 2006). This concept is employed in biofertilizer research. The main idea of biofertilizer is to grow a nitrogen-fixer before planting the desired crop (for releasing fixed nitrogen for uptake by the subsequently desired crop). There are two approaches for releasing nitrogen back to the crop field. The first one is to select nitrogenase-repressed mutants, which excrete ammonia into the selected field (Latorre et al., 1986). Another approach for releasing nitrogen into the selected field is decomposition or mineralization.

Biofertilizers have been studied intensively in rice fields (Dhar et al., 2007). However, the production of biofertilizers from secondarily treated wastewater has rarely been studied. In other words, by growing a nitrogen-fixer in secondarily treated wastewater, to enrich the nitrogen content of wastewater, the wastewater could then be reprocessed for biofuel production. Even though the composition of air contains 79% nitrogen, not all microorganisms can use nitrogen directly from the air. Nitrogen in other forms, such as nitrate, nitrite, ammonia, and urea, can be consumed by microorganisms. The different forms of nitrogen can affect the growth or composition of microalgae. Nitrogen effects among three different kinds of nitrogen sources (i.e. nitrate, urea, and ammonium) have been studied for the growth of *Neochloris oleoabundans*. It has been reported that *N. oleoabundans* shows a higher lipid content and biomass concentration in nitrate medium (Li et al., 2008). However, for *Ellipsoidion* sp., no difference in growth was observed among nitrate, ammonium, and urea nitrogen sources, while a higher lipid content was observed for this species when grown in ammonium media (Xu et al., 2001). Shen et al. (2010) have studied the effect of nitrogen sources on four strains of *Chlorella protothecoides* (i.e. UTEX 25, 31 249, and 255) (Shen et al., 2010). No significant differences in growth are observed in three nitrogen sources (nitrate, urea, and yeast extract) for the microalgae strain UTEX 255. However, UTEX 25 and UTEX 249 grew much better in nitrate. UTEX 31 grows marginally better in urea than the other two nitrogen sources. However, no gene information was reported in this study to indicate that UTEX 255, UTEX 25, UTEX249, and UTEX 255 are the same species even though they are each named to be *C. protothecoides*. When microalgae grow under nitrogen-limited conditions, specific degradation of phycobilisomes is activated for reuse of those nutrients, and bleaching is observed (the algal color is changed from green to yellow-green) (Collier and Grossman, 1992). As nitrogen concentration decreases, the rate of photosynthesis is reduced and metabolism is changed to secondary metabolism, which changes from a protein synthesis pathway to a carbohydrate or lipid synthesis pathway.

Many researchers have reported the significant effect of stress factors like iron, copper, chloramphenicol, nutrient starvation, and light-dark condition on the growth and cellular variability in microalgae. Accumulation of lipids or carbohydrates inside cells under nitrogen starvation conditions depends significantly on the type of microalgae. *Neochloris oleoabundans* were reported to accumulate lipids under nitrogen-limited growth conditions. Upon decreasing KNO_3 concentration from 20 μM to 3 μM, total lipids (w/w) increased from 8% to 40% (Li et al., 2008). Also, *Botryococcus braunii kutz* IPPAS H-252 has been reported to accumulate lipids (w/w) from 9% to 22% (Zhila et al., 2005). β-carotene accumulation by *Dunaliella* cells under nitrogen-limited conditions has also been reported (Ben-Amotz et al., 1982). The effect of N starvation on cellular components like lipid, crude protein, and carbohydrate in *C. pyrenoidosa*, as observed by Dhull et al. (2014), is represented in Figure 20.6. The decrease of N concentration in the culture medium from 1,600 to 0 mg/l resulted in an increase in protein content to 80±1% in N-starved condition in the presence of light, which was about 26.9% higher than the N-rich culture, but it was found that in the absence of light, crude protein content decreased to 29±2%, which was about 54% and 64% less than those present in biomass cultured N-rich and N-starved conditions in the presence of light, respectively. A possible reason for the decrease in protein in N-starved dark cultivation strategy is that the cells may have tarnished the nitrogenous compounds to sustain intracellular protein parts in support of the irregular metabolic function. Further, when photosynthetic microalgae cultivated under N-starved conditions carbon flow from photosynthetic pathways convert into various modes to channel metabolic energy into different energy-rich components like lipid sand carbohydrates (Siaut et al., 2011). The algal storage products (lipid and carbohydrate) are preferred under stress probably for use during adverse conditions for cell survival and division. These are in highly reduced states and being hydrophobic can be bundled in a small chamber of the cell (Courchesne et al., 2009). Figure 20.6 shows that *C. pyrenoidosa* cultivated in N-starved medium significantly increased the lipid content from 13% in N-rich conditions to 63±5% after four days of cultivation in N-deficient conditions in dark.

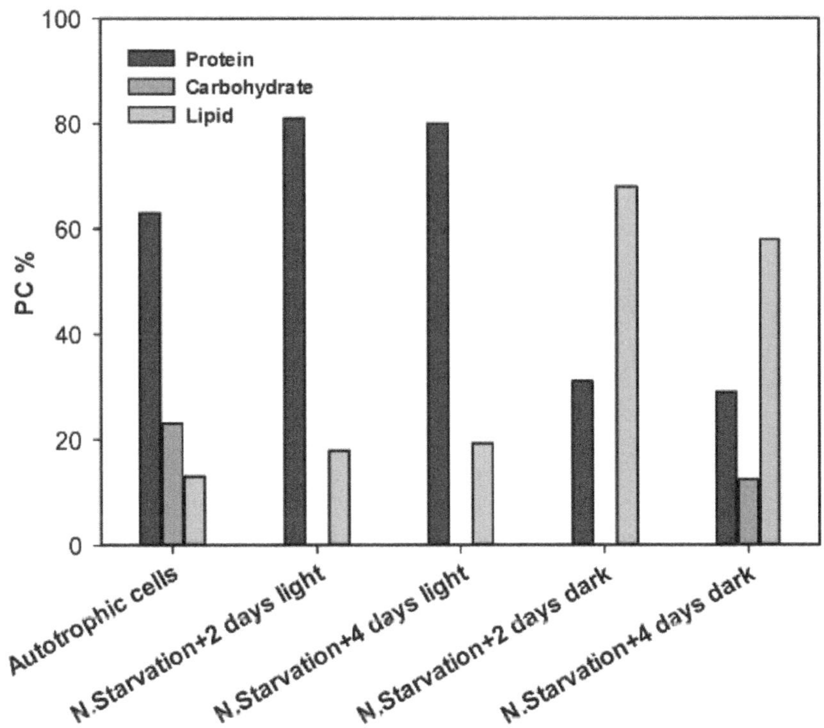

FIGURE 20.6 The cellular variability of *C. pyrenoidosa* in terms of protein, carbohydrate, and lipid content under N stress conditions on day 2 and day 4. PC refers to Principal Components derived from the data analysis of peak area integration from FTIR.

Source: Cited from Dhull et al. (2014)

Table 20.4 represents the changing spectral ratios due to the application of N-starved conditions. The study also revealed that under N-starved cultivation the cells had a negative correlation with the cellular carbohydrate and lipid content. *C. pyrenoidosa* cells grown in N-starved dark medium increased in lipid and decreased in protein content, whereas in the N-starved medium exposed to light, it showed a significant increase in protein whereas no significant increase in lipid content was found (Dhull, 2016). Cells under N-starved light exposure for two days showed the maximum lipid/AmideII shown in Figure 20.6. The study also indicated that four-day light exposure to N-starved culture had slightly less amount of lipid as observed in two days of exposed culture, while the protein content almost remained the same. However, there was no accumulation of total carbohydrates indicating that *C. pyrenoidosa* is probably a lipid accumulating microalgae under N-stress. Total removal of N from the culture medium resulted in the highest lipid (68%) and protein content (81%), making it a potential feed stock for biodiesel land value-added food supplement as single-cell production.

20.4.5 Phosphorus Source

Phosphorous is one of the major nutrients and plays an important role in every biological system. Biomolecules containing phosphorus can be used for storing energy, such as ATP, NADP, and phospholipids or can be used for synthesizing other important biological components, such as coenzymes, proteins, DNA, and RNA. Furthermore, the supply of phosphorus affects the composition of biomass, such as lipids content and carbohydrates (Borowitzka et al., 1991). Also, the nitrogen

TABLE 20.4
The Characteristic Peak Areas of Lipids, Proteins, and Carbohydrates Resulted from the Absorbance Band of FTIR Spectra from Dried algal Cells of *C. pyrenoidosa* (Cited from Dhull, 2016)

Different Stress Conditions	Lipid[a]	Protein[b] (AmideII)	Carbohydrate[c]	Lipid/AmideII[b]	Carbohydrate/ AmideII[b]
Autotrophic cells	0.2853	1.7227	0.1906	0.165612	0.11064
N. Starvation +2 days light	0.3131	1.5839	–0.4680	0.197677	–0.29547**
N. Starvation +4 days light	0.3051	1.4426	–0.4618	0.211493*	–0.32012
N. Starvation +2 days dark	0.9432	5.3889	–0.1349	0.175026	–0.02503
N. Starvation +4 days dark	0.4745	2.9462	0.1417	0.161055	0.048096

*maximum lipid content.

**(-) shows absence of carbohydrate content, corresponding wave number ranges: a- 2,984–2,780 cm^{-1}, b- 1,590 and 1,477 cm^{-1}, c- 1,180 and 1,133 cm^{-1}.

phosphorus (N:P) ratio not only controls potential productivity but also may significantly affect the dominance of specific species in a mixed culture (Miller et al., 2005). A lower N:P ratio has been observed to promote the domination of N-fixing species under high light intensity; however, dominant species shifted to green algae when the N:P ratio was high under high light intensity. In nature, there are several phosphorus sources (Barsanti and Gualtieri, 2006). Phosphorus can be released as phosphate ions from the rock during weathering. Also, the decomposition of dead biological creature, such as animals, plants, or microalgae, can release phosphorus compounds. Phosphorus can be introduced into the biological system from human activities, such as improperly disposed of animal wastes, and agricultural or lawn fertilizers. There are three forms of phosphate: orthophosphate, metaphosphate (a benzyl ring-like structure), and organically bound phosphate.

Orthophosphate is a preferred form for feeding microalgae. When phosphorus is rich, algae store excess phosphorus in the form of polyphosphate. This polyphosphate disappears under phosphorus-deficient conditions (Powell and Hill, 2009). In this condition, the content of chlorophylla is decreased, but carbohydrate is increased. Enhancement of lipid accumulation of *Mundus subterraneus* was reported under a phosphorus-deficient condition (Khozin-Goldberg et al., 2006). With a decrease in K_2HPO_4 concentration from 175 μM to 0 μM, the total lipid content is increased and its composition is changed. The proportion of triacylglycerol in the total lipid is increased from 6.5% (w/w) to 39.3% (w/w); however, the proportion of phospholipids is decreased from 8.3% (w/w) to 1.4% (w/w).

Dhull et al. (2014) statistically designed an autotrophic medium for maximum biomass production by *Chlorella pyrenoidosa* using response surface methodology. After evaluating one factor at a time approach, KH_2PO_4, KNO_3, $MgSO_4.7H_2O$, and $NaHCO_3$ were preferred over the other components of the Fog's medium as most critical autotrophic medium components. The study showed that the maximum biomass yield was achieved while the concentrations of $MgSO_4.7H_2O$, K_2HPO_4, KNO_3, and $NaHCO_3$ were 0.409g/l, 0.24g/l, 1.033g/l, and 3.265g/l, respectively. The study reported that the biomass productivity of *C. pyrenoidosa* improved from 0.14g/l in defined Fog's medium to 1.40 g/l in modified Fog's medium, resulting in a tenfold increase. The biochemical composition biosynthesis of *C. pyrenoidosa* was altered using nitrogen-limiting stress, bringing about 5.23-fold increase in lipid content more than control (cell without stress), as analyzed by the FTIR integration method.

20.5 CONCLUSION AND FUTURE PERSPECTIVE

There is an ever-increasing demand for transportation fuels around the world. This demand is abnormally affecting developing countries and can only be mitigated by producing ethanol in the scenario of shrinking economic and energy resources. Despite laboratory-based bioethanol success stories, the production of fuel ethanol at a plant scale remains a challenging issue. A positive solution to this issue could bring economic advantage not only for the fuel and power industry but also benefit the environmental rehabilitation. The lignocellulosic feedstocks are abundant, cost-effective, and renewable, yet the production cost of the second-generation bioethanol is high because of the pretreatment of the complex structure of lignocellulosic raw materials.

Therefore, the algal biomass as a raw material for bioethanol production is undoubtedly a sustainable and eco-friendly source for renewable biofuel production, as it leads to decreased emission of GHGs and can combat the current energy crisis. Like any other industrial process, the commercialization of algae bioethanol is also dependent on the capitals of the process. Before the commercialization of large-scale bioethanol production, problems arising in the technology of algae bioethanol should be overcome. Additionally, from a basic research point of view, one approach that is receiving more attention is the study of the effects of enzymatic hydrolysis at a more fundamental level. This knowledge should be applied to achieve an integrated and efficient biomass conversion process to ethanol. The hydrolysis method should be simple, devoid of expensive chemicals, and high energy. About 33% of the total cost in the production of algae bioethanol is accounted for by pretreatment and hydrolysis methods. Likewise, carbohydrates from the dried algae biomass must be hydrolyzed directly without its degradation or it will cause the production of fermentation inhibitors. It is essential to minimize sugar losses, to increase high possible solids concentration, and to keep low the cost of reactors and associated equipment, to decrease the pretreatment cost in the production of bioethanol. The next significant technical barrier is the cost of enzymes. Efforts should be made to increase the effectiveness and specificity of enzymes. The development of novel technology for degradation of algae biomass and recovery of enzymes to reduce the enzyme cost by several folds is the need of the hour. Co-biofuel or expanded biofuel production from microalgae are essential for recovering the total energy balance of third-generation biofuels. A biorefinery approach can be developed in the production chain of algae bioethanol. After extraction of useful chemicals from the algae and residue to contained rich starch that can be utilized as raw material for bioethanol production. Even after the ethanol production, the leftover residue still contains a good amount of organic matter and useful minerals and eventually can be used as biofertilizers. One of the successful examples is to use the microalgae biomass (after lipid extraction) for bioethanol production because high concentrations of carbohydrates remain in the biomass. This is a win-win strategy in reutilizing the waste to produce another source of energy, which greatly amplifies the sustainability of microalgae biofuels.

BIBLIOGRAPHY

Andrade, R.C.E., Vera, B.A.L., Cardenas, L.C.H., and Morales, A.L.E.D. 2009. Biomass production of microalga *Scenedesmus* sp. with wastewater from fishery. *Revista Tecnica de la Facultad de Ingenieria Universidad del Zulia*, 32: 126–134.

Avagyan, A. 2008. A contribution to global sustainable development: inclusion of microalgae and their biomass in production and biocycles. *Clean Technologies and Environmental Policy*, 10: 313–317.

Azad, A.K., Ameer U.S.M., and Alam, M.M. 2012. A comprehensive study of DI diesel engine performance with vegetable oil: an alternative biofuel source of energy. *International Journal of Automotive and Mechanical Engineering*, 5: 576–586.

Banerjee, C., Ghosh, S., Sen, G., Mishra, S., Shukla, P., and Bandopadhyay, R. 2013. Study of algal harvesting using cationic guar gum from the natural plant source as flocculant. *Carbohydrate Polymers*, 92: 675–681.

Banerjee, S., Mudliar, S., Giri, B., Satpute, D., Chakrabarti, T., and Pandey, R.A. 2010. Commercializing lignocellulosic bioethanol: technology bottlenecks and possible remedies. *Journal of Bioresources and Bioproducts*, 4: 77–93.

Barsanti, L., and Gualtieri, P. 2006. *Algae: Anatomy, biochemistry, and biotechnology.* London, New York: Taylor and Francis Group, Boca Raton.

Bates, S.S., Tessier, A., Campbell, P.G.C., and Buffle, J. 1982. Zinc adsorption and transport by *Chlamydomonas varuiabilis* and *Scenedsmus subspicatus* (Chlorophyceae) grown in semi continuous culture. *Journal of Phycology*, 18: 521–529.

Beer, L.L., Boyd, E.S., Peters, J.W., and Posewitz, M.C. 2009. Engineering algae for biohydrogen and biofuel production. *Current Opinion in Biotechnology*, 20: 264–271.

Ben-Amotz, A., Katz, A., and Avron, M. 1982. Accumulation of β-carotene in halotolerant alge: purification and characterization of β-carotene-rich globules from *Dunaliella bardawil (Chlorophyceae)* 1. *Journal of Phycology*, 18: 529–537.

Borowitzka, M.A. 2016. Algal physiology and large-scale outdoor cultures of microalgae. In: *The physiology of microalgae*. Cham: Springer, pp. 601–651.

Borowitzka, M.A., Huisman, J.M., and Osborn, A. 1991. Culture of the astaxanthin-producing green alga *Haematococcus pluvialis* 1. Effects of nutrients on growth and cell type. *Journal of Applied Phycology*, 3: 295–304.

Brennan, L., and Owende, P. 2010. Biofuels from microalgae: a review of technologies for production, processing, and extractions of biofuels and co-products. *Renewable and Sustainable Energy Reviews*, 14: 557–577.

Castro, Y., Ellis, J.T., Miller, C.D., and Sims, R.C. 2015. Optimization of waste water microalgae saccharification using dilute acid hydrolysis for acetone, butanol, and ethanol fermentation. *Applied Energy*, 140: 14–19.

Chae, S.R., Hwang, E.J., and Shen, H.S. 2006. Single Cell protein production of *Euglena gracilis* and carbon dioxide fixation in an innovative photo-bioreactor. *Bioresource Technology*, 97: 322–329.

Chakinala, A.G., Brilman, D.W.F., van Swaaij, W.P.M., and Kersten, S.R.A. 2010. Catalytic and non-catalytic supercritical water gasification of microalgae and glycerol. *Industrial and Engineering Chemistry Research*, 49: 1113–1122.

Chen, G., Zheng, Z., Yang, S., Fang, C., Zou, X., and Luo, Y. 2010. Experimental digestion of corn stalk and vermicompost to improve biogas production. *Waste Management*, 30: 1834–1840.

Cheng, Y.S., Zheng, Y., Labavitch, J.M., Vander heynst, J.S. 2013. Virus infection of *Chlorella variabilis* and enzymatic saccharification of algal biomass for bioethanol production. *Bioresource Technology*, 137: 326–331.

Chinnasamy, S., Ramakrishnan, B., Bhatnagar, A., and Keshav, C.D. 2009. Biomass production potential of a wastewater alga *Chlorella vulgaris* ARC 1 under elevated levels of CO_2 and temperature. *International Journal of Molecular Sciences*, 10: 518–532.

Chisti, Y. 2007. Biodiesel from Microalgae. *Biotechnology Advances*, 25: 294–306.

Choi, S., Nguyen, M.T., and Sim, S.J. 2010. Enzymatic pretreatment of *Chlamydomonas reinhardtii* biomass for ethanol production. *Bioresource Technology*, 101: 5330–5336

Collier, J.L., and Grossman, A.R. 1992. Chlorosis induced by nutrient deprivation in *Synechococcus* sp. strain PCC 7942: Not all bleaching is the same. *Journal of Bacteriology*, 174: 4718–4726.

Courchesne, N.M.D., Parisien, A., Wang, B., and Lan, C.Q. 2009. Enhancementof lipid production using biochemical, genetic and transcription factor engineering approaches. *Journal of Biotechnology*, 141: 31–41.

Daneshvar, N., Ayazloo, M., Khataee, A.R., and Pourhassan, M. 2007. Biological decolorization of dye solution containing Malachite Green by microalgae *Cosmarium* sp. *Bioresource Technology*, 98: 1176–1182.

De Godos, I., Guzman, H.O., Soto, R., García-Encina, P.A., Becares, E., and Muñoz, R. 2011. Coagulation/flocculation-based removal of algal bacterial biomass from piggery wastewater treatment. *Bioresource Technology*, 102: 923–927.

Demirbas, A. 2009. Thermochemical coversion of mosses and algae to gaseous products. *Energ Sources Part A-Recovery Utilization and Environmental Effects*, 31: 746–753.

Dhar, D.W., Prasanna, R., and Singh, B.V. 2007. Comparative Performance of three carrier based blue green algal biofertilizers for sustainable rice cultivation. *Journal of Sustainable Agriculture*, 30: 41–50

Dhull, N. 2016. *Optimization of cultivation conditions for Chlorella pyrenoidosa and bioconversion of its biomass into third generation ethanol*. Ph.D. thesis, Panjab University, Department of Microbiology, Chandigarh, India.

Dhull, N.P., Gupta, K., Soni, R., Rahi, D.K., and Soni, S.K. 2014. Statistical optimization of medium components for biomass production of *Chlorella pyrenoidosa* under autotrophic conditions and evaluation of its biochemical composition under stress conditions. *International Journal of Biological, Veterinary, Agricultural and Food Engineering*, 8: 908–915.

Doan, Q.C., Moheimani, N.R., Mastrangelo, A.J., and Lewis, D.M. 2012. Microalgal biomass for bioethanol fermentation : Implications for hypersaline systems with an industrial focus. *Biomass & Bioenergy*, 46: 79–88.

Elmore, B.B. 1987. *The effect of agitation on the high solids digestion of corn stover to methane*. M. D. thesis, University of Arkansas, Chemical Engineering, Fayetteville, AR, May.

Eriksen, N., Poulsen, B., and Lønsmann Iversen, J. 1998. Dual sparging laboratory-scale photobioreactor for continuous production of microalgae. *Journal of Applied Phycology*, 10: 377–382.

Galbe, M., and Zacchi, G. 2007. Pretreatment of lignocellulosic materials for efficient bioethanol production. *Advances in Biochemical Engineering/Biotechnology*, 108: 41–65.

Ganesh, I. 2014. Conversion of carbon dioxide into methanol-A potential liquid fuel: fundamental challenges and opportunities (a review). *Renewable & Sustainable Energy Reviews*, 31: 221–257.

Girio, F.M., Fonseca, C., Carvalheiro, F., Duarte, L., Marques, S., and Bogel Lukasik, R. 2010. Hemicelluloses for fuel ethanol: a review. *Bioresource Technology*, 101: 4775–4800.

Gouveia, L. 2011. *Microalgae as a feedstock for biofuels, in microalgae as a feedstock for biofuels*. Berlin, GDR: Springer, 1–69.

Gupta, V.K., and Rastogi, A. 2008. Biosorption of lead from aqueous solutions by green algae *Spirogyra* species: Kinetics and equilibrium studies. *Journal of Hazardous Materials*, 152: 407–414.

Hammouda, O., Gaber, A., and Abdel-Raouf, N. 1995. Microalgae and wastewater treatment. *Ecotoxicology and Environmental Safety*, 31: 205–210.

Harris, L., Tozzi, S., Wiley, P., Young, C., Richardson, T.M.J., Clark, K., and Trent, J.D. 2013. Potential impact of biofouling on the photobioreactors of the Offshore Membrane Enclosures for Growing Algae (OMEGA) system. *Bioresource Technology*, 144: 420–428.

Harun, R., Danquah, M.K., and Forde, G.M. 2010. Microalgal biomass as a fermentation feedstock for biomethanol production. *Journal of Chemical Technology & Biotechnology*, 85: 199–203.

Hsieh, C.H., and Wu, W.T. 2009. Cultivation of microalgae for oil production with a cultivation strategy of urea limitation. *Bioresource Technology*, 100:3921–3926

Ho, S.H., Huang, S-W., Chen, C-Y., Hasunuma, T., Kondo, A., and Chang, J-S. 2013. Characterization and optimization of carbohydrate production from an indigenous microalga *Chlorella vulgaris* FSP-E. *Bioresource Technology*, 135: 157–165.

Hossain, M., ZabedAkter, S., Yun, J., Zhang, G., Zhang, Y., and Qi, X. 2020. Biogas from microalgae: Technologies, challenges and opportunities. *Renewable and Sustainable Energy Reviews*, 117: 109503.

Jacinto, M.L.J.A.J., David, C.P.C., Perez, T.R., and Jesus, B.R.D. 2009. Comparative efficiency of algal biofilters in the removal of chromium and copper from wastewater. *Ecological Engineering*, 35: 856–860.

John, R.P., Anisha, G.S., Nampoothiri, K.M., and Pandey, A. 2011. Micro and macroalgal biomass: a renewable source for bioethanol. *Bioresource Technology*, 102: 186–193.

Khan, A., Singh, P., and Srivastava, A. 2020. Microbial biofuels: An economic and eco-friendly approach. In: Kumar N. (eds) *Biotechnology for biofuels: a sustainable green energy solution*. Springer, Singapore. https://doi.org/10.1007/978-981-15-3761-5_7

Khozin-Goldberg, I., Cohen, Z., Pimenta-Leibowitz, M., Nechev, J., and Zilberg, D. 2006. Feeding with arachidonic acid-rich triacylglycerols from the microalga *Parietochloris incisa* improved recovery of guppies from infection with *Tetrahymena* sp. *Aquaculture*, 255(1–4): 142–150.

Kim, K.H., Choi, I.S., Kim, H.M., Wi, S.G., and Bae, H-J. 2014. Bioethanol production from the nutrient stress-induced microalga *Chlorella vulgaris* by enzymatic hydrolysis and immobilized yeast fermentation. *Bioresource Technology*, 153: 47–54.

Kuyucak, N., and Volesky, B. 1989. The mechanism of cobalt biosorption. *Biotechnology and Bioengineering*, 33: 823–831.

Larsdotter, K., Lacourjanse, J., and Dalhammar, G. 2007. Biologically mediated phosphorus precipitation in wastewater treatment. *Environmental technology*, 28(9): 953–960.

Latorre, C., Lee, J.H., Spiller, H., and Shanmugam, K.T. 1986. Ammonium ion-excreting cyanobacterial mutant as a source of nitrogen for growth of rice: A feasibility study. *Biotechnology Letters, 8:* 507–512.

Lee, Y.K. 1997. Commercial production of microalgae in the Asia-Pacific rim. *Journal of Applied Phycology*, 9: 403–411.

Li, K., Liu, S., and Liu, X. 2014. An overview of algae bioethanol production. *International Journal of Energy Research*, 38: 965–977.

Li, Y., Horsman, M., Wu, N., Lam, C.Q., and Dubois-Calero, N. 2008. Biofuels from microalgae. *Biotechnology Progress*, 24: 815–820.

Li, Y., Zhou, W., Hu, B., Min, M., Chen, P., and Ruan, R.R. 2011. Integration of algae cultivation as biodiesel production feedstock with municipal wastewater treatment: Strains screening and significance evaluation of environmental factors. *Bioresource Technology*, 102: 10861–10867.

Lin, C.Y., and Lu, C. 2021. Development perspectives of promising lignocellulose feedstocks for production of advanced generation biofuels: A review. *Renewable and Sustainable Energy Reviews*, 136:110445.

Lin, Y., and Tanaka, S. 2006. Ethanol fermentation from biomass resources: Current state and prospects. *Applied Microbiology and Biotechnology*, 69:627–642.

Lopez-Linares, J.C., Romero, I., Cara, C., Ruiz, E., Castro, E., and Moya, M. 2014. Experimental study on ethanol production from hydrothermal pretreated rapeseed straw by simultaneous saccharification and fermentation. *Journal of Chemical Technology & Biotechnology*, 89: 104–110.

Lu, C., and Vonshak, A. 1999. Photoinhibition of outdoor *Spirulina platensis* cultures assured by polyphasic chlorophyll fluorescence transients. *Journal of Applied Phycology*, 11: 355–359.

Maity, J.P., Bundschuh, J., Chen, C.Y., and Bhattacharya, P. 2014. Microalgae for third generation biofuel production, mitigation of greenhouse gas emissions and wastewater treatment: Present and future perspectives – A mini review. *Energy*, 78: 104–113.

Menger-Krug, K.E. Hollenberg, J.N., Hillenbrand, T., and Hiessel, H. 2012. Integration of microalgae systems at municipal wastewater treatment plants: implications for energy and emission balances. *Environmental Science & Technology*, 46: 11505–11514.

Miller, T.E., Burns, J.H., Munguia, P., Walters, E.L., Kneitel, J.M., Richards, M.P., Mouqet, N., and Buckley, H.L. 2005. A critical review of twenty years' use of the resource-ratio theory. *The American Naturalist*, 165: 439–448.

Molina Grima, E. 1999. Microalgae, mass culture methods. In: Flickinger MC, Drew SW (eds.) *Encyclopedia of bioprocess technology: fermentation, biocatalysis, and bioseparation*, New York, NY: John Wiley & Sons, pp. 1753–1769.

Nigam, P., and Singh, A. 2010. Production of liquid biofuels from renewable resources. *Progress in Energy and Combustion Science*. 37: 52–68.

Oncel, M.S., Muhcu, A., Demirbas, E., and Kobya, M. 2013. A comparative study of chemical precipitation and electrocoagulation for treatment of coal acid drainage wastewater. *Journal of Environmental Chemical Engineering*, 1: 989–995.

Oslaj, M., Mursec, B., and Vindis, P. 2010. Biogas production from maize hybrids. *Biomass & Bioenergy*, 34: 1538–1545.

Osundeko, O., Davies, H., and Pittman, J.K. 2013. Oxidative stress-tolerant microalgae strains are highly efficient for biofuel feedstock production on wastewater. *Biomass & Bioenergy*, 56: 284–294.

Pan, P., Hu, C., Yang, W., Li, Y., Dong, L., Zhu, L., Tong, D., Qing, R., and Fan, Y. 2010. The direct pyrolysis and catalytic pyrolysis of *Nannochloropsis* sp residue for renewable bio-oils. *Bioresource Technology*, 101: 4593–4599.

Peng, W., Wu, Q., and Tu, P. 2000. Effects of temperature and holding time on production of renewable fuels from pyrolysis of *Chlorella protothecoides*. *Journal of Applied Phycology*, 12: 147–152.

Poulsen, B.R., and Iversen, J.J.L. 1999. Membrane sparger in bubble column, airlift, and combined membrane-ring sparger bioreactors. *Biotechnology and Bioengineering*, 64: 452–458.

Powell, E.E., and Hill, G.A. 2009. Chemical engineering research and design economic assessment of an integrated bioethanol – biodiesel – microbial fuel cell facility utilizing yeast and photosynthetic algae. *Chemical Engineering Research and Design*. 7: 1340–1348.

Prussi, M., Buffi, M., Casini, D., Chiaramonti, D., Martelli, F., Carnevale, M., Tredici, M.R., and Rodolfi, L. 2014. Experimental and numerical investigations of mixing in raceway ponds for algae cultivation. *Biomass & Bioenergy*, 67: 390–400.

Pulz, O., and Gross, W. 2010. Valuable products from biotechnology of microalgae. *Applied Microbiology and Biotechnology*, 65: 635–648.

Puspawati, S., Ainuri, M., and Nugraha, D.A. 2015. The production of bioethanol fermentation substrate from *Eucheuma cottonii* seaweed through hydrolysis by cellulase enzyme. *Agriculture and Agricultural Science Procedia*, 3: 200–205.

Rangarao, A., Sarada, R., and Ravishankar, G.A. 2007. Influence of CO_2 on Growth and hydrocarbon production in *Botryococcus branuii*. *Journal of Microbiology and Biotechnology*, 17: 414–419.

Richmond, A. 2000. Microalgal Biotechnology at the turn of the millennium-a personal view. *Journal of Applied Phycology*, 12: 441–451.

Rodolfi, L., Zittelli, G.C., Bassi, N., Padovani, G., Biondi, N., and Bonini, G. 2008. Microalgae for oil: strain selection, induction of lipid synthesis and outdoor mass cultivation in a low-cost photobioreactor. *Biotechnology and Bioengineering*, 102: 100–112.

Roncarati, A., Meluzzi, A., Acciarri, S., Tallarico, N., and Paolo, M. 2004. Fatty acid composition of different microalgae strains (*Nannochloropsis* sp. *Nannochloropsis oculata* (Droop) hibberd, *Nannochloris atomus* butcher and *Isochrysis* sp) according to the culture phase and the carbon dioxide concentration. *Journal of the World Aquaculture Society*, 35: 401–411.

Ross, A.B., Biller, P., Kubacki, M.L., Li, H., Lea-Langton, A., and Jones, J.M. 2010. Hydrothermal processing of microalgae using alkali and organic acids. *Fuel*, 89: 2234–2243.

Schenk, P.M., Thomas-Hall, S.R., Stephens, E., Marx, U.C., Mussgnug, J.H., Posten, C., Kruse, O., and Hankamer, B. 2008. Second generation biofuels: high-efficiency microalgae for biodiesel production. *BioEnergy Research*, 1: 20–43.

Schlesinger, A., Eisenstadt, D., Bar-Gil, A., Carmely, H., Einbinder, S., and Gressel, J. 2012. Inexpensive non-toxic flocculation of microalgae contradicts theories; overcoming a major hurdle to bulk algal production. *Biotechnology Advances*, 30: 1023–1030.

Schubert, C. 2006. Can biofuels finally take center stage in nature? *Biotechnology*, 24: 777–784.

Schwab, A.W., Dykstra, G.J., Selke, E., Sorenson, S.C., and Pryde, E.H. 1988. Diesel fuel from thermal decomposition of soybean oil. *Journal of the American Oil Chemists' Society*, 65: 1781–1786.

Seo, Y.H., Cho, C., Lee, J.Y., and Han, J.I. 2014. Enhancement of growth and lipid production from microalgae using fluorescent paint under the solar radiation. *Bioresource Technology*, 173: 193–197.

Shen, Y., Yuan, W., Pei, Z., and Mao, E. 2010. Heterotrophic culture of *Chlorella protothecoides* in various nitrogen sources for lipid production. *Biotechnology and Applied Biochemistry*, 160: 1674–1684.

Siaut, M., Cuine, S., Cagnon, C., Fessler, B., Nguyen, M., Carrier, P., Beyly, A., Beisson, F., Triantaphylides, C., Li-Beisson, Y.H., and Peltier, G. 2011. Oil accumulation in the model green alga *Chlamydomonas reinhardtii*: characterization, variability between common laboratory strains and relationship with starch reserves. *BMC Biotechnology*, 11:7–21.

Soni, S.K. 2007a. Microbes and biodiesel production. In: Soni SK (ed.) *Microbes: a source of energy for 21st century*, New delhi: New India Publishing House, 359–390.

Soni, S.K. 2007b. Microbes and methane production. In: Soni SK (ed.) *Microbes: a source of energy for 21st century*, New delhi: New India Publishing House, 392–422.

Subramaniam, Y., and Masron, T.A. 2020. The impact of economic globalization on biofuel in developing countries. *Energy Conversion and Management*, 100064.

Sunda, W.G., Price, N.M., and Morel, F.M. 2005. Trace metal ion buffers and their use in culture studies. *Algal Culturing Techniques*, 4: 35–63.

Tan, I.S., and Lee, K.T. 2014. Enzymatic hydrolysis and fermentation of seaweed solid wastes for bioethanol production: An optimization study. *Energy*, 78: 53–62.

Torzillo, G., Pushparaj, B., Masojidek, J., and Vonshak, A. 2003. Biological constraints in algal Biotechnology. *Biotechnology and Bioprocess Engineering*, 8: 338–348.

Tredici, M.R. 2004. Mass production of microalgae: photobioreactors. In: Richmond A (eds.) *Handbook of microalgal culture: biotechnol app phycol*. Oxford: Blackwell Science, pp. 178–214.

Trivedi, N., Reddy, C.R.K., Radulovich, R., and Jha, B. 2015. Solid state fermentation (SSF)- derived cellulase for saccharification of the green seaweed *Ulva* for bioethanol production. *Algal Research*, 9: 48–54.

United Nations. 2011. *World population prospects: the 2010 revision, highlights and advance tables*. Available at: http://esa. un. org/unpd/wpp/Documentation/pdf/WPP2010_Highlights. Pdf

Vakulchuk, R., Overland, I., and Scholten, D. 2020. Renewable energy and geopolitics: A review. *Renewable and Sustainable Energy Reviews*, 122: 109547.

Wang, H., Ji, C., Bi, S., Zhou, P., Chen, L., and Liu, T. 2014. Joint production of biodiesel and bioethanol from filamentous oleaginous microalgae *Tribonema* sp. *Bioresource Technology*, 172: 169–173.

Widjaja, A., Chien, C.C., and Ju, Y.H. 2009. Study of increasing lipid production from fresh water microalgae *Chlorella vulgaris*. *Journal of the Taiwan Institute of Chemical Engineers*, 40: 13–20.

Woertz, I., Feffer, A., Lundquist, T., and Nelson, Y. 2009. Algae grown on dairy and municipal wastewater for simultaneous nutrient removal and lipid production for biofuel feedstock. *Journal of Environmental Engineering*, 135: 1115–1122.

Wong, P.T.S., Maguire, R.J., Chau, Y.K., and Kramar, O. 1984. Uptake and accumulation of inorganic tin by alga, *Ankistrodesmus falcatus*. *Canadian Journal of Fisheries and Aquatic Sciences*, 41: 1570–1574.

Xu, N., Zhang, X., Fan, X., Han, L., and Zeng, C. 2001. Effects of nitrogen source and concentration on growth rate and fatty acid composition of *Ellipsoidion* sp. *Journal of Applied Phycology*, 13: 363–469.

Zhang, L.A., Chen, Q.A., Jin, Y.L., Xue, H.L., Guan, J.F., Wang, Z.Y., and Zhao, H. 2011. Energy saving direct ethanol production from viscosity reduction mash of sweet potato at very high gravity (VHG). *Fuel Processing Technology*, 91: 1845–1850.

Zhila, N.O., Kalacheva, G.S., and Volova, T.G. 2005. Effect of nitrogen limitation on the growth and lipid composition of green alga *Botryococcus braunii* Kutz IPPAS H-252. *Russian Journal of Plant Physiology*, 52: 311–319.

Zinoviev, S., Muller Langer, F., Das, P., Bertero, N., Fornasiero, P., and Kaltschmitt, M. 2010. Next generation biofuels: survey of emerging technologies and sustainability issues. *Chem Sus Chem*, 3: 1106–1133.

21 Biomedical Applications of Fungal Enzymes

Deepak K Rahi, Sonu Rahi, Maninderjeet Kaur and Ekta Chaudhary

CONTENTS

21.1 INTRODUCTION

The word 'enzyme' was first used by the German physiologist Wilhelm Kühne in 1878, when he was describing the ability of yeast to produce alcohol from sugars, and it is derived from the Greek words *en* (meaning 'within') and *zume* (meaning 'yeast'). Enzymes are macromolecules

DOI: 10.1201/9781003220831-21

excreted by living organisms that, when brought into a reaction or system, can help to speed up biological/biochemical reactions. They are mainly composed of long chains of amino acids which are connected by amide bonds with molecular mass ranging from kilodalton to megadalton (Singh *et al.*, 2016). They help in changing substrates into useful end products, as they give suitable conditions for reactions to occur (Lehninger *et al.*, 2004; Karigar and Rao, 2011). They are usually proteins which within mild conditions of temperature and pH carry out various chemical reactions at an amazingly high rate with remarkable efficiency and specificity. Enzymes are typically named after the reaction that they catalyse with the suffix *-ase* (e.g. oxidase, dehydrogenase, carboxylase), however some proteolytic enzymes generally have the suffix *-in* (e.g. trypsin, chymotrypsin, papain). Enzymes are also named after the substrate on which they act (e.g. glucose oxidase, alcohol dehydrogenase, pyruvate decarboxylase). The International Union of Biochemistry set up the Enzyme Commission, which provided a systematic approach to the naming of enzymes. Within this system, all enzymes are described by a four-part Enzyme Commission (EC) number. The first part of the EC number refers to the reaction that the enzyme catalyzes. The remaining digits have different meanings according to the nature of the reaction identified by the first digit (Robinson, 2015). All the known enzymes are categorized into one of the six main classes according to the type of reaction that they catalyze. The six categories of enzymes include oxidoreductases, transferases, hydrolases, lyases, isomerases, and ligases (Lehninger *et al.*, 2004). *The mechanism of action of enzymes* depends on the ability of enzymes to accelerate the reaction rate by decreasing the *activation energy*. During the course of the reaction, the enzyme (E) binds to the substrate/s (S) and forms a transient enzyme–substrate *complex* (ES). At the end of the reaction, the product/s are formed, the enzyme remains unchanged, can bind another substrate and can be reused many times. Humans recognized the importance of enzymes thousands of years ago; clarification and filtration of wines and beer being the earliest examples of the application of industrial enzymes. Enzymes have been used in brewing, baking, and alcohol production since prehistoric times; however, they did not call them enzymes. Today, enzymes are envisaged as the bread and butter of biotechnology because they are the main tools for several biotechnological techniques (gene restriction, ligation, and cloning, etc.), bioprocesses (fermentation and cell culture), and in analytics in human and animal therapy as medicines or as drug targets. Furthermore, they find applications in several other industries, such as food and feed, textiles, effluent and waste treatment, paper, tannery, baking, brewing, dairy, pharmaceuticals, confectionary, and so on. The enzymes utilized today are also found in animals (pepsin, trypsin, pancratin, and chimosin) and plants (papain, bromelain, and ficin), but most of them are of microbial origin, such as glucoamylase, α-amylase, pectinases, and so on. (Patel *et al.*, 2017).

21.1.1 Microbial Sources of Enzymes

The bio-catalytic potential of microorganisms has been employed for centuries to produce various products such as wine, bread, and so on. Microorganisms are considered to be the primary source of enzymes. The enzymes produced by them are known to be more superior enzymes, due to their active and stable nature, than the enzymes obtained from plant and animal (Anbu *et al.*, 2013; Gopinath *et al.*, 2013). Many enzymes from microbial sources (bacteria, fungi, and yeasts) have been studied worldwide and are already being used in various commercial processes. The microbial enzymes have gained interest for their widespread use due to their stability, catalytic activity, and ease of production as compared to the plant and animal enzymes. Their use in various industries, be it food, agriculture, or pharmaceutical, is increasing mainly due to their reduced processing time, cost effectiveness, and their non-toxic and eco-friendly behavior. Moreover, with recombinant technology the microbes can be easily manipulated to produce high yields of enzyme in order to meet the increasing demand (Liu *et al.*, 2013). Table 21.1 summarizes the microbial sources of some major enzymes.

TABLE 21.1
Microbial Sources of Enzymes

Enzyme	Producer Organism	References
Xylanase	*Clostridium acetobutylicum*; *Thermoascus aurantiacus*; *Paenibacillus polymyxa* CKWX1; *Streptomyces* T-7; *Cellulosimicrobium cellulans*; *Bacillus* sp. AB16; *Aspergillus* sp.; *Coriolus versicolor*; *Fusarium* sp.; *Phanerochaetae* sp.; *Trichoderma* sp.	Dhillon *et al.* (2000), Sizova *et al.* (2011), Walia *et al.* (2013), Singh (2019)
Proteases	*Bacillus subtilis*; *Bacillus licheniformis*; *Bacillus thuringiensi*; *Aspergillus* sp.; *Penicillium* sp.; *Rhizopus* sp.; *Mucor* sp.; *Humicola* sp.	Rehman *et al.* (2017)
Amylase	*Bacillus amyloliquefaciens*; *Bacillus licheniformis*; *Bacillus stearothermophilus*; *Geobacillus bacterium*; *Nesterenkonia* sp.; *Pseudoalteromonas* sp.; *Nocardiopsis aegyptia*; *Aspergillus niger*; *Aspergillus oryzae*; *Aspergillus awamori*; *Aspergillus flavus*	Abou-Elela *et al.* (2009), Prakash and Jaiswal (2010), Sundarram and Murthy (2014), Gazali and Suwastika (2018)
Cellulase	*Aspergillus niger*; *Fusarium solani*; *Humicola insolens*; *Trichoderma viride*; *Trichoderma reesei*; *Phanerochaete chrysosporium*; *Sporotrichum thermophile*; *Trametes versicolor*; *Acinetobacter junii*; *Bacillus subtilis*; *Pseudomonas cellulose*; *Thermomonospora fusca*	Kuhad *et al.* (2011)
Laccase	*Streptomyces lavendulae*; *Marinomonas mediterranea*; *Phanerochaete chrysosporium*; *Trametes versicolour*; *Monocillium indicum*; *Phlebia radiate*; *Pleurotus pulmonarius*	Palmieri *et al.* (2000), Arias *et al.* (2003), Kiiskinen *et al.* (2004)
Peroxidase	*Bacillus sphaericus*; *Pseudomonas* sp.; *Citrobacter* sp.; *Phanerochaete chrysosporium*; *Anabaena* sp.; *Candida krusei*; *Thermobifida fusca*; *Coprinopsis cinerea*	Bansal and Kanwar (2013)

21.1.2 Fungal Sources of Enzymes

Currently more than half of the industrial enzymes are derived from fungi, which have been used widely in various industrial processes and products (Kango *et al.*, 2019). Some of the well-known areas where the fungal enzymes are utilized include pulp and paper, textiles, detergent, food, nutraceuticals, and therapeutics. The total species of fungi which exist on earth are estimated to be 5.1 million, based on the high-throughput sequencing methods (Blackwell, 2011), of which only a handful are known to humankind and have been explored for the production of various industrial enzymes. Of all the known species *Aspergillus* belonging to the division Ascomycota is the most exploited one. The ascomycete phylum has a large potential for the production of industrially important enzymes such as laccases, amylases, proteases, cellulases, xylanases etc. Although numerous enzymatic profiles have already been unraveled, the research has been covering only a limited number of species and genera, thus leaving many ascomycetes to be analyzed (Hansen *et al.*, 2015). Another group of fungi that is largely explored for their enzyme producing potential are the basidiomycetes. Fungi are the key decomposers of plant polymers like cellulose, hemicelluloses, and lignin. The foremost rapid degraders in this group are basidiomycetes (Sánchez, 2009), which secrete greater amounts of lignocellulolytic enzymes and convey about the degradation of complex macromolecules like cellulose, hemicellulose, chitin, lignin, lipid, and protein within the substrates (Kuforiji and Fasidi, 2008). Wood-rotting basidiomycetes fungi are usually divided into white-rot, brown-rot, and litter decomposing fungi. White-rot basidiomycetes fungi grow in clusters on dead wood throughout the year. Most of the white-rot fungi are wood rotting basidiomycetes, are capable of decaying lignin-based polymers and several other variable recalcitrant environmental pollutants because of its inherent ability to secrete cellulolytic and ligninolytic enzymes like lignin peroxidase, manganese peroxidase and laccases, which aid the decaying process of dyes and inorganic pollutants. Lignin peroxidase, polyphenol oxidase, and laccases produced by white-rot fungi degrade many aromatic compounds (Singh and Pakshirajan, 2010). Lignocellulolytic fungi actively secrete different sets of oxidative and hydrolytic enzymes along with low molecular-mass effectors that are employed within the discoloration and detoxification of wastewaters, in bio-pulping and bio-bleaching of paper pulps, and within the assembly of bio-based chemicals (Ruiz-Dueñas and Martínez, 2009). Table 21.2 summarizes the various enzymes produced by two major groups of fungi: Ascomycetes and Basidiomycetes and their applications.

TABLE 21.2
Enzyme Producing Ascomycetes and Basidiomycetes Fungi and Their Applications

Fungal Species	Enzyme Produced	Reference
Ascomycetes		
Aspergillus niger	Cellulases, exo-β-1,4-glucan cellobiohydrolase, hemicellulases, Glucose oxidase, Catalse, Lipase, Pectinase	Cianchetta *et al.* (2012), Singhania *et al.* (2013)
Aspergillus oryzae	Protease, Lipase, Glucoamylase, alpha-amylase	De Castro and Sato (2014)
Aspergillus fumigatus	Exo-glucanases, Xylanase, Amylase	Mahmood *et al.* (2013), Sarkar and Aikat (2014)
Fusraium spp.	Pectinases, Amylases, Beta-glucosidases and Cellulases	Kwon *et al.* (2007)
Myceliophthora thermophile	Laccase, Cellulase	Østergaard and Olsen (2011)
Penicillium simplicissimum	Cellulase and Xylanase	Østergaard and Olsen (2011)
Talaromyces pinophilus	Cellulases, exo-β-1,4-glucan cellobiohydrolase, hemicellulases, Amuloglucosidase	Krogh and Symington (2004), Cianchetta *et al.* (2012), Singhania *et al.* (2013)
Trichoderma viride	Hemicellulase, Protease, Xylanase, Beta-glucanase	Haltrich *et al.* (1996)
Trichoderma reesei	Cellulases, exo-β-1,4-glucan cellobiohydrolase, hemicellulases	Krogh and Symington (2004), Cianchetta *et al.* (2012), Singhania *et al.* (2013)
Basidiomycetes		
Cerrena unicolor	Laccase, Manganese peroxidase,	Michniewicz *et al.* (2003)
Dichomitus squalens	Laccase, Manganese peroxidase	Arora and Gill (2001)
Fomes fomentarius	Laccase	Arora and Sharma (2010), Forootanfar and Faramarzi (2015)
Ganoderma lucidium	Laccase, Manganese peroxidase	Postemsky *et al.* (2017)
Irpex lacteus	Laccase, Manganese peroxidase	Kasinath *et al.* (2003)
Phlebia radiate	Laccase, Manganese peroxidase, Lignin peroxidase	Arora and Sharma (2010), Forootanfar and Faramarzi (2015)
Pleurotus eryngii	Laccase, Peptidase, Lipase	Camarero *et al.* (2004)
Pleurotus oryzae	Laccase, Manganese peroxidase	Carunchio *et al.* (2001)
Pleurotus ostreatus	Laccase, Manganese peroxidase	Hou *et al.* (2004), Palmieri *et al.* (2005)
Phanerochaete chrysosporium	Lignin peroxidase	Kang *et al.* (2004)
Trametes versicolor	Laccase, Lignin peroxidase, Cellulase, Xylanase	Claus *et al.* (2002), Camarero *et al.* (2004), Sun *et al.* (2004)
Trametes villosa	Laccase, Manganese peroxidase	Baborová *et al.* (2006), Arora and Sharma (2010), Forootanfar and Faramarzi (2015)

21.2 LIGNOCELLULOLYTIC ENZYMES

Lignocellulolytic enzymes are biocatalysts engaged with the breakdown of lignin and cellulosic materials into their constituents for further hydrolysis into useful products. Sometimes called lignocellulases, they have hydrolytic enzymes that degrade recalcitrant lignocellulose, a component of plant biomass. Studies show that lignocellulolytic enzymes can be considered as a large group of mainly extracellular proteins, which include ligninolytic enzymes (peroxidases and oxidases) and hydrolytic enzymes (cellulases, hemicellulases, pectinases, chitinases, amylases, proteases, esterases, and mannanases) (Mtui, 2012). Figure 21.1 presents various lignocellulolytic enzymes produced by fungi.

21.2.1 Sources of Lignocellulolytic Enzymes

21.2.1.1 Fungi

White-rot fungi, having their capability to produce lignocellulolytic enzymes, are significant resources in enzyme technology. A few investigations have uncovered their role in the advancement of degradation of lignin, when compared with different living beings. However, despite the mineralization of lignin to CO_2 and H_2O, white-rot fungi are unfit for use as a carbon and fuel source. But, they acquire energy from metabolism of cellulose (Arora and Sharma, 2010). The phyla Ascomycota and Basidiomycota have very crucial species of wood-decaying fungi. Their characteristic name is initially derived from their capability to decay wood and making decayed wood appear white or yellow. Their capacity to oxidize the recalcitrant lignin substrates implies their significance in enzyme technology. Lignin degradation by white-rot fungi is reliant on a ligninolytic complex, containing manganese peroxidase, lignin peroxidase, laccase, and enzymes associated with generation of hydrogen peroxide (i.e., glucose oxidase) (Deacon, 2006). Wood rot is brought about by some fungal species, in light of their capacity to degrade lignin. The species that hydrolyze lignin are called white rot, because of their capacity to lighten the color of rotted wood. Brown-rot fungi hydrolyze cellulose and hemicelluloses, but no lignin. Also, some soft-rot fungi degrade lignin in angiosperm wood (Sánchez, 2009). Among the wood-rot fungi, white-rot fungi are the most notable organisms and have been continually investigated for mycoremediation, due their ability for degradation of complex chemical compounds. Organisms including *Ceriporiopsis subvermispora, Coriolus versicolor, Dichomitus squalens, Irpex lacteus, Phanerochaete chrysosporium, Phanerochaete carnosa, Phlebiopsis gigantean, Pleurotus ostreatus,* and *Trametes versicolor* are some examples of white-rot fungi, which have discovered applications in biotechnological research (Dashtban *et al.*, 2009; Vaithanomsat *et al.*, 2010; Couturier *et al.*, 2015; Montoya *et al.*, 2015).

21.2.1.2 Bacteria

A number of bacteria biodegrade lignocellulose (Bandounas *et al.*, 2011). Members from the family *Bacillus* have been accounted to have either ligninolytic or cellulolytic activities or both. These include *B. pumilus, B. brevis, B. subtilis, B. cereus, B. licheniformis, B. circulans, B. halodurans, B. tequilensis, and B. polymyxa* (Bandounas *et al.*, 2011; Dubey *et al.*, 2014; Prasad *et al.*, 2014; Behera *et al.*, 2014). *Klebsiella variicola* has lignocellulolytic activities and can use different sorts of sugars (Jiménez *et al.*, 2014; Dantur *et al.*, 2015); *Acinetobacter calcoaceticus* furthermore, other Acinetobacter, use lignin as carbon sources (Ghodake *et al.*, 2009; Ponnambalam *et al.*, 2011), *A. calcoaceticus* (*A. pittii*), *A. baumannii, K. variicola, Chryseobacterium gleum, Bacillus* spp., and *Stenotrophomonas maltophilia* have also been found to degrade both lignin and cellulose (Ahmed *et al.*, 2018).

21.2.1.3 Actinomycetes

Actinomycetes have a place with order Actinomycetales, a different taxonomic group inside domain bacteria (Chaudhary *et al.*, 2013). They are Gram-positive microscopic organisms, basically oxygen

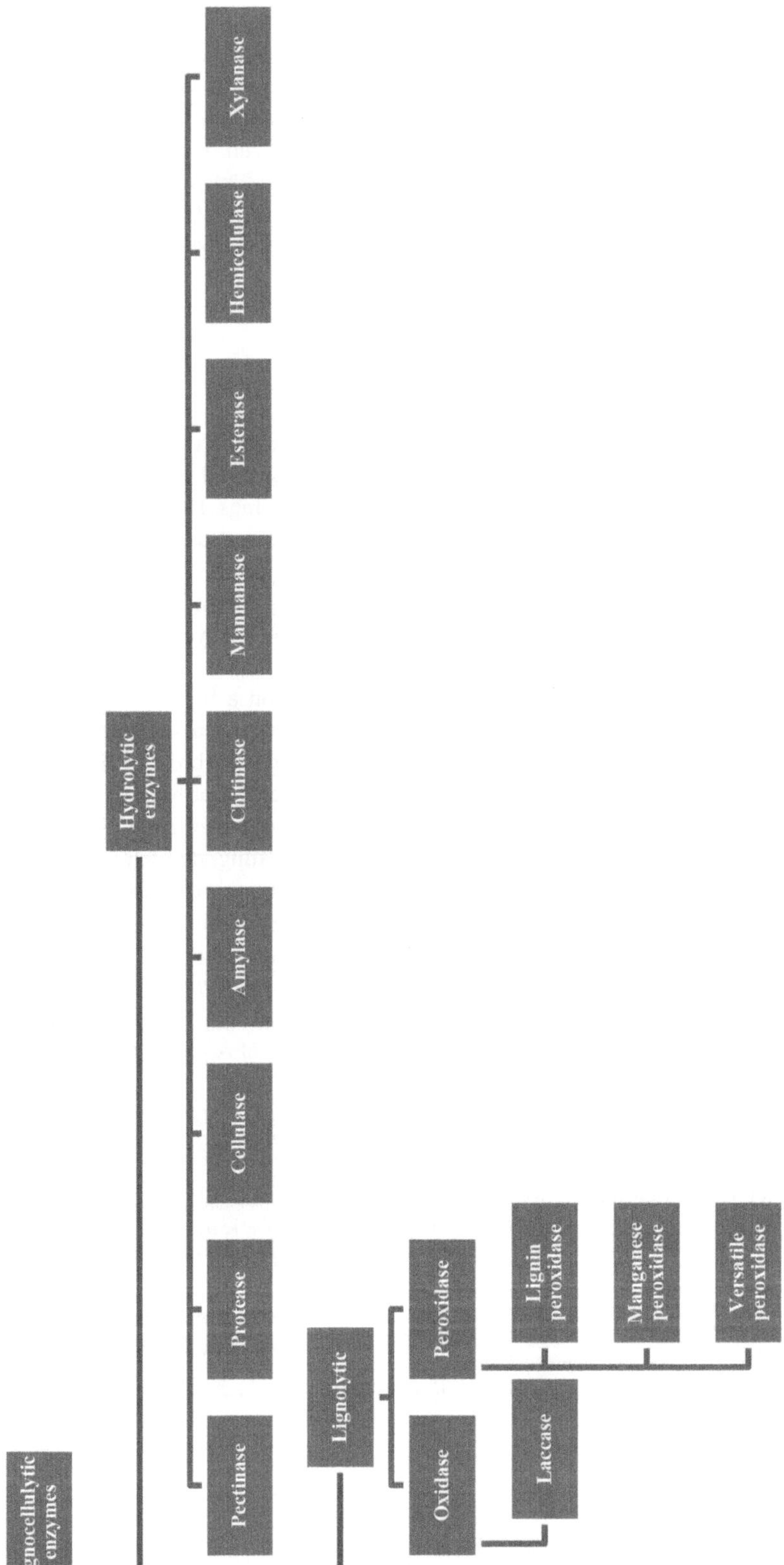

FIGURE 21.1 Various lignocellulolytic enzymes grouped according to their enzymatic systems.

consuming and sporulating (Jeffrey, 2008). They have filamentous development and produce airborne or substrate mycelium, features they share with fungi. Being both mesophilic and thermophilic (Wilson, 1992), they have a wide scope of living habitats and are universal in nature, present in both amphibian territories, for example, mangroves and ocean residue, just as terrestrial environments (Veiga *et al.*, 1983; Chaudhary *et al.*, 2013). Actinomycetes have been accounted to produce cellulases, hemicellulases, and lignolytic enzymes. Hyper cellulase activity has been reported by various investigations (Kluepfel *et al.*, 1986; Saini *et al.*, 2015). Hemicellulase activity, chiefly xylanase in actinomycetes, has been the most broadly studied over the most recent forty years, particularly in streptomyces (McCarthy *et al.*, 1985; Kluepfel *et al.*, 1986; Morosoli *et al.*, 1986; Ball and McCarthy, 1988; Beg *et al.*, 2000; Rifaat *et al.*, 2005; Maryandani, 2007; Bhosale *et al.*, 2011).

21.3 BIOMEDICAL APPLICATIONS OF FUNGAL ENZYMES

The field involving the application of enzymes is expanding every day. We now find that enzymes are used in many branches of our economy, be it food, textiles, pulp and paper, electronics, medicine, or chemistry. The recent growth in the industrial and medical applications of enzymes has resulted from greatly improved methods for their production, especially from the microbial sources. The processing of enzymes for their biomedical use is now an important aspect of today's pharmaceutical industry. Nowadays enzymes are being used routinely in the medical field, such as in the diagnosis and treatment of various diseases. For example, the immobilized antibodies, receptors, or enzymes are used in biosensors; enzymes are also used in bioreactors for the removal of waste metabolites and correction of inborn metabolic deficiency. The biomedical applications of various enzymes produced by fungi are described in the following section.

21.4 BIOMEDICAL APPLICATIONS OF ENZYMES

21.4.1 Xylanase

The enzyme xylanase has recently been used widely for the production of xylooligosaccharides (XOs). The action of xylanase results in the formation of xylose, arabinose, and xylooligosaccharides, which are known to possess many beneficial biomedical and health benefits (Yang *et al.*, 2005).

The xylooligosaccharides produced by the action of xylanase have found applicability in various fields like pharmaceuticals, feed formulations, agricultural purposes, and food applications (Vazquez *et al.*, 2000). As a food additive the XOs are known to possess prebiotic activity since they are found to be capable of improving the intestinal function by increasing the number of healthy *Bifidobacteria* (Rycroft *et al.*, 2001; Fooks and Gibson, 2002; Izumi and Kojo, 2003). These XOs have an acceptable odor, are noncarcinogenic, and are also known to reduce the risk of colon cancer (Whitehead and Cotta, 2001). Some studies have also reported the low-calorific value of these XOs, thus making them suitable for use in anti-obesity diet products (Taeko *et al.*, 1998; Toshio *et al.*, 1990).

21.4.2 Cellulases and Hemicellulase

A bezoar is a mass discovered in the gastrointestinal system. Phytobezoars, as its name recommends, are made out of indigestible plant material (e.g., cellulose). In other words, it is a gastric solidification made by vegetable fibers, seeds, and skins of fruits, and at times starch granules and fat globules caught inside the gastrointestinal tract. This is a typical issue frequently found in patients with disturbed digestion and decreased gastric motility. Sometimes medical procedures are needed to eliminate these substances; certain minor conditions can be treated by cellulases. The normal application reported is with the use of cellulases. As there are no such reports about use of cellulases, research can be directed to discover the use of potential cellulolytic bacterial cellulases to treat this condition. However, what is most significant is that any of these individual enzymes or enzyme mixes should not unfavorably influence healthy body cells. Another direct use of cellulase in medicine is degradation of cell walls of pathogenic organisms. *Acanthamoeba* is a Protista that causes keratitis, an uncommon as well as serious corneal disease that prompts visual deficiency. This amoebic keratitis is an intense sight-threatening corneal infection related with contact lens misuse (Rinaudo, 2006). This has two phases in its life cycle: the blister and trophozoite. Both of these are found when the eye is affected by this organism. The cyst wall is similar to a plant cell wall. It is conceivable that amoebae emit cellulose since its cell wall is fundamentally made out of cellulose. There is proof that has been found to demonstrate this matter about the presence of cellulose (Pillai *et al.*, 2009). Consequently, it is possible to apply cellulases in controlling the microbe that causes this infection. Cellulases can be utilized to break the cell wall and control the microbe. But, it needs a ton of trials prior to utilizing cellulase as a treatment for the eye.

Another potential application of cellulase is their role in the degradation of biofilms. A biofilm is a group of microbial cells that is irreversibly related (not eliminated by delicate rinsing) with a surface and is encased in an extracellular polymeric substance (EPS) framework. Typically, pathogenic microorganisms form this sort of assemblage. They might be found on a wide scope of surfaces, including living tissues and inhabiting clinical gadgets. Artificial hip prosthesis, focal venous catheter, prosthetic heart valve, intrauterine gadget, and urinary catheter are a few examples for inhabiting clinical gadgets that are usually connected with biofilms. The vast majority of the microorganisms produce extracellular polymeric substances composed of backbone structures that contain 1,3- or 1,4-β-connected hexose residues and will in general be more rigid, less deformable, and, in specific cases, poorly soluble. This is the specific linkage present in cellulose polymer. Therefore, cellulases can be further studied for their effective application in eliminating this type of biofilm from medical devices.

21.4.3 Esterase

21.4.3.1 Application in Tumor and Disease Treatment

Different deadly types of cancer are known, and lung cancer is one of them. Recent chemotherapeutic techniques against lung cancer need expansive specificity and effectiveness. A biocompatible and biodegradable poly (ethylene glycol)-b-poly (D, L-26 lactic corrosive) (PEG-b-PLA) micelle-conveyed β-lap-dC3 and 28-dC6 prodrugs nanotherapeutics were changed over to β-lapachone (β-lap) utilizing porcine liver esterase (PLE). The β-lapachone prodrug showed antitumor adequacy and long-term endurance utilizing cytotoxicity tests on A549 and H596 lung cancer (Ma *et al.*, 2014). In a neural stem cells (NSCs)-derived enzyme/prodrug treatment (NDEPT), antitumor impact of NSCs communicating carboxylesterase (CE) was demonstrated to treat primary lung cancer or metastatic lung cancers in the brain (Yi *et al.*, 2014). Neural stem cells (NSCs) communicating Hare carboxylesterase (F3. CE) could repress the development of A549 human non-small-cell lung adenocarcinoma cells in vitro and in vivo, and accordingly delivering therapeutic genes to brain tumors has been considered as a powerful therapy for cellular breakdown in the lungs brain metastases (Hong *et al.*, 2013).

21.4.3.2 Application in Biosensor Creation

In the development of a cholesterol-based biosensor, cholesterol esterase related to cholesterol oxidase and peroxidase was immobilized onto polyaniline films and has been utilized as detecting components in assessment of cholesterol fixation and brought about improved time span of usability of biosensor terminals (Singh *et al.*, 2006). The assessment of metabolites like glucose, urea, lactate, and cholesterol in the blood test is significant in clinical diagnostics and avoidance of various clinical problems like hypertension, arteriosclerosis, cerebral apoplexy, and coronary heart sicknesses

(Zicha *et al.*, 1999; Forster *et al.*, 2000). Recently, highly delicate fluorogenic esterase tests from the far-red fluorophore 7-hydroxy-9H- (1, 3 dichloro-9, 9-dimethylacridin-2-one) (DDAO) have been utilized to recognize low picogram measures of PLE movement in mycobacterial lysates during various phases of tuberculosis disease (Tallman and Beatty, 2015). Mycobacterium tuberculosis esterases have an unmistakable part in pathogenesis (Singh *et al.*, 2010) and it has been hard to portray them in light of the fact that most structure incorporation bodies in heterologous hosts (Deb *et al.*, 2006). In a new report, esterase has been utilized with ferrocene-covered gold nanoparticle for cholesterol discovery in blood tests (Davis-Lorton, 2015).

21.4.3.3 HAE Treatment

HAE is an autosomal dominant disease, described by episodes of painful abundant liquid development, normally influencing the bowels, face and upper airway, body trunk, and genitalia. HAE is caused by an insufficiency in C1 esterase inhibitor (C1-INH) without any distinctions in prevalence dependent on sex or race (Bork, 2014). The management of HAE may incorporate treatment for acute attacks, or short- or long-haul prophylaxis for anticipation of HAE attacks (Pham *et al.*, 2014). One account of a patient with HAE that included hypovolemic shock, ascites, extreme sepsis from nosocomial pneumonia, renal and respiratory disappointment was given a trial of everyday intravenous mixture of Human plasma-inferred C1 esterase inhibitor (pdC1-INH) concentrate; his clinical status was discovered to be improved, especially renal capacity (Cowan *et al.*, 2001). Subsequently, human plasma-inferred C1 esterase inhibitor (pdC1-INH) is shown for the therapy of both HAE attacks and counteraction of HAE episodes in patients going through clinical treatment.

21.4.3.4 Application in Gene Treatment

Gene therapy includes replacing a defective gene in the body with a sound one. This should be possible by eliminating cells from the body utilizing genetic engineering procedures to change defective sequences in the DNA, at that point reinserting the cells instead of utilizing medications or medical procedure. A substitution treatment has additionally been intended for raising serum levels of PON1 paraoxonase/aryl esterase chemicals by utilizing a gene delivery vector for hydrolytic destruction of organophosphate before it can enter the brain and cause poisonousness (Gordon *et al.*, 1981).

21.4.3.5 Application in Treatment of Hypercholesterolemia

Hypercholesterolemia is portrayed by extremely higher degrees of cholesterol in the blood and it is a set up hazard factor for atherosclerosis and coronary illness (CHD) in people (Heidrich *et al.*, 2004). In a report while examining the activity of cholesterol esterase it was discovered that cholesterol esterase focused on inhibitors could be valuable therapeutics for restricting cholesterol assimilation (Ellidag *et al.*, 2013).

21.4.4 Mannanase

Mannooligosaccharides (MOS) are a sort of prebiotics and have been accounted for to help the development of gainful gut microbiota like *Lactobacilli* and *Bifidobacteria*. These microscopic organisms or probiotics have numerous gainful capacities. Different scientists have contemplated MOS creation by hydrolytic activity of β-mannanases from various microorganisms (*Pradeep et al.*, 2016; Yopi *et al.*, 2017; Bååth *et al.*, 2018). Mannooligosaccharides can be utilized as anti-tumor agents. Colon cancer in Western nations is the second-most basic cancer type (Fernández *et al.*, 2015). The utilization of prebiotics could prompt more positive outcomes on the well-being compared with commercial anti-cancer agents. Prebiotics have been accounted to have anti-cancer properties and detoxify gastrointestinal genotoxins. Ongoing investigations have shown that prebiotics, for example, MOS, have amazing anti-tumor properties. In vitro cytotoxicity examine of MOS (500 μg/mL) on human epithelial colorectal adenocarcinoma cell line (HT-29) shows, after

48 hours, a 60% decrease in the viability of cells (Ghosh *et al.*, 2015). Thus, for colon cancer treatment, MOS can be acceptable therapeutics specialists. Mannooligosaccharides can be utilized as weight-reduction agents. Clinical investigations have shown that MOS utilization may lessen body weight and consequently can be utilized as a guide to a weight reduction routine. Salinardi *et al.* (2010) report that in overweight men, utilization of MOS-containing drink as a feature of a weight keeping up diet decreases absolute body volume and body weight. Also, St-Onge *et al.* (2012) have additionally seen that adding MOS to a weight-reduction diet can speed up fat tissue and weight reduction in men, proposing a possible potential function of MOS in fat tissue dispersion and weight management. However, more exploration is important to additionally comprehend the system by which MOS improves the composition of the body and to explain the impact of gender.

21.4.5 Chitinase

Chitinases are used along with antifungal medications to expand the productivity in treating the parasitic diseases. Chitinases are utilized as added substances in skin moisturizers and antifungal creams (Dahiya *et al.*, 2006). The antifungal action and highly biocompatible quality make chitinase and its derivatives especially valuable for biomedical applications, like injury healing, drug delivery, ligament tissue designing, and nerve generation (Manivasagan *et al.*, 2014). Chitin and chitosan are utilized as film for drug delivery and furthermore in tissue designing. This expands the need of exceptionally purified enzymes. Chitinases are also utilized in anti-cancer treatment.

21.4.5.1 Allergic Inflammation

In spite of the fact that vertebrates can't integrate or process chitin, various genuine chitinolytic compounds like chitotriosidases, acidic mammalian chitinase (AMCase), or chitin-restricting proteins or chitinase-like proteins (CLPs), for example, Ym-1, Ym-2, and bosom relapse protein 39 were found in vertebrates. Subsequently, regardless of whether the chitotriosidase in people has similar immunologic movement in unfavorably susceptible reactions as AMCase in mouse is a fascinating inquiry that should be researched in additional examinations (Lee, 2009).

21.4.6 Amylase

α-amylases would be possibly helpful in the pharmaceutical and fine chemical compounds industries if enzymes with appropriate properties could be prepared (Fogarty and Kelly, 1980). Curiously, the first enzyme created industrially was an amylase from a fungal source in 1894, which was utilized as a drug for the treatment of stomach-related problems (Pandey *et al.*, 2000). Engineered and normal biodegradable polymers have been a significant focal point of interest in drug research. The biodegradable polymers are utilized to control the drug release rate from parenteral-controlled delivery systems (Dumoulina *et al.*, 1999). For drugs with limited solubility or for certain drugs for which dissolvability can be affected by variety of gastrointestinal pH, a system is needed to speed up drug release. Polysaccharide biodegradable frameworks are of interest since the degradation of a characteristic item like starch happens normally in the human body (Kost and Shefar, 1990). Alpha-starch (pregelatinized starch) and cross-linked starch have been utilized as hydrogels. It has been reported that increasing the level of cross-linking of starch decreases the drug-release rate. The delivery rate is also reliant on α-amylase action contained in the dissolution media. The addition of α-amylase to cross-linked amylose (CLA) tablets can modulate the delivery action of drugs (Dumoulina *et al.*, 1999).

21.4.7 Protease

Proteases are an expanding class of drugs that have found wide applicability in the medical field, and some of the protease-based therapies have already been approved by U.S.F.D.A. In medication, various equations, like dressing, nonwoven tissues, and ointment composition containing alkaline proteases delivered by *B. subtilis* show promising therapeutic properties (Sen *et al.*, 2011). Certain lytic enzyme deficiencies are analyzed to be supported by an oral administration of alkaline proteases (Gupta and Khare, 2007; Joshi and Satyanarayana, 2013). It has been accounted for that fibrin degradation has been accomplished by alkaline fibrinolytic proteases. The utilization of this fibrinolytic enzyme suggests its future application as an anti-cancer drug and in thrombolytic treatment (Jaouadi *et al.*, 2011, 2012). Slow delivery dose form preparation containing collagenases with alkaline proteases is broadly utilized in therapeutic applications. The hydrolysis of collagen by the enzyme liberates low-atomic weight peptides with no amino acid delivery for therapeutic use (Romsomsa *et al.*, 2010; Suwannaphan *et al.*, 2017). For the treatment of different diseases, like burns, carbuncles, furuncles, and wounds, a preparation of elastoterase immobilized on a bandage is utilized (Davidenko, 1999; Palanivel *et al.*, 2013).

21.4.8 Pectinase

21.4.8.1 Tablet Formulation

In tablet formulations, presently, gelatin hydrogels are generally utilized as binding agents that go about as a controlled delivering agent, which are mostly utilized in colon cancer treatment medicines. Hydrolyzed pectin beads are made by adopting the ionotropic gelation strategy and are utilized as a supported drug delivery system. The modulation of low-methoxy gelatin by esterification or execution of calcium pectinate gel beads in formulations likewise upholds the administration of a drug. Fundamentally, calcium pectinate is utilized in light of the improvement of insoluble hydrophilic coatings that further connect with one another in the host cell and helps in the smooth arrival of focused medication (Wong *et al.*, 2011; Butte *et al.*, 2014). Besides, a report also recommended the viability and controlled release of oral plans with compacting the pectic substances that were obtained from orange and mango strips. These are having an amazing contact with useful polymers and have remarkable restricting properties during the formulation of different tablets (Liu *et al.*, 2007; Murata *et al.*, 2019).

21.4.9 Laccase

21.4.9.1 Laccase as a Virulence Factor and a Marker of Fitness

The capacity to produce melanin pigments was emphatically related with virulence in early investigations of cryptococcal infection. The principal hereditary proof that laccase was significant came from studies in which the gene encoding the laccase was deleted in a strain of *Cryptococcus neoformans* var. *neoformans*. Laccase has been related with the production of immunomodulatory catecholamines from brain, just as protection from antifungals and oxidative products of macrophages. Furthermore, laccase action has been discovered to be a marker of stress, as acceptance of laccase connected with substrate starvation and the presence of possibly poisonous metals. It might be because of an extra capacity of laccase to act as a hereditary marker for cellular processes that are significant for virulence (Panepinto and Williamson, 2006).

21.4.9.2 Herbal Treatment

Laccase is produced from shellfish mushroom (*Pleurotus ostreatus*) and is being utilized as a natural medication to hinder the HCV (hepatitis C infection) replication rate (Munir *et al.*, 2010).

21.4.10 Lignin Peroxidases (LiPs)

Nowadays, different uses of lignin peroxidase (LiP) have seen dramatic increase in use in numerous areas. Woo *et al.* (2004) considered LiP purified from *P. chrysosporium* that was utilized in the decolorization of melanin (engineered), and furthermore its utilization in the use of new beauty lightening products. It exhibited skin-lightening results compared to that of hydroquinone, without an observable antagonistic outcome, and with prevalence in pores and skin surface and harshness (Draelos, 2015). The ability of LiP enyzmes to oxidize a tremendous range of basically particular substrates makes them a suitable candidate for the oxidation of melanin that is fundamentally similar as lignin. This enzyme with melanocytic capacity has the potential for utility inside the cosmetic manufacturing industries (Falade *et al.*, 2017). Depigmentation and pores and skin-lightening products, which have been being used for quite a while in Asian countries where skin whiteness is the main esthetic model, are particularly esteemed by the Western populace, who endure sun based skin harm with extended pores and skin spots (Couteau and Coiffard, 2016).

21.4.11 Versatile Peroxidases (VPs)

Endocrine Disrupting Chemicals (EDCs) are a group of anthropogenic or engineered synthetic mixtures present in the environment that cause unfavorable impacts on the well-being and behavior of animals and furthermore on other living organisms by changing the elements of the endocrine system. These synthetic compounds include bisphenol, triclosan, and so forth, which are unsafe even at small amounts. The regular strategies for treatment are not powerful in complete removal of these EDCs from the environment. Recently, the ability of versatile peroxidase to degrade the EDCs and decrease in activity was illustrated. Versatile peroxidase is exceptionally liked in interruption of these synthetic compounds inferable from their high redox potential and autonomy of redox arbiters. Two phase enzymatic framework for evacuation of EDCs at high and at natural focuses was considered and answered to be effective (Taboada-Puig *et al.*, 2015).

21.5 CONCLUSION AND FUTURE PROSPECTS

In the past few decades, significant progress has been made to understand the enzymes produced by fungi and their enormous industrial potential. The fungi are a major source of various biotechnologically valued enzymes, like cellulases, amylases, peroxidases, xylanases, laccases, and peroxidases. The enzymes produced by fungi have a broad substrate specificity and are functionally more effective compared to their synthetic counterparts, thus leading to the development of sustainable technologies. Their vast diversity, specific range of action, and property of being active over a very wide range of temperature and pH have attracted the attention of biotechnologists worldwide. In addition, the use of enzymes in industry and medicine is highly ethical and socially desirable, which represents an advance in modern technological processes. The science of using enzymes has evolved many folds, which has resulted in opportunities for new applications in many industries. The future of fungal enzymes in the biomedical field is very promising since they have successfully replaced many conventional chemical-catalyst-based processes, for example, as enzyme therapy for the treatment of various metabolic disorders and genetic diseases such as SCID, Gaucher's disease, Parkinson's disease, PKU, and so on. Recently enzymes have found their applicability in cancer treatment and treatment of infectious diseases, where antibiotics are no longer useful. However, despite the current progress, there is still ample scope for fungal enzymes in the biomedical sector. A large proportion of fungal enzymes are still to be harnessed, and extensive research is thus required in this field to fully utilize their potential.

BIBLIOGRAPHY

Abou-Elela, G M, El-Sersy, N A, & Wefky, S H (2009). Statistical optimization of cold adapted α-amylase production by free and immobilized cells of Nocardiopsis aegyptia. *Journal of Applied Sciences Research*, 5(3): 286–292.

Ahmed, M A, Sanaullah, M, Blagodatskaya, E, Mason-Jones, K, Jawad, H, Kuzyakov, Y, & Dippold, M A (2018). Soil microorganisms exhibit enzymatic and priming response to root mucilage under drought. *Soil Biology and Biochemistry*, 116: 410–418.

Anbu, P, Gopinath, S C, Cihan, A C, & Chaulagain, B P (2013). Microbial enzymes and their applications in industries and medicine. *BioMed Research International,* 2017: 1–14.

Arias, M E, Arenas, M, Rodríguez, J, Soliveri, J, Ball, A S, & Hernández, M (2003). Kraft pulp biobleaching and mediated oxidation of a nonphenolic substrate by laccase from Streptomyces cyaneus CECT 3335. *Applied and Environmental Microbiology*, 69(4): 1953–1958.

Arora, D S, & Gill, P K (2001). Effects of various media and supplements on laccase production by some white rot fungi. *Bioresource Technology*, 77(1): 89–91.

Arora, D S, & Sharma, R K (2010). Ligninolytic fungal laccases and their biotechnological applications. *Applied Biochemistry and Biotechnology*, 160(6): 1760–1788.

Bååth, J A, Martínez-Abad, A, Berglund, J, Larsbrink, J, Vilaplana, F, & Olsson, L (2018). Mannanase hydrolysis of spruce galactoglucomannan focusing on the influence of acetylation on enzymatic mannan degradation. *Biotechnology for Biofuels*, 11: 1–15.

Baborová, P, Möder, M, Baldrian, P, Cajthamlová, K, & Cajthaml, T (2006). Purification of a new manganese peroxidase of the white-rot fungus Irpex lacteus, and degradation of polycyclic aromatic hydrocarbons by the enzyme. *Research in Microbiology*, 157(3): 248–253.

Ball, A S, & McCarthy, A J (1988). Sacchariiication of straw by actinomycete enzymes. *Microbiology*, 134(8): 2139–2147.

Bandounas, L, Wierckx, N J, de Winde, J H, & Ruijssenaars, H J (2011). Isolation and characterization of novel bacterial strains exhibiting ligninolytic potential. *BMC Biotechnology*, 11(1): 1–11.

Bansal, N, & Kanwar, S S (2013). Peroxidase (s) in environment protection. *The Scientific World Journal*, 2013: 1–10.

Beg, Q K, Bhushan, B, Kapoor, M, & Hoondal, G S (2000). Production and characterization of thermostable xylanase and pectinase from Streptomyces sp. QG-11-3. *Journal of Industrial Microbiology and Biotechnology*, 24(6): 396–402.

Behera, S, Arora, R, Nandhagopal, N, & Kumar, S (2014). Importance of chemical pretreatment for bioconversion of lignocellulosic biomass. *Renewable and Sustainable Energy Reviews*, 36: 91–106.

Bhosale, H J, Sukalkar, S R, Uzma, S M Z, & Kadam, T A (2011). Production of xylanase by Streptomyces rameus grown on agricultural wastes. *Biotechnol Bioinf Bioeng*, 1(4): 505–512.

Blackwell, M (2011). The Fungi: 1, 2, 3… 5.1 million species? *American Journal of Botany*, 98(3): 426–438.

Bork, K (2014). Pasteurized and nanofiltered, plasma-derived C1 esterase inhibitor concentrate for the treatment of hereditary angioedema. *Immunotherapy*, 6: 533–551.

Butte, K, Momin, M, & Deshmukh, H (2014). Optimisation and *in vivo* evaluation of pectin based drug delivery system containing curcumin for colon. *International Journal of Biomaterials*: 924278. doi: 10.1155/2014/924278

Camarero, S, Garcıa, O, Vidal, T, Colom, J, del Rıo, J C, Gutiérrez, A, . . . & Martınez, Á T (2004). Efficient bleaching of non-wood high-quality paper pulp using laccase-mediator system. *Enzyme and Microbial Technology*, 35(2–3): 113–120.

Carunchio, F, Crescenzi, C, Girelli, A M, Messina, A, & Tarola, A M (2001). Oxidation of ferulic acid by laccase: identification of the products and inhibitory effects of some dipeptides. *Talanta*, 55(1): 189–200.

Chaudhary, H S, Soni, B, Shrivastava, A R, & Shrivastava, S (2013). Diversity and versatility of actinomycetes and its role in antibiotic production. *Journal of Applied Pharmaceutical Science*, 3(8): S83–S94.

Cianchetta, S, Galletti, S, Burzi, P L, & Cerato, C (2012). Hydrolytic potential of Trichoderma sp. strains evaluated by microplate-based screening followed by switchgrass saccharification. *Enzyme and Microbial Technology*, 50(6–7): 304–310.

Claus, H, Faber, G, & König, H J A M (2002). Redox-mediated decolorization of synthetic dyes by fungal laccases. *Applied Microbiology and Biotechnology*, 59(6): 672–678.

Couteau, C, & Coiffard, L (2016). Overview of skin whitening agents: drugs and cosmetic products. *Cosmetics*, 3(3): 27.

Couturier, M, Navarro, D, Chevret, D, Henrissat, B, Piumi, F, Ruiz-Dueñas, F J, & Rosso, M N (2015). Enhanced degradation of softwood versus hardwood by the white-rot fungus Pycnoporus coccineus. *Biotechnology for biofuels*. 8(1): 1–16.

Cowan, J, Sinton, C M, Varley, A W, Wians, F H, Haley, R W, & Munford, R S (2001). Gene therapy to prevent organophosphate intoxication. *Toxicology and Applied Pharmacology*, 173: 1–6.

Dahiya, N, Tewari, R, & Hoondal, G S (2006). Biotechnological aspects of chitinolytic enzymes: a review. *Applied Microbiology and Biotechnology*, 71(6): 773–782.

Dantur, K I, Enrique, R, Welin, B, & Castagnaro, A P (2015). Isolation of cellulolytic bacteria from the intestine of *Diatraea saccharalis* larvae and evaluation of their capacity to degrade sugarcane biomass. *Amb Express*, 5(1): 1–11.

Dashtban, M, Schraft, H, & Qin, W (2009). Fungal bioconversion of lignocellulosic residues; opportunities & perspectives. *International Journal of Biological Sciences*, 5(6): 578.

Davidenko, T I (1999). Immobilization of alkaline protease on polysaccharides of microbial origin. *Pharmaceutical Chemistry Journal*, 33(9): 487–489.

Davis-Lorton, M (2015). An update on the diagnosis and management of hereditary angioedema with abnormal c1 inhibitor. *Journal of Drugs in Dermatology*, 14: 151–157.

Deacon, J (2006). *Fungal Biology*. Oxford: Wiley-Blackwell Publishing Ltd.

Deb C, Daniel J, Sirakova T D, Abomoelak B, Dubey V S, & Kolattukudy P E (2006). A novel lipase belonging to the hormone-sensitive lipase family induced under starvation to utilize stored triacylglycerol in Mycobacterium tuberculosis. *Journal of Biological Chemistry*, 281: 3866–3875.

de Castro, R J S, & Sato, H H (2014). Production and biochemical characterization of protease from Aspergillus oryzae: an evaluation of the physical – chemical parameters using agroindustrial wastes as supports. *Biocatalysis and Agricultural Biotechnology*, 3(3): 20–25.

Dhillon, A, Gupta, J K, Jauhari, B M, & Khanna, S (2000). A cellulase-poor, thermostable, alkalitolerant xylanase produced by Bacillus circulans AB 16 grown on rice straw and its application in biobleaching of eucalyptus pulp. *Bioresource Technology*, 73(3): 273–277.

Draelos, Z D (2015). A split-face evaluation of a novel pigment-lightening agent compared with no treatment and hydroquinone. *Journal of the American Academy of Dermatology*, 72: 105107.

Dubey, S K, Meena, R K, Sao, S, Patel, J, Thakur, S, & Shukla, P (2014). Isolation and characterization of cellulose degrading bacteria from biogas slurry and their RAPD profiling. *Current Research in Microbiology and Biotechnology*, 2(4): 416–421.

Dumoulina, Y Cartiliera, L, & Mateescub, M (1999). Cross-linked amylose tablets containing a-amylase: an enzymatically-controlled drug release system. *Journal of Controlled Release*, 60: 161–167.

Ellidag, H Y, Eren, E, Aydin, O, Yıldırım, M, Sezer, C, & Yilmaz, N (2013). Multiple myeloma: relationship to antioxidant esterases. *Medical Principles and Practice*, 23: 18–23.

Falade, A O, Nwodo, U U, Iweriebor, B C, Green, E, Mabinya, L V, & Okoh, A I (2017). Lignin peroxidase functionalities and prosp pective applications. *Microbiologyopen*, 6: 114.

Fernández, J, Redondo-Blanco, S, Miguélez, E M, Villar, C J, Clemente, A, & Lombó, F (2015). Healthy effects of prebiotics and their metabolites against intestinal diseases and colorectal cancer. *AIMS Microbiology*, 1: 48. doi: 10.3934/microbiol.2015.1.48

Fogarty, W M, & Kelly, C T (1980). *In Economic Microbiology, Microbial Enzymes and Bioconversions* (pp. 115–1170). New York: Academic Press.

Fooks, L J, & Gibson, G R (2002). In vitro investigations of the effect of probiotics and prebiotics on selected human intestinal pathogens. *FEMS Microbiology Ecology*, 39(1): 67–75.

Forootanfar, H, & Faramarzi, M A (2015). Insights into laccase producing organisms, fermentation states, purification strategies, and biotechnological applications. *Biotechnology Progress*, 31(6): 1443–1463.

Forster, R J, Cassidy, E, & Donoghue, O (2000). Electrochemical diagnostic strip device for the total cholesterol and its subfractions. *Electroanalysis*, 12: 716–721.

Gazali, F M, & Suwastika, I N (2018). Thermostable α-amylase activity from thermophilic bacteria isolated from Bora hot spring, Central Sulawesi. *Journal of Physics: Conference Series*, 979: 012001.

Ghodake, G, Jadhav, S, Dawkar, V, & Govindwar, S (2009). Biodegradation of diazo dye direct brown MR by *Acinetobacter calcoaceticus* NCIM 2890. *International Biodeterioration & Biodegradation*, 63(4): 433–439.

Ghosh, A, Verma, A K, Tingirikari, J R, Shukla, R, & Goyal, A (2015). Recovery and purification of oligosaccharides from copra meal by recombinant endo-β-mannanase and deciphering molecular mechanism involved and its role as potent therapeutic agent. *Molecular Biotechnology*, 57: 111–127. doi: 10.1007/s12033-014-9807-4

Gopinath, S C, Anbu, P, Lakshmipriya, T, & Hilda, A (2013). Strategies to characterize fungal lipases for applications in medicine and dairy industry. *BioMed Research International*, 2013: 1–11.

Gordon, T, Kannel, W B, Castelli, W P, & Dawber, T R (1981). Lipoproteins, cardiovascular disease and death: the framingham study. *Archives of Internal Medicine*, 141: 1128–1131.

Gupta, A, & Khare, S K (2007). Enhanced production and characterization of a solvent stable protease from solvent tolerant *Pseudomonas aeruginosa* PseA. *Enzyme and Microbial Technology*, 42(1): 11–16.

Haltrich, D, Nidetzky, B, Kulbe, K D, Steiner, W, & Župančič, S (1996). Production of fungal xylanases. *Bioresource Technology*, 58(2): 137–161.

Hansen, G H, Lübeck, M, Frisvad, J C, Lübeck, P S, & Andersen, B (2015). Production of cellulolytic enzymes from ascomycetes: comparison of solid state and submerged fermentation. *Process Biochemistry*, 50(9): 1327–1341.

Heidrich, J E, Contos, L M, Hunsaker, L A, Deck, L M, & Vander Jagt, D L (2004). Inhibition of pancreatic cholesterol esterase reduces cholesterol absorption in the hamster. *BMC Pharmacology*, 4: 1–9.

Hong, S H, Lee, H J, An, J, Lim, I, Borlongan, C, Aboody, K S, *et al.* (2013). Human neural stem cells expressing carboxyl esterase target and inhibit tumor growth of lung cancer brain metastases. *Cancer Gene Therapy*, 20: 678–682.

Hou, H, Zhou, J, Wang, J, Du, C, & Yan, B (2004). Enhancement of laccase production by Pleurotus ostreatus and its use for the decolorization of anthraquinone dye. *Process Biochemistry*, 39(11): 1415–1419.

Izumi, Y, & Kojo, A (2003). Long-chain xylooligosaccharide compositions with intestinal function-improving and hypolipemic activities, and their manufacture. *JP Patent*, 2003048901.

Jaouadi, B, Abdelmalek, B, Jaouadi, N Z, & Bejar, S (2011). The bioengineering and industrial applications of bacterial alkaline proteases: the case of SAPB and KERAB. *INTECH Open Access Publisher*, 20: 445–466.

Jaouadi, NZ, Jaouadi, B, Aghajari, N, & Bejar, S (2012). The overexpression of the SAPB of *Bacillus pumilus* CBS and mutated sapB-L31I/T33S/N99Y alkaline proteases in Bacillus subtilis DB430: new attractive properties for the mutant enzyme. *Bioresource Technology*, 105: 142–151.

Jeffrey, L S H (2008). Isolation, characterization and identification of actinomycetes from agriculture soils at Semongok, Sarawak. *African Journal of Biotechnology*, 7(20): 3700–3705.

Jiménez, N, Esteban-Torres, M, Mancheño, J M, de Las Rivas, B, & Muñoz, R (2014). Tannin degradation by a novel tannase enzyme present in some *Lactobacillus plantarum* strains. *Applied and Environmental Microbiology*, 80(10): 2991–2997.

Joshi, S, & Satyanarayana, T (2013). Biotechnology of cold-active proteases. *Biology*, 2(2): 755–783.

Kang, S W, Park, Y S, Lee, J S, Hong, S I, & Kim, S W (2004). Production of cellulases and hemicellulases by Aspergillus niger KK2 from lignocellulosic biomass. *Bioresource Technology*, 91(2): 153–156.

Kango, N, Jana, U K, & Choukade, R (2019). Fungal enzymes: sources and biotechnological applications. In *Advancing Frontiers in Mycology & Mycotechnology* (pp. 515–538). Singapore: Springer.

Karigar, C S, & Rao, S S (2011). Role of microbial enzymes in the bioremediation of pollutants: a review. *Enzyme Research*, 2011.

Kasinath, A, Novotný, Č, Svobodová, K, Patel, K C, & Šašek, V (2003). Decolorization of synthetic dyes by Irpex lacteus in liquid cultures and packed-bed bioreactor. *Enzyme and Microbial Technology*, 32(1): 167–173.

Kiiskinen, L L, Rättö, M, & Kruus, K (2004). Screening for novel laccase-producing microbes. *Journal of Applied Microbiology*, 97(3): 640–646.

Kluepfel, D, Shareck, F, Mondou, F, & Morosoli, R (1986). Characterization of cellulase and xylanase activities of Streptomyces lividans. *Applied Microbiology and Biotechnology*, 24(3): 230–234.

Kost, J, & Shefer, S (1990). Chemically-modified polysaccharides for enzymatically-controlled oral drug delivery. *Biomaterials*, 11: 695–698.

Krogh, B O, & Symington, L S (2004). Recombination proteins in yeast. *Annual Review of Genetics*, 38(1): 233–271.

Kuforiji, O O, & Fasidi, I O (2008). Enzyme activities of Pleurotus tuber-regium (Fries) Singer, cultivated on selected agricultural wastes. *Bioresource Technology*, 99(10): 4275–4278.

Kuhad, R C, Gupta, R, & Singh, A (2011). Microbial cellulases and their industrial applications. *Enzyme Research*, 2013: 1–11

Kwon, H W, Yoon, J H, Kim, S H, Hong, S B, Cheon, Y, & Ko, S J (2007). Detection of Extracellular Enzyme Activities in Various Fusarium spp. *Mycobiology*, 35(3): 162–165.

Lee, C G (2009). Chitin, chitinases and chitinase-like proteins in allergic inflammation and tissue remodeling. *Yonsei Medical Journal*, 50(1): 22.

Lehninger, A L (2004). *Lehninger Principles of Biochemistry: David L. Nelson, Michael M. Cox*. New York: Recording for the Blind & Dyslexic.

Lehninger, A L, Nelson, D L, & Cox, M C (2004). *Cell Biochemistry and Function. Lehninger: Principles of Biochemistry* (p. 1119), 4th ed., Nelson, D L, Cox, M C, Eds. W. H. Freeman & Co.: New York, NY, USA, Volume 23.

Liu, L, Fishman, M L, & Hicks, K B (2007). Pectin in controlled drug delivery – a review. *Cellulose*, 14: 15–24. doi: 10.1007/s10570-006-9095-7

Liu, Y, Du, J, Yan, M, Lau, M Y, Hu, J, Han, H, . . . & Lu, Y (2013). Biomimetic enzyme nanocomplexes and their use as antidotes and preventive measures for alcohol intoxication. *Nature Nanotechnology*, 8(3): 187–192.

Ma, X, Huang, X, Moore, Z, Huang, G, Kilgore, J A, Wang, Y, *et al.* (2014). Esterase-activatable β-lapachone prodrug micelles for NQO1-targeted lung cancer therapy. *Journal of Controlled Release*, 200: 201–211.

Mahmood, R T, Asad, M J, Mehboob, N, Mushtaq, M, Gulfraz, M, Asgher, M, . . . & Hadri, S H (2013). Production, purification, and characterization of exoglucanase by Aspergillus fumigatus. *Applied Biochemistry and Biotechnology*, 170(4): 895–908.

Manivasagan, P, Venkatesan, J, Sivakumar, K, & Kim, S K (2014). Pharmaceutically active secondary metabolites of marine actinobacteria. *Microbiological Research*, 169(4): 262–278.

Maryandani, A (2007). Characterization of xylanase from Streptomyces sp. strain C1-3. *HAYATI Journal of Biosciences*, 14(3): 115–118.

McCarthy, A J, Peace, E, & Broda, P (1985). Studies on the extracellular xylanase activity of some thermophilic actinomycetes. *Applied Microbiology and Biotechnology*, 21(3): 238–244.

Michniewicz, A, Ledakowicz, S, Jamroz, T, Jarosz-Wilkolazka, A, & Leonowicz, A (2003). Decolorization of aqueous solution of dyes by the laccase complex from Cerrena unicolor. *Biotechnologia*, 4: 194–204.

Montoya, S, Sánchez, Ó J, & Levin, L (2015). Production of lignocellulolytic enzymes from three white-rot fungi by solid-state fermentation and mathematical modeling. *African Journal of Biotechnology*, 14(15): 1304–1317.

Morosoli, R, Bertrand, J L, Mondou, F, Shareck, F, & Kluepfel, D (1986). Purification and properties of a xylanase from Streptomyces lividans. *Biochemical Journal*, 239(3): 587–592.

Mtui, G Y S (2012). Lignocellulolytic enzymes from tropical fungi: types, substrates and applications. *Scientific Research and Essays*, 7: 1544–1555.

Munir, S, Saleem, S, Idrees, M, Tariq, A, Butt, S, Rauff, B, *et al.* (2010). Hepatitis C treatment: current and future perspectives. *Journal of Virology*, 7: 296–304.

Murata, Y, Maida, C, & Kofuji, K (2019). Drug release profiles and disintegration properties of pectin films. *Materials*, 12: 355. doi: 10.3390/ma12030355

Østergaard, L H, & Olsen, H S (2011). Industrial applications of fungal enzymes. In *Industrial applications* (pp. 269–290). Berlin, Heidelberg: Springer.

Palanivel, P, Ashokkumar, L, & Balagurunathan, R (2013). Production, purification and fibrinolytic characterization of alkaline protease from extremophilic soil fungi. *International Journal of Pharma and Bio Sciences*, 4: 101–110.

Palmieri, G, Cennamo, G, & Sannia, G (2005). Remazol Brilliant Blue R decolourisation by the fungus Pleurotus ostreatus and its oxidative enzymatic system. *Enzyme and Microbial Technology*, 36(1): 17–24.

Palmieri, G, Giardina, P, Bianco, C, Fontanella, B, & Sannia, G (2000). Copper induction of laccase isoenzymes in the ligninolytic fungus Pleurotus ostreatus. *Applied and Environmental Microbiology*, 66(3): 920–924.

Pandey, A, Nigam, P R, Sccol, C T, Soccol, V, Singh, D, & Mohan, R (2000). Advances in microbial amylases. *Journal of Biotechnology*, 31: 135–152.

Panepinto, J C, & Williamson, P R (2006). Intersection of fungal fitness and virulence in Cryptococcus neoformans. *FEMS Yeast Research*, 6: 489–498.

Patel, H, Royall, P G, Gaisford, S, Williams, G R, Edwards, C H, Warren, F J, . . . & Butterworth, P J (2017). Structural and enzyme kinetic studies of retrograded starch: inhibition of α-amylase and consequences for intestinal digestion of starch. *Carbohydrate Polymers*, 164: 154–161.

Pham, H, Santucci, S, & Yang, W H (2014). Successful use of daily intravenous infusion of C1 esterase inhibitor concentrate in the treatment of a hereditary angioedema patient with ascites, hypovolemic shock, sepsis, renal and respiratory failure. *The Journal of Allergy and Clinical Immunology*, 10: 1–5.

Pillai, C K S, Paul, W, & Sharma, C P (2009). Chitin and chitosan polymers: chemistry, solubility and fiber formation. *Progress in Polymer Science*, 34: 641–678.

Ponnambalam, A S, Deepthi, R S, & Ghosh, A R (2011). Qualitative display and measurement of enzyme activity of isolated cellulolytic bacteria. *Biotechnology, Bioinformatics and Bioengineering*, 1(1): 33–37.

Postemsky, P D, Bidegain, M A, González-Matute, R A M I R O, Figlas, N D, & Cubitto, M A (2017). Pilot-scale bioconversion of rice and sunflower agro-residues into medicinal mushrooms and laccase enzymes through solid-state fermentation with Ganoderma lucidum. *Bioresource Technology*, 231: 85–93.

Pradeep, G, Cho, S S, Choi, Y H, Choi, Y S, Jee, J-P, Seong, C N, *et al.* (2016). An extremely alkaline mannanase from *Streptomyces sp.* CS428 hydrolyzes galactomannan producing series of mannooligosaccharides. *World Journal of Microbiology and Biotechnology*, 32: 84.

Prakash, O, & Jaiswal, N (2010). α-Amylase: an ideal representative of thermostable enzymes. *Applied Biochemistry and Biotechnology*, 160(8): 2401–2414.

Prasad, S, Manasa, P, Buddhi, S, Tirunagari, P, Begum, Z, Rajan, S, & Shivaji, S (2014). Diversity and bioprospective potential (cold-active enzymes) of cultivable marine bacteria from the subarctic glacial fjord, Kongsfjorden. *Current Microbiology*, 68(2): 233–238.

Rehman, Z U, Kamran, J, Abd El-Hack, M E, Alagawany, M, Bhatti, S A, Ahmad, G, . . . & Ding, C (2017). Influence of low-protein and low-amino acid diets with different sources of protease on performance, carcasses and nitrogen retention of broiler chickens. *Animal Production Science*, 58(9): 1625–1631.

Rifaat, H M, Hassanein, S M, El-Said, O H, Saleh, S A, & Selim, M S (2005). Purification and characterisation of extracellular neutral protease from Streptomyces microflavus. *Arab J Biotechnol*, 9: 51–60575.

Rinaudo, M (2006). Chitin and chitosan: properties and applications. *Progress in Polymer Science*, 31: 603–632.

Robinson, P K (2015). Enzymes: principles and biotechnological applications. *Essays in Biochemistry*, 59: 1.

Romsomsa, N, Chim-anagae, P, & Jangchud, A (2010). Optimization of silk degumming protease production from Bacillus subtilis C4 using Plackett-Burman design and response surface methodology. *Science Asia*, 36: 118–124.

Ruiz-Dueñas, F J, & Martínez, Á T (2009). Microbial degradation of lignin: how a bulky recalcitrant polymer is efficiently recycled in nature and how we can take advantage of this. *Microbial Biotechnology*, 2(2): 164–177.

Rycroft, C E, Jones, M R, Gibson, G R, & Rastall, R A (2001). A comparative in vitro evaluation of the fermentation properties of prebiotic oligosaccharides. *Journal of Applied Microbiology*, 91(5): 878–887.

Saini, A, Aggarwal, N K, Sharma, A, & Yadav A (2015). Actinomycetes: a source of lignocellulolytic enzymes. *Enzyme Research*, 2015: 1–16

Salinardi, T C, Rubin, K H, Black, R M, & St-Onge, M-P (2010). Coffee mannooligosaccharides, consumed as part of a free-living, weight-maintaining diet, increase the proportional reduction in body volume in overweight men. *Journal of Nutrition*, 140: 1943–1948. doi: 10.3945/jn.110.128207

Sánchez, C (2009). Lignocellulosic residues: biodegradation and bioconversion by fungi. *Biotechnology Advances*, 27(2): 185–194.

Sarkar, N, & Aikat, K (2014). Aspergillus fumigatus NITDGPKA3 provides for increased cellulase production. *International Journal of Chemical Engineering*, 2014: 1–10.

Sen, P, Gopalakrishnan, R, Kothari, H, Keshava, S, Clark, C A, Esmon, C T, Rao, L V M (2011). Factor VIIa bound to endothelial cell protein C receptor activates protease activated receptor-1 and mediates cell signaling and barrier protection. *Blood, The Journal of the American Society of Hematology*, 117(11): 3199–3208.

Singh Arora D, & Kumar Sharma, R (2010). Ligninolytic fungal laccases and their biotechnological applications. *Applied biochemistry and biotechnology*. 160(6): 1760-1788.

Singh, B (2019). Production, characteristics, and biotechnological applications of microbial xylanases. *Applied Microbiology and Biotechnology*, 103(21): 8763–8784.

Singh, G, Singh, G, Jadeja, D, & Kaur, J (2010). Lipid hydrolizing enzymes in virulence: *Mycobacterium tuberculosis* as a model system. *Critical Reviews in Microbiology*, 36: 259–269.

Singh, R, Kumar, M, Mittal, A, & Mehta, P K (2016). Microbial enzymes: industrial progress in 21st century. *3 Biotech*, 6(2): 1–15.

Singh, S, & Pakshirajan, K (2010). Enzyme activities and decolourization of single and mixed azo dyes by the white-rot fungus Phanerochaete chrysosporium. *International Biodeterioration & Biodegradation*, 64(2): 146–150.

Singh, S, Pratima, R, Solanki, M K, & Pandey, B D (2006). Malhotra cholesterol biosensor based on cholesterol esterase, cholesterol oxidase and peroxidase immobilized onto conducting polyaniline films. *Sensors & Actuators, B*, 115: 534–541.

Singhania, R R, Patel, A K, Sukumaran, R K, Larroche, C, & Pandey, A (2013). Role and significance of beta-glucosidases in the hydrolysis of cellulose for bioethanol production. *Bioresource Technology*, 127. 500–507.

Sizova, M V, Izquierdo, J A, Panikov, N S, & Lynd, L R (2011). Cellulose-and xylan-degrading thermophilic anaerobic bacteria from biocompost. *Applied and Environmental Microbiology*, 77(7): 2282–2291.

St-Onge, M P, Salinardi, T, Herron-Rubin, K, & Black, R M (2012). A weight-loss diet including coffee-derived mannooligosaccharides enhances adipose tissue loss in overweight men but not women. *Obesity*, 20: 343–348. doi: 10.1038/oby.2011.289

Sun, X, Zhang, R, & Zhang, Y (2004). Production of lignocellulolytic enzymes by Trametes gallica and detection of polysaccharide hydrolase and laccase activities in polyacrylamide gels. *Journal of Basic Microbiology: An International Journal on Biochemistry, Physiology, Genetics, Morphology, and Ecology of Microorganisms*, 44(3): 220–231.

Sundarram, A, & Murthy, T P K (2014). α-amylase production and applications: a review. *Journal of Applied & Environmental Microbiology*, 2(4): 166–175.

Suwannaphan, S, Fufeungsombut, E, Promboon, A, & Chim-Anage, P (2017). A serine protease from newly isolated Bacillus sp. for efficient silk degumming, sericin degrading and colour bleaching activities. *International Biodeterioration & Biodegradation*, 117: 141–149.

Taboada-Puig, R, Lu-Chau, T A, Eibes, G, *et al.* (2015). Continuous removal of endocrine disruptors by versatile peroxidase using a two-stage system. *Biotechnology Progress*, 31(4): 908–916.

Taeko, I, Koichi, N, Yasushi, N, Akira, K, & Yoshinobu, K (1998). Food and drink effective in anti-obesity. *Japanese Patent JP, 10290681.*

Tallman K R, & Beatty K E (2015). Far-red fluorogenic probes for esterase and lipase detection. *ChemBioChem*, 16: 70–75.

Toshio, I, Noriyoshi, I, Toshiaki, K, Toshiyuki, N, & Kunimasa, K (1990). Production of xylobiose. *Japanese Patent JP*: 2119790.

Vaithanomsat, P, Apiwatanapiwat, W, Petchoy, O, & Chedchant, J (2010). Production of ligninolytic enzymes by white-rot fungus Datronia sp. KAPI0039 and their application for reactive dye removal. *International Journal of Chemical Engineering*, 2010: 1–7.

Vazquez, M J, Alonso, J L, Domınguez, H, & Parajo, J C (2000). Xylooligosaccharides: manufacture and applications. *Trends in Food Science & Technology*, 11(11): 387–393.

Veiga, M, Esparis, A, & Fabregas, J (1983). Isolation of cellulolytic actinomycetes from marine sediments. *Applied and Environmental Microbiology*, 46(1): 286–287.

Walia, A, Mehta, P, Chauhan, A, & Shirkot, C K (2013). Optimization of cellulase-free xylanase production by alkalophilic Cellulosimicrobium sp. CKMX1 in solid-state fermentation of apple pomace using central composite design and response surface methodology. *Annals of Microbiology*, 63(1): 187–198.

Whitehead, T R, & Cotta, M A (2001). Identification of a broad-specificity xylosidase/arabinosidase important for xylooligosaccharide fermentation by the ruminal anaerobe Selenomonas ruminantium GA192. *Current Microbiology*, 43(4): 293–298.

Wilson, D B (1992). Biochemistry and genetics of actinomycete cellulases. *Critical Reviews in Biotechnology*, 12(1–2): 45–63.

Wong, T W, Colombo, G, & Sonvico, F (2011). Pectin matrix as oral drug delivery vehicle for colon cancer treatment. *AAPS PharmSciTech*, 12: 201–214. doi: 10.1208/s12249-010-9564.

Woo, S H, Cho, J S, Lee, B S, & Kim, E K (2004). Decolorization of melanin by lignin peroxidase from Phanerochaete chrysosporium. *Biotechnol. Bioprocess Engineering*, 9: 256260.

Yang, R, Xu, S, Wang, Z, & Yang, W (2005). Aqueous extraction of corncob xylan and production of xylooligosaccharides. *LWT-Food Science and Technology*, 38(6): 677–682.

Yi, B R, Kim, S U, & Choi, K C (2014). Co-treatment with therapeutic neural stem cells expressing carboxyl esterase and CPT-11 inhibit growth of primary and metastatic lung cancers in mice. *Oncotarget*, 5: 12835–12848.

Yopi, Rahmani, N, Jannah, A M, Nugraha, I P, & Ramadana, R M (2017). "Production of mannanase from Bacillus Subtilis LBF-005 and its potential for manno-oligosaccharides production," in *Proceedings of the AIP Conference Proceedings*, (College Park, MD: AIP Publishing LLC), 020022.

Zicha, J, Kunes, J, & Devynck, M A (1999). Abnormalities of membrane function and lipid metabolism in hypertension. *American Journal of Hypertension*, 12: 315–331.

Index

A

B

S

T

U

V